Developments in
Dynamic Soil-Structure Interaction

NATO ASI Series

Advanced Science Institutes Series

A Series presenting the results of activities sponsored by the NATO Science Committee, which aims at the dissemination of advanced scientific and technological knowledge, with a view to strengthening links between scientific communities.

The Series is published by an international board of publishers in conjunction with the NATO Scientific Affairs Division

A Life Sciences **B Physics**	Plenum Publishing Corporation London and New York
C Mathematical and Physical Sciences **D Behavioural and Social Sciences** **E Applied Sciences**	Kluwer Academic Publishers Dordrecht, Boston and London
F Computer and Systems Sciences **G Ecological Sciences** **H Cell Biology** **I Global Environmental Change**	Springer-Verlag Berlin, Heidelberg, New York, London, Paris and Tokyo

NATO-PCO-DATA BASE

The electronic index to the NATO ASI Series provides full bibliographical references (with keywords and/or abstracts) to more than 30000 contributions from international scientists published in all sections of the NATO ASI Series.
Access to the NATO-PCO-DATA BASE is possible in two ways:

– via online FILE 128 (NATO-PCO-DATA BASE) hosted by ESRIN, Via Galileo Galilei, I-00044 Frascati, Italy.

– via CD-ROM "NATO-PCO-DATA BASE" with user-friendly retrieval software in English, French and German (© WTV GmbH and DATAWARE Technologies Inc. 1989).

The CD-ROM can be ordered through any member of the Board of Publishers or through NATO-PCO, Overijse, Belgium.

Series C: Mathematical and Physical Sciences - Vol. 390

Developments in Dynamic Soil-Structure Interaction

edited by

Polat Gülkan

Department of Civil Engineering and
Earthquake Engineering Research Center,
Middle East Technical University,
Ankara, Turkey

and

Ray W. Clough

Department of Civil Engineering,
University of California,
Berkeley, California, U.S.A.

Springer Science+Business Media, B.V.

Proceedings of the NATO Advanced Study Institute on
Developments in Dynamic Soil-Structure Interaction
Kemer, Antalya, Turkey
July 8–16, 1992

ISBN 978-94-010-4770-8 ISBN 978-94-011-1755-5 (eBook)
DOI 10.1007/978-94-011-1755-5

Printed on acid-free paper

TABLE OF CONTENTS

PREFACE

This book is a collection of papers which were presented at the NATO Advanced Study Institute on "Developments in Dynamic Soil-Structure Interaction" held at the Hotel Bilkent Paradise in Kemer, Antalya, Turkey during 8-16 July 1992.

A broad overview of the soil-structure interaction phenomenon in a dynamic environment was performed at the Institute. For the last couple of decades it has been well recognized that the foundation material on which a structure is constructed may interact dynamically with the structure during its response to earthquake or other form of disturbance to the extent that stresses and deflections in the system are modified, sometimes significantly, from the values which would have been developed if it were on a rigid foundation. Treatment of the subject matter fell into five major areas. This treatment is also the way in which this book is organized. Six papers dealt with what may be described as basic theory. Soil-structure interaction methods were classified, and time and frequency domain solutions were compared. While relatively recent approaches such as the boundary element method, critical assessment of transmitting boundaries or effect of incoherent ground motions on spatially extended structures received attention, the unifying formulation in terms of well-known structural engineering concepts was also cited.

Given the vigorous development of computational techniques oriented towards analysis, the idea of checking the effectiveness and correctness of these procedures through correlation studies using field tests was a natural sequel. Several papers addressed observations, data and correlative analysis in soil-structure systems, generally with a slant towards nuclear facilities. Not surprisingly, the SSI validation project in Lotung, Taiwan co-located with the SMART-1 array found frequent reference during the Institute.

A wide variety of soil-structure interaction problems may be formulated and solved quite accurately using simplified models in lieu of rigorous procedures. Papers dealing with such models emphasized the fact that for end-user purposes a clear understanding of the fundamental concepts not overshadowed by the mathematics of the exact solutions can be achieved. The success of any soil-structure analysis rests on the accuracy with which the dynamic behavior of the soil domain can be modeled. Two papers addressed this problem from the computational and physical characterization viewpoints. Practical problems were illustrated by analysis of tunnel linings and stiffness and damping of pile groups. Finally, design code provisions and estimation and generation of design input motions completed the overview of soil-structure interaction in engineering practice.

We acknowledge gratefully the award granted by the NATO Scientific Affairs Committee which enabled us to organize this Advanced Study Institute. Other supporters and contributors were: Middle East Technical University, Turkish Atomic Energy Authority, STFA Construction Company, KAS Construction Company, and the Turkish Electricty Authority.

We wish to express our appreciation to Nesrin Başöz for her outstanding job as administrative secretary and to Dr. Haluk Sucuoğlu, Uğurhan Akyüz, Alphan Nurtuğ and Oğuzhan Bayrak for their assistance. Our final words of thanks go to the contributors for their involvement, and to the participants who made the whole effort worthwhile.

Polat Gülkan and Ray W. Clough
November 1992

NATO Advanced Study Institute on
Developments in Dynamic Soil-Structure Interaction
Kemer, Antalya, Turkey
8-16 July 1992

LIST OF CONTRIBUTORS

Prof. Polat Gülkan	Department of Civil Engineering Middle East Technical University 06531 Ankara, Turkey
Prof. Ray W. Clough	Department of Civil Engineering University of California, Berkeley Berkeley, California 94720 USA
Prof. Milos Novak	Department of Civil Engineering University of Western Ontario London, Ontario Canada
Dr. Jethro W. Meek	Fr. Holst Construction Company Ellerholzweg 14 D-2102 Hamburg 93 Germany
Prof. Dimitri E. Beskos	Department of Civil Engineering University of Patras, GR-26110 Patras, Greece
Prof. Panayotis G. Carydis	Laboratory for Earthquake Engineering Polytechnic Campus-Zografos 15700 Athens Greece
Prof. Makato Watabe	Senior Manager, Shimizu Corporation Seavans South No. 2-3 Shibaura 1-chome Minato-ku, Tokyo 105-07 Japan
Prof. John P. Wolf	Laboratoire de Constructions Hydrauliques Institut d'Hydraulique et d'Energie Department de Genie Civil, EPFL CH-1015 Lausanne Switzerland

Dr. Nuray Aydınoğlu	Engineering Manager STFA Construction Company Tophanelioğlu Cad. No. 19 Altunizade, Çamlıca İstanbul, Turkey
Prof. Yalçın Mengi	Department of Engineering Sciences Middle East Technical University 06531 Ankara, Turkey
Prof. Joseph Penzien	International Civil Engineering Consultants, Inc. 1995 University Avenue, Suite 119 Berkeley, California 94704 USA
Prof. Anestis S. Veletsos	Department of Civil Engineering Rice University P.O. Box 1892 Houston, Texas 77251 USA
Dr. James J. Johnson	EQE Engineering Consultants 44 Montgomery St., Suite 3200 San Francisco, California 94507 USA
Dr. Aldo Castoldi	ISMES Viale G. Cesare 29 24150 Bergamo Italy
Prof. Denis Aubry	Ecole Centrale de Paris Grande Voie des Vignes 92295 Chatenay Malabry Cedex France

LIST OF PARTICIPANTS

Dr. Aybars Gürpınar*	IAEA Wagramerstr. 5 P.O. Box 100 A-1400 Vienna Austria
Geert Degrande	Department of Civil Engineering K.U. Luevan Willem de Croylaan B-3001 Heverlee Belgium
Dr. Christian Legrand	Head of Geotech. Div. CSTC/WTCB 53 rue d'Arlon B-1000 Brussels Belgium
Dr. Kiril Hadjiyski	Bulgarian Academy of Sciences 1113 Sofia Acad. G. Bonchev Str. Block 3, CLSMEE Bulgaria
Dr. Yvonna Kolekova*	Institute of Construction and Architecture Slovak Academy of Sciences Dubravska Cesta 9 84220 Bratislava CSFR
Prof. Sheldon Cherry	Department of Civil Engineering University of British Columbia Vancouver, B.C. V6PIZ4 Canada
Prof. Gary Sabin	Faculty of Engineering and Applied Sciences Memorial University of Newfoundland St. John's, NF A1B3X5 Canada
Mauri Koskinen	Tampere University of Technology P.O. Box 600 Tampere, Finland
Prof. Stavroula Pantazopoulou*	Department of Civil Engineering University of Toronto Toronto, Canada

Prof. Sudhir K. Jain — Department of Civil Engineering
Indian Institute of Technology
Kanpur 208016
India

Prof. Moshe Eisenberger — Civil Engineering Department
Technion City 3200
Israel

Dr. Azm Al-Hamoud — Civil Engineering Department
Jordan University of Sciences & Tech.
P.O. Box 3030
Irbid, Jordan

Prof. Dimitar Petrovski — IZIIS
Inst. of Earthquake Engineering and Eng. Seismology
P.O. Box 101
91000 Skopje
Republic of Macedonia

Dr. Robert Gajewski — Warsaw University of Technology
00-637 Warszawa
Aleja Armii Ludowej 16
Poland

Prof. Rodrigues Carvalho — Universidade Nova de Lisboa
2825 Monte de Caparica
Portugal

Dr. Alexandra Chaves — Universidade Nova de Lisboa/Seccao
Autonoma de Geotecnica
2825 Monte de Caparica
Portugal

Manuel Nunes — LNEC
Avenue do Brasil 101
1799 Lisboa Codex
Portugal

Joao Paulo B. Serra — LNEC
Av. Brasil 101
1799 Lisboa Codex
Portugal

Jose Freire Lucas* — Escola Superior Agraria
Quinta da Senhora de Mercules
6000 Castelo Branco
Portugal

Altan Abdulamit	Bucharest Inst. of Civil Engineering 124, Lacul Tei Bd. Sector 2 Bucharest 72302 Romania
Dr. Anton Chirica	Bucharest Inst. of Civil Engineering Boul. Lacul Tei 124, Sector 2 Bucharest 72302 Romania
Prof. Jacob Eisenberg	Department of Structural Earthquake Resistance 2nd Institutskay St. 2 6 TsNIISK Moscow 109428 Russia
Tadeusz Szczesiak	HIL E 13.2 ETH-Hönggerberg Institute of Structural Engineering 8093 Zürich Switzerland
Uğurhan Akyüz	Department of Civil Engineering Middle East Technical University 06531 Ankara, Turkey
Prof. Gökhan Baykal	Department of Civil Engineering Boğaziçi University 80815 Bebek, İstanbul Turkey
Yusuf Calayır	Department of Civil Engineering Karadeniz Technical University 61080 Trabzon Turkey
Halil Ceylan	Dokuzeylül University Civil Engineering Department Bornova 35100, İzmir Turkey
Dr. Ayfer Erken	İstanbul Technical University Civil Engineering Faculty 80626 Maslak, İstanbul Turkey

Dr. Necmettin Gündüz
İstanbul Technical University
Civil Engineering Faculty
80686 Maslak, İstanbul
Turkey

Prof. Zeki Hasgür
İstanbul Technical University
Civil Engineering Faculty
Ayazağa, İstanbul
Turkey

Prof. Engin Karaesmen
Department of Engineering Sciences
Middle East Technical University
06531 Ankara, Turkey

Aydın Kavak
Boğaziçi University
Department of Civil Engineering
P.K. 2 Bebek, İstanbul
Turkey

Dr. Arif Ş. Kayalar
Dokuzeylül University
Faculty of Engineering and Architecture
Bornova, İzmir
Turkey

Dr. Nevzat Kıraç
Anadolu University
Faculty of Engineering and Architecture
Department of Civil Engineering
26030 Eskişehir
Turkey

Gonca Kutluay
Department of Engineering Sciences
Middle East Technical University
06531 Ankara, Turkey

Alphan Nurtuğ
Department of Civil Engineering
Middle East Technical University
06531 Ankara, Turkey

Prof. Semih Tezcan
Boğaziçi University
Department of Civil Engineering
Bebek, İstanbul
Turkey

Prof. Mehmet Utku
Department of Civil Engineering
Middle East Technical University
06531 Ankara, Turkey

Prof. Yıldız Wasti
Department of Civil Engineering
Middle East Technical University
06531 Ankara, Turkey

Prof. C.V.G. Vallabhan	Texas Tech University Civil Engineering Department Lubbock, Texas 74409 USA
Adel Younan	Department of Civil Engineering Rice University P.O. Box 1892 Houston, Texas 77251 USA
Nesrin Başöz	Graduate Research Assistant Department of Civil Engineering Stanford University Stanford, CA 94305 USA
Dr. Erdal Şafak*	USGS Reston, Virginia 20420 USA

* Unable to attend

SURVEY AND CLASSIFICATION OF COMPUTATIONAL APPROACHES IN SOIL-STRUCTURE INTERACTION: COMPARISON OF TIME- AND FREQUENCY-DOMAIN ANALYSES

JOHN P. WOLF
Institute of Hydraulics and Energy, Department of Civil Engineering,
Swiss Federal Institute of Technology Lausanne, CH-1015 Lausanne, Switzerland

ABSTRACT: The various methods to perform soil-structure-interaction analysis are classified. The first classification uses as criterion the behaviour (linear or nonlinear) of the structure and of the unbounded soil which determines for the analysis either the familiar time domain, the natural choice for a dynamic analysis, or the frequency domain, originally developed for machine foundation calculations. The second classification distinguishes between the direct method and the substructure method, which do not necessarily lead to identical results. Within each method, however, the various procedures are mathematically equivalent. In the substructure method the dynamic stiffness representing the interaction forces of the unbounded soil is determined based on the boundary element method in the time or frequency domain. In the latter case various so-called realizations in the time domain are distinguished using the extent of the frequency-domain calculations as a criterion. The temporal global coupling of the interaction force-displacement relationship can be eliminated with these realizations.

1. Introduction

The various methods to analyse dynamic soil-structure interaction appear at first sight to be quite different from each other, although they all model the same feature: *wave propagation in the unbounded soil towards infinity*. As shown is this paper, certain differences between the results of the so-called direct method and of the substructure method do actually exist. Within each of these two methods, however, the various approaches, which are based on different concepts, are mathematically equivalent. They are thus not truly different and independent procedures. Using certain criteria, such as e.g. which steps are performed in the frequency domain and which in the time domain, a classification of the methods to analyse soil-structure interaction results.

To establish a firm base for the classification of the methods, the objective and the significant features of soil-structure-interaction analysis described in the two text books [1, 2] are summarized in the following. The dynamic system whose response is to be determined consists of two distinct parts with different properties: the (generalized) *structure with bounded dimensions* consisting of the actual structure and possibly an irregular adjacent soil region, and the *unbounded soil extending to infinity* (Fig. 1). The (generalized) structure is modelled straightforwardly with finite elements (the word "generalized" is dropped for the sake of conciseness in the following). To analyse the semi-infinite domain of the unbounded soil numerically, a surface forming the boundary of the numerical model is chosen which encloses the structure. The properties associated

P. Gülkan and R. W. Clough (eds.), Developments in Dynamic Soil-Structure Interaction, 1–23.

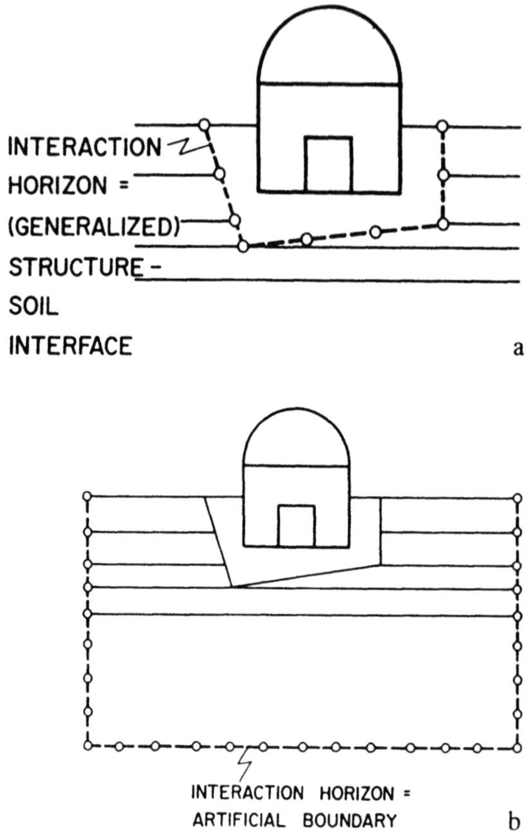

Fig. 1 Dynamic models for (a) Substructure method. (b) Direct method.

with the degrees of freedom of the nodes on this so-called *interaction horizon* [3] represent the significant dynamic features of the unbounded domain located on the exterior of this surface. In particular, the *radiation condition* [4] has to be enforced which states that for a unique solution *no energy may be radiated from infinity into the dynamic system* for a load applied to the structure or for the scattered motion in the case of e.g. seismic excitation. A certain arbitrariness exists when selecting the location of the interaction horizon, which actually has no physical significance. The interaction horizon can coincide with the structure-soil interface leading to the *substructure method*; or it can be identical to an artificial boundary up to which the soil is modelled with, for example, finite elements, which results in the *direct method.* If the same rigorous radiation condition were formulated in both methods, the two methods would actually be the same, leading to identical results.

These concepts can be applied to other dynamic structure-medium interactions, such as fluid-structure interaction and structural acoustics.

In Section 2 it is shown that the radiation condition stating that energy radiates towards infinity does not exclude "reflected" waves from propagating with the phase velocity in the opposite direction. In Section 3 the damping-solvent extraction method to determine the dynamic stiffness of an unbounded domain is discussed. In this procedure the dynamic stiffness of a bounded soil domain with artificially introduced material damping is first calculated and then the influence of the damping is extracted.

Two procedures of classification are then discussed. The first procedure examined in the Section 4 is based on the linear versus nonlinear behaviour of the structure and of the unbounded soil. The second procedure distinguishes between the direct method (Section 5) and the substructure method (Section 6).

Not all methods which have been developed can be discussed in this paper. Preference is given to those with which the author has had some contact. It is also not possible to provide an accurate historical review of the development of soil-structure-interaction analysis. The cited literature is restricted to a selection of historical references, some review papers which are still worthwhile to consult today and recent articles describing the latest progress.

2. Radiation Condition

To examine wave propagation in the unbounded domain towards infinity, the dynamic-stiffness coefficients of simple one-dimensional cases of increasing complexity are addressed. First, the *semi-infinite prismatic rod* is examined (Fig. 2a). For the axial motion with the displacement u of a rod with area A, modulus of elasticity E and density ρ (rod velocity $c_\ell = \sqrt{E/\rho}$), applying a displacement u_0 at x=0 leads to the outwardly propagating wave

$$u(x,t) = u_0\left(t - \frac{x}{c_\ell}\right) \tag{1}$$

The force $P_0 = -EAu_{0,x}$ follows as

$$P_0(t) = C\,\dot{u}_0(t) \tag{2}$$

with the damping coefficient

$$C = \rho c_\ell A \tag{3}$$

In the frequency domain with factor $e^{+i\omega t}$, the displacement amplitude

$$u(x,\omega) = u_0(\omega)e^{-i\omega x/c_\ell} \tag{4}$$

and the dynamic-stiffness coefficient $S(\omega) = P_0(\omega)/u_0(\omega)$ equals

$$S(\omega) = i\omega\rho c_\ell A \tag{5}$$

Second, the *semi-infinite rod on elastic foundation* (Fig. 2b) with the static spring stiffness per unit length of the foundation k_g is discussed. This rod model is much more important than expected, as it represents exactly a single mode of an infinite modal expansion of the layer built-in at its base. The equation of motion in the frequency domain equals

$$EAu(x,\omega)_{,xx} + (-k_g + \omega^2\rho A)u(x,\omega) = 0 \tag{6}$$

The solution with the outwardly propagating wave equals

$$u(x,\omega) = u_0(\omega)e^{-i\omega x/c} \tag{7}$$

where the phase velocity

$$c = \frac{c_\ell}{\sqrt{1 - \left(\frac{\omega_c}{\omega}\right)^2}} \tag{8}$$

and with the natural frequency of the rigid bar supported by the elastic foundation in rigid-body motion $\omega_c = \sqrt{k_g/(\rho A)}$ (which is also equal to the cutoff frequency). The dynamic stiffness coefficient is found by evaluating the force amplitude $P_0(\omega) = -EAu_0(\omega)_{,x}$, yielding

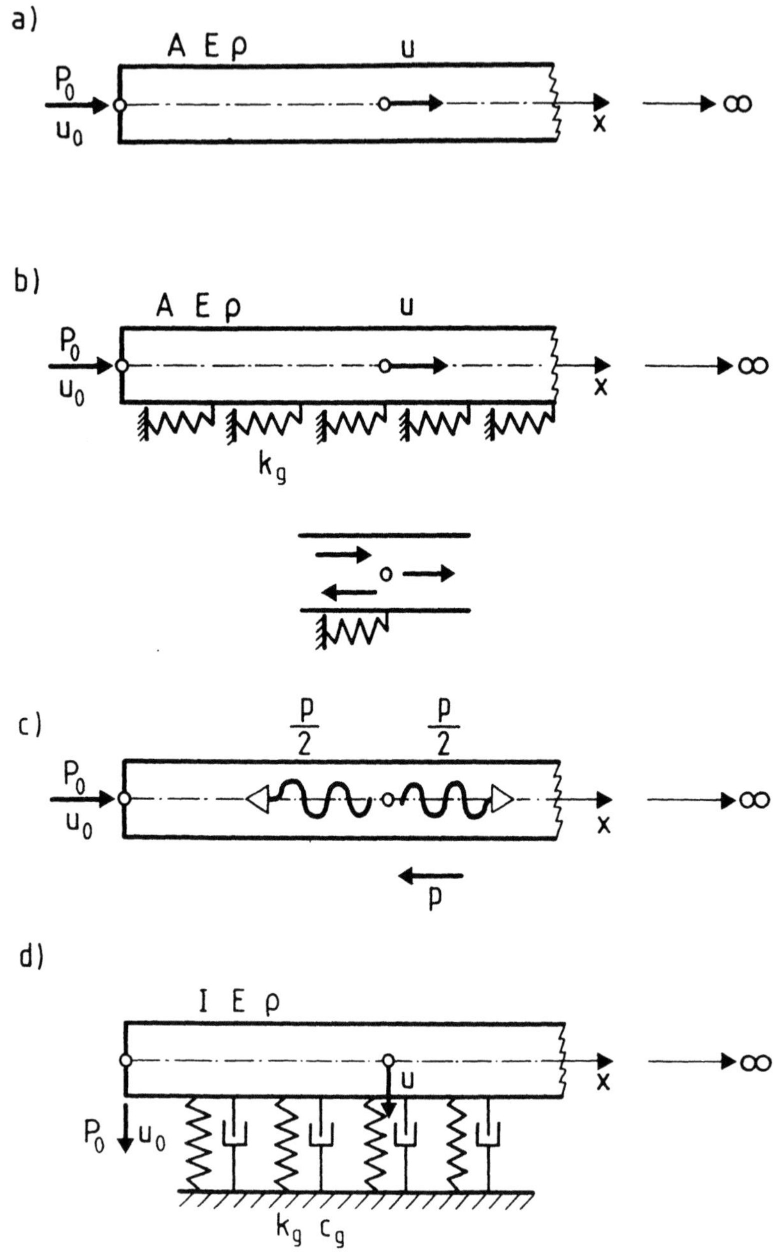

Fig. 2 Wave propagation towards infinity in one-dimensional systems.
- a) Prismatic rod.
- b) Rod on elastic foundation with reflected and refracted waves on microscopic scale.
- c) Wave pattern caused by elastic reaction.
- d) Beam on visco-elastic foundation.

$$S(\omega) = i\omega\rho\frac{c_\ell^2}{c}A = i\sqrt{EAk_g}\sqrt{\frac{\omega^2}{\omega_c^2} - 1} \tag{9}$$

For $\omega > \omega_c$, i.e. above the cutoff frequency where wave propagation actually occurs, the wave does indeed propagate outwardly with the phase velocity c in this dispersive system (Eq. 7) and the rate of energy transmission which is proportional to the imaginary part of the dynamic-stiffness coefficient is positive (Eq. 9), indicating that energy is radiated towards infinity (radiation condition). The wave patterns of the prismatic rod (Eq. 4) follows from that of the rod on an elastic foundation (Eq. 7) by replacing the frequency-dependent phase velocity c by the rod velocity c_ℓ. The same relationship holds also for the dynamic-stiffness coefficients (see middle term in Eq. 9).

On a «microscopic» scale, the only wave velocity is c_ℓ, a material constant. When in any point along the rod an incident wave propagating outwardly with c_ℓ encounters a spring of the elastic foundation, a reflected wave propagating inwardly with c_ℓ and a refracted wave propagating outwardly with c_ℓ are generated (Fig. 2b). After superposition, the net effect on a «macroscopic» scale of this very complicated wave pattern is a dispersive wave propagating outwardly with the phase velocity c. Thus on a microscopic scale, also inwardly propagating waves exist. This concept can be used to calculate the dynamic-stiffness coefficient based solely on wave-propagation techniques [5]. A distributed force p arising from the spring of the elastic foundation at some location x (Fig. 2c) induces a dilatational wave in tension with amplitude p/2 which propagates towards infinity and a dilatational wave in compression with amplitude p/2 which propagates in the negative x-direction, reaching the beginning of the bar at time x/c_ℓ later and being reflected there with double amplitude. The force at the beginning of the bar is therefore given by

$$P_0(t) = \rho c_\ell A\,\dot{u}_0(t) + \int_0^\infty p\left(x, t - \frac{x}{c_\ell}\right) dx \tag{10}$$

The first term on the right-hand side follows from Eqs. 2 and 3. The distributed force along the rod $p(x,t - x/c_\ell)$ is the product of k_g and the local displacement $u(x,t - x/c_\ell)$.

$$P_0(t) = \rho c_\ell A\,\dot{u}_0(t) + k_g \int_0^\infty u\left(x, t - \frac{x}{c_\ell}\right) dx \tag{11}$$

or in the frequency domain

$$P_0(\omega) = \rho c_\ell A i\omega\, u_0(\omega) + k_g \int_0^\infty u(x,\omega)\, e^{-i\omega x/c_\ell}\, dx \tag{12}$$

Substituting $u(x,\omega)$ (Eqs. 7, 8) the integral in Eq. 12 can be evaluated (it vanishes at $x = \infty$).

$$k_g \int_0^\infty u_0(\omega)\exp\left\{-i\omega\left(\frac{1}{c} + \frac{1}{c_\ell}\right)x\right\}dx = k_g u_0(\omega) \lim_{d\to\infty} \int_0^\infty \exp\left\{-\left[\frac{1}{d} + i\omega\left(\frac{1}{c} + \frac{1}{c_\ell}\right)\right]x\right\}dx$$

$$= k_g u_0(\omega) \lim_{d\to\infty} - \left.\frac{e^{-x/d}\, e^{-i\omega(1/c + 1/c_\ell)x}}{\frac{1}{d} + i\omega\left(\frac{1}{c} + \frac{1}{c_\ell}\right)}\right|_0^\infty = \frac{1}{i\omega\left(\frac{1}{c} + \frac{1}{c_\ell}\right)}\, k_g u_0(\omega) \tag{13}$$

The dynamic-stiffness coefficient follows as $S(\omega) = P_0(\omega)/u_0(\omega)$, leading to the same expression as specified in Eq. 9.

Third, as a slight extension, *hysteretic material damping* with the damping ratio ζ is introduced in the rod. Based on the correspondence principle, the equation of motion (Eq. 6) is transformed to

$$E(1 + 2\zeta i)\, Au(x,\omega)_{,xx} + (-k_g + \omega^2 \rho A)\, u(x,\omega) = 0 \tag{14}$$

The solution is still specified by Eq. 7, but in Eq. 8 c_ℓ is replaced by $c_\ell \sqrt{1+2\zeta i}$, yielding with $1/(c_\ell \sqrt{1+2\zeta i}) \cong (1-\zeta i)/c_\ell$

$$u(x,\omega) = u_0(\omega) \exp\left[\frac{-i\omega x(1-\zeta i)}{c_\ell / \sqrt{1-(\omega_c/\omega)^2}}\right] \tag{15}$$

The case of an excitation ω below the cutoff frequency of the undamped system ($\omega<\omega_c$) is investigated. Eq. 15 is transformed to

$$u(x,\omega) = u_0(\omega) \exp\left[-\frac{\omega x}{c_\ell / \sqrt{(\omega_c/\omega)^2 - 1}}\right] \exp\left[+i\,\frac{\omega x \zeta}{c_\ell / \sqrt{(\omega_c/\omega)^2 - 1}}\right] \tag{16}$$

The first exponential function describes a decay along the axis, leading to zero at $x = \infty$, the second an inwardly propagating wave. (The other solution of Eq. 7 with a sign change in the argument of the exponential function cannot be used, as in this case, the first exponential function goes to infinity for $x \to \infty$). The dynamic-stiffness coefficient follows from $P_0(\omega) = -E(1+2\zeta i)Au_0(\omega)_{,x}$ as

$$S(\omega) = \sqrt{EAk_g} \sqrt{1 - \frac{\omega^2}{\omega_c^2}(1 + \zeta i)} \tag{17}$$

For $\omega<\omega_c$, the imaginary part of the dynamic-stiffness coefficient and thus the rate of energy transmission are positive, indicating that no energy is radiated from infinity. *The radiation condition is satisfied, although the wave propagates inwardly!*

Fourth, an example is addressed where the exact solution with outward radiation of energy exhibits an inwardly propagating wave for all frequencies. The *semi-infinite beam* with moment of inertia I *on a visco-elastic foundation* (Fig. 2d) with the damping coefficient per unit length of the foundation c_g can be used to model a pile in lateral motion. Following Ref. [6], the equation of motion in the frequency domain equals

$$EIu(x,\omega)_{,xxxx} + (k_g + i\omega c_g - \omega^2 \rho A)\, u(x,\omega) = 0 \tag{18}$$

which can also be written as

$$u(x,\omega)_{,xxxx} + 4R^4 e^{i\theta} u(x,\omega) = 0 \tag{19}$$

with

$$R = \left[\frac{(k_g - \omega^2 \rho A)^2 + \omega^2 c_g^2}{(4EI)^2}\right]^{1/8} \tag{20}$$

$$\theta = \arctan \frac{\omega c_g}{k_g - \omega^2 \rho A} \tag{21}$$

As $\omega c_g>0$, $0<\theta<\pi$. With $u(x,\omega) = e^{\lambda z}$, Eq. 19 yields

$$\lambda^4 + 4R^4 e^{i\theta} = 0 \tag{22}$$

Its solution equals

$$\lambda_1 = R\,(b + ia) \tag{23a}$$

$$\lambda_2 = -R\,(b + ia) \tag{23b}$$

$$\lambda_3 = -R\,(a - ib) \tag{23c}$$

$$\lambda_4 = R\,(a - ib) \tag{23d}$$

with

$$a = \sin\frac{\theta}{4} + \cos\frac{\theta}{4} \tag{24a}$$

$$b = -\sin\frac{\theta}{4} + \cos\frac{\theta}{4} \tag{24b}$$

As $0<\theta<\pi$, $a>b>0$.
As the displacement amplitude cannot become infinite at $x = \infty$, the exponential functions with λ_1 and λ_4 will not appear in the solution

$$u(x,\omega) = c_1 e^{\lambda_2 x} + c_2 e^{\lambda_3 x} \tag{25}$$

$$= c_1 e^{-Rbx}\, e^{-iRax} + c_2 e^{-Rax}\, e^{iRbx} \tag{26}$$

where c_1 and c_2 are the integration constants. Enforcing the boundary conditions $u(x{=}0,\omega)=u_0(\omega)$, $u(x{=}0,\omega)_{,x}=0$ leads to

$$u(x,\omega) = \frac{u_0(\omega)}{2}\left[(1+i)\, e^{-Rbx}\, e^{-iRax} + (1-i)\, e^{-Rax}\, e^{iRbx}\right] \tag{27}$$

The first term on the right-hand side corresponds to an outwardly propagating wave, but the second term with the factor e^{iRbx} to an inwardly propagating wave (in the negative x-direction) for the beam loaded at x=0.

The dynamic-stiffness coefficient is calculated by evaluating the force amplitude $P_0(\omega) = -EIu_0(\omega)_{,xxx}$, yielding

$$S(\omega) = EIR^3\,[-a^3 - b^3 + 3a^2b + 3ab^2 + i(a^3 - b^3 + 3a^2b - 3ab^2)] \tag{28}$$

As $a>b$, the imaginary part of $S(\omega)$ is positive and thus also the rate of energy transmission.

In Ref. [7] it is mentioned, that for Rayleigh waves occurring in a layered medium, the directions of the phase velocity and of the energy propagation are sometimes opposed.

The radiation condition [4] guaranteeing uniqueness of the solution states that the forces P_0 applied at x=0 must be sources, not sinks of energy. The energy must propagate towards infinity or formulated differently, no energy may be radiated from infinity. As the rate of energy transmission is proportional to the imaginary part of the dynamic-stiffness coefficient, the sign of the latter can be checked. The radiation condition is satisfied for these simple one-dimensional systems. This does, however, not mean that an inwardly travelling «reflected» wave cannot exist propagating with the phase velocity from infinity towards the applied load. This aspect is important when constructing a transmitting boundary to be used in the direct method of analysis.

3. Damping-Solvent Extraction Method

To extract salt from the ground, water can be injected into the ground. The salt dissolves in the water which can then be pumped to the surface. After evaporation of the water, the salt remains. A similar procedure, where a chemical solvent is added and at the end extracted again, can be applied in oil exploration. The same concept can be applied to calculate the dynamic-stiffness in the frequency domain of the undamped unbounded soil, relating the displacement amplitudes to the interaction force amplitudes on the interaction horizon (structure-soil interface in the substructure method).

This so-called *damping-solvent extraction method* is described and evaluated in depth in Ref. [8]. It is explained and illustrated using a rectangular foundation embedded in an elastic halfplane

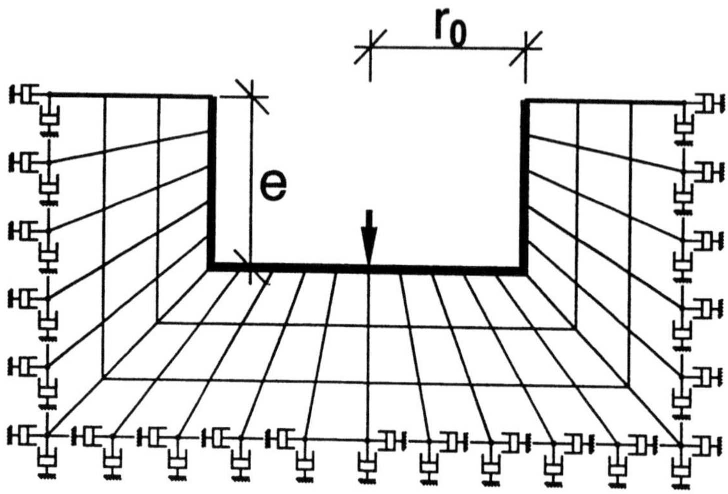

Fig. 3 Foundation embedded in half-plane with finite-element mesh of adjacent bounded soil region with injected material damping and viscous dampers.

(Fig. 3). Its dynamic-stiffness matrix is to be determined. In the first step, the dynamic-stiffness matrix of an adjacent *bounded soil region with artificially injected material damping* is calculated. The number of finite elements in the radial direction can be chosen as small (in the range of 3 to 6). The ratio of the hysteretic damping is selected as quite large (in the range $0.2<\zeta<0.6$) to diminish the energy of the reflected waves reaching the structure-soil interface. This is achieved to a larger extent by placing in addition the well-known viscous dampers [9] at the outer boundary. The analysis of this bounded soil domain with hysteretic damping is straightforward applying the correspondence principle, which multiplies the Lamé constants by $(1+2i\zeta)$. The coefficients of the viscous dashpots will also become complex. As a good approximation, the *dynamic-stiffness matrix of this highly damped bounded soil domain can be assumed to be equal to that of the unbounded damped soil domain.* In the second step, the influence of the artificially injected hysteretic damping on the dynamic-stiffness matrix of the unbounded domain is eliminated, i.e. the *effect of damping* (the solvent) *is extracted.* This operation can be performed for each coefficient independently of the others. The dynamic-stiffness coefficient of the undamped unbounded domain $S(a_0)$, which is proportional to its static value, is a function of the dimensionless frequency $a_0 = \omega r_0/c_s$ (r_0 = characteristic length, c_s = shear-wave velocity). Introducing hysteretic material damping affects the dynamic-stiffness coefficient in two ways: first, the static-stiffness coefficient is multiplied by $(1 + 2i\zeta)$ and second, a_0 is replaced by a_0^*

$$a_0^* = \frac{\omega r_0}{c_s^*} = \frac{\omega r_0}{c_s\sqrt{1+2i\zeta}} \tag{29}$$

The dynamic-stiffness coefficient of the damped unbounded domain $S(a_0^*)$ divided by $(1+2i\zeta)$ is thus a function of a_0^* (which now describes the only dependence on ζ), that of the undamped unbounded domain $S(a_0)$ is the same function, but of a_0. To evaluate this function at a_{0j} corresponding to a specific ω_j, a Taylor expansion at $a_0 = a_{0j}^*$ is formulated

$$S(a_{0j}) = \frac{S(a_{0j}^*)}{1+2i\zeta} + \left.\frac{S(a_0)_{,a_0}}{1+2i\zeta}\right|_{a_0 = a_{0j}^*} (a_{0j} - a_{0j}^*) \tag{30}$$

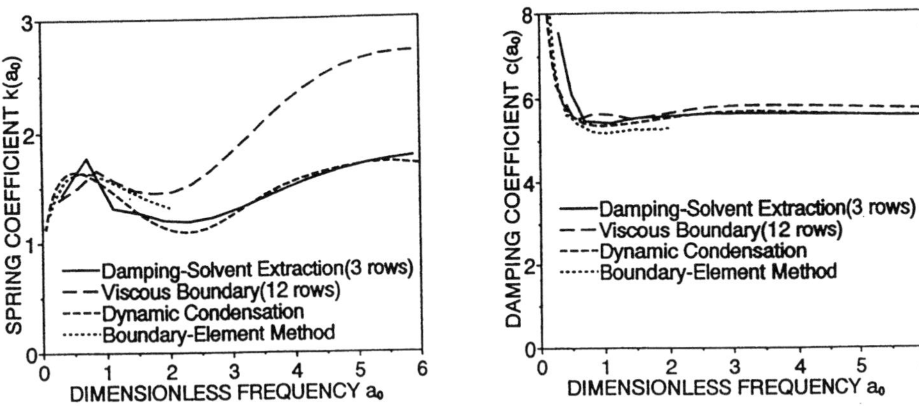

Fig. 4 Vertical dynamic-stiffness coefficient of rigid rectangular foundation embedded in half-plane applying damping-solvent extraction method.

with $a_{0j} - a^*_{0j} = a_{0j}(1 - 1/\sqrt{1+2i\zeta})$. The partial derivative $S(a_0),_{a_0}$ at $a_0 = a^*_{0j}$ is calculated numerically using two neighbouring values of ω_j (and thus of a^*_{0j}). The extraction of the damping can thus be performed for each frequency independent of the others.

For illustration, the vertical dynamic-stiffness coefficient of the rigid rectangular foundation with half-length r_0 embedded with depth e in a halfplane is calculated. $e = r_0$ and Poisson's ratio = 0.25. The mesh of the adjacent bounded soil (3 rows of finite elements in the radial direction) with the viscous dampers is shown in Fig. 3. The dynamic-stiffness coefficient of this two-dimensional problem is non-dimensionalized as

$$S(a_0) = G[k(a_0) + ia_0c(a_0)] \tag{31}$$

with G representing the shear modulus. This is a stringent test, as $k(a_0)$ and $c(a_0)$ start for $a_0 = 0$ at zero and infinity, respectively (static case). The applied hysteretic damping ratio of the solvent equals $\zeta = 0.3$. The results of the damping-solvent extraction method (Fig. 4) agree well with the boundary-element solution [10] and those determined with dynamic condensation which is based on finite elements [11]. The results using only the viscous dampers (and no material damping) as a transmitting boundary at the artificial boundary deviate significantly, although the number of finite elements in the radial direction is selected as 12, leading to a finite-element mesh which is 4 times as large.

Summarizing, the damping-solvent extraction method permits an efficient calculation of the dynamic-stiffness matrix of a foundation embedded in an unbounded soil, analysing the bounded adjacent soil domain only, which exhibits the same dynamic characteristics as the (bounded) structure. The introduced hysteretic material damping can be extracted at the end of the analysis for each coefficient of the dynamic-stiffness matrix and for each frequency concisely.

4. Linear Versus Nonlinear Structure and Unbounded Soil

The first classification uses as criterion the behaviour (linear or nonlinear) of the structure and of the unbounded soil (Table 1). In the *first case both the structure and the unbounded soil remain linear*, which applies to many analyses of nuclear power plants and machine foundations. Quite surprisingly, also for a transient excitation such as an earthquake loading, the analysis for this total

STRUCTURE	UNBOUNDED SOIL	CALCULATIONAL DOMAIN
LINEARLINEAR NONLINEAR (NON) LINEAR	FREQUENCY (TIME) LINEAR NONLINEAR	 TIME TIME

Table 1 Calculational domain determined by behaviour of structure and unbounded soil

linear system is routinely performed in the frequency domain and only exceptionally in the time domain. This is due to the fact that the formulation of the radiation condition is as an analytical expression better known in the frequency than in the time domain (and to a lesser extent as it is straightforward to incorporate hysteretic material damping in the frequency domain). But equivalent formulations in the time domain of the radiation condition do, of course, exist. Reviews of this so-called complex response analysis working in the frequency domain exist [12, 13, 14], as do general-purpose computer programs [15, 16]. This first case which has reached a very high level of development is not addressed any further.

In the *second case the structure can behave nonlinearly while the unbounded soil will remain linear*. The latter is normally justified, as for the three-dimensional spreading of the waves when propagating away from the structure the amplitudes decay. Examples are structures which perform in the nonlinear range for high seismic excitation; base-isolation systems with friction plates exhibiting strong nonlinear characteristics which have to be considered in design; local nonlinearities such as the partial uplift of the basemat and the separation occurring between the sidewalls of the base and the neighbouring soil in the case of embedded structures; and the highly nonlinear soil behaviour arising adjacent to the basemat. The analytical methods working in the time domain (with possibly certain steps performed in the frequency domain) are summarized in a text book [2].

In the *third case the total dynamic system will be nonlinear* (with as a special case linear behaviour of the structure), which is analysed in the time domain. The nonlinearity of the unbounded soil can be caused by e.g. the two-dimensional propagation of surface waves, for which no decay of the amplitudes occurs. This case has hardly been addressed. An exception is Ref. 17 where as an approximation the far field is modelled based on one-dimensional elasto-plastic wave propagation with one stress component.

The direct and substructure methods discussed below will concentrate on the second case; i.e. the structure will be nonlinear (with as a special case linear behaviour) while the unbounded soil will behave linearly.

5. Direct Method

How to formulate the radiation condition of the unbounded soil is the key issue in the analysis of soil-structure interaction. If the radiation condition is formulated at infinity, the rigorous boundary condition results, which is *global in space and time* (Fig. 5). That is, all degrees of freedom of the nodes located on the interaction horizon from the start of the excitation contribute to the interaction forces. The substructure method (with the structure-soil interface coinciding with the interaction horizon) is based on this concept. In the direct method approximations are introduced. The radiation condition is formulated directly on the interaction horizon (= artificial boundary) in such a

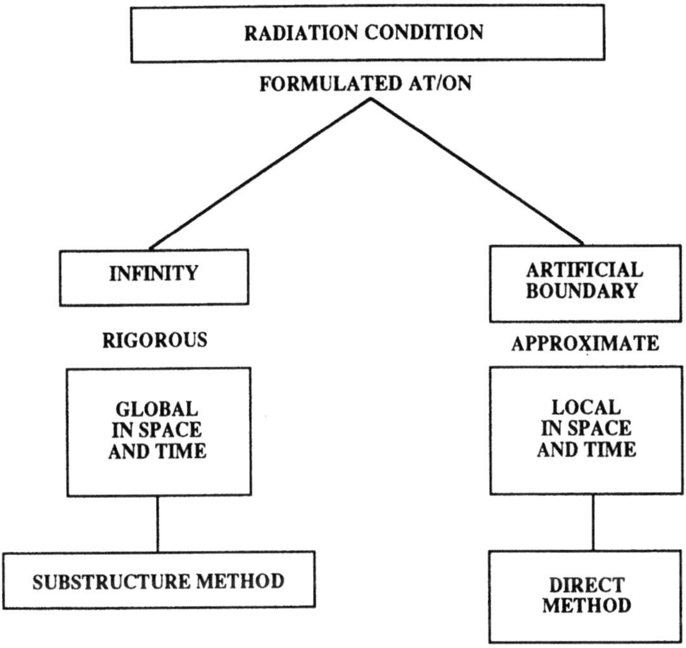

Fig. 5 Differences in formulating the radiation condition for substructure and direct methods.

way that a *(frequency independent)* highly absorbing boundary condition results which is *local in space and time*. These *transmitting boundaries* thus use information only from the node being addressed or from the nearly region of the mesh at the current time station or, at most, at a few recent time stations.

Based on various mathematical principles many apparently different local transmitting boundaries have been developed: the viscous damper [9], the paraxial approximation [18, 19], the extrapolation algorithm [20] and the superposition boundary [21, 22] to name just a few. Ref. [23] contains a detailed evaluation of some of these formulations. In Ref. [24] it is shown that all the transmitting boundaries mentioned above, although they appear to be vastly different from each other, are actually all mathematically equivalent and thus are essentially alternative realizations of one and the same boundary mechanism. It is also demonstrated that the transmitting boundaries can be classified as being of first order, second order, etc. whereby higher-order schemes which would result in higher accuracy may lead to dynamic instabilities and can thus not be used.

A real challenge exists to develop a transmitting boundary of higher accuracy which is local in time and space and which can handle (approximately) all types of waves without restrictions on the geometry and on the material properties such as Poisson's ratio. This is definitively the area of soil-structure-interaction analysis where the research efforts should be concentrated.

To demonstrate that improvement is possible, it is shown in Ref. [25] that for a one-dimensional dynamic system which is dispersive and exhibits a cutoff frequency, starting from the dispersion relation, a lumped-parameter model with a few internal degrees of freedom and frequency-independent coefficients of the springs, dashpots and masses can be developed.

6. Substructure Method

6.1 DYNAMIC STIFFNESS

The dynamic-stiffness matrix relating the displacements in the nodes on the structure-soil interface to the interaction forces of the unbounded soil (or in a computational algorithm the interaction forces at a specific time station) are routinely calculated based on the ***boundary element method***. Ref. [26] provides a review of this very effective numerical method, which is well suited to represent semi-infinite domains such as the unbounded soil. For this case the advantages of the boundary element method become significant, i.e. the discretization is performed only on the surface of the domain (reduction of the dimensionality of the problem by one) and the radiation condition can be enforced exactly at infinity by incorporating it in the fundamental solution.

The first time-domain boundary-element formulation for the unbounded soil was presented in Refs. [27] and [28], whereby in the latter a nonlinear application is included. For restricted cases such as the out-of-plane motion, earlier work e.g. Ref. [29] can be mentioned. Many different boundary-element formulations in the time domain exist (Refs. [30, 26]). Besides the direct method, which is applied widely, indirect methods, weighted residual procedures and others have been used.

As an alternative, the so-called *multi-cell cloning algorithm* [11] can be used to calculate the dynamic-stiffness matrix of the unbounded domain. This is a stand-alone procedure based entirely on the finite-element method [31]. Starting with the dynamic-stiffness matrix of a bounded cell of finite elements, that of the unbounded domain can be determined. The *damping-solvent extraction method* is also very attractive to calculate the dynamic-stiffness matrix, using the bounded adjacent soil domain.

6.2 DISPLACEMENT, VELOCITY AND ACCELERATION CONVOLUTIONS AND THEIR RECURSIVE EVALUATION

The interaction force-displacement relationship formulated in the frequency domain as the product of the frequency-domain dynamic-stiffness matrix and the displacement amplitudes is written in the time-domain as the convolution of the time-domain dynamic-stiffness matrix and the displacements. Besides this displacement convolution, corresponding velocity and acceleration convolutions leading to the same interaction forces can be used. The following is based on Ref. [32], whereby velocity and acceleration convolutions for the unbounded soil are also addressed in Ref. [33].
In the frequency domain the interaction force (amplitudes $\{R(\omega)\}$) - displacement (amplitudes $\{u(\omega)\}$) relationship is formulated with the dynamic-stiffness matrix $[S(\omega)]$ as

$$\{R(\omega)\} = [S(\omega)]\ \{u(\omega)\} \tag{32}$$

To determine its Fourier transformation, $[S(\omega)]$ is decomposed into a singular part, which is equal to its asymptotic value at $\omega = \infty$, $[K_\infty] + i\omega[C_\infty]$, and the remaining regular part $[K_r(\omega)]$, which is absolutely integrable over the ω-axis [2]

$$[K_r(\omega)] = [S(\omega)] - [K_\infty] - i\omega[C_\infty] \tag{33}$$

The interaction force-displacement relationship in the time domain involving a *displacement convolution* is then equal to

$$\{R(t)\} = \int_0^t [K_r(t-\tau)]\ \{u(\tau)\}\, d\tau + [K_\infty]\ \{u(t)\} + [C_\infty]\ \{\dot{u}(t)\} \tag{34}$$

with $[K_r(t)]$ denoting the inverse Fourier transform of $[K_r(\omega)]$.
Alternatively, Eq. 32 can be rewritten as

$$\{R(\omega)\} = \frac{[S(\omega)]}{i\omega}\, i\omega\{u(\omega)\} \tag{35}$$

where $i\omega\{u(\omega)\}$ are the velocity-amplitudes. $[S(\omega)]/(i\omega)$ is again decomposed into its singular part consisting of $[C_\infty]$ and of its real term at $\omega = 0$, $[K_0]/(i\omega)$, and the remaining regular part $[C_r(\omega)]$. $[K_0]$ is the static-stiffness matrix.

$$[C_r(\omega)] = \frac{[S(\omega)]}{i\omega} - \frac{[K_0]}{i\omega} - [C_\infty] \tag{36}$$

Eq. 34 can then be reformulated involving a *velocity convolution* with $[C_r(t)]$ denoting the inverse Fourier transform of $[C_r(\omega)]$ as

$$\{R(t)\} = \int_0^t [C_r(t-\tau)]\ \{\dot{u}(\tau)\}\, d\tau + [K_0]\ \{u(t)\} + [C_\infty]\ \{\dot{u}(t)\} \tag{37}$$

$[C_r(t)]$ is also the indefinite time integral of $[K_r(t)]$.

Finally, Eq. 32 can be specified as

$$\{R(\omega)\} = \frac{[S(\omega)]}{(i\omega)^2}\, (i\omega)^2\ \{u(\omega)\} \tag{38}$$

The singular part of $[S(\omega)]/(i\omega)^2$ determined by its behaviour at $\omega = 0$ equals $[K_0]/(i\omega)^2$ + $[C_0]/(i\omega)$. The Fourier transform $[M_r(t)]$ of the regular part

$$[M_r(\omega)] = \frac{[S(\omega)]}{(i\omega)^2} - \frac{[K_0]}{(i\omega)^2} - \frac{[C_0]}{i\omega} \tag{39}$$

allows the interaction forces involving an *acceleration convolution* to be formulated as

$$\{R(t)\} = \int_0^t [M_r(t-\tau)]\ \{\ddot{u}(\tau)\}\, d\tau + [K_0]\ \{u(t)\} + [C_0]\ \{\dot{u}(t)\} \tag{40}$$

Again, $[M_r(t)]$ is the indefinite time integral of $[C_r(t)]$.

The formulation involving the velocity convolution (Eq. 37) is especially attractive, as for the calculation of the singular term only the static-stiffness matrix $[K_0]$ and the asymptotic value at $\omega = \infty$ of the damper $[C_\infty]$ are needed. The contribution of a unit area to $[C_\infty]$ equals the product of the mass density and the wave- propagation velocity.

The two alternative formulations (Eqs. 37 and 40) can also be derived in the time domain starting from Eq. 34 and using integration by parts, whereby the singularities arising at the limits of integration have to be taken into account.

As an example a very simple one-dimensional case is examined, the spherical cavity with radius a embedded in a full space (shear modulus G, mass density ρ, dilatational-wave velocity c_p) with symmetric waves occurring caused by a uniform pressure p acting on the cavity's wall (Fig. 6). The dynamic-stiffness coefficient $S(a_0)$ relating the amplitude of the radial wall displacement $u(a_0)$ to $p(a_0)$ with $a_0 = \omega a/c_p$ equals [2]

$$S(a_0) = K_0 \left[1 - \frac{c_p^2}{4c_s^2} + i a_0 \frac{c_p^2}{4c_s^2} + \frac{c_p^2}{4c_s^2} \frac{1}{1 + i a_0} \right] \tag{41}$$

where $K_0 = 4G/a$. With $K_\infty = K_0 (1 - c_p^2/(4c_s^2))$, $C_\infty = K_0 a c_p/(4c_s^2) = \rho c_p$ and with ($\bar{t} = tc_p/a$)

$$K_r(\bar{t}) = K_0 \frac{c_p^3}{4ac_s^2} e^{-\bar{t}} \tag{42}$$

the interaction force-displacement relationship (Eq. 34) equals

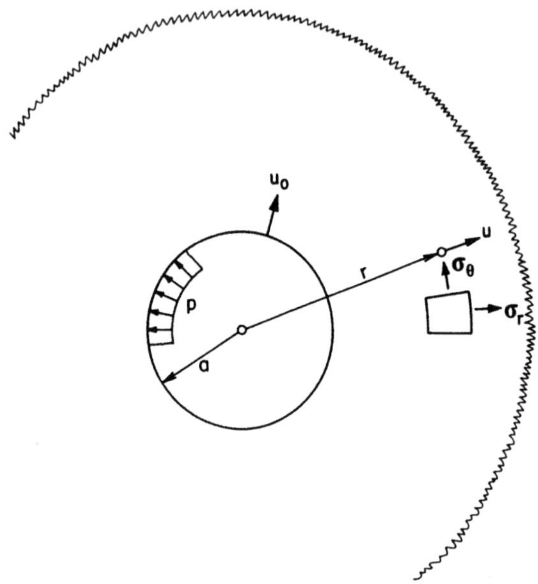

Fig. 6 Spherical cavity with uniform pressure.

$$p(t) = K_0 \frac{c_p^3}{4ac_s^2} \int_0^t e^{-c_p/a(t-\tau)} u_0(\tau)d\tau + K_0 \left(1 - \frac{c_p^2}{4c_s^2}\right) u_0(t) + \rho c_p \dot{u}_0(t) \tag{43}$$

With

$$C_r(\bar{t}) = -K_0 \frac{c_p^2}{4c_s^2} e^{-\bar{t}} \tag{44}$$

Eq. 37 with the velocity convolution is formulated as

$$p(t) = -K_0 \frac{c_p^2}{4c_s^2} \int_0^t e^{-c_p/a(t-\tau)} \dot{u}_0(\tau)d\tau + K_0 u_0(t) + \rho c_p \dot{u}_0(t) \tag{45}$$

With $C_0 = 0$ and with

$$M_r(\bar{t}) = K_0 \frac{ac_p}{4c_s^2} e^{-\bar{t}} \tag{46}$$

Eq. 40 with the acceleration convolution results in

$$p(t) = K_0 \frac{ac_p}{4c_s^2} \int_0^t e^{-c_p/a(t-\tau)} \ddot{u}_0(\tau)d\tau + K_0 u_0(t) \tag{47}$$

The interaction forces $\{R(t)\}$ expressed in Eqs. 34, 37 or 40 have to be calculated at each time station. The evaluation of the convolution integrals is computionally expensive: the total number of operations is proportional to the square of the number of time steps and in addition the total time histories of the displacements, velocites or accelerations have to be stored. As discussed in the next subsection, however, the *recursive evaluation* of the convolution integrals makes this time-domain analysis using the substructure method computationally competitive, as for a typical seismic excitation a reduction of one to three orders of magnitude results (Refs. [34, 35, 36, 37]). For instance, the recursive evaluation of the convolution integral in Eq. 34 at time $t_n = n\Delta t$ leads to

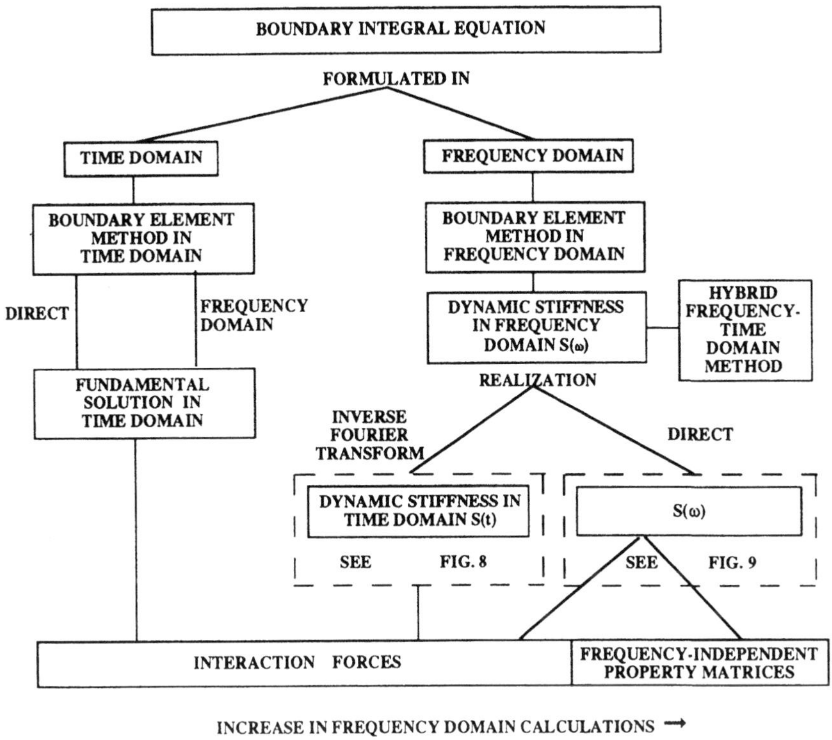

Fig. 7 Classification of computational procedures of substructure method to model unbounded soil.

$$\{R_r\}_n = \int_0^{t_n} [K_r(t-\tau)]\ \{u(\tau)\}\ d\tau \tag{48}$$

$$\{R_r\}_n = \sum_{i=1}^{M} [a]_i\ \{R_r\}_{n-i}\ + \Delta t \sum_{i=0}^{L}\ [b]_i\ \{u\}_{n-i} \tag{49}$$

where $[a]_i$ and $[b]_i$ are matrices which are independent of the time step index n, but are a function of Δt. The interaction forces $\{R_r\}_n$ are thus computed from the n-th displacements $\{u\}_n$ and the M and L most recent past values of the forces and displacements, respectively. The decrease in computational effort (the number of operations is proportional to the number of time steps) and storage requirement is especially large for dynamic-stiffness coefficients which do not return to zero immediately after the waves have passed. The velocity and acceleration convolution integrals, which are not addressed any further in this paper, can be treated analogously.

6.3 COMPUTATIONAL ALGORITHMS

A classification of the various methods to model the contribution of the unbounded soil to the equations of motion is shown in Figs. 7, 8 and 9. The further to the right a procedure is placed in Fig. 7, the more calculations are performed in the frequency domain. Many other methods with slight differences also exist.

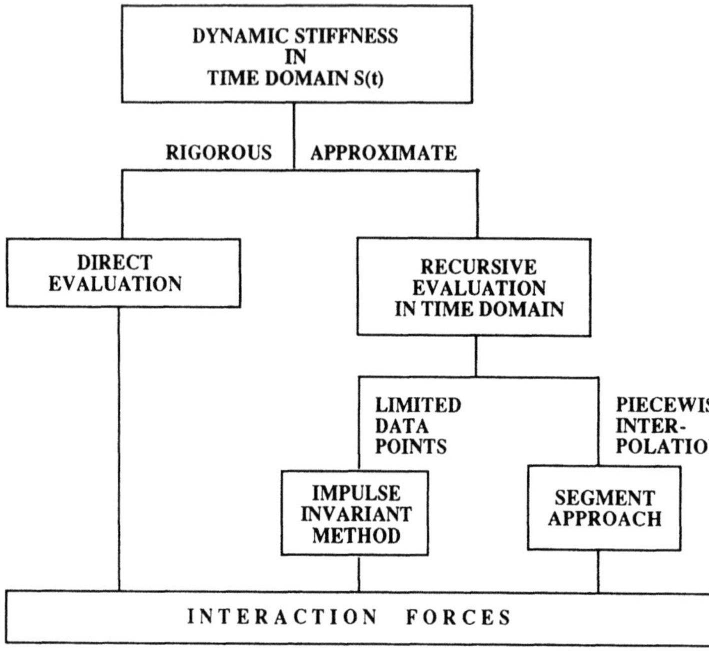

Fig. 8 Interaction forces determined starting from dynamic stiffness in time domain.

As already pointed out, the unbounded soil is modelled using the boundary element method (Ref. [26]), which is based on a boundary integral equation. The latter e.g. in the form of a reciprocity relationship or of an application of the superposition principle can either be formulated in the time domain or in the frequency domain.

The *time-domain boundary-element method* is addressed first (left part of Fig. 7). The fundamental solution (Green's function) of the full space is specified directly in the time domain, which leads to an additional discretization of the free surface of the site [27, 38]. Alternatively, as recommended for layered sites, the fundamental solution is first determined in the frequency domain and then using the inverse Fourier transformation calculated in the time domain [39]. In this case the discretization is limited to the structure-soil interface. The interaction forces of the unbounded soil acting in the nodes on the structure-soil interface can then be calculated, which is performed for each time station. A recursive evaluation of the convolution integrals appearing in the boundary integral equation should be possible. The time-domain boundary element method using the fundamental solution of the full space avoids all calculations in the frequency domain.

The *frequency-domain boundary element method* is examined next (right part of Fig. 7). The procedure leads to the dynamic-stiffness matrix in the frequency domain [S(ω)] of the unbounded soil referred to the nodes located on the structure-soil interface. This transfer function matrix describes the displacement amplitude (input) - interaction force amplitude (output) relationship in the frequency domain. The corresponding relationship in the time domain is called a *realization*, whereby many possibilities exist for a specific dynamic-stiffness matrix in the frequency domain [40]. It is appropriate to distinguish between those which, by first performing an inverse Fourier transformation, are based on the dynamic-stiffness matrix in the time domain [S(t)] and those which start directly from [S(ω)]. The former, which lead to the interaction forces at a specific time station, are classified as shown in Fig. 8. The latter which can result in addition in a set of

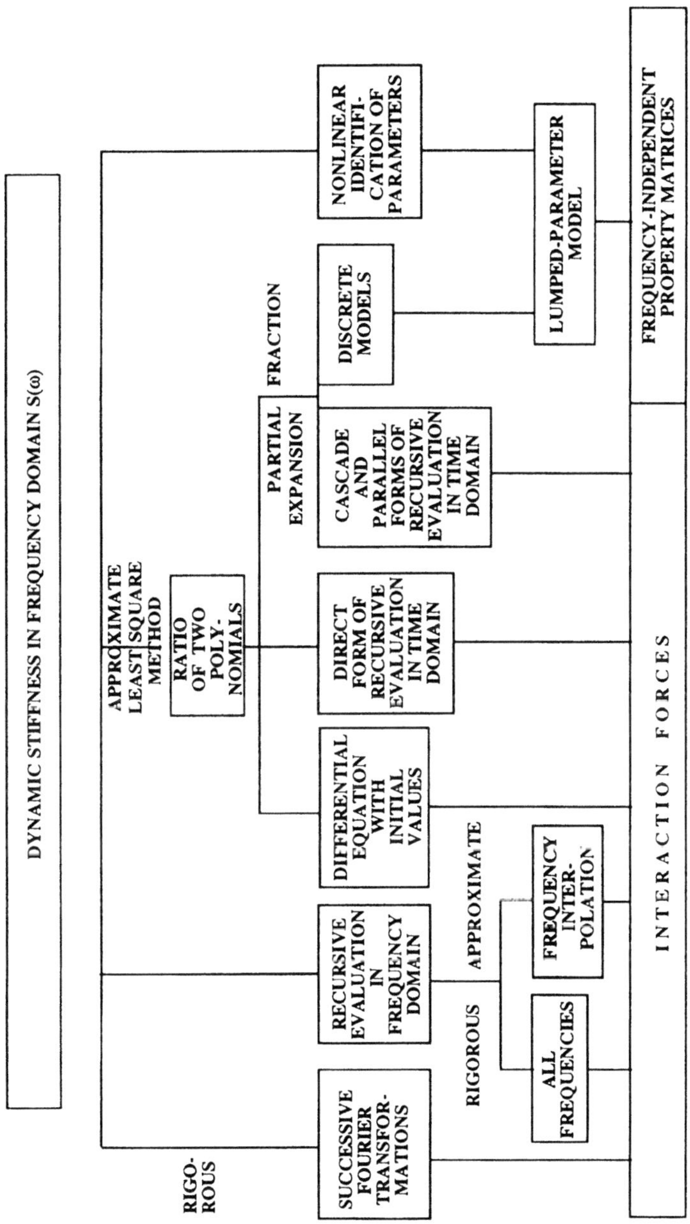

Fig. 9 Interaction forces and property matrices determined from dynamic stiffness in frequency domain.

differential equations for the interaction forces with initial values or in frequency-independent property matrices are classified as in Fig. 9. Finally, $[S(\omega)]$ of the unbounded soil can be assembled with the dynamic-stiffness matrix of the structure and the total dynamic system solved in the frequency domain. This leads to the *hybrid frequency-time domain method* [41, 42, 43], where a series of linear analyses are performed in the frequency domain iteratively with pseudo-loads taking the nonlinearities into account.

Turning to the *realizations which start from the dynamic-stiffness matrix in the time domain* $[S(t)]$ (Fig. 8), the convolution integral can be evaluated directly non-recursively. The equations for the displacement, velocity and acceleration convolution are specified in Eqs. 34, 37 and 40. The recursive formulation (Eq. 49) represents, in general, an approximation. Actually, the dynamic-stiffness matrix in the time domain $[S(t)]$ is approximated in some way. The choice of a recursive equation is not unique, and many possibilities exist. Two options are developed in Ref. [36]. The first called the *impulse-invariant method* [34] sets the approximate dynamic-stiffness matrix corresponding to the recursive formulation equal to the exact one at specified points in a certain time range. This results in a system of equations with the unknown $[a]_i$ and $[b]_i$. The approximate dynamic-stiffness matrix will, in general, deviate from the exact one in the other time ranges. In the second procedure, the *segment approach*, the dynamic-stiffness coefficients in the time domain are interpolated piecewise. Applying the so-called z-transformation then results in an explicit recursive equation without solving a system of equations.

The *realizations which work directly from the dynamic-stiffness matrix in the frequency domain* $[S(\omega)]$ and thus avoid the calculation of $[S(t)]$ are classified in Fig. 9. Various possibilities exist. The first quite inefficient procedure consists of performing at each time step a Fourier transformation of the displacement time history $\{u(t)\}$, which leads to $\{u(\omega)\}$. The interaction forces in the frequency domain $\{R(\omega)\}$ then follow as the product of $[S(\omega)]$ and $\{u(\omega)\}$; those in the time domain $\{R(t)\}$ are equal to the inverse Fourier transform. This procedure, consisting of successive Fourier transformations, is discussed in Ref. [44]. In the same reference a recursive evaluation of the amplitudes of the displacements in the frequency domain at the time station $n\Delta t$, $\{u(\omega)\}_n$, is described, using only the amplitudes of the previous time step $\{u(\omega)\}_{n-1}$ and the displacements at $n\Delta t$, $\{u\}_n$, and at $(n - 1)\Delta t$, $\{u\}_{n-1}$. An alternative derivation based on the z-transformation is possible which is addressed in Ref. [37]. It is important to stress that the recursive evaluation calculated for all frequency components is rigorous. It corresponds to the exact calculation of the convolution integral in the frequency domain. It is customary in the standard complex response analysis performed for a total linear system in the frequency domain not to solve the system of equations for all frequencies, but also to make use of *interpolation schemes*. The same concept can of course also be used in the recursive evaluation which will now, however, only be approximate.

As an alternative each coefficient of the dynamic-stiffness matrix in the frequency domain $[S(\omega)]$ can be approximated as a *ratio of two polynomials in* $i\omega$ using a curve-fitting technique based on the least-squares method which leads to the solution of a system of linear equations (right-hand side of Fig. 9). No other approximation is introduced. It is possible to transform the ratios of the two polynomials to ordinary differential equations which constant coefficients for the interaction forces together with the initial conditions, which can be solved directly [37]. Using the z-transformation, the so-called direct form of the recursive evaluation of the convolution integral can be derived [37]. Applying the partial fraction expansion to the ratio of the two polynomials and using the z-transformation the cascade [45] and parallel forms [37] of the recursive evaluation in the time domain of the interaction forces are derived. Alternatively, each term of the partial-fraction expansion can be rigorously represented by a discrete model consisting of frequency-independent springs, dashpots and masses. They form the *lumped-parameter model* [46] which can be directly incorporated in a general-purpose computer program. For important cases, such as a rigid disk foundation on the surface of or a rigid cylindrical foundation embedded in a soil layer resting on rigid rock, the coefficients of the springs, dashpots and masses have been tabulated for practical use [47, 48]. Alternatively, the corresponding frequency-independent property matrices (stiffness, damping, mass) [49] can be used as input. In this case the interaction force-displacement relationship follows as the realization. The latter can also be derived from [S(ω)] by a nonlinear identification of the parameters of the lumped-parameter model [50, 51].

As already mentioned, as an alternative to the boundary-element method, the cloning algorithm can be applied, which is not limited to the calculation of the dynamic-stiffness matrix in the frequency domain. In multi-cell cloning [11], the dynamic-stiffness coefficients can be expressed as a ratio of two polynomials in iω directly in the algorithm. The corresponding recursive evaluation and the lumped-parameter model (Fig. 9) thus follow directly from the cloning algorithm.

The various realizations shown in Figs. 8 and 9 eliminate the global coupling of the dynamic-stiffness matrix with respect to time. The spatial coupling remains, which is also present in a complex-response analysis in the frequency domain and in a static calculation.

7. Conclusions

The first classification uses as criterion the behaviour (linear or nonlinear) of the structure and of the unbounded soil. When they both remain linear, unified highly-developed analysis procedures exist, which work mostly in the frequency domain. When the unbounded soil exhibits nonlinear behaviour with the structure being linear or nonlinear, the only available analysis procedure is based on one-dimensional elasto-plastic wave propagation with one stress component in the far field. The analysis is performed in the time domain. In the remaining case the unbounded soil will behave linearly with the structure exhibiting nonlinear behaviour (with linearity as a special case). For this case the second classification distinguishes between the direct method and the substructure method. It has been demonstrated before that all available local transmitting boundaries of the direct method working in the time domain are at least in the limit of the continuum formulation mathematically equivalent. In the substructure method the dynamic stiffness representing the interaction forces of the unbounded soil is calculated based on the boundary element method (or with the multi-cell cloning algorithm) either in the time domain or in the frequency domain. In the former case, it is possible to formulate the entire procedure in the time domain. In the latter case, various realizations in the time domain are distinguished using the extent of the frequency-domain calculations as a criterion. The dynamic stiffness in the frequency domain can either be transformed to the time domain or used directly. Either interaction forces calculated recursively or frequency-independent property matrices corresponding to a lumped-parameter model of the unbounded soil are determined, which eliminate the global temporal coupling present in the interaction-force relationship. As an alternative, multi-cell cloning can lead directly to the recursive evaluation and the lumped-parameter model.

REFERENCES

[1] Wolf, J.P., «*Dynamic Soil-Structure Interaction*», Prentice-Hall, Englewood Cliffs, NJ, 1985.

[2] Wolf, J.P., «*Soil-Structure-Interaction Analysis in Time Domain*», Prentice-Hall, Englewood Cliffs, NJ, 1988.

[3] Wong, F.S. and Weidlinger P., «*Dynamic Soil-Structure Interaction and the Design of Underground Shelters*», Proceedings, Design of Protective Structures, Bundesakademie für Wehrverwaltung und Wehrtechnik, Germany, 1982.

[4] Sommerfeld, A., «*Partial Differential Equations in Physics*», Section 28, Academic Press, New York, 1949.

[5] Meek, J.W. and Wolf, J.P., «*Insights on Cutoff Frequency for Foundation on Soil Layer*», Earthquake Engineering and Structural Dynamics, 1991, vol. 20, 651-665.

[6] Makris, N, and Gazetas G., «*Dynamic Pile-Soil-Pile Interaction. Part II: Lateral and Seismic Response*», Earthquake Engineering and Structural Dynamics, 1992, vol. 21, 145-162.

[7] Waas, G., «*Analysis Methods for Footing Vibrations through Layered Media*», Ph.D. Thesis, Department of Civil Engineering, University of California, Berkeley, 1972.

[8] Song, Ch. and Wolf, J.P., «*Dynamic Stiffness of Unbounded Medium Based on Damping-Solvent Extraction*», submitted to Earthquake Engineering and Structural Dynamics for review and possible publication.

[9] Lysmer, J. and Kuhlemeyer, R.L., «*Finite Dynamic Model for Infinite Media*», Journal of Engineering Mechanics, ASCE, 1969, vol. 95, 859-877.

[10] Wang, Y. and Rajapakse, R.K.N.D., «*Dynamics of Rigid Strip Foundations Embedded in Orthotropic Elastic Soils*», Earthquake Engineering and Structural Dynamics, 1991, vol. 20, 927-947.

[11] Wolf, J.P. and Song, Ch., «*Dynamic-Stiffness Matrix of Unbounded Soil by Finite-Element Cloning*», submitted to Earthquake Engineering and Structural Dynamics for review and possible publication.

[12] Lysmer, J., «*Analytical Procedures in Soil Dynamics*», Proceedings Specialty Conference on Earthquake Engineering and Soil Dynamics, Geotechnical Engineering Division, ASCE, Pasadena, CA, 1978, vol. 3, 1267-1316.

[13] Roesset, J.M., «*Stiffness and Damping Coefficients of Foundations*», Dynamic Response of Pile Foundations: Analytical Aspects, M.W. O'Neil and R. Dobry, eds., Proceedings of the Geotechnical Engineering Division, ASCE, October 1980, 1-30.

[14] Luco, J.E., «*Linear Soil-Structure Interaction: A Review*», Applied Mechanics Division, ASME, 1982, vol. 53, 41-57.

[15] Luco, J.E., «*CLASSI, a Program for Analysis of Soil-Structure Interaction*», Personal Communication, University of California, San Diego, CA.

[16] Lysmer, J., Tabatabaie, M., Tajirian, F., Vahdani, S. and Ostadan, F., «*SASSI, a System for Analysis of Soil-Structure Interaction*», UCB/GT, 81-02, University of California, Berkley, CA, 1981.

[17] Wolf, J.P. and Paronesso, A., «*One-Dimensional Modelling of the Non-Linear Far-Field in Soil-Structure-Interaction Analysis*», Earthquake Engineering and Structural Dynamics, 1992, vol. 21, November.

[18] Engquist, B. and Majda A., «*Absorbing Boundary Conditions for the Numerical Simulation of Waves*», Mathematics of Computation, 1977, vol. 31, 629-651.

[19] Clayton, R. and Engquist, B., *«Absorbing Boundary Conditions for Acoustic and Elastic Wave Equations»*, Bulletin of the Seismological Society of America, 1977, vol. 67, 1529-1540.
[20] Liao, Z.P. and Wong, H.L., *«A Transmitting Boundary for the Numerical Simulation of Elastic Wave Propagation»*, Soil Dynamics and Earthquake Engineering, 1984, vol. 3, 174-183.
[21] Smith, W.D., *«A Nonreflecting Plane Boundary for Wave Propagation Problems»*, Journal of Computational Physics, 1974, vol. 15, 492-503.
[22] Cundall, P.A., Kunar, R.R., Carpenter, P.C. and Marti, J., *«Solution of Infinite Dynamic Problems by Finite Modelling in the Time Domain»*, Proceedings of the 2nd International Conference on Applied Numerical Modelling, Madrid, Pentech Press London, 1974, 339-351.
[23] Cohen, M. and Jennings, P.C., *«Silent Boundary Methods for Transient Analysis»*, in Computational Methods for Transient Analysis, Belytschko, T. and Hughes, T.J.R., eds. Elsevier Science Publishers, Amsterdam, 1983, 301-360.
[24] Kausel, E., *«Local Transmitting Boundaries»*, Journal of Engineering Mechanics, ASCE, 1988, vol. 114, 1011-1027.
[25] Wolf, J.P., «Classification of Analysis Methods for Dynamic Soil-Structure Interaction», Proceedings 2nd Internatoinal Conference on Recent Advances in Geotechnical Earthquake Engineering and Soil Dynamics, S. Prakash, ed., St. Louis, Missouri, March 1991, vol. II, 1821-1832.
[26] Beskos, D.E., *«Boundary Element Methods in Dynamic Analysis»*, Applied Mechanics Review, 1987, vol. 40, 1-23.
[27] Karabalis, D.L. and Beskos, D.E., *«Dynamic Response of 3-D Rigid Surface Foundations by Time Domain Boundary Element Method»*, Earthquake Engineering and Structural Dynamics, 1984, vol. 12, 73-94.
[28] Wolf, J.P. and Obernhuber, P., *«Nonlinear Soil-Structure-Interaction Analysis Using Dynamic Flexibility of Soil for Impulse Forces»*, Proceedings, 8th World Conference on Earthquake Engineering, San Francisco, CA, July 1984, vol. 3, 969-976.
[29] Cole, D.M., Kosloff, D.D. and Minster, J.B., *«A Numerical Boundary Integral Equation Method for Elastodynamics»*, Bulletin of Seismological Society of America, 1978, Vol. 68, 1331-1357.
[30] Wolf, J.P. and Darbre, G.R., *«Time-Domain Boundary Element Method in Visco-Elasticity with Application to a Spherical Cavity»*, Soil Dynamics and Earthquake Engineering, 1986, vol. 5, 138-148.
[31] Dasgupta, G., *«A Finite Element Formulation for Unbounded Homogeneous Continua»*, Journal of Applied Mechanics, ASME, 1982, vol. 49, 136-140.
[32] Motosaka, M. and Nagano, M., *«Basic Investigation on the Evaluating of Interaction Forces of Unbounded Soil in the Time-Domain Substructure Method»*, Proceedings, 8th Japan Earthquake Engineering Symposium, 1990 and Summary of Technical Papers of Annual Meeting of Architectural Institute of Japan, 1990 (in Japanese).
[33] Wolf, J.P. and Obernhuber, P., *«Nonlinear Soil-Structure-Interaction Analysis Using Dynamic-Stiffness or Flexibility of Soil in the Time Domain»*, Earthquake Engineering and Structural Dynamics, 1985, vol. 13, 195-212.
[34] Verbic, B., *«Analysis of Certain Structure-Foundation Systems»*, Ph.D. Dissertation, Department of Civil Engineering, Rice University, 1973.
[35] Meek, J.W., *«Recursive Analysis of Dynamical Phenomena in Civil Engineering»*, Bautechnik, 1990, vol. 67, 205-210 (in German).

[36] Wolf, J.P. and Motosaka, M, *«Recursive Evaluation of Interaction Forces of Unbounded Soil in the Time Domain»*, Earthquake Engineering and Structural Dynamics,1989, vol. 18, 345-363.

[37] Wolf, J.P. and Motosaka, M, *«Recursive Evaluation of Interaction Forces of Unbounded Soil in the Time Domain from Dynamic-Stiffness Coefficients in the Frequency Domain»*, Earthquake Engineering and Structural Dynamics,1989, vol. 18, 365-376.

[38] Antes, H., *«A Boundary Element Procedure for Transient Wave Propagation in Two-Dimensional Isotropic Elastic Media»*, Finite Elements in Analysis and Design, 1985, vol. 1, 313-322.

[39] Wolf, J.P. and Obernhuber, P., *«Nonlinear Soil-Structure-Interaction Analysis Using Green's Function of Soil in the Time Domain»*, Earthquake Engineering and Structural Dynamics, 1985, vol. 13, 213-223.

[40] Chen, C.T., *«Linear System Theory and Design»*, Holt, Rinehart and Winston, New York, 1984.

[41] Kawamoto, J.D., *«Solution of Nonlinear Dynamic Structural Systems by a Hybrid Frequency-Time Domain Approach»*, Research Report R 83-5, Department of Civil Engineering, Massachusetts Institute of Technology, Cambridge, MA, 1983.

[42] Darbre, G.R. and Wolf, J.P., *«Criterion of Stability and Implementation Issues of Hybrid Frequency-Time Domain Procedure for Nonlinear Dynamic Analysis»*, Earthquake Engineering and Structural Dynamics, 1988, vol.18, 569-581.

[43] Darbre, G.R., *«Seismic Analysis of Non-Linearly Base-Isolated Soil-Structure Interacting Reactor Building by Way of the Hybrid Frequency-Time Domain Procedure»*, Earthquake Engineering and Structural Dynamics, 1990, vol.19, 725-738.

[44] Mohasseb, S.K. and Wolf, J.P., *«Recursive Evaluation of Interaction Forces of Unbounded Soil in Frequency Domain»*, Soil Dynamics and Earthquake Engineering, 1989, vol. 8, 176-188 (also in Soil-Structure Interaction, Cakmak, A.S., ed., Developments in Geotechnical Engineering, vol. 43, Elsevier Science, Amsterdam, 1987, 41-55).

[45] Oppenheim, A.V. and Willsky, A.S., *«Signals and Systems»*, Prentice-Hall, Englewod Cliffs, NJ, 1983.

[46] Wolf, J.P., *«Consistent Lumped-Parameter Models for Unbounded Soil: Physical Representation»*, Earthquake Engineering and Structural Dynamics, 1991, vol. 20, 11-32.

[47] Wolf, J.P. and Paronesso, A., *«Lumped-Parameter Model for Foundation on Layer»*, Proceedings 2nd International Conference on Recent Advances in Geotechnical Earthquake Engineering and Soil Dynamics, S. Prakash, ed., St. Louis, Missouri, March 1991, vol. I, 895-906.

[48] Wolf, J.P. and Paronesso, A., *«Lumped-Parameter Model for Rigid Cylindrical Foundation Embedded in Soil Layer on Rigid Rock»*, Earthquake Engineering and Structural Dynamics, 1992, vol. 21, October.

[49] Wolf, J.P., *«Consistent Lumped-Parameter Models for Unbounded Soil: Frequency-Independent Stiffness, Damping and Mass Matrices»*, Earthquake Engineering and Structural Dynamics, 1991, vol. 20, 33-41.

[50] De Barros, F.C.P. and Luco, J.E., *«Discrete Models for Vertical Vibrations of Surface and Embedded Foundations»*, Earthquake Engineering and Structural Dynamics, 1990, vol. 19, 289-303.

[51] Jean, W.Y., Lin, T.W. and Penzien, J., *«System Parameters of Soil Foundations for Time Domain Dynamic Analysis»*, Earthquake Engineering and Structural Dynamics, 1990, vol. 19, 541-553.

[52] Veletsos, A.S., «*Basic Response Functions for Elastic Foundations*», Journal Engineering Mechanics, ASCE, 1974, vol. 100, 189-202..

[53] Hayashi, Y., Fukuwa, N., Nakai, S. and Koyanagi, Y., «*Earthquake Response Analysis Considering Soil-Structure Separation Using Contact Elements and Dynamic Flexibility of Soil in Time Domain*», Proceedings, 9th World Conference on Earthquake Engineering, Tokyo-Kyoto, August 1988, vol. VIII, 265-270.

Final note

This paper for the course, which was originally scheduled to be held two years earlier, was prepared in parallel to Ref. 25. It has since been updated and Sections 2 and 3 have been added.

DEVELOPMENT OF ANALYTICAL TECHNIQUES IN SOIL-STRUCTURE INTERACTION

M.NURAY AYDINOĞLU
STFA Construction Co.,
81190 Altunizade, Çamlıca, Istanbul
Turkey

ABSTRACT. The development of soil-structure interaction during the last quarter of a century has followed two distinctive paths yielding the Direct and Substructure Methods of analysis. In current state-of-the-art both methods are represented by quite similar models with differences being restricted to the dynamic boundary conditions defined along the boundary of the finite soil region considered. Therefore achieving a common formulation equally applicable to both methods is possible. Within the framework of Substructure Method, the Boundary Element Method has been very effective in handling geometrically irregular conditions at the soil-structure interface. For the stiffness problem of embedded foundations, the determination of Green functions in unexcavated, virgin soil condition and the subsequent subtraction of the excavation zone are the two main steps of the analysis concept called Analytical Excavation. Effective input motions of flexible and rigid foundations are easily obtained once the dynamic stiffness and free-field response of virgin soil are known. Examples presented are the stiffness and damping influence coefficients as well as the components of effective base input motion of a rectangular foundation embedded into a damped half plane.

1. Introduction

Over the last quarter of a century the dynamic soil-structure interaction has attracted an intensive interest among researchers and engineers dealing with structural dynamics, wave mechanics and soil dynamics. This interest may be attributed, on one hand to the practical significance of the subject in the design of massive and rigid structures such as nuclear power plants, and on the other hand to the challenging nature of the problem which involves various analytical techniques and unaccustomed modelling requirements of soil medium with boundaries at infinity. The latter which is caused by the radiation of dynamic energy purely due to unbounded geometry of soil medium is, in fact, a unique feature of dynamic soil-structure interaction and constitutes the most difficult and complex part of the problem. It may even be stated that soil-structure interaction would not have been counted beyond the classical scope of structural dynamics had the radiation phenomenon were not so much significant.

In the course of development of soil-structure interaction, modelling and analysis of soil medium has evolved through various stages but always in two distinctive directions, namely Direct Method and Substructure Method. Although conceived and developed in different ways, both however used the same analysis domain, that is the frequency domain, for quite a long time which inevitably restricted soil-structure interaction with linear system behaviour. Nonlinearity of soil was taken into account only in an approximate manner through equivalent linear analysis procedure in which dynamic soil parameters were adjusted in accordance with peak or average strains during iterative linear solutions of the system in frequency domain. The structure, whose

P. Gülkan and R. W. Clough (eds.), Developments in Dynamic Soil-Structure Interaction, 25–42.

analysis is the main objective of soil-structure interaction, had to be assumed being linear.

On the other hand, as Professor R.W.Clough rightly stated in his foreword to Ref. [11], "The other disadvantage is that the typical structural analyst is not accustomed to working in the frequency domain; his natural approach is to consider the sequence of developments from one time to the next-that is to apply the time domain concept. This mental obstacle has greatly limited the general acceptance of analytical procedures using frequency domain formulations."

In spite of what is mentioned above, the development in frequency domain was bound to be experienced as an evolution stage of soil-structure interaction. The significance of the problem has been well understood and numerous practical applications have been realized. The efficiency in frequency domain analysis and computational procedures should not be disregarded. Simple expressions of equations motion in frequency domain for discretized parts of the system made it possible to manipulate them as easy as static equilibrium equations leading to various analytical formulations particularly in Substructure Method. In addition, hysteretic nature of soil damping was easily incorporated into formulation of the problem. On computational side, availability of Fast Fourier Transform algorithm coupled with interpolation schemes have made the numerical procedures highly efficient. Therefore it can be concluded that it is the maturity gained in frequency domain analysis that eventually has led to the initiation of efforts during the last decade to attack the ultimate problem-that is the nonlinear response analysis of soil-structure systems which can be handled in the time domain only.

2. Evaluation of Direct Method and Substructure Method

As mentioned above, modelling and formulation of soil-structure interaction conceived and developed in two different directions yielding Direct Method and Substructure Method. A brief evaluation of developments in both methods and interaction in between could be interesting and instructive.

2.1. A BRIEF EVALUATION OF PAST DEVELOPMENT

Throughout their simultaneous development, Direct and Substructure Methods continuously influenced each other such that at every main step one had achieved what the other was lacking for or vice versa. The Direct Method initiated with standard finite element modelling of soil underlain by bedrock where embedment of structure and irregularities in adjacent soil region were easily incorporated (Fig 1a). Apparently Direct Method was intended to work in the time domain using well established procedures of structural dynamics. However at this stage Direct Method failed to model radiation effect which was inherently considered by Substructure Method in its simple soil-structure model where readily available continuum solutions involving steady-state response of rigid surface foundations were utilized to represent the dynamic stiffness of the soil medium (Fig 1b). Thus, inevitably Substructure Method had to start with working in frequency domain. In response, on the Direct Method side, recognition of radiation effect led to the development of artificial "Transmitting Boundaries" by which reflections of waves emanating from the structure and adjacent soil would be avoided. In this respect the introduction of "viscous boundaries" occured relatively in early years [6]. Being time or frequency independent, these simple and uncoupled boundaries looked very convenient and attractive for Direct Method analysis in time domain, but it was immediately discovered that they were insufficient to represent the real wave pattern within the soil medium (Fig 2a). In the meantime, Substructure Method with its simple model allowed the realization of parametric studies to demonstrate the significance and consequences of soil-structure interaction [8]. At a particular stage, Direct Method practically used the substructuring concept by the development of frequency dependent "Consistent Boundaries" [9]. As the name implies, these boundaries were able to consistently represent the general wave

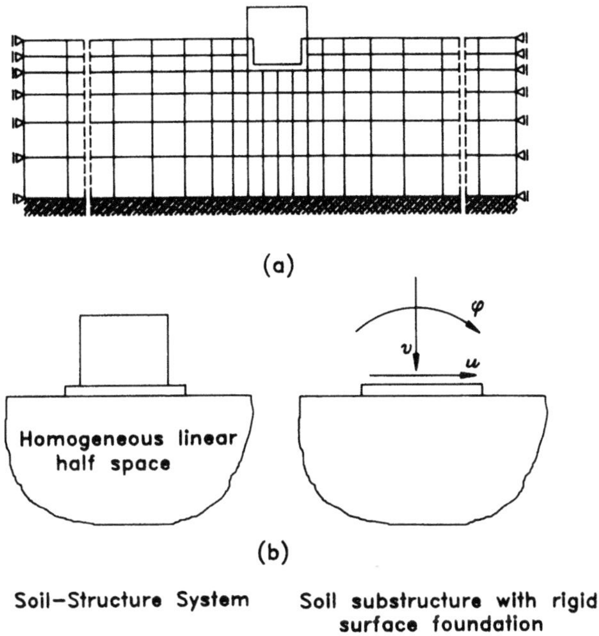

Fig.1. Early models of (a) Direct Method and (b) Substructure Method

pattern in layered soils underlain by a rigid bedrock. In fact, the far-field soil zones represented by consistent boundaries were perfect substructures with their frequency dependent and coupled force-displacement relationships. Consequently, consistent boundaries were able to be placed just near the structure and adjacent irregular soil zone leading to a drastic reduction in number of degrees of freedom in soil-structure system (Fig 2b). By their rigorous formulation consistent boundaries effectively made the Direct Method enter into the frequency domain. This situation was incidentally strengthened by the findings in soil dynamics which indicated the hysteretic nature of soil damping, an energy dissipation mechanism well suited to frequency domain analysis.

This development in Direct Method using consistent boundaries coupled with hysteretic soil damping led to general computer codes [7] but hardly affected the proponents of Substructure Method who had virtually committed themselves to continuum approach. On that side the attention had focused on embedment problem which had meant a further complication of the mixed-boundary value problem, now to be solved for also geometrically irregular boundary conditions. Gradually the concept of discretizing the soil-structure interface was introduced [3] which initiated the efforts to deal with embedment problem within the framework of Substructure Method. However the effect of embedment on free-field response as well as interrelationships between and equivalence of two methods remained as serious problems. The well known concept of seperating soil-structure interaction into two phases, namely Kinematic Interaction and Inertial Interaction (see Ref.[10]) has been introduced not only to provide an engineering sense of understanding to the problem, but at the same time to show through kinematic interaction that the free- field response with embedment was drastically different from that used for surface supported structures. Eventually it was in late 70's that the effect of embedment in Substructure Method was clarified [4] followed by an attempt to unify the formulation of soil-structure interaction in Ref.[1]

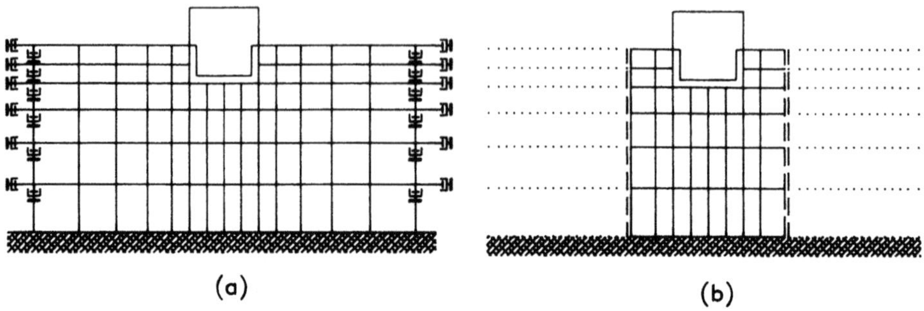

Fig.2. Schematic representation of Direct Method with (a) viscous , (b) consistent boundaries.

where it was shown that governing equations of motion of Substructure Method can be derived for various combinations of total and relative response components utilizing the equations of "Complete Soil-Structure System" and certain "Reference Soil Systems" through which free-field soil response is defined (for more detailed treatment of Ref [1] see chapter 8 of Ref [10]).

Through the last decade both Direct and Substructure methods developed towards maturity as evidenced and well documented in two textbooks published [10,11]. Substructure Method overcame the difficulties associated with embedded foundations by the rapid development of Boundary Element Method with which virtually any type of geometrical boundary conditions can be simulated in both frequency and time domains by discrete modelling performed merely at the soil-structure interface. In this method, rigorous treatment of unbounded soil was achieved through the derivation of discretized Green functions. In order to consider an irregular soil zone adjacent to the structure, this zone was treated as part of the structure with finite element modelling [2,11].

The Direct Method, on the other hand, is still being developed to fulfill its primary goal, i.e., creating efficient transmitting boundaries in both frequency and time domains [5,11].

2.2. EVALUATION OF CURRENT MODELS

Time domain analysis is the new era for both methods whose current models are shown in Fig.3. Each consists of an irregular (inner) soil zone and a regular (outer) soil zone divided by what is called "Interaction Horizon" [11]. Interaction horizon, through which the seismic input motion is applied, corresponds to Transmitting Boundaries in Direct Method and to "Irregular-Regular Soil Interface" in Substructure Method. Irregular soil zone, which is modelled with finite elements, is intended to consider all geometrical, mechanical irregularities and, material and geometrical nonlinearities. In both models regular (outer) zone is taken as linear whereas superstructure may behave nonlinearly.

In analytical terms, transmitting boundaries of Direct Method actually serve for the determination of interaction forces to enforce the dynamic equilibrium conditions along the interaction horizon. In fact the same is realized in Substructure Method through dynamic stiffness matrix of regular soil zone. The difference in terminology corresponds to the fact that in Direct Method dynamic properties of transmitting boundaries are required to be "local", that is uncoupled both in space and time. One can intuitively conclude that this requirement would lead to approximate analytical expressions for transmitting boundaries as opposed to rigorous and fully coupled force-displacement relationships in Substructure Method.

In Substructure Method dynamic stiffness matrix of unbounded regular soil zone in time domain can be obtained either by working in the same domain or from frequency domain solutions using convolution theorem of Fourier Transform. However what is actually needed in time domain

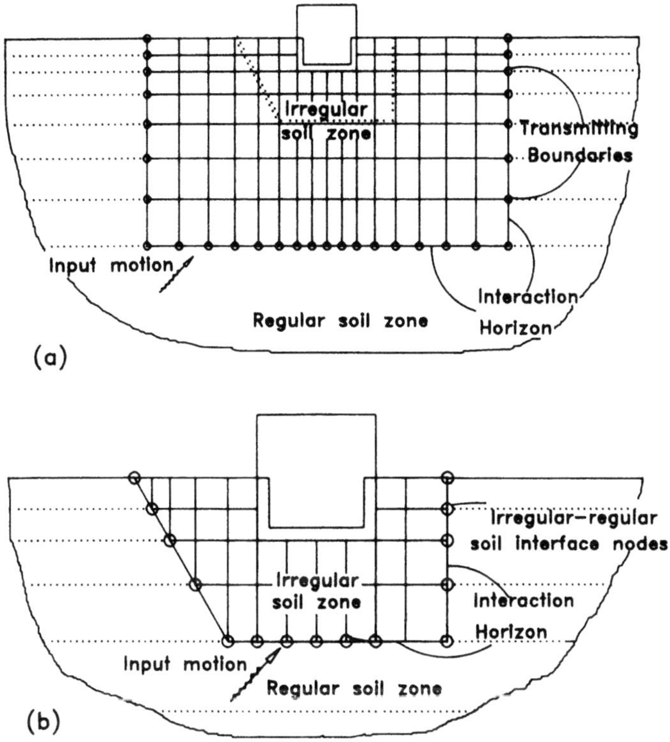

Fig.3. Current models of (a) Direct Method and (b) Substructure Method

analysis is to determine the so-called interaction forces which are equal to time integration of the product of dynamic stiffness matrix and relative response vector (interaction displacements) calculated along the interaction horizon (See Eq. 12a in next section). According to Fourier convolution theorem this time integral can also be evaluated from its counterpart in the frequency domain for each time step. Certain recursive calculation techniques have been developed [11] to evaluate convolution integrals in both frequency and time domain in order to reduce the number of numerical operations involved. It is worth to mention that determination of time dependent interaction forces via frequency domain analysis is limited with the regular soil zone only and should not be regarded as a backward trend to the frequency dependency of the solution process in soil-structure interaction.

The location of interaction horizon is an important problem for both methods. In Direct Method it is usually located farther away from the structure than Substructure Method to reduce the adverse effects that may arise from the approximate nature of transmitting boundaries. It can be such that the outer parts of the irregular soil zone may be far away from the real irregularities and nonlinearities and thus may behave linearly. The Substructure Method has no such problem and in the absence of geometrical or mechanical irregularities and nonlinearities, interaction horizon can be coincided directly with the soil-structure interface. Substructure Method can consider the material nonlinearity of soil within the irregular zone due to response caused by the structure (i.e., inertial interaction) but fails to consider soil nonlinearity, if any, occuring in the free-field response of soil when formulation based on unexcavated soil response is used [10,11]. Both types of nonlinearity may affect the location of interaction horizon out of which the latter will further be addressed in the next section.

Regarding the future prospects of Direct and Substructure Methods in time domain analysis, a straightforward evaluation may be premature. The Direct Method may become more attractive to engineers as long as sufficiently accurate transmitting boundaries are developed which, as stated in Ref.[12], is "definitely the area of soil-structure interaction analysis where the research efforts should be concentrated". However resulting transmitting boundaries may not be as simple as attaching certain masses, springs and dashpots directly to interaction horizon and may come up with sophisticated couplings [13,14]. The Substructure Method, on the other hand, with its rigorous formulation, could see its chance for practical applications if efficient numerical procedures are further developed for the evaluation of convolution integrals.

3. A Common Formulation for Direct and Substructure Methods

Considering the similar modelling of Direct and Substructure Methods (Fig.3) except possibly the size of irregular soil zone and definition of boundary conditions along interaction horizon, one can expect that a common formulation applicable to both methods should be achievable. This is indeed possible as given in the following.

For convenience, following formulation is first given in frequency domain and then converted to time domain. In the course of derivation in frequency domain certain observations are made for soil nonlinearity which, of course, can be handled with time domain version of the formulation. Since the aim is to reach a common formulation for both methods, one can start with the Substructure formulation of the system shown in Fig 4a as follows;

$$\begin{bmatrix} \bar{K}_{ss} & \bar{K}_{sb} & 0 & 0 \\ \bar{K}_{bs} & (\bar{K}^{s}_{bb}+\bar{K}^{i}_{bb}) & \bar{K}_{bi} & 0 \\ 0 & \bar{K}_{ib} & \bar{K}_{ii} & \bar{K}_{ih} \\ 0 & 0 & \bar{K}_{hi} & (\bar{K}^{i}_{hh}+\bar{S}^{r}_{hh}) \end{bmatrix} \begin{Bmatrix} \bar{r}^{t}_{s} \\ \bar{r}^{t}_{b} \\ \bar{r}^{t}_{i} \\ \bar{r}^{t}_{h} \end{Bmatrix} = \begin{Bmatrix} 0 \\ 0 \\ 0 \\ \bar{P}_{h} \end{Bmatrix} \tag{1}$$

in which subscripts and/or superscripts s,b,i,h,r refer to structure, its base, irregular soil zone , interaction horizon and regular soil zone, respectively. The dynamic stiffness submatrices on the left-hand side are expressed, in general, either as

$$\bar{K} = K+i\omega C-\omega^2 M \tag{2a}$$

or

$$\bar{K} = K(1+2i\beta)-\omega^2 M \tag{2b}$$

Eq.2a or Eq.2b can be used for systems with viscous or hysteretic damping, respectively. In the latter, β indicates the hysteretic damping ratio. **K**, **C**, **M** represent the dynamic property matrices and ω is the frequency of excitation. Overbars used for all terms in Eq.1 indicate the frequency domain formulation. Submatrix $\bar{S}^{r}_{hh}$ is the dynamic stiffness matrix of unbounded regular soil zone.The response vector is represented by total displacements and indicated by superscript t.

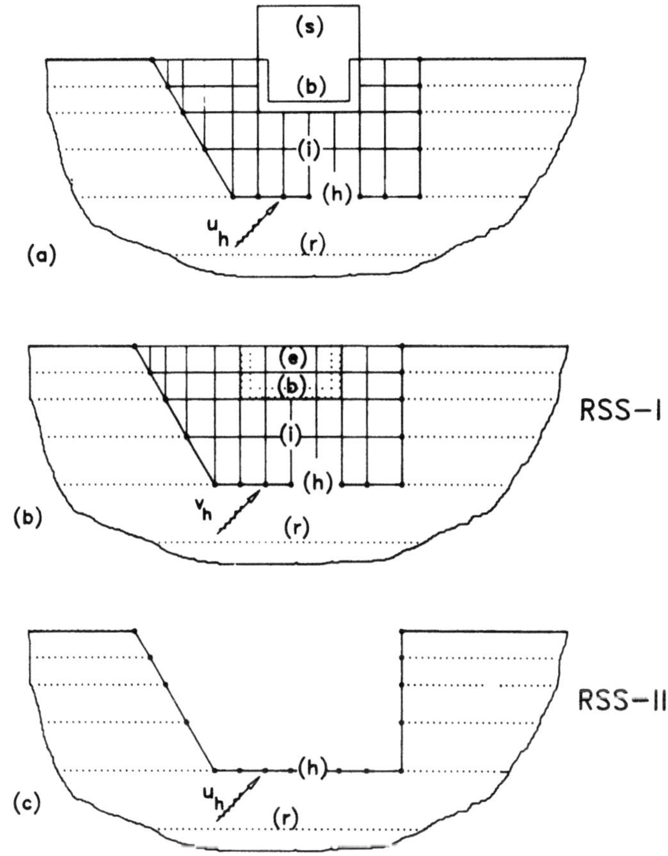

Fig.4. Modelling of (a) Soil-Structure System , (b) Reference Soil System I (RSS-I), (c) Reference Soil System II (RSS-II)

The nonzero effective force vector component on the rigt-hand side of Eq.1 depends on the choice of Reference Soil System (RSS) used in order to define the free-field response [1]. In Fig.4b and Fig.4c two reference soil systems, named RSS-I and RSS-II, respectively, are shown. The first system represents the so-called virgin soil condition whereas the second one corresponds to an imaginary condition as if the soil is excavated down to interaction horizon. Based on RSS-II the nonzero effective force vector component of Eq.1 can be expressed as;

$$\bar{P}_h = \bar{S}^r_{hh}\, \bar{u}_h \tag{3}$$

where $\bar{u}_h$ represents the free-field motion for RSS-II (Same expression can be found, with notation differences, in Refs [1,10] for the case where interaction horizon coincides with soil-structure interface). Effective force vector can also be expressed in terms of free-field motion of RSS-I (virgin soil). In order to achieve this, first equations of motion of RSS-I are written simply by replacing structural degrees of freedom in Eq.1 with those of soil zone to be excavated for embedment (as denoted by e). Thus,

$$\begin{bmatrix} \bar{K}_{ee} & \bar{K}_{eb} & 0 & 0 \\ \bar{K}_{be} & (\bar{K}^{e}_{bb}+\bar{K}^{i}_{bb}) & \bar{K}_{bi} & 0 \\ 0 & \bar{K}_{ib} & \bar{K}_{ii} & \bar{K}_{ih} \\ 0 & 0 & \bar{K}_{hi} & (\bar{K}^{i}_{hh}+\bar{S}^{r}_{hh}) \end{bmatrix} \begin{Bmatrix} \bar{v}_{e} \\ \bar{v}_{b} \\ \bar{v}_{i} \\ \bar{v}_{h} \end{Bmatrix} = \begin{Bmatrix} 0 \\ 0 \\ 0 \\ \bar{P}_{h} \end{Bmatrix} \tag{4}$$

in which response vector components now represent the free-field motion for RSS-I. It is evident that effective force vector component of Eq.4 is the same as that of Eq.1 as given in Eq.3. Physically they correspond to the reaction forces (fixed-end forces) acting along the outer face of interaction horizon due to incident and reflected waves travelling through regular soil zone, when interaction horizon is held fixed in space. There are two possibilities to relate Eq.4 to Eq.1 in order to express the effective force vector of Eq.1 in terms of free-field response. The first is to eliminate all degrees of freedom in Eq.4 except those on interaction horizon which yields to

$$(\bar{S}^{i}_{hh}+\bar{S}^{r}_{hh})\ \bar{v}_{h} = \bar{P}_{h} \tag{5}$$

in which $\bar{S}^{i}_{hh}$ is the condensed stiffness submatrix representing the whole soil portion inside the interaction horizon. Eq.5 can be rewritten as

$$\bar{P}_{h} = \bar{S}_{hh}\ \bar{v}_{h} \tag{6}$$

where $\bar{S}_{hh}$ (without any superscript) being the sum of two submatrices on the left-hand side of Eq.5 represents the stiffness matrix of virgin soil (RSS-I) defined along the nodes on interaction horizon. Substituting Eq.6 into Eq.1 thus allows the use of free-field motion of virgin soil defined only along interaction horizon for the formulation of Substructure Method as given in Refs.[10,11]. However, this formulation is valid for only linear free-field response of irregular soil zone because of the condensation applied in Eq.5.

The second possibility to relate Eq.4 to Eq.1 is directly to extract the last row of Eq.4 to define the effective force vector component as used in Ref [1], i.e.,

$$\bar{P}_{h} = \bar{S}^{r}_{hh}\ \bar{v}_{h} + (\ \bar{K}^{i}_{hh}\ \bar{v}_{h} + \bar{K}_{hi}\ \bar{v}_{i}\ ) \tag{7}$$

Regarding the physical interpretation of effective force vector, the first term at the right-hand side represents the reaction forces acting along the outer face of interaction horizon in case of RSS-I while the second term in paranthesis corresponds to the forces along the inner face when (imaginarily) interaction horizon is held fixed in space.

Substituting Eq.7 into Eq.1 results in a second version of Substructure formulation in terms of free-field response of virgin soil. It is interesting to note that the expression of Eq.7 is valid also for nonlinear free-field response of irregular soil zone except for the nodes immediately inside the interaction horizon as coupling submatrix $\bar{K}_{hi}$ relates only those nodes to the ones on interaction horizon which is linear by definition. It is to be mentioned that in free-field response of virgin soil (either by downward back analysis from a control point on the surface or by upward analysis for a rock outcrop input) the soil amplification may lead to nonlinearities in the upper parts which would also affect the linear response below. Thus locating the interaction horizon such that upper boundary of linear part coincides with the nodes immediately inside the interaction horizon would allow an indirect consideration of nonlinear effects occuring in the free-field response of virgin soil.

Eq.1 with load vector component given in Eq.7 is, in principle, applicable to both Substructure and Direct Methods. However this formulation can be further simplified by defining the response vector component related to interaction horizon as relative to the free-field response of virgin soil, i.e., the total response component can be written as,

$$\bar{r}_h^t = \bar{v}_h + \bar{r}_h^\Delta \tag{8}$$

in which the second term refers to the relative response. Since the nodes on the interaction horizon (and those immediately inside) are linear by definition, the superposition principle as applied in Eq.8 is perfectly valid. Substituting Eq.8 into Eq.1 and considering Eq.7 result in;

$$\begin{bmatrix} \bar{K}_{ss} & \bar{K}_{sb} & 0 & 0 \\ \bar{K}_{bs} & (\bar{K}_{bb}^s + \bar{K}_{bb}^i) & \bar{K}_{bi} & 0 \\ 0 & \bar{K}_{ib} & \bar{K}_{ii} & \bar{K}_{ih} \\ 0 & 0 & \bar{K}_{hi} & (\bar{K}_{hh}^i + \bar{S}_{hh}^r) \end{bmatrix} \begin{Bmatrix} \bar{r}_s^t \\ \bar{r}_b^t \\ \bar{r}_i^t \\ \bar{r}_h^\Delta \end{Bmatrix} = \begin{Bmatrix} 0 \\ 0 \\ -\bar{K}_{ih}\bar{v}_h \\ \bar{K}_{hi}\bar{v}_i \end{Bmatrix} \tag{9}$$

Eq.9 is the ultimate equation of motion of the Common Formulation of Direct and Substructure Methods in the frequency domain. This equation can be looked upon as the generalization of Direct Method, however it is equally applicable to Substructure Method. It is interesting to observe that as relative displacements on interaction horizon approach to zero, which physically corresponds to augmenting of irregular soil zone such that interaction horizon approaches to infinity, upper three rows of Eq.9 yield to traditional formulation of Direct Method [1,10]. Furthermore Eq.9 allows both methods for indirect consideration of material nonlinearity within the irregular soil zone regarding nonlinear response of the free-field. Of course, nonlinearity can be worked out in the time domain for which corresponding equations of motion can be obtained from Eq.9 as,

$$\begin{bmatrix} M_{ss} & M_{sb} & 0 & 0 \\ M_{bs} & (M_{bb}^s + M_{bb}^i) & M_{bi} & 0 \\ 0 & M_{ib} & M_{ii} & M_{ih} \\ 0 & 0 & M_{hi} & M_{hh}^i \end{bmatrix} \begin{Bmatrix} \ddot{r}_s^t(t) \\ \ddot{r}_b^t(t) \\ \ddot{r}_i^t(t) \\ \ddot{r}_h^\Delta(t) \end{Bmatrix} + \begin{Bmatrix} Q_s(t) \\ Q_b(t) \\ Q_i^i(t) \\ 0 \end{Bmatrix} + \begin{Bmatrix} 0 \\ 0 \\ Q_i^h(t) \\ Q_h^h(t) + R_h(t) \end{Bmatrix} = \begin{Bmatrix} 0 \\ 0 \\ P_i(t) \\ P_h(t) \end{Bmatrix} \tag{10}$$

where the first term on the left-hand side represents the inertial forces whereas $Q_s(t)$, $Q_b(t)$ and $Q^i_i(t)$ refer to nonlinear internal forces in corresponding parts of the system (the last one with superscript i represents the nonlinear forces on the nodes of irregular soil zone except those immediately inside the interaction horizon). $Q^h_h(t)$ and $Q^h_i(t)$ are linear internal forces acting on the nodes along the interaction horizon and immediately inside of it, respectively, which can be expressed as

$$Q_i^h(t) = K_{ih}\, r_h^{\Delta}(t) + C_{ih}\, \dot{r}_h^{\Delta}(t) \tag{11a}$$

$$Q_h^h(t) = K_{hi}\, r_i^t(t) + C_{hi}\, \dot{r}_i^t(t) + K_{hh}^i\, r_h^{\Delta}(t) + C_{hh}^i\, \dot{r}_h^{\Delta}(t) \tag{11b}$$

The interaction forces can now be obtained directly using relative displacements calculated along the interaction horizon. In case of Substructure Method they can be expressed as,

$$R_h(t) = \int_0^t S_{hh}^r(t-\tau)\, r_h^{\Delta}(\tau)\, d\tau \tag{12a}$$

in which $S^r_{hh}(t)$ represents the dynamic stiffness matrix of unbounded regular soil zone in time domain. Eq.12a corresponds to the convolution integral mentioned in Section 2 above which can also be derived from its counterpart in the frequency domain. For the Direct Method, the expression of interaction forces depends on the definition of artifical transmitting boundaries. If, for example, they can be defined by a set of masses, springs and dashpots for each node on interaction horizon, then uncoupled interaction force for j th node can be written as,

$$R_h^j(t) = K_{hh}^j\, r_h^{\Delta j}(t) + C_{hh}^j\, \dot{r}_h^{\Delta j}(t) + M_{hh}^j\, \ddot{r}_h^{\Delta j}(t) \tag{12b}$$

As a special case of transmitting boundaries defined as viscous boundaries, the second term only would be sufficient to represent the interaction force.

Finally effective force vector components on the right-hand side of Eq.10 can be given as,

$$P_i(t) = -K_{ih}\, v_h(t) - C_{ih}\, \dot{v}_h(t) - M_{ih}\, \ddot{v}_h(t) \tag{13a}$$

$$P_h(t) = K_{hi}\, v_i(t) + C_{hi}\, \dot{v}_i(t) + M_{hi}\, \ddot{v}_i(t) \tag{13b}$$

The common time domain formulation of both methods is thus given by Eqs.10 through 13.

4.Stiffness and Effective Base Input Motion of Embedded Foundations in Frequency Domain

In determination of dynamic stiffness of embedded foundations the Boundary Element Method has been proven to be very effective. The method, in principle, is based on the calculation of Green functions, i.e., in structural mechanics terms, the dynamic flexibility influence coefficients which define the displacement field due to application of unit tractions. If the structural system is discrete in nature, then the stiffness problem can be defined as the determination of unknown (redundant) forces due to prescibed unit displacements satisfying the compatibility conditions. In this respect Green functions are the auxiliary functions that relate the unknown forces to the displacements to be prescribed. In continuum problems the system is discretized along where the stiffness is to be determined. In this case compatibility requirements are satisfied in average sense on the basis of approximations made regarding the distributions of unknown stresses and displacements between discrete points. Various forms of Boundary Element Method [11] are actually the results of different averaging techniques used.

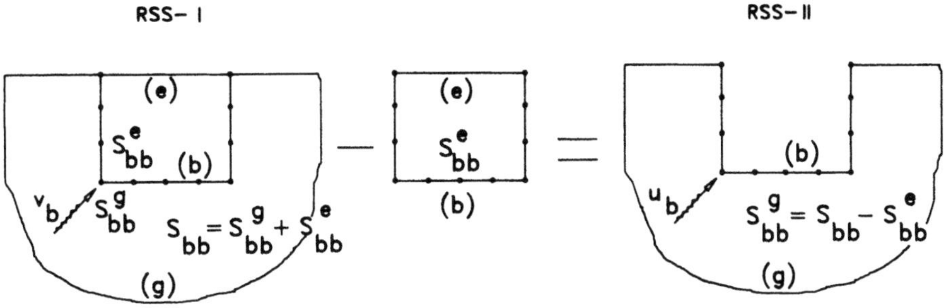

Fig.5. The concept of "Analytical Excavation"

4.1. CONCEPT OF ANALYTICAL EXCAVATION

In embedment problems the stiffness matrix is conveniently determined by using the concept what the author calls "Analytical Excavation". It is based on a simple superposition procedure as shown in Fig.5. (The overbars used to denote frequency domain analysis in previous section are dropped from here on for convencience). Unexcavated virgin soil is discretized along the lines denoted by letter b which later would consitute the excavation boundary of the soil medium. The virgin soil shown in Fig.5 is actually the same as shown in Fig.4b (RSS-I) except that whole outer soil zone is denoted by letter g as in Ref [1]. Inner soil zone to be excavated is again indicated by e. S^g_{bb} and S^e_{bb} are condensed dynamic stiffness matrices of outer and inner soil zones, respectively, whereas their sum is represented by S_{bb} (without superscript).

The total stiffness matrix, S_{bb}, of virgin soil is appropriately determined by using Boundary Element Method, and Finite Element Method can be effectively used for stiffness matrix S^e_{bb}, of the bounded soil zone to be excavated [1,10]. However there is one another possibility to extract S^g_{bb} without condensation of inner soil zone to be excavated [1].Both possibilities can be worked out by starting with the following relationship written for the virgin soil as,

$$\begin{bmatrix} K_{ee} & K_{eb} \\ K_{be} & (K^e_{bb}+S^g_{bb}) \end{bmatrix} \begin{bmatrix} F_{ee} & F_{eb} \\ F_{be} & F_{bb} \end{bmatrix} = \begin{bmatrix} I_{ee} & 0 \\ 0 & I_{bb} \end{bmatrix} \qquad (14)$$

in which the matrices on the left-hand side are the stiffness and flexibility matrices (Green functions) of virgin soil (RSS-I) ,respectively, defined in terms of degrees of freedom of excavation boundary (b) and inner soil zone (e) only. (Note that in this case condensation has not been applied to inner zone). I_{bb} and I_{ee} are unit matrices of corresponding orders. Using the identity of

$$S_{bb} = F^{-1}_{bb} \qquad (15)$$

the result is obtained from the lower diagonal element of the matrix product on the left-hand side of Eq.14 as [1],

$$\mathbf{S}^{g}_{bb} = \mathbf{S}_{bb} - (\mathbf{K}^{e}_{bb} + \mathbf{K}_{be}\,\mathbf{F}_{eb}\,\mathbf{S}_{bb}) \tag{16}$$

On the other hand, condensation applied to the first matrix on left-hand side of Eq.14 and consideration of Eq.15 results in

$$\mathbf{S}^{g}_{bb} = \mathbf{S}_{bb} - (\mathbf{K}^{e}_{bb} - \mathbf{K}_{be}\,\mathbf{K}^{-1}_{ee}\,\mathbf{K}_{eb}) \tag{17}$$

It can be concluded that the second terms in parantheses in Eq.16 and Eq.17 are identical and equal to condensed stiffness matrix, $\mathbf{S}^{e}_{bb}$, of the soil zone to be excavated. A probable advantage of the former is the partial discretization required, as coupling submatrices $\mathbf{K}_{be}$ and $\mathbf{F}_{eb}$ are related to the nodes only on the excavation boundary and their immediate neighbours inside. However in this case, flexibility submatrix $\mathbf{F}_{eb}$ should also be calculated in addition to $\mathbf{F}_{bb}$ (Eq.15) which is required in both cases.

Dynamic stiffness matrix of a rigid foundation embedded in damped half plane is determined using Green functions and Eq.17 will be presented in Section 4.4 below as an example.

4.2. EFFECTIVE BASE INPUT MOTION

As it is shown in Section 3 above, the effective force vector component defined along the interaction horizon can be determined in two ways, i.e., using the free-field response of either unexcavated virgin soil (RSS-I, Fig.4b) or excavated soil (RSS-II, Fig.4c). When interaction horizon coincides with the soil-structure interface (or excavation boundary), RSS-I and RSS-II are to be defined as shown in Fig.5. With appropriate notation modifications Eq.3 and Eq.6 can be rewritten, respectively, as

$$\mathbf{P}_{b} = \mathbf{S}^{g}_{bb}\,\mathbf{u}_{b} \tag{18a}$$

$$\mathbf{P}_{b} = \mathbf{S}_{bb}\,\mathbf{v}_{b} \tag{18b}$$

where $\mathbf{u}_b$ and $\mathbf{v}_b$ represent base input motions corresponding to RSS-II and RSS-I, respectively. Considering the expressions given in Fig.5 and utilizing Eqs.18, effective base input motion can be obtained in terms of free-field response of virgin soil as

$$\mathbf{u}_{b} = \mathbf{v}_{b} + \mathbf{S}^{g^{-1}}_{bb}\,\mathbf{S}^{e}_{bb}\,\mathbf{v}_{b} \tag{19}$$

This expression is not used in the formulation of Substructure Method, however it is quite instructive to demonstrate the effect of embedment which is represented by the second term.

4.3. RIGID FOUNDATIONS

In the previous sections all formulations presented are developed for flexible foundations, i.e., no constraints are imposed regarding their rigidity. In practice most foundations are relatively rigid with respect to adjacent soil. In case of infinitely rigid foundation assumption, all elements related to foundation (base) degrees of freedom have to be transformed into rigid body degrees of freedom of the base. In general, base response vector is defined as

$$\mathbf{r}_b = \mathbf{T}_{bo}\,\mathbf{r}_o \tag{20}$$

where $\mathbf{r}_o$ represents the rigid base response and $\mathbf{T}_{bo}$ is the associated kinematic transformation matrix. Effective base input motion must also satisfy the rigid body motion conditions, i.e., similar to Eq.20,

$$\mathbf{u}_b = \mathbf{T}_{bo}\,\mathbf{u}_o \tag{21}$$

Relevant transformations on various forms of equations of motion of soil-structure systems are given in Ref. [1]. In the following, transformed expressions for dynamic stiffness matrix and effective base input motion only are given.

Stiffness matrix of rigid foundation is expressed as,

$$\mathbf{S}^g_{oo} = \mathbf{T}^T_{bo}\,\mathbf{S}^g_{bb}\,\mathbf{T}_{bo} \tag{22}$$

and utilizing Eqs.21,22 transformed forms of Eqs.18 result in

$$\mathbf{P}_o = \mathbf{T}^T_{bo}\,\mathbf{S}^g_{bb}\,\mathbf{T}_{bo}\,\mathbf{u}_o = \mathbf{S}^g_{oo}\,\mathbf{u}_o \tag{23a}$$

$$\mathbf{P}_o = \mathbf{T}^T_{bo}\,\mathbf{S}_{bb}\,\mathbf{V}_b \tag{23b}$$

from which effective base input motion of rigid foundation can be obtained in terms of free-field response of virgin soil as,

$$\mathbf{u}_o = \mathbf{S}^{g^{-1}}_{oo}\,\mathbf{T}^T_{bo}\,\mathbf{S}_{bb}\,\mathbf{v}_b \tag{24}$$

In the special case of surface foundations subject to vertically indicent waves, free-field motion inherently satisfies the rigid body motion condition, i.e.,

$$\mathbf{v}_b = \mathbf{T}_{bo}\,\mathbf{v}_o \tag{25}$$

In this case considering Eqs.22,24,25 results in

$$\mathbf{u}_o = \mathbf{v}_o \tag{26}$$

which indicates that only in this special case effective base input motion is identical to the free-field motion of virgin soil.

4.4. EXAMPLE :RIGID RECTANGULAR FOUNDATION EMBEDDED IN HALF PLANE

The concept of Analytical Excavation explained above for the determination of dynamic stiffness matrix and effective base input motion is applied to rigid rectangular foundations embedded in half plane [2]. First the fundamental problem is solved for the unexcavated half plane to obtain

Green functions for unit harmonic point load as shown in Fig.6. The solution is performed by the application of superposition procedure as described in Fig.7 for horizontal load only. The infinite plane is solved for point load yielding to non-zero stresses along x axis which later applied onto

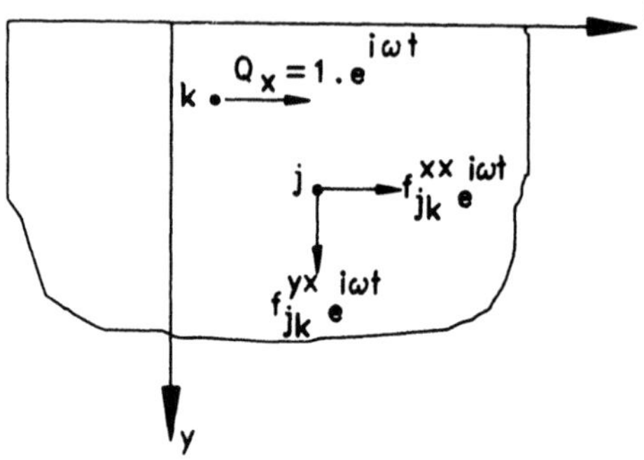

Fig.6. Definition of Green Functions for unit harmonic point load.(Horizontal load indicated only)

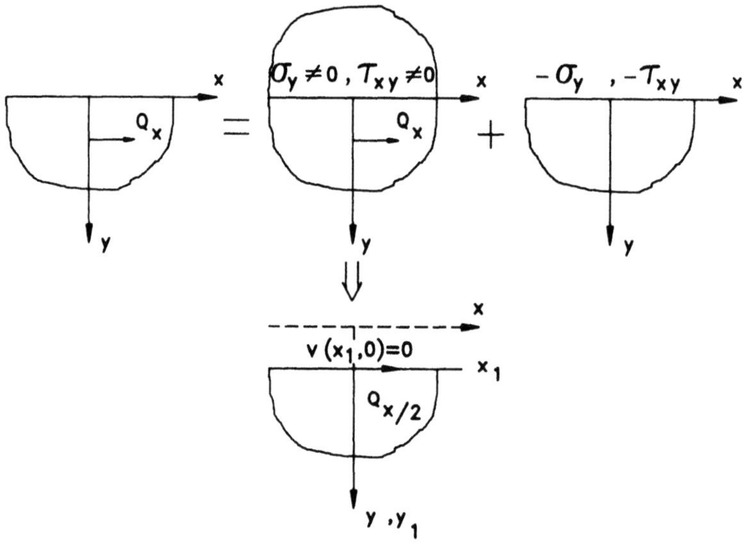

Fig.7. Superposition procedure applied to the solution of half-plane for unit harmonic point load

the surface of half plane in opposite direction. Superposition of two solutions yield the object solution. The solution of the infinite plane, on the other hand, is also converted to that of the half plane as shown in the lower part of Fig.7 yielding to the solution of the problem as the combination of two half plane solutions. Resulting Green functions in terms of inverse Fourier integrals are presented in the Appendix.

Regarding the discretization for rectangular embedded foundations (before the excavation), the load distribution is taken piecewise constant between consecutive nodal points whereas displacements vary linearly. With these assumptions, procedure for the derivation of stiffness matrix of unexcavated soil is equivalent to the application of Indirect Boundary Element Method [11] and therefore details are not given here.

The second step in the implementation of concept of analytical excavation is the extraction of the soil zone to be excavated (Fig.5, Eqs.16,17). Here Eq.17 is applied with finite element modelling of excavation zone. Resulting stiffness matrix can be used for both flexible and rigid foundations. In case of the latter the stiffness matrix is transformed into rigid body degrees of freedom of the foundation by using Eq.22. Dynamic force-displacement relationship can be written as

$$\begin{Bmatrix} P_1/G \\ P_2/(Ga) \\ P_3/G \end{Bmatrix} = \begin{bmatrix} K_{11} & K_{12} & 0 \\ K_{21} & K_{22} & 0 \\ 0 & 0 & K_{33} \end{bmatrix} \begin{Bmatrix} r_1 \\ ar_2 \\ r_3 \end{Bmatrix} \tag{27}$$

where stiffness influence coefficients are expressed in non-dimensional form. In Eq.27, G represents the shear modulus of the soil and (a) is half width of rigid foundation. Each coefficient in Eq.27 can be written as

$$K_{ij} = k_{ij} + i\, a_o\, c_{ij} \tag{28}$$

in which k_{ij} and c_{ij} correspond to frequency-dependent stifness and damping coefficients. a_o represents the non-dimensional frequency which is defined as $a_o = \omega\, a / C_s$ where ω is the frequency and C_s represents the shear wave velocity of the soil medium.

In example problem, material damping of soil is incorporated into the solution with hysteretic damping ratio being β=0.05. Poisson's ratio is taken as υ =1/3. The half width of the base of foundation is divided into six elements. Solutions are obtained for embedment ratios of h/a=1 and 1/2 as well as for a surface foundation. Six elements and three elements, respectively, are taken along the walls of the foundation for specified embedment ratios. Figs.8,9 show the non-dimensional stiffness and damping coefficients defined in Eqs.27 and 28 with respect to nondimensional frequency, a_o.

The effective base input motion of the same foundation is obtained by using Eq.24 for vertically incident S and P waves. Fig.10 shows non-dimensional input motion components corresponding to rigid body degrees of freedom defined above and normalized with respect to horizontal and vertical components of free-field surface motion of virgin soil, v_x for S wave and v_y for P wave, respectively.

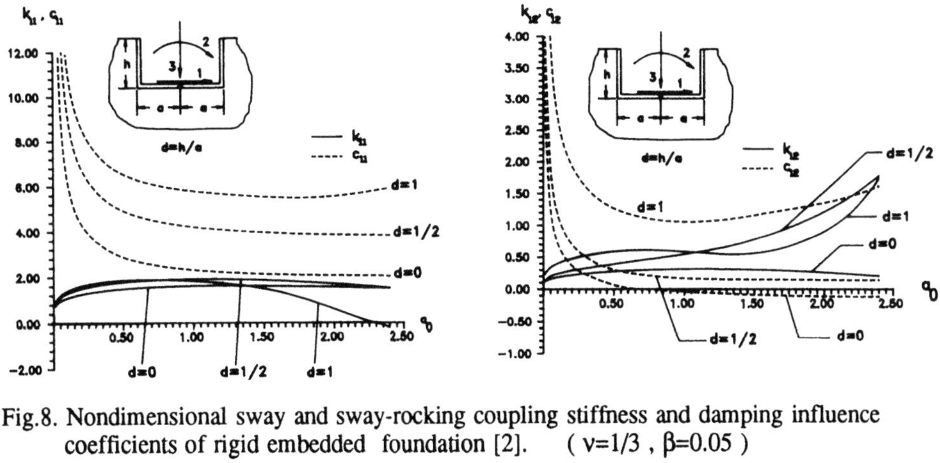

Fig.8. Nondimensional sway and sway-rocking coupling stiffness and damping influence coefficients of rigid embedded foundation [2]. (ν=1/3 , β=0.05)

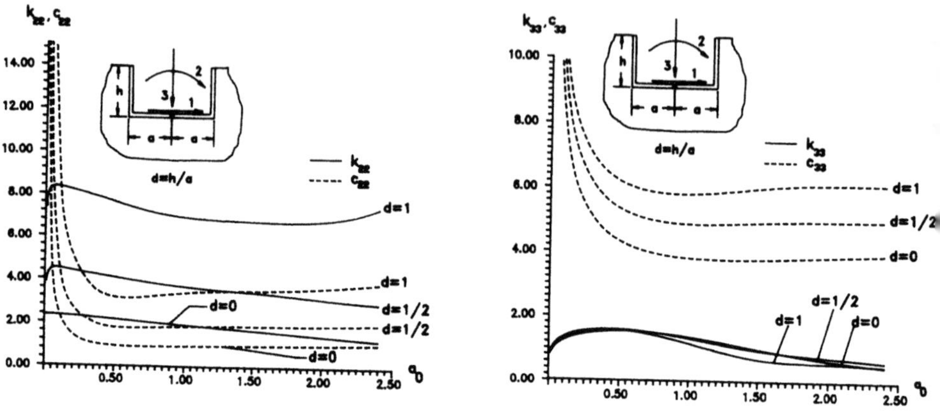

Fig.9. Nondimensional rocking and vertical stiffness and damping influence coefficients of rigid embedded foundation [2]. (ν=1/3 , β=0.05)

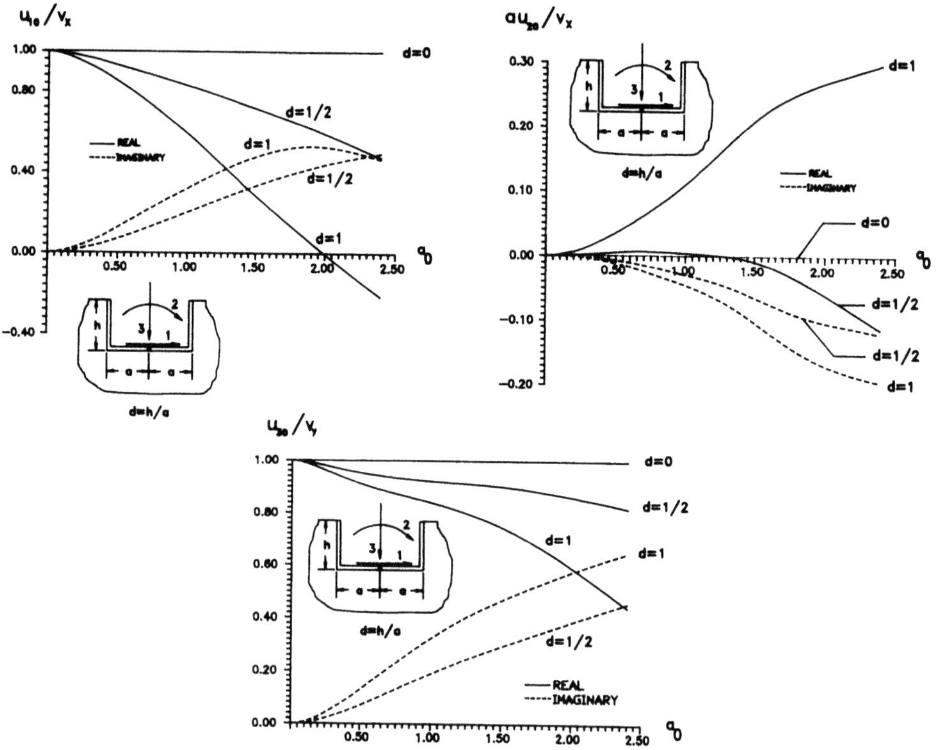

Fig.10. Effective horizontal , rocking and vertical input motions of rigid embedded foundation normalized with respect to horizontal and vertical components of free-field surface motion [2]. (ν=1/3 , β=0.05)

5. Conclusions

The exciting adventure of soil-structure interaction during the last quarter of a century is briefly reviewed with particular emphasis given to head to head development of Direct Method and Substructure Method. Current physical and analytical modelling of both methods are evaluated and an attempt is made to reach a common formulation equally applicable to both methods. It is shown that the common formulation makes it possible for Substructure Method to indirectly consider the nonlinear soil behaviour in free-field response in addition to that caused by the structure.

The procedures for determining stiffness and effective base input motion of embedded foundations utilizing the concept of Analytical Excavation are reviewed and results of an example problem of rectangular foundation embedded into damped half plane are presented.

References

[1] Aydınoğlu, M.N. (1980) 'Unified Formulations for Soil-Structure Interaction', Proc. 7th World Conference on Earthquake Engineering, Vol.6, pp. 121-128.

[2] Aydınoğlu M.N. (1981) 'General Formulation of Dynamic Soil-Structure Interaction and a Substructure Method for Embedded Structures' Unpublished report submitted to Istanbul Technical University for the partial fulfillment of Assoc. Professorship degree (in Turkish).

[3] Gutierrez, J.A. (1976) 'A Substructure Method for Earthquake Analysis of Structure-Soil Interaction', Report No. EERC 76-9, University of California, Berkeley.

[4] Kausel, E., Whitman, R.V., Morray, J.P., Elsabee, F. (1978) 'The Spring Method for Embedded Foundations', Nuclear Engineering and Design, Vol.48, pp. 377-392.

[5] Kausel, E. (1988) 'Local Transmitting Boundaries', Journal of Engineering Mechanics, ASCE, Vol.114, pp. 1011-1027.

[6] Lysmer, J., Kuhlemeyer, R.L. (1969) 'Finite Dynamic Model for Infinite Media', Journal of Engineering Mechanics Division, ASCE, Vol.95, pp. 859-877.

[7] Lysmer,J., Udaka, T., Tsai, C.F., Seed, H.B. (1975) 'FLUSH-Computer Program for Approximate 3-D Analysis of Soil-Structure Interaction Problems', Report No. 75-30, University of California, Berkeley.

[8] Veletsos, A.S., Meek, J.W. (1974) 'Dynamic Behaviour of Building-Foundation Systems', Int. J. Earthquake Engineering and Structural Dynamics, Vol.3, pp. 121-138.

[9] Waas, G. (1972) 'Earth Vibration Effects and Abatement for Military Facilities-Analysis Method for Footing Vibrations through Layered Media', Technical Report S-71-14, U.S. Army Engineer Waterways Experiment Station, Vicksburg, Missisipi.

[10] Wolf, J.P. (1985) Dynamic Soil-Structure Interaction, Prentice-Hall, Englewood Cliffs, N.J.

[11] Wolf, J.P. (1988) Soil-Structure Interaction in Time Domain, Prentice-Hall, Englewood Cliffs, N.J.

[12] Wolf, J.P. (1991) 'Classification of Analysis Methods for Dynamic Soil-Structure Interaction (State of the Art Paper)', Proc. Second International Conference on Recent Advances in Geotechnical Earthquake Engineering and Soil Dynamics, S.Prakash, ed., St. Louis, Missouri, pp. 1821-1832.

[13] Wolf, J.P. (1991) 'Consistent Lumped-Parameter Models for Unbounded Soil: Physical Representation', Earthquake Engineering and Structural Dynamics, Vol.20, pp. 11-32.

[14] Wolf, J.P. (1991) 'Consistent Lumped-Parameter Models for Unbounded Soil: Frequency Independent Stiffness, Damping and Mass Matrices', Earthquake Engineering and Structural Dynamics, Vol.20, pp. 33-42.

Appendix : Green Functions for Unit Harmonic Point Load

Green Functions are presented below for unit harmonic point load in terms of inverse Fourier transform integrals. Notations are as defined in Fig.6. Auxiliary expressions are given first as;

$s_1=(\xi^2-\kappa_1^2)^{1/2}$; $s_2=(\xi^2-\kappa_2^2)^{1/2}$; $w=2\xi^2-\kappa_2^2$

$r=4\xi^2 s_1 s_2-w^2$; $g=(4\xi^2 s_1 s_2+w^2)/r$; $A=1/(2\pi G\kappa_2^2)$

$\Delta_1=y_j-y_k$; $\Delta_2=y_j+y_k$; $\Delta_3=x_j-x_k$

$t_1=\exp(-s_1|\Delta_1|)$; $t_2=\exp(-s_1\Delta_2)$; $t_3=\exp(-s_2|\Delta_1|)$

$t_4=\exp(-s_2\Delta_2)$; $t_5=\exp(-s_1y_k-s_2y_j)$; $t_6=\exp(-s_2y_k-s_1y_j)$

in which

$\kappa_1=\omega/C_p$ and $\kappa_2=\omega/C_s$

where ω is the excitation frequency, C_p and C_s are dilatational and shear wave velocities, respectively.

Expressions for Green Functions are given below utilizing auxiliary expressions defined above:

$$f_{jk}^{xx}=A\int_0^\infty[\xi^2 t_1/s_1-s_2t_3+gt_2\xi^2/s_1+gs_2t_4-4\xi^2 s_2 w(t_5+t_6)/r]\cos(\xi\Delta_3)\,d\xi$$

$$f_{jk}^{yx}=A\int_0^\infty[(t_1-t_3)\operatorname{sign}(\Delta_1)-g(t_2+t_4)-4s_1s_2wt_6/r-4\xi^2wt_5/r]\xi\sin(\xi\Delta_3)\,d\xi$$

$$f_{jk}^{xy}=A\int_0^\infty[(t_1-t_3)\operatorname{sign}(\Delta_1)-g(t_2+t_4)+4s_1s_2wt_5/r-4\xi^2wt_6/r]\xi\sin(\xi\Delta_3)\,d\xi$$

$$f_{jk}^{yy}=A\int_0^\infty[\xi^2 t_3/s_2-s_1t_1+gs_1t_2+gt_4\xi^2/s_2-4\xi^2 s_1 w(t_5+t_6)/r]\cos(\xi\Delta_3)\,d\xi$$

As a special case where $y_j=y_k=0$ Green Functions reduce to the following:

$$f_{jk}^{xx}=\kappa_2^2/(\pi G)\int_0^\infty(s_2/r)\cos(\xi\Delta_3)\,d\xi$$

$$f_{jk}^{yx}=-f_{jk}^{xy}=1/(\pi G)\int_0^\infty[(2s_1s_2-w)\xi/r]\sin(\xi\Delta_3)\,d\xi$$

$$f_{jk}^{yy}=\kappa_2^2/(\pi G)\int_0^\infty(s_1/r)\cos(\xi\Delta_3)\,d\xi$$

COMPUTATIONAL SOIL DYNAMICS AND SOIL-STRUCTURE INTERACTION

DENIS AUBRY
Ecole Centrale de Paris
92295 Chatenay Malabry Cedex, France

ABSTRACT. Review of the governing equations for the earthquake analysis of saturated porous media and the cyclic constitutive model for the soil skeleton are described. The mathematical formulation is implemented with consideration of the boundary and interface conditions. An example is presented for the analysis of an earth dam subjected to seismic loading.

1. INTRODUCTION

The analysis of seismic wave propagation in soils can be considered as one of the major fields of research in civil engineering in the past years [2]. The advent of numerical methods in the late sixties has made it possible to analyze the response of soils in a severe environment. Numerous constitutive models have been developed and proposed rather recently and many of them have been able to perform good predictions. In this presentation a cyclic constitutive model developed at Ecole Centrale de Paris is used and some of its merits will be explained below [3].

Besides the non-linear aspects of the soil behavior the pore water pressure which modifies the effective stress path is essential to the analysis of both stiffness and strength. One of the most surprising manifestations of the non-linear behavior of soils combined with the presence of two phases is the phenomenon of liquefaction. The numerical simulation of soil liquefaction has only been performed very recently, as in addition to a suitable formulation of the two-phase behavior, a performant constitutive model is required [2][7][8].

In the following sections, a brief review of the governing equations and the cyclic constitutive model for the soil skeleton will be given for the analysis of saturated porous media. Then, some simplifications introduced for numerical modelling in earthquake engineering practices will be presented. The third section is dedicated to the numerical implementation of the mathematical formulation of the problem and the boundary and interface conditions. Finally an example of an earthdam submitted to seismic loadings will be presented in order to illustrate the different aspects of the whole analysis.

2. GOVERNING EQUATIONS

In this section the conservation equations, the constitutive model and boundary and interface conditions are briefly presented. As usual in soil mechanics the total stress tensor σ is first decomposed into the effective stress tensor σ' and the pore pressure p (positive traction):

$$\sigma = \sigma' - p\,\boldsymbol{I}$$

P. Gülkan and R. W. Clough (eds.), Developments in Dynamic Soil-Structure Interaction, 43–60.

The displacement of the solid phase $\mathbf{u}_s$, fluid phase $\mathbf{u}_f$ and the relative displacement of the fluid phase $\mathbf{u}_{rf}$ given by [2]:

$$u_{rf} = n\,(u_f - u_s)$$

in addition to the pore water pressure p, will be kept as the primary unknowns. In the above relation n represents the porosity of the solid skeleton.

The momentum conservation of the two phases is given by:

$$\boldsymbol{Div}\ \sigma' + \rho\,\mathbf{g} = \rho\,\ddot{u}_s + \rho_f\,\ddot{u}_{rf}$$

where the density is:

$$\rho = (1-n)\,\rho_s + n\,\rho_f$$

ρ_s and ρ_f are respectively the density of the solid and the fluid phases and $\mathbf{g}$ the acceleration of gravity vector. By introducing the total stress decomposition the following expression is obtained:

$$\boldsymbol{Div}\ \sigma' - \boldsymbol{grad}\ p + \rho\,\mathbf{g} = \rho\,\ddot{u}_s + \rho_f\,\ddot{u}_{rf}$$

The mass conservation for the solid and fluid phases are given by:

$$div\,(\,(1-n)\rho_s\,\dot{u}_s\,) + \partial_t(\,(1-n)\,\rho_s) = 0$$

$$n\,(\dot{\rho}_f/\rho_f) + n\ div\ \dot{u}_f + \dot{n} = 0$$

By adding the two last equations and by invoking the decomposition of the total stress, the mass conservation for the two phases may finally be written as:

$$div\ \dot{u}_{rf} + div\ \dot{u}_s = -\,\dot{p}\,/\,Q$$

with the compressibility parameter Q given by:

$$1/Q = n\,/\,K_f + (1-n)\,/\,K_s$$

The above conservation equations must, of course, be augmented with an equation which describes the movement of one phase with respect to the other. This equation is given by a generalized Darcy law:

$$\dot{u}_{rf} = \mathbf{K}\,\{\,-\boldsymbol{grad}\ p + \rho_f(\,\mathbf{g} - \ddot{u}_s - \ddot{u}_{rf}\,/\,n)\,)\}$$

where $\mathbf{K}$ the permeability tensor ($\mathbf{K} = \mathbf{k}\ /\,\rho_f\,g$). Biot in his original work introduced a further inertial coupling term between the two phases in the Darcy equation:

$$\dot{u}_{rf} = \mathbf{K}\,\{\,-\boldsymbol{grad}\ p + \rho_f(\,\mathbf{g} - \ddot{u}_s - \ddot{u}_{rf}\,/\,n)) - \rho_a\,\ddot{u}_{rf}\,\} \qquad \rho_a \geq 0$$

This parameter depends on the porosity and the geometry of the pores. This inertia coupling term will be neglected in this paper. In summary, the three sets of equations which describe the (u_s-u_{rf}-p) model are given by:

$$\begin{cases} Div\ \sigma' - \mathbf{grad}\ p + \rho\, \mathbf{g} = \rho\ \ddot{\mathbf{u}}_s + \rho_f\ \ddot{\mathbf{u}}_{rf} \\ div\ \dot{\mathbf{u}}_s - div\ [\mathbf{K}\ (\mathbf{grad}\ p)] + \dot{p}\ /\ Q = -\ div\ [\mathbf{K}\ (\ \rho_f (\ \mathbf{g} - \ddot{\mathbf{u}}_s - (\Xi/n)\ \ddot{\mathbf{u}}_{rf}))\ )] \\ p = -\ Q\ (div\ \mathbf{u}_{rf} + div\ \mathbf{u}_s\ ) \end{cases}$$

The preceeding model is interesting for higher frequencies. However, for the frequency range suited in earthquake engineering it seems that a simplified version would be more cost-effective. In this model, the relative acceleration of the fluid is neglected so that the momentum conservation and Darcy's law may be written as:

$$Div\ \sigma' - \mathbf{grad}\ p + \rho\, \mathbf{g} = \rho\ \ddot{\mathbf{u}}_s$$

$$\dot{\mathbf{u}}_{rf} = \mathbf{K}\ \{\ -\mathbf{grad}\ p + \rho_f (\ \mathbf{g} - \ddot{\mathbf{u}}_s\ )\ \}$$

The main consequence of this assumption is that now by using the conservation of mass, the relative displacement of the fluid may be eliminated:

$$div\ \dot{\mathbf{u}}_s - div(\mathbf{K}\ \mathbf{grad}\ (p - \rho_f \mathbf{g} \cdot \mathbf{x})) - div(\mathbf{K} \rho_f\ \ddot{\mathbf{u}}_s) + \dot{p}/Q = 0$$

so that the displacement and the pore pressure are the only unknowns. In soil mechanics Q mainly depends on the compressibilty of the water especially in the case of partially saturated soil and the following simplified assumptions may be used:

$$Q = K_f\ /\ n$$

In summary, the two sets of equations together with the constitutive equation for the skeleton which describe the (u_s-p) model are given by:

$$\begin{cases} Div\ \sigma' - \mathbf{grad}\ p + \rho\, \mathbf{g} = \rho\ \ddot{\mathbf{u}}_s \\ div\ \dot{\mathbf{u}}_s - div(\mathbf{K}\ \mathbf{grad}\ (p - \rho_f \mathbf{g} \cdot \mathbf{x})\ ) - div(\mathbf{K} \rho_f\ \ddot{\mathbf{u}}_s) + n\ \dot{p}\ /K_f = 0 \end{cases}$$

Corresponding to this simplified model the following boundary conditions are employed. The first partition of the boundary Γ is considered: Γ_σ and Γ_{us}. On Γ_{us} the displacement is imposed while the total traction boundary condition is retained on Γ_σ. For the fluid flow boundary conditions another partition is considered : Γ_p and Γ_φ . On Γ_p there is a pore pressure boundary condition while on Γ_φ there is a normal velocity condition, weaker than in the full (u_s-u_{rf}-p) model:

$$\dot{\mathbf{u}}_{rf}.\mathbf{n} = \varphi \qquad , X \in \Gamma_\varphi$$

Again in this case on the interface between a saturated porous material and a one-phase medium the continuity of absolute solid dispacements as well as the total stress vector **T** is required. The condition on the fluid is now restricted to the vanishing of the fluid flow through the boundary:

$$\begin{cases} \mathbf{u}_s^{(two\text{-}phase)} = \mathbf{u}_s^{(one\text{-}phase)} \\ \mathbf{K}\ \mathbf{grad}\ p^{(two\text{-}phase)}\ .\ \mathbf{n} = 0 \\ \mathbf{T}^{(two\text{-}phase)} + \mathbf{T}^{(one\text{-}phase)} = 0 \end{cases}$$

or a prescribed pore pressure on the interface:

$$\left| \begin{array}{l} u_s^{(two\text{-}phase)} = u_s^{(one\text{-}phase)} \\ p^{(two\text{-}phase)} = p^{(one\text{-}phase)} (prescribed) \\ T^{(two\text{-}phase)} + T^{(one\text{-}phase)} = 0 \end{array} \right.$$

3. CYCLIC EFFECTIVE SOIL BEHAVIOUR

The general scheme of soil behaviour is now recalled and then quantified by using a simple numerical model. The dynamic or rather cyclic properties of soils are basically described starting from static soil properties. The following key factors are fundamental to the description of soil behaviour:

- the effective stresses are able to describe essential soil properties and should be used at the beginning;
- the effective confinement affects very strongly the stiffness and the strength (Figure 1)
- the porosity (or void ratio or dry density) at a given confinement level modifies highly the behaviour (Figure 2). The dilatancy, ie the ability to produce volumetric strains even with no effective pressure variation, depends both on the porosity and the ratio between the shear and the normal stress.
- under undrained and saturated conditions, the behaviour is fully accounted by the effective stress principle and the dilatancy and the grains and water incompressibilities (Figure 3)
- the frequency has some influence because it is related to the strain rate. It will not be considered because in seismic loadings and more particularly for sands it can be neglected.
- For larger shear strain amplitudes the stress-strain loops are modified at each cycle due either to the densification of the material under drained conditions or to the pore pressure increase under undrained conditions (Figure 4). But the classical (G-γ) and (ξ,γ) are essentially understood by the creation of plastic deformations.
- The undrained cyclic behaviour, liquefaction and cyclic mobility are the consequences of the preceeding observations (Figure 5 and 6)

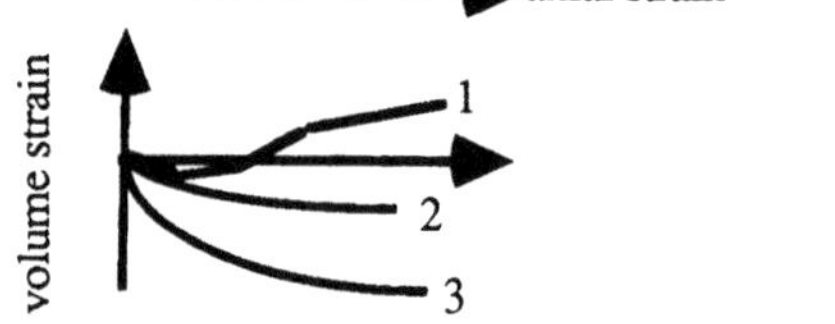

Figure 1: Influence of effective confinement on drained soil behaviour

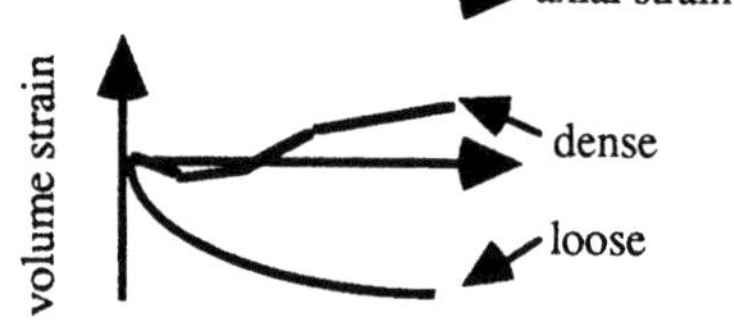

Figure 2: Influence of the porosity in drained triaxial tests

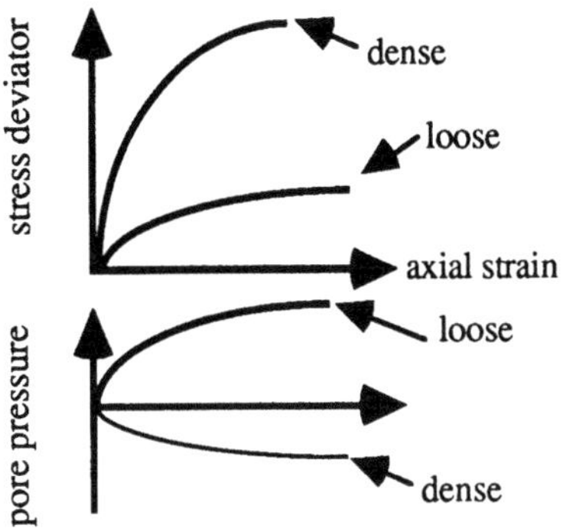

Figure 3: Influence of initial density on undrained soil behaviour

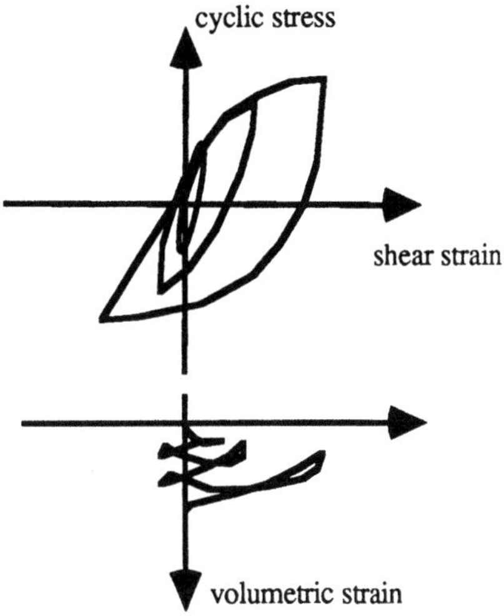

Figure 4: Drained cyclic soil behaviour (dense or loose)

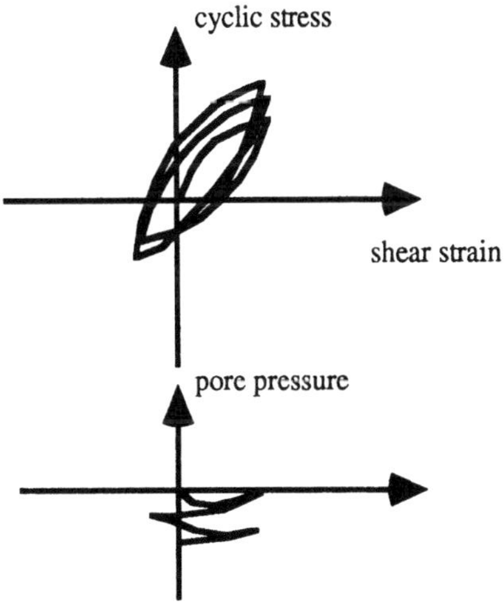

Figure 5: Undrained cyclic soil behaviour for a dense sand

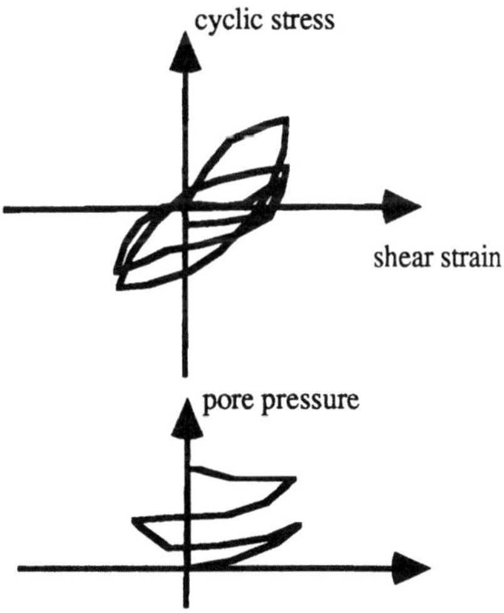

Figure 6: Undrained cyclic soil behaviour for a loose sand

The level of plastic shear strain is an indicator of the type of behavior because it has a major influence on cyclic soil properties and because in seismic analysis the strain range may be extremely wide.

- It is recalled that experimental results show that when the cyclic shear strain amplitude is less than 10^{-6}, the soil response is approximately reversible and non-linear elasticity due to the effects of the confinement may describe the stress-strain curve quite well.

- Between 10^{-6} and 10^{-4} irreversible deformations begin to appear but the cycles are rather stable. In such cases, an approach like the linear equivalent moduli with hysteretic damping may be of some value.
- For cyclic shear strain amplitude larger than 10^{-3} only a full cyclic elastoplastic model can provide accurate response even from a qualitative standpoint.

The above considerations will now be illustrated by using simple equations, using only one dimensional effective stresses and strains: the shear stress τ, the shear strain γ, the effective normal stress σ'_n, the normal strain ε_n. First, the role of the confinement is essentially due to the frictionnal characteristic of granular material, at failure, and a Mohr-Coulomb strength with friction angle ϕ is assumed:

$$\tau = - \sigma'_n \tan \phi$$

To arrive at a stress-strain relationship, it will be assumed that the plastic shear strain γ^p parametrizes the current shear stress with a hyperbolic curve with the parameter a, so that:

$$\tau = - \sigma'_n \tan \phi (\gamma^p / (a + \gamma^p)$$

Then the role of the initial density is accounted by introducing a Cam-Clay type equation and the critical stress σ_c, which is the effective stress at which no normal strain is activated. σ_{c0} is the initial critical stress and β a soil parameter:

$$\left| \begin{array}{l} \tau = - (\sigma'_n \tan \phi)\ F(\sigma'_n, \sigma_c(\varepsilon^p{}_n))\ \ (\gamma^p / (a + \gamma^p)) \\ F(\sigma'_n, \sigma_c(\varepsilon^p{}_n)) = 1 - Log(\sigma'_n / \sigma_c(\varepsilon^p{}_n), \qquad \sigma_c(\varepsilon^p{}_n) = \sigma_{c0} \exp(- \beta\, \varepsilon^p{}_n) \end{array} \right.$$

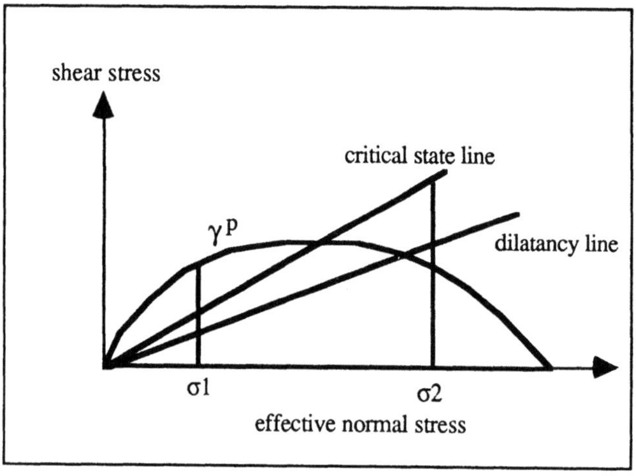

Figure 7: Yield locus in effective stress space

When σ_{c0} is larger than σ'_n the material is overconlidated. When the material is contractant, $\varepsilon^p{}_n$ is negative and decreases. Then σ_c increases and for a given σ'_n the strength of the material is increased. On the contrary, when the material is dilatant, ε^p_n is positive and increases. Then σ_c decreases and for a given σ'_n the strength of the material is decreased.

The evolution of ε^p_n is now the key supplementary equation. It is given by a Roscoe type dilatancy equation with ψ the dilatancy angle:

$$\dot{\varepsilon}^p_n = -(\tan\psi + (\tau / \sigma'_n))\, \dot{\gamma}^p$$

The final equation will give the total shear strain γ through the elastic shear modulus G:

$$\dot{\gamma} = \dot{\gamma}_p + \dot{\tau} / G$$

Very often for soils, the elastic strain is small and may be neglected.

Undrained loading

In the case of undrained loading condition, the total normal strain is constant:

$$0 = \dot{\varepsilon}_n = \dot{\varepsilon}^e_n + \dot{\varepsilon}^p_n$$

The evolution of the effective stress can be computed from this assumption with the bulk modulus K:

$$\dot{\sigma}'_n = -K\, \dot{\varepsilon}^p_n$$

This value is then incremented in the evaluation of the shear stress τ. The pore pressure p is computed from:

$$\sigma_n = \sigma'_n - p = cst$$

Cyclic loading

The cyclic loading can be easily approached by the concept of Masing rule, with (γ^{pc}, τ_c) being the the corresponding values at the load reversal time, and a_c a cyclic soil parameter:

$$\tau = \tau_c - h\,(\sigma'_n \tan\phi)\ (1 - Log(\sigma'_n/\sigma_c(\varepsilon^p_n)))\ \ (\gamma^p - \gamma^{pc} / (a_c + \gamma^p - \gamma^{pc}))$$

if $\dot{\gamma}^p \geq 0$, *then* $h=+1$, $h=-1$ *otherwise*

To account for pore pressure increase being dependant on the strain level the dilatancy rule is modified slightly:

$$\dot{\varepsilon}^p_n = -\zeta(\gamma^p)\,(\tan\psi + (\tau / \sigma'_n))\, \dot{\gamma}^p$$

$\zeta(\gamma^p)$ is defined as:

$\zeta(\gamma^p) = 0$, *if* $\gamma^p \leq \gamma^{pe}$ $\qquad$ $\zeta(\gamma^p) = 1$, *if* $\gamma^p \geq \gamma^{pm}$

$\zeta(\gamma^p)$ *increases smoothly from zero to one, if* $\gamma^{pe} \leq \gamma^p \leq \gamma^{pm}$

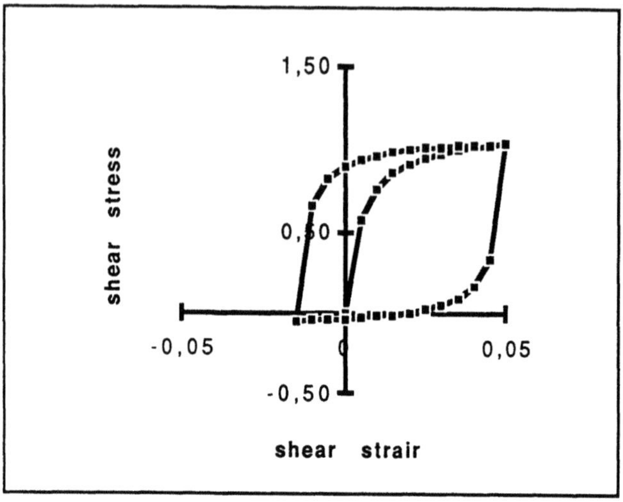

Figure 8: Drained cyclic stress-strain relationship

Generalization to the 3D case

The prediction of the pore pressure increase and of the permanent deformations can only be made by using an incremental constitutive equation in the framework of the elastoplasticity theory with due account taken of the evolution of the internal variables like the density of the material. The elastoplasticity theory with some adaptation is able to account for the main aspects of cyclic soil behaviour. Let us recall that the theory is based on the following assumptions:

- decomposition of the strain into elastic and plastic parts:

$$\varepsilon = \varepsilon^e + \varepsilon^p$$

- rate of effective stresses related to rate of elastic strains through the elastic stiffness:

$$\dot{\sigma}' = \boldsymbol{D} : \dot{\varepsilon}^e$$

- rate of plastic strains triggered by the yield function and developped by the flow rule for the plastic strain and internal parameter like plastic shear strain and volumetric strain:

$$\dot{\varepsilon}^p = \dot{\lambda}^p \left(\tau_n \otimes \boldsymbol{n} + \zeta(\varepsilon^{pn}) \left(\tan\psi + (\| \tau_n \| / \sigma'_{nn}) \right) \boldsymbol{I} \right)$$

In such an approach, effective stress increments are related to strain increments; the relationship being driven by a Mohr-Coulomb type yield function such as the following:

$$f(\sigma', \gamma^p, \varepsilon^{pn}, \boldsymbol{n}, \boldsymbol{r}_H, \boldsymbol{n}_H) = \| \tau_n \| + \sigma'_{nn} \tan\phi \, F(\sigma'_{nn}, \varepsilon^{pn}) \, (\gamma^{pc} / (a + \gamma^{pc}))$$

$$\sigma'_{nn} = \boldsymbol{n} \, \sigma' \, \boldsymbol{n}$$

$$\tau_n = \sigma' \boldsymbol{n} - \sigma'_{nn} \boldsymbol{n} + (\boldsymbol{r}_H - (\gamma^{pc} / (a_c + \gamma^{pc})) \, . \, \boldsymbol{n}_H) \, \sigma'_{nn} \, F(\sigma'_{nn}, \varepsilon^{pn}) \; tg\,\phi$$

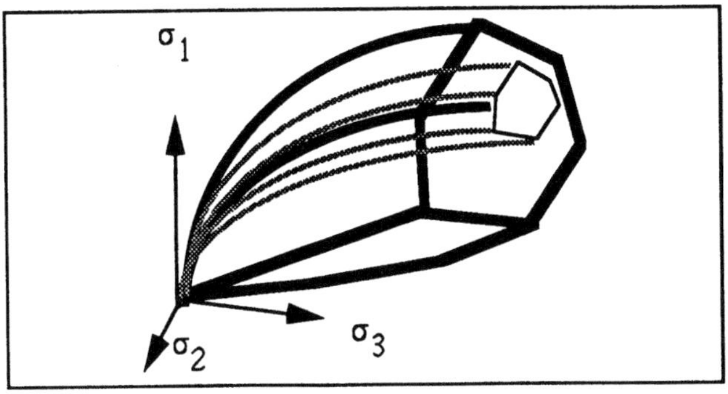

Figure 9: Cyclic and monotonous yield surfaces

where **n** stand respectively for the unit normal to a plane in the case of directional properties.

Several soils have been described with such a constitutive equation including sands and clays and a good correlation with observed experimental investigations is generally obtained. This model has been successfully used by the authors for the determination of the well-known G-γ and ξ-γ curves commonly used in engineering practice (linear equivalent method). As mentionned at the very beginning of this section, a more detailed discussion of these topics is not possible here and the reader is referred to Aubry et al.[3].

The primary dependant variables in the soil dynamics problem are the displacements $\mathbf{u}_S$, the pore pressure p, the effective stresses σ' and the internal parameters. In some cases their initial values are easily estimated by hand. However in most pratical cases, a true static computation must be performed prior to the dynamic simulation. Let us take several examples:

Building and foundation problem

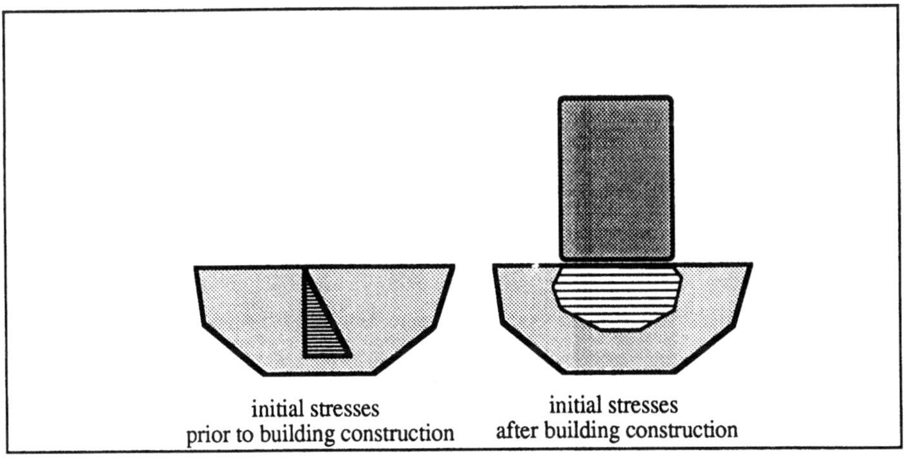

Figure 10: Effective initial stresses before dynamic analysis

Dam construction

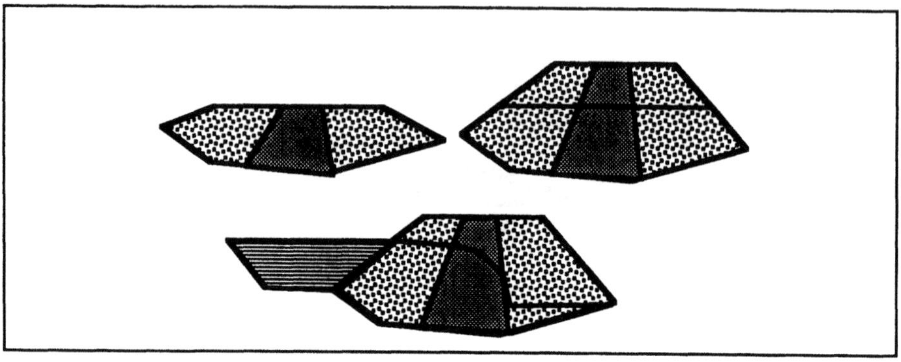

Figure 11: Effective stresses and pore pressures computation s
for Dam construction and reservoir filling

Excavation in front of retaining walls

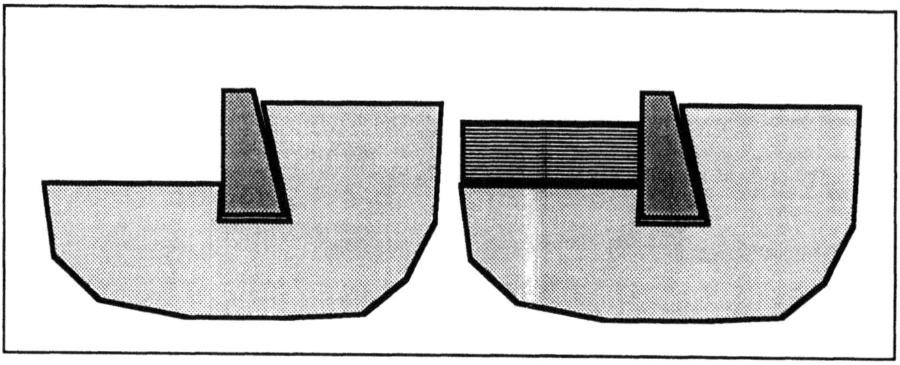

Figure 12: Excavation followed by water filling behind a retaining wall

The boundary conditions will obviously be more complex than for one-phase material and must be decomposed into boundary conditions relative to the bulk material and to the fluid flow. The boundary Γ of the domain Ω is thus decomposed into two first parts: Γ_σ and Γ_{us}. On Γ_{us} a displacement boundary condition is selected, while on Γ_σ a traction boundary condition is applied. Then a second partition of the boundary is introduced to deal with the fluid flow : Γ_p and Γ_{urf}. On Γ_{urf} a no-drainage type boundary condition is applied, and on Γ_p a pressure boundary condition is considered.

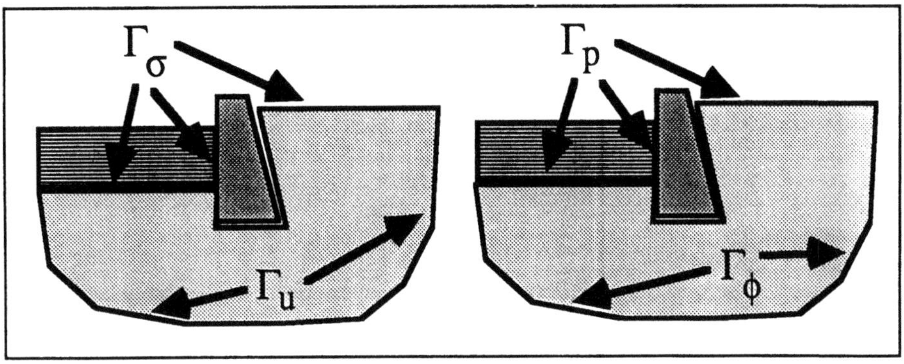

Figure 13: Example of boundary conditions

On the interface between a saturated porous material and a one-phase medium, the continuity of absolute solid dispacements as well as the total stress vector has to be verified and the relative displacement of the fluid phase must vanish. At the interface between structures or foundations and the soil, contact and sliding boundary have to be considered.

4. THE EQUIVALENT SEISMIC LOADING

The seismic event is defined by an incident displacement field $\mathbf{u}_i$ which is assumed to be known beforehand and which is compatible with the outer domain Ω_E. The inner domain Ω_I is assumed to be bordered by an elastic domain where a simplified version of wave propagation is assumed. On the interface Σ between these two domains, the continuity equations between the displacement $\mathbf{u}$ and stress vectors $\mathbf{t}$ are given by :

$$\mathbf{u}_I = \mathbf{u}_E \quad \text{and} \quad \mathbf{t}_I + \mathbf{t}_E(\mathbf{u}_E) = 0$$

At the vicinity of the boundary Σ, a linear elastic behaviour is assumed. The paraxial equation is an approximation to the wave equation for waves which have a preferred direction of propagation. The performance of the resulting boundary elements is connected to the approximation order. At zeroth order, these boundaries correspond to consistant dampers. If the incident field is to be accounted for, the continuity relations between inner and outer domains introduce the stress and velocity vectors of the incident field in the equilibrium equation.

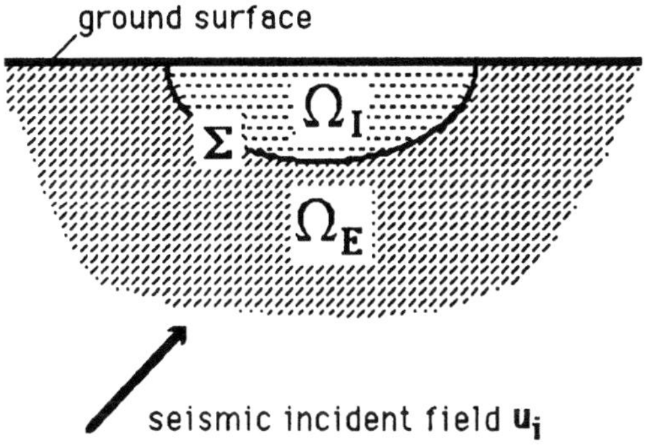

Figure 14: Schematic representation of the problem

With u_n, the displacement in the direction **n** normal to the boundary (Figure 14) , ρ the mass density, c_p and c_s respectively the dilatational and transversal body wave velocities, at *zeroth order*, the stress vector on the boundary is given by the following relation for any outgoing displacements field $\mathbf{u}_r$:

$$\boldsymbol{t}_r = \rho\, c_p\, \partial_t u_{rn}\, \boldsymbol{n} + \rho\, c_s\, \partial_t \boldsymbol{u}_{rT} = A_0(\partial_t \boldsymbol{u}_r)$$

which is precisely the transient impedance of consistant dampers distributed along the boundary Σ, commonly used in numerical modelling of dynamic problems.

Variational Formulation

We recall that in the inner domain, at the vicinity of the boundary Σ, the constitutive material is supposed to be linear elastic. At the infinity, the total displacement field tends to the incident displacement field. On the boundary Σ, the total displacement vector is splitt into the incident and diffracted displacement fields:

$$\boldsymbol{u}_E = \boldsymbol{u}_i + \boldsymbol{u}_r$$

The continuity condition for the stress vector over Σ and the linearity hypotheses at its vicinity enables us to write :

$$\boldsymbol{t}_I = -\, \boldsymbol{t}_E(\boldsymbol{u}_E) = -\, \boldsymbol{t}_E(\boldsymbol{u}_i) - \boldsymbol{t}_E(\boldsymbol{u}_r)$$

$$\boldsymbol{t}_I = -\, \boldsymbol{t}_E(\boldsymbol{u}_i) - A_0(\partial_t \boldsymbol{u}_r) \;=\; -\, \boldsymbol{t}_E(\boldsymbol{u}_i) + A_0(\partial_t \boldsymbol{u}_i) - A_0(\partial_t \boldsymbol{u}_I)$$

Now, let **w** be a kinematically admissible virtual displacement field over Ω_I, we can write the virtual work principle for every **w** :

$$(\rho \partial_{tt} \boldsymbol{u}_I\, ,\, \boldsymbol{w}\,)_{\Omega_I} + (\sigma_I\, ,\, \varepsilon(\boldsymbol{w})\,)_{\Omega_I} - \langle \boldsymbol{t}_I, \boldsymbol{w} \rangle_\Sigma = (\, \rho\, \boldsymbol{g}\, ,\, \boldsymbol{w}\,)_{\Omega_I}$$

in which $(\ ,\ )_{\Omega I}$ denotes the integration over the domain Ω_I and $<\ ,\ >_\Sigma$ designates the integration on its boundary Σ. The above equation then becomes :

$$(\rho\, \partial_{tt} \boldsymbol{u}_I\, ,\, \boldsymbol{w}\,)_{\Omega_I} + (\, \sigma_I\, ,\, \varepsilon(\boldsymbol{w})\,)_{\Omega_I} + \langle A_0(\partial_t \boldsymbol{u}_I)\, ,\boldsymbol{w} \rangle_\Sigma$$
$$= (\, \rho\, \boldsymbol{g}\, ,\, \boldsymbol{w})_{\Omega_I} \;+ \langle -\, \boldsymbol{t}_E\, (\boldsymbol{u}_i) + A_0(\partial_t \boldsymbol{u}_i)\, ,\boldsymbol{w} \rangle_\Sigma$$

The preceeding equation shows clearly how the seismic loading can be described with at the same time a damping matrix and a right hand side where both the distribution of the incident displacement and the incident traction vectors are to be known.

Definition of the Seismic Incident Field

The definition of the seismic incident field presents a difficult issue in Earthquake Engineering. In almost every case, the seismic loading (acceleration vector) is supposed to be known at the ground surface at some (usually one) points called control points. For modelling purposes, due to theoretical limitations, this control point has to be the farthest possible from the surface or buried irregularities.

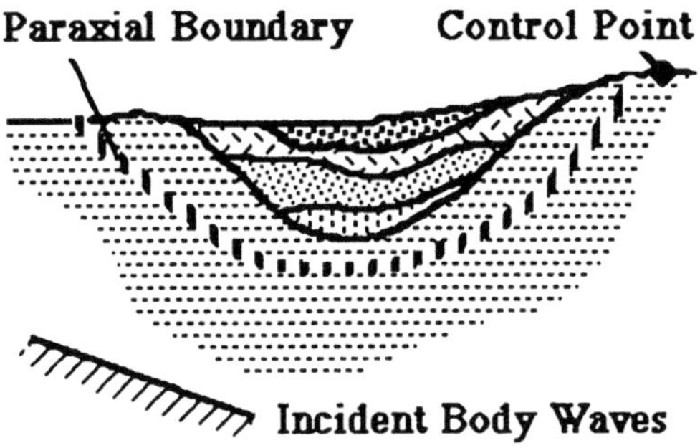

Figure 15: Schematic representation of a sediment-filled valley with a homogeneous substratum

Now, if the outer domain can be assumed as a homogeneous linear elastic domain (figure 15), the seismic incident field can be evaluated at the artificial interface directly by a simple deconvolution of the acceleration at the control point for any incident angle. As a direct time integration procedure is used, a base line correction is necessary to determine velocity and stress vectors due to the input acceleration vector. In any other cases (for example a multilayer substratum shown in figure 4), the deconvolution must be performed using other techniques (for example Thomson-Haskell method for a multilayer substratum) and then the time history of the stress and velocity fields can be entered for along the paraxial boundary.

5. NUMERICAL IMPLEMENTATION

The weak form of the governing equations may be obtained using a variational formulation. Then a finite element approximation of $\mathbf{u}_s(t)$ and p(t) defined by $\mathbf{u}_{sh}(t)$ and $p_h(t)$ should be computed by restricting the test and trial functions to finite dimensional spaces for the displacement of the solid, and of the pressure. The finite element spaces provide the usual finite element shape functions : $w_{sJ}(\mathbf{x})\cdot\mathbf{e}_j$, and $w_{pJ}(\mathbf{x})$ where J stands for a node number, $\mathbf{e}_j$ is the unit basis vector of the euclidean space. In this manner the following expansion may be written:

$$u_{sh}(x,t) = \sum_{Ii} u_{sIi}(t)\cdot w_{sI}(x)\, e_i \qquad p_h(x,t) = \sum p_I(t)\cdot w_{pI}(x)$$

Correspondingly the test functions will be chosen successively equal to each member of the basis, that is:

$$w_{sh} = w_{sJ}\cdot e_j\,, \qquad w_{ph} = w_{pJ}$$

The not mandatory special choice of the same test function for both $\mathbf{w}_{sh}$ and p is now assumed:

$$w_J \;=\; w_{sJ} \;=\; w_{pJ}$$

The following notations are introduced for either two vector fields ($\mathbf{v}_1$, $\mathbf{v}_2$), or two tensor fields (τ_1,τ_2) defined on the domain Ω or its boundary Γ:

$$\begin{aligned}
(\tau_1,\tau_2)_\Omega &= \int_\Omega \tau_1 : \tau_2 d\Omega &&= \sum_{ij}\int_\Omega \tau_{1ij}\,.\,\tau_{2ij}\, d\Omega\\
(v_1, v_2)_\Omega &= \int_\Omega v_1\,.\,v_2 d\Omega &&= \sum_i \int_\Omega v_{1i}\,.\,v_{2i}\, d\Omega\\
\langle v_1, v_2\rangle_\Gamma &= \int_\Gamma v_1\,.\,v_2 d\Gamma &&= \sum_i \int_\Gamma v_{1i}\,.\,v_{2i}\; d\Gamma
\end{aligned}$$

With the following matrices defined by:

$$(m_{ss})_{IiJj} = \delta_{ij}\ (\rho\, w_I\, , w_J\,)_\Omega \qquad (m_{ps})_{IiJ} = (\,div(\rho_f \mathbf{K} w_I\, \mathbf{e}_i\,), w_J)_\Omega$$

$$(c_{ps})_{IiJ} = (div\,(w_I\, \mathbf{e}_i), w_J)_\Omega \qquad (c_{pp})_{IJ} = (n/K_f\, w_I\, ,w_J\,)_\Omega$$

$$(B\sigma')_{Jj} = (\sigma', \varepsilon(\, w_J \cdot \mathbf{e}_j\,))_\Omega$$

$$(k_{sp})_{IJj} = -\,(\, w_I\, , div\,(w_J\, \mathbf{e}_j)\,)_\Omega$$

$$(k_{pp})_{IJ} = (\mathbf{K.grad}\ w_I\, , \mathbf{grad}\ w_J\,)_\Omega$$

$$(f_s)_{Jj} = (\rho\, \mathbf{g}, w_J \cdot \mathbf{e}_j\,)_\Omega + <\mathbf{T}, w_J \cdot \mathbf{e}_j>_{\Gamma\sigma}$$

$$(f_p)_J = -<\varphi\, , w_J>_{\Gamma\varphi} + (\mathbf{Kgrad}\,(\rho_f\, \mathbf{g}\, \mathbf{x}), \mathbf{grad}\ w_J\,)_\Omega$$

the governing equations written in the matrix form become:

$$m_{ss}\ \ddot{u}_s + B\sigma' + k_{sp}\ p = f_s$$

$$m_{ps}\ \ddot{u}_s + c_{ps}\ \dot{u}_s + c_{pp}\ \dot{p} + k_{pp}\ p = f_p$$

For the time integration, the predictor-corrector Newmark scheme is used with the parameters β and γ. If at time step number n, the following quantities are already known ($\mathbf{u}_{sn}$, p_n, $\mathbf{v}_{sn}$, $\mathbf{a}_{sn}$) then the Newmark predictor is given by:

$$\mathbf{u}_{n+1}^* = \mathbf{u}_n + \Delta t\ \mathbf{v}_n + \Delta t^2 (1 - \beta/2)\, \mathbf{a}_n$$

$$\mathbf{v}_{n+1}^* = \mathbf{v}_n + \Delta t\,(1 - \gamma)\, \mathbf{a}_n$$

$$\sigma'_{sn+1}{}^* = \sigma'_{sn} + \Delta t\ \mathbf{D}^{ep} : \varepsilon(\, \mathbf{u}_{n+1}^* - \mathbf{u}_n\,)$$

The Newmark corrector is, on the other hand, given by:

$$\mathbf{u}_{n+1} = \mathbf{u}_{n+1}^* + \Delta t^2\, \beta\, \mathbf{a}_{n+1}$$

$$\mathbf{v}_{n+1} = \mathbf{v}_{n+1}^* + \Delta t\ \gamma\, \mathbf{a}_{n+1}$$

$$\sigma'_{sn+1} = \sigma'_{sn} + \Delta t\ \mathbf{D}^{ep} : \varepsilon(\, \mathbf{u}_{n+1} - \mathbf{u}_n\,)$$

These expressions are then introduced into the finite element equations described above and a step by step procedure is initiated.

6. NUMERICAL APPLICATIONS

The preceeding theoretical considerations are now applied to a complete example: the dynamic analysis of the El Infiernillo dam. The El Infiernillo dam has been analyzed for a Benchmark organized by the International Committee for Large Dams. The construction scenario and the reservoir filling have been modelled to get the initial conditions before the input of the seismic event.

The figure 16 shows the general layout of the geometry of this earth-dam. On the figure 17 a simulation of the (G-γ) curves for the rockfill material shows clearly the capability of the proposed model.

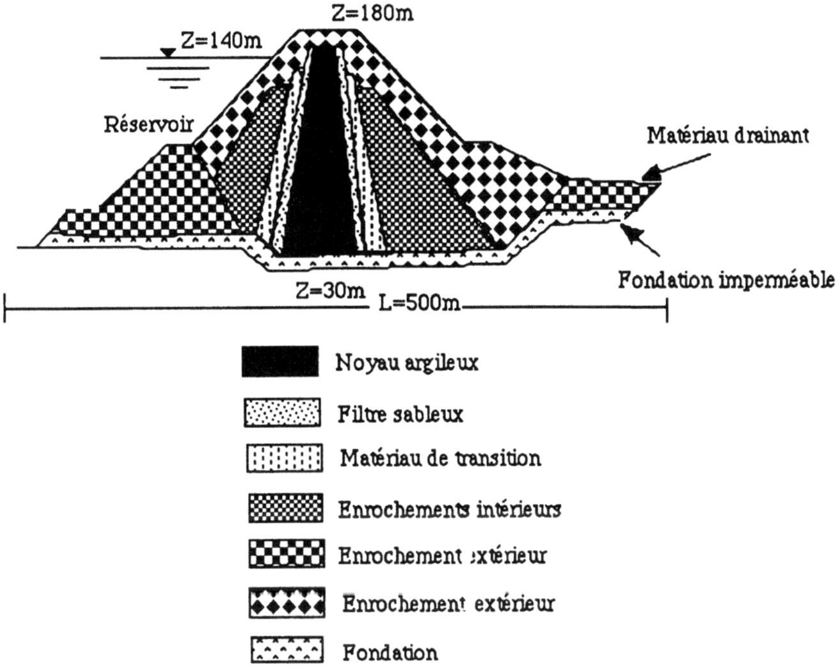

Figure 16: General layout of the dam

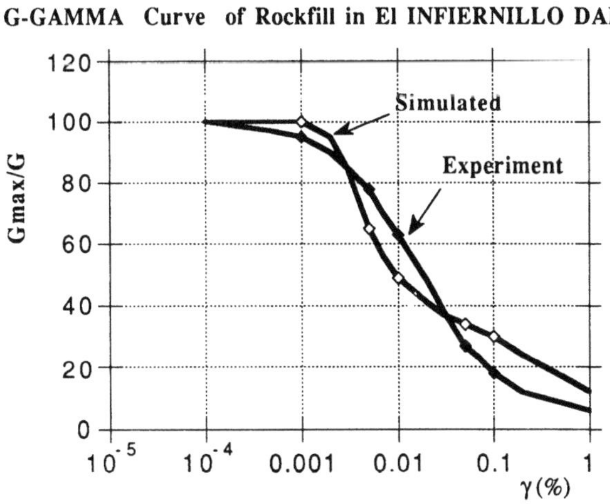

Figure 17: Simulated (G-γ) curves for the rockfill material

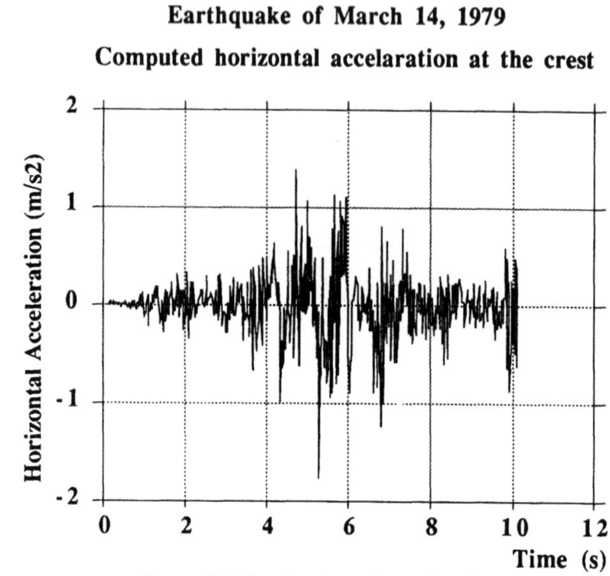

Figure 18: Crest horizontal acceleration

The figure 18 gives an example of the obtained accelaration at the crest and the figure 19 gives the history of the horizontal displacements in the core. The amount of permanent displacement is well visualized. On the figure 20 the permanent settlements are quantitatively compared with the experimental measurements. A rather satisfactory agreement is obtained.

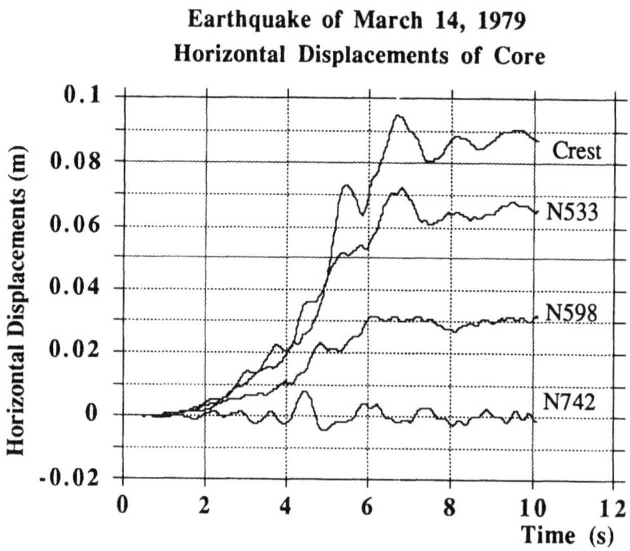

Figure 19: History of the horizontal displacements

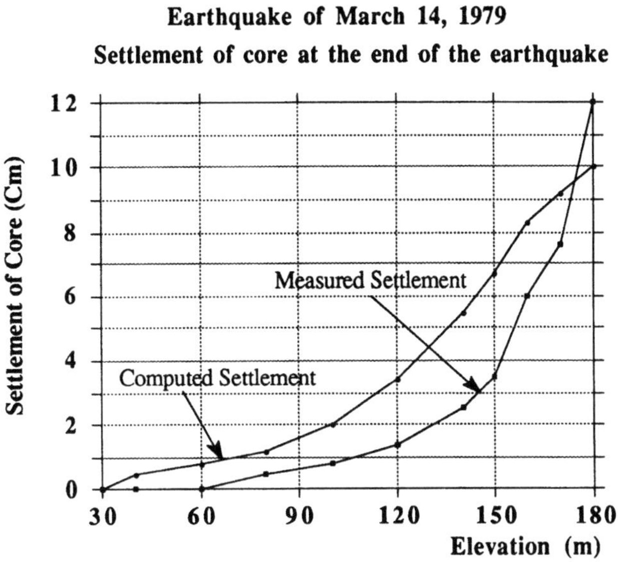

Figure 20: Vertical component of the permanent displacements in the core due to the seismic input

Finally the evolution and the rise of the pore pressure with time is shown on figure 21. This evolution can obviously be obtained only with the use of an adequate constitutive model.

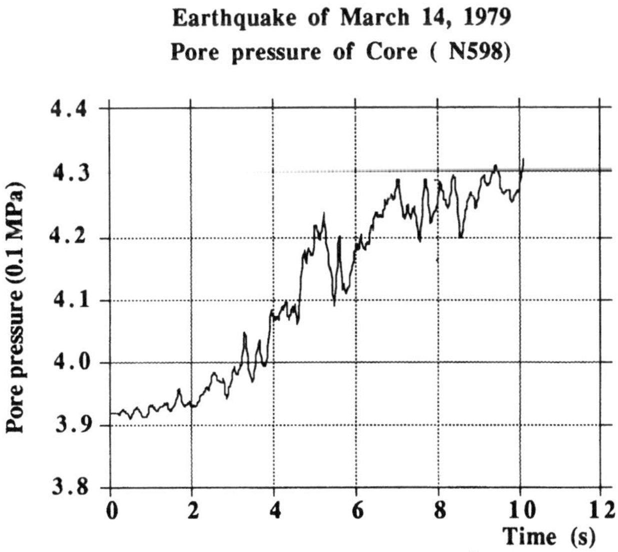

Figure 21: History of the pore pressure at node 598 during the seismic event

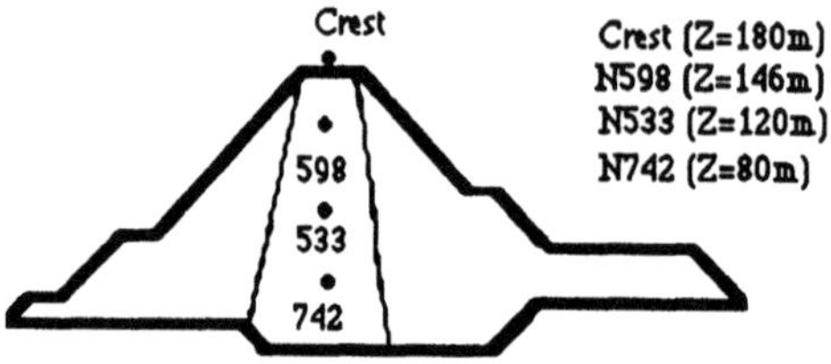

Figure 22: Position of the output points

7. CONCLUSIONS

A formulation for the two-phase soil combined with a suitable cyclic constitutive model has been presented for the analysis of wave propagation in saturated soils. This formulation is well-adapted for earthquake engineering and for soil dynamics. The implementation of this model in a finite element code constitutes an attractive tool for the prediction of soil behavior, permanent settlement or liquefaction. The unified approach which is however affordable at least in two dimensions allows to have now a consistant appreciation of the behaviour of large earth structures [1] when submitted to low or strong events.

8. REFERENCES

1 Seed H.B., Landslides during earthquakes due to liquefaction, J. Soil Mech. and Found. Div., ASCE, 94:SM5, pp. 1053-1122, (1968).

2 Zienkiewicz O.C. & Shiomi T., Dynamic behaviour of saturated porous media ; The generalized Biot formulation and its numerical solution. Int. Jour. Num. and Anal. Meth. Geomech. vol 8, 71-96 (1984).

3 Aubry D., Hujeux J.C., Lassoudière F. & Meimon Y., A double memory model with multiple mechanisms for cyclic soil behaviour, Int. Symp. Num. Mod. Geomech., Balkema.3-13 (1982).

4 Hughes T.J.R., Liu I.K., Implicit explicit finite elements in transient analysis: stability theory. J. Applied Mechanics. Vol 45, 371-375 (1977).

5 Ghaboussi J. & Dikmen U., Liquefaction analysis of Horizontaly layered Sands, Jou. of the Geotech. Eng. Div. pp 341:356, (1978).

6 Aubry D., Modaressi H., Seismic wave propagation in soils including non linear and pore pressure effects, Recent Advances in Earthquake Enginering and Structural Dynamics, Davidovici Ed, Ouest Edition, 209-224, (1992)

7 Aubry D. , Benzenati I. , Modaressi A., A coupled dynamic strain and pore pressure analysis of embankment dams, Dam Engineering, pp 53-94, (1990)

8 Benzenati I., Coupled nonlinear numerical simulation of the seismic behaviour of earthdams and slopes, Doctoral thesis, Ecole Centrale Paris, 1992.

APPLICATIONS OF THE BOUNDARY ELEMENT METHOD IN DYNAMIC SOIL-STRUCTURE INTERACTION

D.E. BESKOS
Department of Civil Engineering
University of Patras
GR-26110 Patras
Greece

ABSTRACT. The conventional direct frequency or time domain boundary element method as applied to dynamic soil-structure interaction analysis is discussed. Both the structure and the soil are assumed to be homogeneous, isotropic and linear elastic or viscoelastic bodies under two - or three - dimensional conditions. The dynamic disturbances may be externally applied loads or seismic waves of an arbitrary spatial and temporal variation. The effects of inhomogeneity and anisotropy of soil on the system response are also considered. Applications including the dynamics of foundations and above ground structures and the dynamic response of underground structures are described in detail. The general conclusion is that the boundary element method alone, or coupled with the finite element method, is ideally suited for dynamic soil-structure interaction problems from both the accuracy and efficiency viewpoints.

1. Introduction

The two main advantages of the boundary element method (BEM), namely the reduction of the spatial dimensions of the problem by one and the high accuracy of the method are more pronounced in linear elastodynamics, especially when the domain of interest is infinite or semi-infinite. Domain type of numerical methods, such as the finite element method (FEM) and the finite difference method (FDM), require a discretization of both the surface and the interior of the domain and create problems due to wave reflections at artificial boundaries for infinite or semi-infinite domains. Remedies to this problem include large uneconomical meshes or employment of complicated absorbing boundaries. The BEM, on the other hand, requires just a surface discretization and takes automatically into account the radiation condition at infinity. And all these advantages become more important in three spatial dimensions.

Thus, problems associated with wave propagation in soils and dynamic soil-structure interaction under the assumption of linear elastic or viscoelastic soil behavior can be ideally treated by the BEM. This work deals with the conventional direct frequency or time domain BEM as applied to dynamic soil-structure interaction problems involving linear elastic or viscoelastic bodies under two - or three - dimensional conditions. The term "conventional" is used to characterize those BEM's which are based on integral formulations involving the fundamental solution (or Green's function) of the problem and are solved by discretization of the surface of the domain and collocation.

The sources of the dynamic disturbances considered here may be externally applied loads or seismic waves of an arbitrary direction and time variation. The effects of soil inhomogeneity and anisotropy on the response are also considered. Two kinds of applications are discussed in detail. Problems of the dynamic response of foundations and above ground structures placed on or embedded in the half-space soil medium are

P. Gülkan and R. W. Clough (eds.), Developments in Dynamic Soil-Structure Interaction, 61–90.

discussed first. Then the dynamic response of underground structures buried in the full- o half-space soil medium is considered. Other BEM applications in dynamic soil-structur interaction, such as pile dynamics, vibration isolation of machine foundations an unilateral contact analysis are not treated here. The interested reader can consult the book of Manolis and Beskos (1988) and Manolis and Davies (1992) for these and additiona subjects.

Among the existing classical works on elastodynamics and wave propagation that includ integral equation techniques one can mention here the books of Pao and Mow (1973 Achenbach (1973), and Eringen and Suhubi (1975). General reviews on elastodynami analysis by the direct BEM in time or frequency domain can be found in the book o Manolis and Beskos (1988), the article of Beskos (1987), the articles of Kobayashi (1985 1987), Banerjee et al (1987), Nishimura and Kobayashi (1989) and the two books of Wol (1985,1988). Reviews on the BEM as applied to static and dynamic problems i geomechanics and wave propagation in soils have been written by Beskos (1988, 1992) an to dynamic soil-structure interaction by Karabalis and Beskos (1987 a, b), Dominguez an Abascal (1987), Karabalis (1988) and Tassoulas (1989). The aforementioned two books o Wolf (1985, 1988) on soil-structure interaction should also be mentioned.

In connection with the two classes of dynamic soil-structure interaction problems, whic constitute the subject of interest in this work, one can mention, among others, the followin key BEM references: Dominguez (1978 a, b), Alarcon and Dominguez (1980) Ottenstreue and Schmid (1981), Abascal and Dominguez (1984b), Karabalis and Beskos (1984, 1985 1986) and Spyrakos and Beskos (1986 a, b) for foundation dynamics and Kobayashi an Nishimura (1982), Manolis and Beskos (1983), Manolis (1983) and Kitahara et al (1987 for dynamics of underground structures.

The present work, in addition to the introduction - section one, consists of the followin seven sections: The second section provides a brief description of the direct boundar element method in elastodynamics; both the frequency and the time domain formulation are presented. The third section deals with the dynamic analysis of foundations and abov ground structures subjected to external loads or seismic waves; surface, embedded, rigid o flexible foundations on homogeneous or layered soil are considered. The determination o the dynamic response of underground structures to external loads or seismic waves is th subject of the fourth section. Section five presents representative numerical results for al the three cases considered here, while section six summarizes the conclusions. The presen lecture finally closes with a comprehensive list of references in the seventh and last section

2. B.E.M.'S in Elastodynamics

A brief review of conventional boundary element methods in elastodynamics is presente in this section for reasons of completeness. The reader can consult Beskos (1987) an Manolis and Beskos (1988), the basis of the developments in this section, for furthe details.

Consider a homogeneous, isotropic and linearly elastic body of volume V bounded b a surface S. The equations of motion of this body are of the form

$$(c_1^2-c_2^2)\,u_{i,ij}+c_2^2u_{j,ii}+b_j=\ddot{u}_j \tag{1}$$

where $u_i = u_i\,(\underset{\sim}{x},t)$ is the displacement vector at point $\underset{\sim}{x}$ and time t, b_j is the body force vector per unit mass, c_1 and c_2 are the dilational and shear wave velocities, respectively, which are given in terms of the Lamé elastic constants λ and μ as

$$c_1^2=(\lambda+2\mu)/\varrho, \qquad c_2^2=\mu/\varrho \tag{2}$$

with ϱ being the mass density of the body and where Latin indices, e.g., i, j receive the values 1, 2 and 3, summation convention is implied over repeated indices and commas and overdots denote spatial and temporal differentiation, respectively. Stresses t_{ij} are given by

$$t_{ij}=\varrho(c_1^2-2c_2^2)\,u_{m,m}\,\delta_{ij}+\varrho c_2^2(u_{i,j}+u_{j,i}) \tag{3}$$

with δ_{ij} being the Kronecker's delta and are connected with the stress vector (traction) t_i by

$$t_i=t_{ij}n_j \tag{4}$$

with n_j being the outward normal unit vector on S. Displacements and stresses should satisfy the boundary conditions

$$\begin{aligned} u_i(\underset{\sim}{x},t)&=u_{bi}(\underset{\sim}{x},t)\,; \quad \underset{\sim}{x}\,\varepsilon\, S_u \\ t_i(\underset{\sim}{x},t)&=t_{bi}(\underset{\sim}{x},t); \quad \underset{\sim}{x}\,\varepsilon\, S_t \end{aligned} \tag{5}$$

where $S_u \cup S_t = S$ and u_{bi} and t_{bi} are prescribed values, and the initial conditions

$$\begin{aligned} u_i(\underset{\sim}{x},0)&=u_{oi}(x)\,; \quad \underset{\sim}{x}\,\varepsilon\, V \\ \dot{u}_i(x,0)&=\upsilon_{oi}(x)\,; \quad \underset{\sim}{x}\,\varepsilon\, V \end{aligned} \tag{6}$$

where u_{oi} and υ_{oi} are prescribed initial displacements and velocities, respectively.

Two dimensional approximations of the above three-dimensional formulation are a) Anti-plane strain, defined as

$$u_\alpha=0, \quad u_3 = u_3(x_1,x_2,t) \tag{7}$$

where Greek indices, e.g., α receive the values 1 and 2. Use of (7) in (3) and (1) result in the stresses

$$t_{\alpha 3} = \varrho c_2^2 u_{3,\alpha} \tag{8}$$

and the equation

$$c_2^2 u_{3,\alpha\alpha} + b_3 = \ddot{u}_3 \tag{9}$$

b) Plane strain, defined as

$$u_\alpha = u_\alpha(x_1, x_2, t), \quad u_3 = 0 \tag{10}$$

Use of (10) in (3) and (1) and employment of (2) result in the stresses

$$\begin{aligned} t_{\alpha\beta} &= \varrho(c_1^2 - 2c_2^2) u_{\gamma,\gamma} \delta_{\alpha\beta} + \varrho c_2^2 (u_{\alpha,\beta} + u_{\beta,\alpha}) = \\ &= \lambda u_{\gamma,\gamma} \delta_{\alpha\beta} + \mu (u_{\alpha,\beta} + u_{\beta,\alpha}) \\ t_{33} &= \varrho(c_1^2 - 2c_2^2) u_{\gamma,\gamma} = \lambda u_{\gamma,\gamma} \end{aligned} \tag{11}$$

and the equation

$$(\lambda + \mu) u_{\alpha,\alpha\beta} + \mu u_{\beta,\alpha\alpha} + \varrho b_\beta = \varrho \ddot{u}_\beta \tag{12}$$

In many cases it is convenient to formulate and solve an elastodynamic problem not in the real time domain but in the frequency domain with the aid of Fourier (real frequency domain) or Laplace (complex frequency domain) transform with respect to time. The Fourier transform $\tilde{f}(\underset{\sim}{x},\omega)$ of a function $f(\underset{\sim}{x},t)$ is defined as

$$\tilde{f}(x,\omega) = \int_{-\infty}^{\infty} f(\underset{\sim}{x},t) e^{-i\omega t} dt \tag{13}$$

where ω (real number) is the Fourier transform parameter or circular frequency and $i = \sqrt{-1}$. The Laplace transform $\bar{f}(x,s)$ of the same function $f(x,t)$ is defined as

$$\bar{f}(\underset{\sim}{x},s) = \int_{0}^{\infty} f(\underset{\sim}{x},t) e^{-st} dt \tag{14}$$

where s (complex number) is the Laplace transform parameter or complex frequency. Application of Fourier transform with respect to time on Eqs. (1) and (3-6) under zero initial conditions (nonzero initial conditions can be easily treated) leads to

$$(c_1^2 - c_2^2) \tilde{u}_{i,ij} + c_2^2 \tilde{u}_{j,ii} + \tilde{b}_j = -\omega^2 \tilde{u}_j \tag{15}$$

$$\tilde{t}_{ij} = \varrho(c_1^2 - 2c_2^2) \tilde{u}_{m,m} \delta_{ij} + \varrho c_2^2 (\tilde{u}_{i,j} + \tilde{u}_{j,i}) \tag{16}$$

$$\tilde{u}_i(\underset{\sim}{x},\omega)=\tilde{u}_{bi}(\underset{\sim}{x},\omega);\ \underset{\sim}{x}\varepsilon S_u$$
$$\tilde{t}_i(\underset{\sim}{x},\omega)=\tilde{t}_{ij}n_j=\tilde{t}_{bi}(\underset{\sim}{x},\omega);\ \underset{\sim}{x}\varepsilon S_t \tag{17}$$

It can be proved (e.g., Manolis and Beskos 1988) that the above equations hold true also in the Laplace transformed domain if ω is replaced by is. The special cases of anti-plane and plane strain can also be written in the frequency domain without any difficulty.

The conventional boundary element method (BEM) is based on an integral representation of the solution u_i of (1) or $\tilde{u}_i$ of (15), which is achieved in a direct or indirect manner with the aid of the fundamental solution of (1) or (15), respectively. This fundamental solution (or Green's function) is defined as the displacement field $u^*_{ij}(\underset{\sim}{\xi},t)$ or $\tilde{u}^*(\xi,\omega)$ at point $\underset{\sim}{\xi}$ in an infinite body due to a concentrated body force of the form $\varrho b_i(\underset{\sim}{x},t) = \delta(t)\delta(x,\xi)e_i$ or $\varrho \tilde{b}_i(x,\omega) = e^{i\omega t}\ \delta(\underset{\sim}{x},\xi)e_i$ at point $\underset{\sim}{x}$ for the time or frequency domain formulation, respectively, with δ being the Dirac delta function and e_i the direction of the applied force. Use of equations (3) and (4) or (16) and $(17)_2$ can provide the fundamental traction tensor t^*_{ij} or $\tilde{t}^*_{ij}$ in the time or frequency domain, respectively. Explicit expressions for the fundamental solution pair in the time (u^*_{ij}, t^*_{ij}) or the frequency $(\tilde{u}^*_{ij}, \tilde{t}^*_{ij})$ domain for three-dimensional elastodynamics, antiplane strain or plane strain can be found in, e.g., Manolis and Beskos (1988). The main characteristic of these solutions is their singular character. Thus,

$$u^*_{ij}\ or\ \tilde{u}^*_{ij} = O(lnr)$$
$$t^*_{ij}\ or\ \tilde{t}^*_{ij} = O(1/r) \tag{18}$$

for two-dimensions and

$$u^*_{ij}\ or\ \tilde{u}^*_{ij} = O(1/r)$$
$$t^*_{ij}\ or\ \tilde{t}^*_{ij} = O(1/r^2) \tag{19}$$

for three-dimensions, where O means "order of" as $r\rightarrow 0$ and $r = |\underset{\sim}{x} - \xi|$.

A direct integral formulation employs displacements and tractions, while an indirect one fictitious quantities, such as source densities. Only the direct formulation will be briefly presented here in the time and frequency domain under zero body forces and initial conditions. A direct formulation starts either with the reciprocal theorem or the weighted residuals statement and leads to

$$c_{ij}(\underset{\sim}{\xi})u_i(\underset{\sim}{\xi},t) = \int_S \left[u^*_{ij}\circ t_i(\underset{\sim}{x},t)-t^*_{ij}\circ u_i(\underset{\sim}{x},t)\right]dS(\underset{\sim}{x}) \tag{20}$$

$$c_{ij}(\underset{\sim}{\xi})\tilde{u}_i(\underset{\sim}{\xi},\omega)=\int_S \left[\tilde{u}^*_{ij}\tilde{t}_i(\underset{\sim}{x},\omega)-\tilde{t}^*_{ij}\tilde{u}_i(\underset{\sim}{x},\omega)\right]dS(\underset{\sim}{x}) \tag{21}$$

integral representations in the time and frequency domain with the symbol ∘ denoting convolution and the tensor c_{ij} receiving the value of δ_{ij} if $\xi \varepsilon V$ and $(1/2)\ \delta_{ij}$ if $\xi \varepsilon S$ and is smooth. For the numerical solution of (20) the surface S of the body is discretized into a finite number of boundary elements (constant or higher order) and the time axis into a finite number of time steps. Thus for the whole surface of the body (20) yields

$$[U]\{t\}^n = [T]\{u\}^n \tag{22}$$

at the time step n, where [U] and [T] are influence matrices with entries integrals over boundary elements involving u^*_{ij} and t^*_{ij}, respectively and {t} and {u} are the traction and displacement vectors at all nodal points on the surface, respectively. After employment of the boundary conditions and rearrangement, the solution can be obtained through a time marching scheme. The discretized version of (21) requires only a spatial discretization and reads

$$[\tilde{U}]\{\tilde{t}\}=[\tilde{T}]\{\tilde{u}\} \tag{23}$$

After employment of the boundary conditions and rearrangement, the solution of (23) is obtained for a sequence of values of ω in the Fourier transformed domain and the time domain response is finally obtained by a numerical inversion through the fast Fourier algorithm. In case one works in the Laplace transformed domain the numerical inversion algorithm used is that of Durbin. For both inversion algorithms one can consult Manolis and Beskos (1988) for details. Once the solution on the boundary has been found, one can obtain the solution at any interior point ξ by using the discretized versions of (20) and (21) with $c_{ij} = \delta_{ij}$. With known interior displacements, stresses can be computed from (3) or (16).

The preceding developments cover not only the three-dimensional case but the plane strain case as well. The direct integral formulation of the antiplane strain case (equation (9)) under zero body forces and initial conditions takes the form

$$c_{ij}(\underset{\sim}{\xi})u_3(\underset{\sim}{\xi},t) = \int_S \left[u^*_{33} \circ \frac{\partial u_3(\underset{\sim}{x},t)}{\partial n} - \frac{\partial u^*_{33}}{\partial n} \circ u_3(\underset{\sim}{x},t) \right] dS(\underset{\sim}{x}) \tag{24}$$

in the time domain, where u^*_{33} is the fundamental solution of equation (9) and $\partial u_3/\partial n = u_{3,j}\, n_j$. Replacement of t by ω and elimination of the convolution produces the frequency domain formulation of the problem with u^*_{33} being of course function of ω.

All the preceding developments in this section were associated with linear elastic material behavior. Linear viscoelastic soil behavior can be easily incorporated in a BEM frequency domain formulation with the aid of the correspondence principle by replacing the elastic constants λ and μ by $i\omega\, \tilde{\lambda}(\omega)$ and $i\omega\, \tilde{\mu}(\omega)$ (or $s\bar{\lambda}(s)$ and $s\bar{\mu}(s)$), respectively, where λ(t) and μ(t) are given viscoelastic material functions. In practice, however, one simply replaces λ and μ by their complex values

$$\lambda^*=\lambda(1+2i\beta),\ \mu^*=\mu(1+2i\beta) \tag{25}$$

where $i = \sqrt{-1}$ and β is the constant hysteretic damping coefficient. Treatment of viscoelastic soil behavior by the time domain BEM is not easy due to the difficulty in obtaining the time domain viscoelastic Green's function for the general case of viscoelastic material behavior. Tanaka (1985) presented two time domain BEM's for the treatment of viscoelastic problems. The first one is a usual time domain BEM in conjunction with a viscoelastic Green's function for the three-element viscoelastic model (Voight type) under two-dimensional conditions. The second one is a time domain BEM in incremental form (as, e.g., in plasticity) for which the general viscoelastic Green's function can be easily obtained. Wolf (1988) has proposed a numerical evaluation of this function by inverting the frequency domain one for which viscoelastic behavior can be easily incorporated by using (25). Recently, Gaul et al (1991) succeeded in analytically evaluating the time domain viscoelastic Green's function for the special case of the fractional derivative type of viscoelasticity.

3. Dynamics of Foundations

The determination of the dynamic response of rigid or flexibe, surface or embedded, three-or-two dimensional foundations to externally applied loads or seismic waves by the BEM is the main subject of this section. When the foundation is flexible and/or there is a superstructure, one usually employs the BEM for the soil and the FEM for the foundation and/or the superstructure in a way that couples the two methods together. The discussion that follows is based on Karabalis and Beskos (1987 b).

Consider first the dynamic analysis of a three-dimensional, massless, rigid, surface foundation of arbitrary shape resting on a linear elastic, homogeneous and isotropic half-space. Once this kinematic problem has been solved, the mass of the foundation and the superstructure can easily be taken into account. The governing equations of the problem consist of the discretized boundary integral equations (22) or (23) in time or frequency domain, respectively and the compatibility and equilibrium equations at the soil-foundation interface.

The displacement compatibility equations read

$$\{u\}^n = [S]\{D\}^n \tag{26}$$

where $\{u\}^n$ is the 3Mx1 displacement vector corresponding to the centers of the M constant quadrilateral boundary elements and $\{D\}^n$ is the 6x1 rigid body displacement vector given by

$$\{D\}^n = \{\Delta_1\ \Delta_2\ \Delta_3\ \phi_1\ \phi_2\ \phi_3\}^{Tn} \tag{27}$$

with Δ_i and ϕ_i (i = 1,2,3) being the rigid body displacements and small rotations, respectively, at the center of the foundation (Fig. 1). In frequency domain one has again (26) with $\{u\}^n$ and $\{D\}^n$ replaced by their transformed expressions $\{\tilde{u}\}$ and $\{\tilde{D}\}$, respectively.

The equilibrium equations will be formulated now for the two cases of external loads and seismic waves. Equilibrium between external forces P_i - moments M_i and tractions t_{im} (m = 1,2....M) reads

$$\{P\}^n = [K]\{t\}^n, \tag{28}$$

where

$$\{P\}^n = \{P_1\ P_2\ P_3\ M_1\ M_2\ M_3\}^{Tn} \tag{29}$$

and $\{t\}^n$ is a 3Mx1vector with elements t_{im}, the tractions at the center of every element m. Equations (22), (26) and (28) form a system of equations in the time domain, which is solved in a step-by-step fashion marching in time. In frequency domain (28) is again valid with $\{P\}^n$ and $\{t\}^n$ replaced by $\{\tilde{P}\}$ and $\{\tilde{t}\}$, respectively. In this case the solution results in the force-displacement relation

$$\{\tilde{P}\} = [K(\omega)]\{\tilde{D}\}, \tag{30}$$

where $[K(\omega)]$ is the 6x6 impedance (or dynamic stiffness) of the rigid surface foundation. When the elastic half-space is excited by a plane incident wave, as shown in Fig. 1, the placement vector field $\{u_f\}$ on the free surface $x_3 = 0$ is described, in the absence of any foundation, by

$$\{u_f\} = \{u_f(\underset{\sim}{x},t)\} = \{u_{f1},\ u_{f2},\ u_{f3}\}^T \tag{31}$$

where $\{u_f\}$ depends on the vertical and horizontal angles of incidence θ_v and θ_h respectively and the apparent velocity $c = c_{1,2}/\cos\theta_v$, depending on the type (P or S) of the incident wave. In the presense of the foundation, a scattered displacement field $\{u_s(\underset{\sim}{x},t)\}$ is additionally generated so that the total displacement field $\{u(\underset{\sim}{x},t)\}$ under the foundation to be given as

$$\{u\} = \{u_f\} + \{u_s\} \tag{32}$$

The discrtetized in space and time version of (32) involves 3Mx1 vectors for constant elements and time steps. The problem is governed by Eq. (22) for the scattered field, Eq. (26) for the total field and the discretized version of (32), which expresses scattered displacements in terms of the total ones and the known free field ones. The system of these equations is then solved by marching in time.

When the formulation is in the frequency domain,(32) is replaced by its transformed version and the problem is governed by that version as well as Eqs. (23) for $\tilde{u}_s$ and the transformed version of (26). The solution of this system of equations results in

$$\{\tilde{D}\} = [C(\omega)]\{\tilde{u}_f\}, \tag{33}$$

where the 6x3M matrix $[C(\omega)]$ is the input motion matrix of the rigid foundation.

Once the response of a massless rigid foundation has been determined, the dynamics of a massive superstructure supported on this foundation, which is now assumed to possess a mass, can be easily studied by employing superposition of the kinematic and inertia interaction problems. The kinematic problem deals with the determination of the response to seismic waves when the structure-foundation system has only stiffness and zero mass

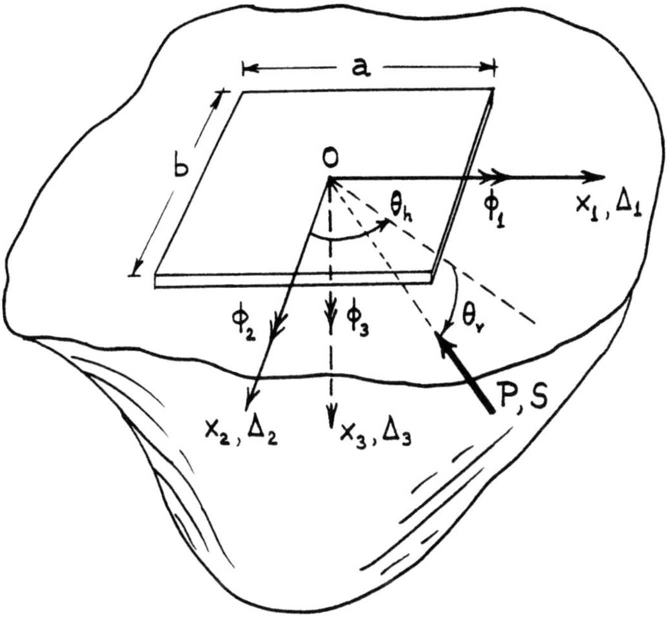

Figure 1. Rectangular rigid surface foundation and its associated coordinate system and displacements-rotations. Incident oblique P and S wave geometry.

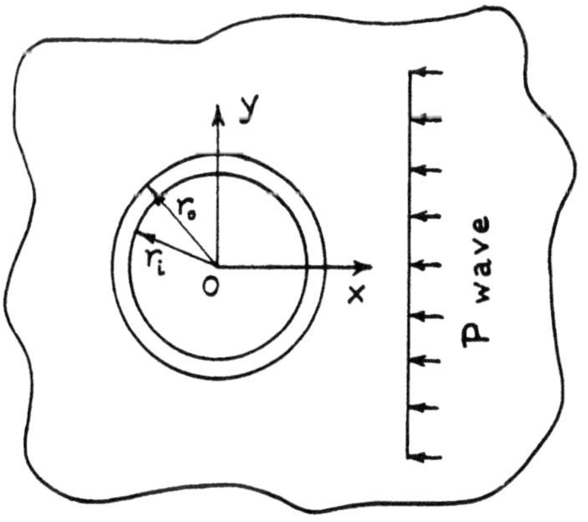

Figure 2. Wave diffraction by a lined tunnel in infinite soil medium.

while the inertial problem with the determination of the response to accelerations on the structure computed from the first problem when the system possesses both stiffness and mass. In the following, a similar superposition approach in the frequency domain due to Luco and Wong (1982) is briefly described.

The rigid body displacements $\{\tilde{D}_K\}$ of a massless surface foundation due to a free field wave excitation $\{\tilde{u}_f\}$ can be obtained by using (33) as

$$\{\tilde{D}_K\} = [C(\omega)]\,\{\tilde{u}_f\} \tag{34}$$

The fictitious inertia forces $\{\tilde{F}_I\}$ corresponding to the total rigid foundation displacements $\{\tilde{D}_T\}$ can be written as

$$\{\tilde{F}_I\} = \omega^2\big[[M_f] + [M_s(\omega)]\big]\{\tilde{D}_T\} \tag{35}$$

where $[M_f]$ is the 6x6 mass matrix of the foundation and $[M_s(\omega)]$ is a 6x6 frequency dependent equivalent mass matrix which accounts for the elastic properties and the mass distribution of the superstructure. The response $\{\tilde{D}_I\}$ of the rigid foundation to these inertia forces can be found by applying Eq. (30) as

$$\{\tilde{D}_I\} = [K(\omega)]^{-1}\{\tilde{F}_I\} \tag{36}$$

Finally the total response $\{\tilde{D}_T\} = \{\tilde{D}_K\} + \{\tilde{D}_I\}$ at the foundation level can be found by adding Eqs (34) and (35), i.e.,

$$\{\tilde{D}_T\} = \Big[[I] - \omega^2[K(\omega)]^{-1}\big[[M_f] + [M_s(\omega)]\big]\Big]^{-1}[C(\omega)]\{\tilde{u}_f\} \tag{37}$$

where [I] is the identity matrix. The response at any level of the structure can be obtained by standard techniques, i.e., through a FEM discretization of the structure, in terms of the total motion at the foundation level.

It should be noted at this point that the approach followed by Karabalis and Beskos (1985) for coupling the BEM, used to model the soil, with the FEM, used to model the flexible foundation, at the soil-foundation interface can also be used to study the response of the complete soil-foundation-structure system thus avoiding the intermediate step that requires the use of the equivalent mass matrix $[M_s(\omega)]$ in Eq. (37). The coupling of the two methods (BEM + FEM) is accomplished by enforcing equilibrium and compatibility at the soil-foundation interface. Since the BEM matrices are fully populated and not symmetric the total system matrix is also fully populated and not symmetric. In addition problems of not complete matching at the interface exist. Early attempts to symmetrize BEM matrices have not been successful (Mang et al 1989). Recently, variationally coupled BEM-FEM schemes for elastostatics (Lu et al 1991) and elastodynamics (Bielak et al 1991, Aubry and Clouteau 1991) have appeared capable of not only creating symmetric matrices but also matching perfectly conditions at the interface of the two domains.

When use is made of the full-space Green's function in the BEM for the soil, then the whole free surface of the half-space and not just the soil-foundation interface have to be discretized. In practice, of course, the free soil surface discretization is confined to a small finite portion around the foundation. For a discussion on the truncation error one can consult Dominguez and Meise (1991). Of course, when use is made of the half-space Green's function only the soil-foundation interface has to be discretized. Half-space Green's function for homogeneous or layered soils are, however, very complicated. For a discussion on these functions as well as relevant references one can consult Beskos (1992). In surface foundation dynamics one usually accepts the so-called relaxed boundary conditions approximation implying that only those contact stresses primarily related to a given vibration mode are monitored while the remaining ones can be neglected. If these conditions are used instead of the complete bonding between soil and foundation, then only the interface has to be discretized. However, when the foundation is embedded, even when relaxed conditions are assumed, a free soil surface discretization around the foundation has to be done. For a discussion on the subject of relaxed boundary conditions one can look at Mohammadi and Karabalis (1990) for details.

When the soil is layered (discretly non-homogeneous) and use is made of the full-space homogeneous Green's function in the BEM, the free soil surface as well as the layer interfaces have to be discretized (finite portions around the region of interest). This approach has the advantage of treating layers of any geometry and not just horizontal ones but is practically restricted to a small number of layers. Another approach is to employ a layered half-space Green's function in which case only the soil-foundation interface has to be discretized. This approach is very well suited for soils with many horizontal layers and continuously nonhomogeneous soils (moduli varying continuously with depth) which are modeled by a system of many thin layers. For a thorough discussion on the subject one can consult Beskos (1992). In case the soil is anisotropic the problem can only be solved for those cases of anisotropy for which half-space Green's functions are available. Thus half-plane Green's functions for orthotropic soil have been reported by Payton (1983) and Rajapakse and Wang (1991) in the time and frequency domain, respectively.

Among the published works dealing with foundation dynamics treated by the direct BEM one can mention those of Dominguez (1978 a, b), Alarcon and Dominguez (1980), Alarcon et al (1980), Abascal and Dominguez (1984 a & b, 1985, 1986), Dominguez and Abascal (1989), von Estorff and Schmid (1984), Huh and Schmid (1984), Schmid et al (1984), Chouw et al (1991 a, b), Tullberg and Wiberg (1985), Reng et al (1986), Wolf and Darbre (1984 a, b), Antes and von Estorff (1989) and Israil and Ahmad (1989) dealing with strip foundations in the frequency domain, Spyrakos and Beskos (1986 a, b), Spyrakos et al (1986), Spyrakos and Antes (1986), Antes and von Estorff (1989) Gallego and Dominguez (1990) and Israil and Banerjee (1990) dealing with strip foundations in the time domain, Dominguez (1978 a, b), Kitamura and Sakurai (1979), Ottenstreuer and Schmid (1981), Ottenstreuer (1982), Sato et al (1983), Huh and Schmid (1984), Ahmad and Manolis (1987), Ahmad et al (1988), Gaul et al (1988), Gonsalves et al (1990), Mohammadi and Karabalis (1990), Karabalis and Mohammadi (1991) and Chouw and Schmid (1991) dealing with three-dimensional rigid or flexible surface and rigid embedded foundations in the frequency domain and Karabalis and Beskos (1984, 1985, 1986), Ahmad and Banerjee (1988), Karabalis and Huang (1989) and Karabalis (1990, 1991) dealing with three-dimensional rigid or flexible surface and rigid embedded foundations in the time domain. The difficult problem of determining the dynamic response of three-dimensional flexible embedded foundations to external forces and seismic waves of harmonic time variation has been solved by Gaitanaros and Karabalis (1988) and Tong and Kuribayashi (1988).

The special case of axisymmetric foundations has been studied in the frequency domain by Gomez-Lera et al (1985) and Emperador and Dominguez (1989) for the surface and embedded foundation cases, respectively, on the basis of specially developed axisymmetric fundamental solutions. Cross interactions between two adjacent three-dimensional surface foundations have been studied in the frequency domain by Ottenstreuer and Schmid (1981) and Chouw and Schmid (1991), while the case of a series of adjacent foundations on a half space connected by a flexible superstructure (ties-railway tracks system) has been investigated by Karabalis and Mohammadi (1991). Extensive comparison studies between frequency and time domain formulations as applied to foundation dynamics have been done by Mohammadi and Karabalis (1990) and Karabalis and Beskos (1990).

In all but one cases of flexible surface or embedded foundations discussed above the flexible foundation was modelled by the FEM while the soil by the BEM. In Ahmad and Banerjee (1988) the foundation was also modelled by the BEM. The FEM is to be preferred in cases there is a superstructure connected to the foundation. Use of the hybrid BEM + FEM scheme in the frequency domain has been used by Huh and Schmid (1986), Kobayashi et al (1986), Kobayashi and Mori (1986), Nishimura et al (1989), Auersch and Schmid (1990) and Wang and Schmid (1992). Use of this hybrid scheme in the time domain has been used by von Estorff and Kausel (1989) and von Estorff and Prabucki (1990).

In connection with the problem of dynamics of foundations the literature is also rich with respect to another method, similar to but less general than the conventional direct BEM discussed in this section. This is actually a Green's function approach, which is restricted to surface foundations under harmonic loading. For more information and pertinent references the interested reader can consult Luco (1982), Luco and Wong (1990), Savidis and Richter (1979) and Whittaker and Christiano (1982 b).

4. Dynamics of Underground Structures

The determination of the dynamic response of underground structures to externally applied dynamic loads (on the free soil surface or inside the structure) or seismic waves of any orientation and time variation is a problem that can be ideally solved by the BEM alone, or in conjunction with the FEM. Here the case of a suddendly applied P-wave diffraction by a lined circular tunnel in the infinite plane as described in Manolis and Beskos (1983, 1988) is briefly discussed for illustration purposes. The case of externally applied dynamic loads for the buried structure can be considered as a special case of the wave diffraction problem.

With reference Fig. 2, the scattered field BEM equations for the soil medium and the liner can be written in the Laplace transformed domain as

$$[\bar{T}]\{\bar{u}^s\} = [\bar{U}]\{\bar{t}^s\} \tag{38}$$

and

$$\begin{bmatrix} \bar{T}_{11} & \bar{T}_{12} \\ \bar{T}_{21} & \bar{T}_{22} \end{bmatrix} \begin{Bmatrix} \bar{u}_o^s \\ \bar{u}_i^s \end{Bmatrix} = \begin{bmatrix} \bar{U}_{11} & \bar{U}_{12} \\ \bar{U}_{21} & \bar{U}_{22} \end{bmatrix} \begin{Bmatrix} \bar{t}_o^s \\ \bar{t}_i^s \end{Bmatrix} \tag{39}$$

respectively, where the superscript s stands for scattered and the subscripts i and o stand for inner and outer surface of the liner. In wave diffraction problems the total field is given by equation (32) and hence the compatibility and equilibrium equations of the problem taking into account that there is no free field in the liner read

$$\begin{gathered} \bar{u}^f + \bar{u}^s = \bar{u}_o^s \\ \bar{t}^f + \bar{t}^s = -\bar{t}_o^s, \quad \bar{t}_i^s = 0 \end{gathered} \tag{40}$$

Thus, one has a system of the six equations (38)-(40) in six unknowns $\bar{u}^s$, $\bar{t}^s$, $\bar{u}_o^s$, $\bar{t}_o^s$, $\bar{u}_i^s$ and $\bar{t}_i^s$ to solve in the Laplace transformed domain and determine the response by a numerical inversion of the transformed solution as discussed in the general section on boundary element methods. Alternatively the problem can be formulated for the total field and solved as described in Manolis and Beskos (1983). Of course, if the structure is embedded in the half-plane (space), then a finite portion of the free soil surface has to be discretized in addition to the cavity or structure discretization provided that the BEM employs the infinite plane (space) Green's function. However, no free soil surface discretization is necessary if the half-plane (space) Green's function is employed.

The frequency domain direct BEM has been used for the determination of the dynamic response of lined circular or horse-shoe type tunnels buried into the infinite plane and subjected to incident harmonic or transient P and SV waves by Kobayashi and Nishimura (1982), Manolis and Beskos (1983) and Kitahara et al (1987) under conditions of plane strain. The case of SH waves being diffracted by circular lined tunnels in the half-plane has been treated by the same method by Kobayashi and Nishimura (1983) and Kobayashi (1985). Vardoulakis et al (1987), Von Estorff et al (1991) and Stamos et al (1992) have computed the dynamic response of rectangular lined tunnels in the half-plane, to externally applied harmonic or transient load by using the direct frequency domain (Laplace transform) BEM. The same problem has also been solved by Von Estorff and Kausel (1989), Von Estorff and Antes (1991), Antes (1990), Von Estorff et al (1991) and Stamos et al (1992) by using the hybrid BEM-FEM time domain scheme mentioned in the previous section. While the literature on the dynamics of two-dimensional under-ground structures is rather rich, this is not the case with three-dimensional underground structures. The three-dimensional problem presents no special difficulty. However, it is very time consuming. One can mention here the axisymmetric direct time domain BEM of Wang and Banerjee (1990), which has been used for the study of the dynamic response of a spherical shell buried into the half-space to a transient surface load and the direct frequency domain (Laplace transform) BEM of Stamos and Beskos (1992 a), which has been used to solve the aforementioned problem in Wang and Banerjee (1990) as well as the problem of P-wave diffraction in the infinite space by a spherical shell. Recently, Stamos and Beskos (1992 b) following the idea of Zhang and Chopra (1991) for the problem of wave diffraction b canyons, developed a direct frequency domain BEM for the study of long cylindrical line tunnels subjected to incident harmonic waves at an angle to the tunnel axis. This metho as in the case of canyons, is capable of reducing the three-dimensional problem to a two dimensional one.

5. Numerical Examples

A number of numerical examples covering some of the subjects treated in the previou sections are presented here in order to illustrate the use of the direct conventional BEM and demonstrate its advantages.

5.1. EXAMPLE 1

Consider a massless, rigid, surface, square foundation of side 2b subjected to an externall applied vertical force P_3 = 100 H (Δt-t) in lbs,, where H is the Heaviside function and Δ = $(1/c_1)(a^2/\pi)^{1/2}$ with a being the side of the square elements in which the contact surfac has been discretized. This example has been taken from Mohammadi and Karabalis (1990) Figure 3 depicts the vertical displacement u_3 (in ft x 10^{-10}) versus time (in sec x 10^{-4}) a computed under relaxed boundary conditions by the frequency domain direct BEM i conjunction with either Fourier or Laplace transform and the time domain direct BEM An 8x8 constant element discretization with a 4x4 point Gaussian quadrature scheme ha been used for the frequency domain calculations, and a 6x6 constant element discretizatio with a 7x7 sub-element mesh per element and a semi-analytical spatial integration using 2 point Gaussian quadrature for the time domain calculations. A number of subelement per element is used in order to take into account the distribution of the travelling wave in an accurate and efficient way. A good agreement between the three BEM versions particularly at the peak values, is observed in Fig. 3. The Fourier and Laplace transform BEM require almost the same amount of computer time, while the time domain BEM requires an additional time by 25%.

5.2. EXAMPLE 2

Consider a 60x60 in. massless square flexible foundation of thickness 11.517 in., modulu of elasticity E_p = 30.004x10^6 lb/in^2, Poisson's ratio ν_p = 0.3 and mass density ϱ_p = 7.3 x 10^{-4} lb.s^2/in^4, resting on the surface of an elastic half-space with modulus of elasticity E = 8.84x10^6 lb/in^2, Poisson's ratio ν_s = 0.3 and mass density ϱ_s = 2.82x10^{-4} lb s^2/in^4. Thi foundation is subjected to a vertical, uniformly distributed rectangular impulse force o intensity 55.55 lb/in^2 and duration 4.12x10^{-5} secs. The history of the vertical deflection o this foundation at its center was first computed by Karabalis and Beskos (1985) using thei time domain BEM for the soil in conjunction with a 5x5 constant element mesh and a 3x3 subelement per element mesh at the soil-foundation interface and the FEM for the flexible foundation in conjunction with square thin plate elements with three degrees-of-freedom per node. Relaxed boundary conditions were assumed in view of the fact that for this case the accuracy is good and the computation very efficient as requiring only an interface discretization (Mohammadi and Karabalis 1990). The same problem was also solved by

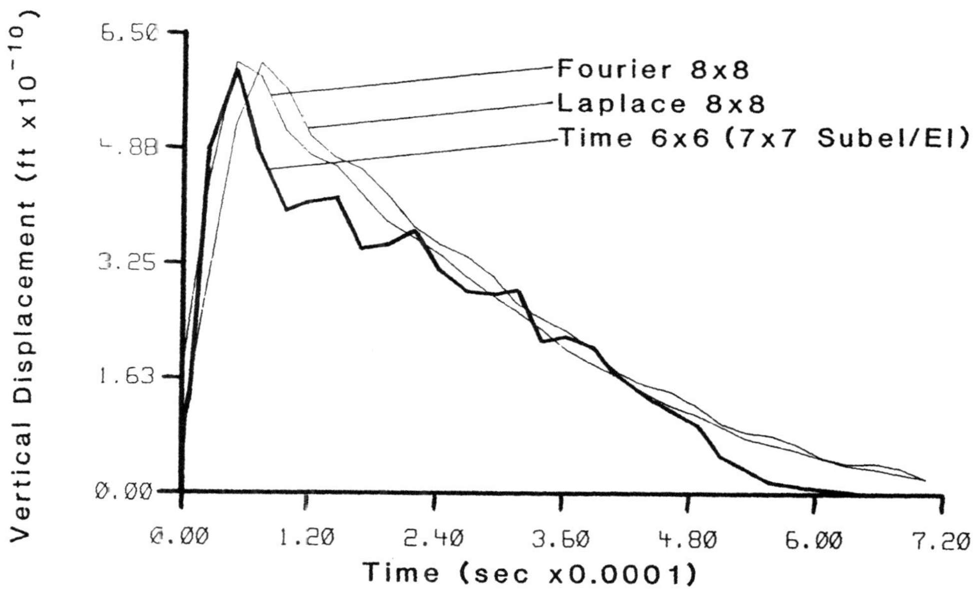

Figure 3. Vertical displacement history of massless, rigid, square foundation subjected to a vertical suddenly applied load as computed by three BEM approaches.

Ahmad and Banerjee (1988) using their time domain BEM both for the soil and the foundation in conjunction with a linear time variation, 12 quadratic elements per quadrant at the soil-foundation interface and 16 at the soil surface around the foundation (non-relaxed boundary conditions).

Figure 4 provides the history of the vertical displacement of the foundation as obtained by Karabalis and Beskos (1985) and Ahmad and Banerjee (1988) by their aforementioned methodologies. The results of these two methods show a considerable difference indicating that one set of them is naturally incorrect. Karabalis and Beskos (1990) through systematic and extensive comparison studies proved that the results of Karabalis and Beskos (1985) are the correct ones. The methodology of Ahmad and Banerjee (1988) is correct, but apparently it was incorrectly applied to this particular example. The results of three other BEM's (A, B, C) are also shown in Fig. 4. Method A employs a 5x5 constant element mesh with a 7x7 subelement mesh per element and works in the time domain, method B employs a 12x12 linear element mesh and works in the frequency domain (Fourier transform) and method C employs a 4x4 quadratic element mesh with a 9x9 subelement and works in the time domain. The plate is discretized by the FEM in all three cases using plate elements in methods A and B and solid elements in method C. Additional support to the arguments of Karabalis and Beskos (1990) is provided by Fig. 5, which depicts the amplitude of the vertical displacement of the foundation versus frequency ($a_o = \omega b/c_2$; b = half-side foundation length) due to harmonic time variation of the load. The results were obtained by convoluting the responses shown in Fig. 4 with a harmonic vertical force of the same intensity as that of Fig. 4, apart from the results of Whittaker and Christiano (1982 a), which were obtained by a Green's function approach based on Lamb's solution and requiring only a discretization of the soil-foundation interface and the results of method D obtained by an 8x8 quadratic element mesh and solid plate finite elements in the frequency domain. The static solution of Timoshenko and Goodier (1970), which constitutes here an upper bound solution for low frequencies, is also shown in Fig. 5. It is again apparent that while the results of Karabalis and Beskos (1985), methods A, B and D and Whittaker and Christiano (1982 a) are very close and nicely approach the static value as the frequency approaches zero, those of Ahmad and Banerjee (1988) are far apart from all the other results.

5.3. EXAMPLE 3

Consider an elastic T-shaped bridge pier embedded in a viscoelastic half-space and subjected to a time harmonic SV wave of oblique incidence, as shown in Fig. 6a. The incident wave has a wavelength L = 100 m and its direction of propagation forms angles of $\theta_v = 20^o$ and $\theta_h = 30^o$, as shown in Fig. 6a. The material constants for the bridge pier (denoted by subscripts f) and the ground (denoted by subscripts g) are as follows: $\lambda_f = 2\nu_f\mu_f/(1\text{-}2\nu_f)$, $\nu_f = 1/6$, $\nu_g = 1/3$, $\varrho_f = 2400$ Kg/m^3, $\varrho_g = (5/6)\varrho_f$, $\lambda_g^* = \mu_g^* 2\nu_g/(1\text{-}2\nu_g)$, $100(1+\nu_g)\mu_g^* = (1+\nu_f)\mu_f(1\text{-}0.2i)$, where λ_g^* and μ_g^* correspond to the hysteretic damping model and have the form of (25) with β replaced by $-(\beta'/2)$. The response of this bridge pier to the incident SV wave has been computed by Kobayashi et al (1986) using a FEM (for the structure) - BEM (for the ground) hybrid scheme in the frequency domain. A part of the free soil surface having a radius of about twice the incident wavelength has also been discretized in addition to the soil-structure interface. Figures 6 b & c provide the BEM and FEM mesh patterns. The real (imaginary) part of the solution shown in Fig. 6 b (6 c)

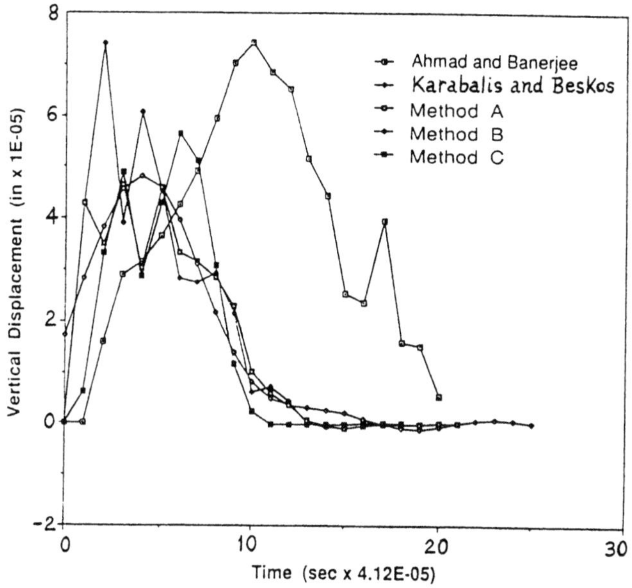

Figure 4. Vertical displacement history at the center of massless, flexible, square foundation subjected to a vertical impulsive load as computed by various BEM approaches.

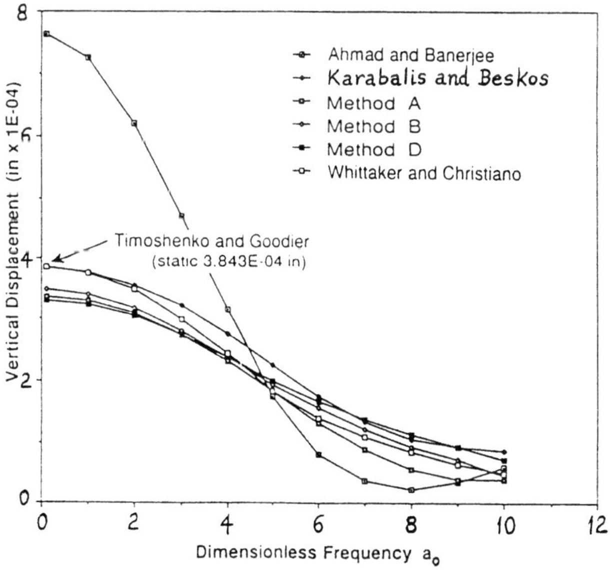

Figure 5. Vertical displacement amplitude versus frequency at the center of massless, flexible, square foundation as computed by various BEM approaches.

indicates the deformation at the moment when the peak (node) of the incident displacement wave reaches the origin of the co-ordinates. The normalizing factor u_o in Figs 6 b,c represents the amplitude of the incident wave.

5.4. EXAMPLE 4

Consider two identical adjacent chimneys supported by identical rigid embedded square foundations, as shown in Fig. 7 a. The material of the structures has a modulus of elasticity $E = 2.6x10^{10}$ N/m^2, a Poisson's ration $v = 0.33$, a mass density $\varrho = 7200$ kg/m^3 and a hysteretic damping coefficient $\beta = 0.05$. The corresponding material constants for the soil are $E = 2.1x10^8$ N/m^2, $v = 0.33$, $\varrho = 1800$ Kg/m^3 and $\beta = 0.05$. The two chimneys are modelled as beams of height L = 10.16 m and annular cross-section with outer radius R = 1.0 m and inner radius r = 0.75 m, while the foundations have dimensions 4.0 m x 4.0 m x 2.0 m and the horizontal distance between their centers is d = 6.0 m. The constant boundary element discretization involves not only the soil-foundation interfaces but a portion of the free soil surface as well. The maximum size of the soil elements is selected to be one fifth of the shear wave length. The structure on the left (excited) is subjected to a horizontal load P of unit amplitude and harmonic time variation and the response of this structure as well as that on the right (passive) are computed by a hybrid FEM (for structures) - BEM (for soil) hybrid scheme in the frequency domain, as described in Wang and Schmid (1992) from where this example is taken. Figs 7 b & c depict the displacement amplitudes U_x and U_z at nodes 1 and 2 of the two chimneys versus frequency for the cases of the load P acting at the base and at the top of the excited structure, respectively. It is evident that the interaction effect is greatly affected by the resonant frequencies of the two structures and is more pronounced when the load is at the base of the structure. When the load is applied at the top of one structure, the interaction occurs only near the resonant frequencies of the structures.

5.5. EXAMPLE 5

Consider a spherical cavity at a depth of 8 ft in the half-space lined with a concrete shell of thickness h = 5 ft and having an inner radius of 25 ft as shown in Fig. 8 a. The soil material constants are $E_s = 4.5 \times 10^6$ psf, $v_s = 0.33$ and $\varrho_s = 3.23$ lb s^2 ft^{-4} and the shell material constants $E_c = 4.5 \times 10^8$ psf, $v_c = 0.15$ and $\varrho_c = 4.66$ lb s^2 ft^{-4}. The soil surface above the buried shell is loaded axisymmetrically by a load having spatial and temporal variation as shown in Fig. 8 a. The dynamic response of this buried shell was computed in Stamos and Beskos (1992 a) by the direct BEM in conjunction with Laplace transform used both for the shell and the soil. A finite portion of the free soil surface had to be discretized due to the use of the full-space Green's function in the formulation. However, use of symmetry with respect to two planes reduced considerably the size of the problem. Thus 1/4 of every surface of the shell was discretized into 28 quadratic quadrilateral boundary elements and 1/4 of the free soil surface into 56 elements in a sectorial area with 427 ft radius. The numerical Laplace transform was done with 20 values of the transform parameter.

Figure 8 b depicts the history of the vertical displacement at point A of the shell as obtained by the above method and also by the axisymmetric time domain BEM of Wang

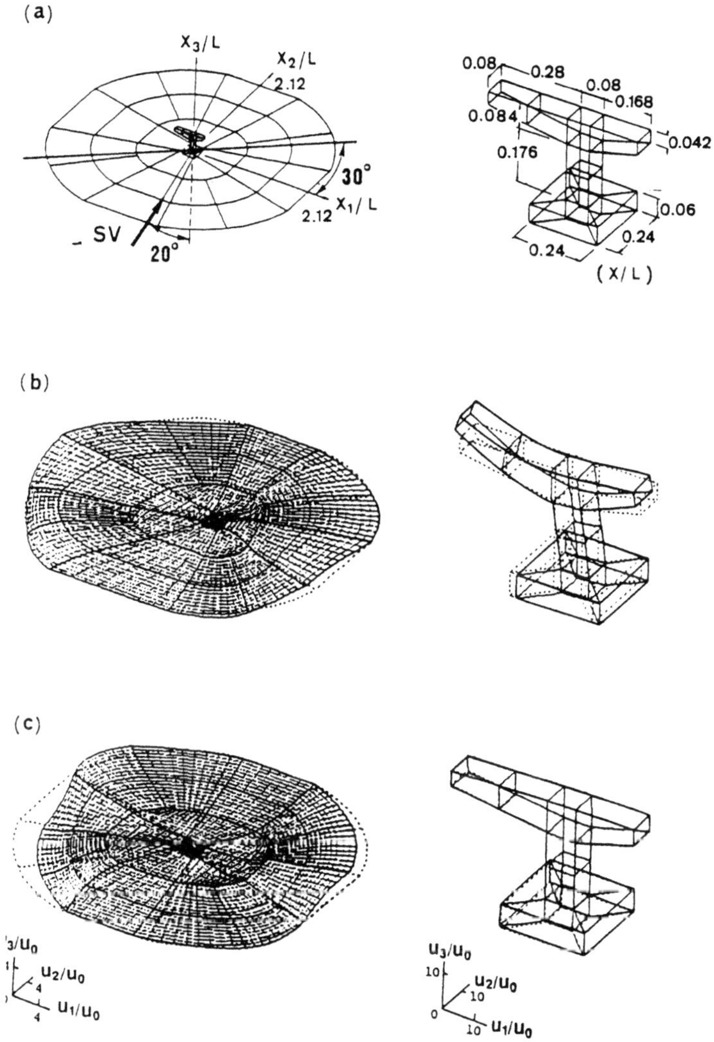

Figure 6. Response of an elastic bridge pier founded in viscoelastic soil to an oblique, harmonic SV wave: (a) geometry of system; (b) real part of response; (c) imaginary part of response.

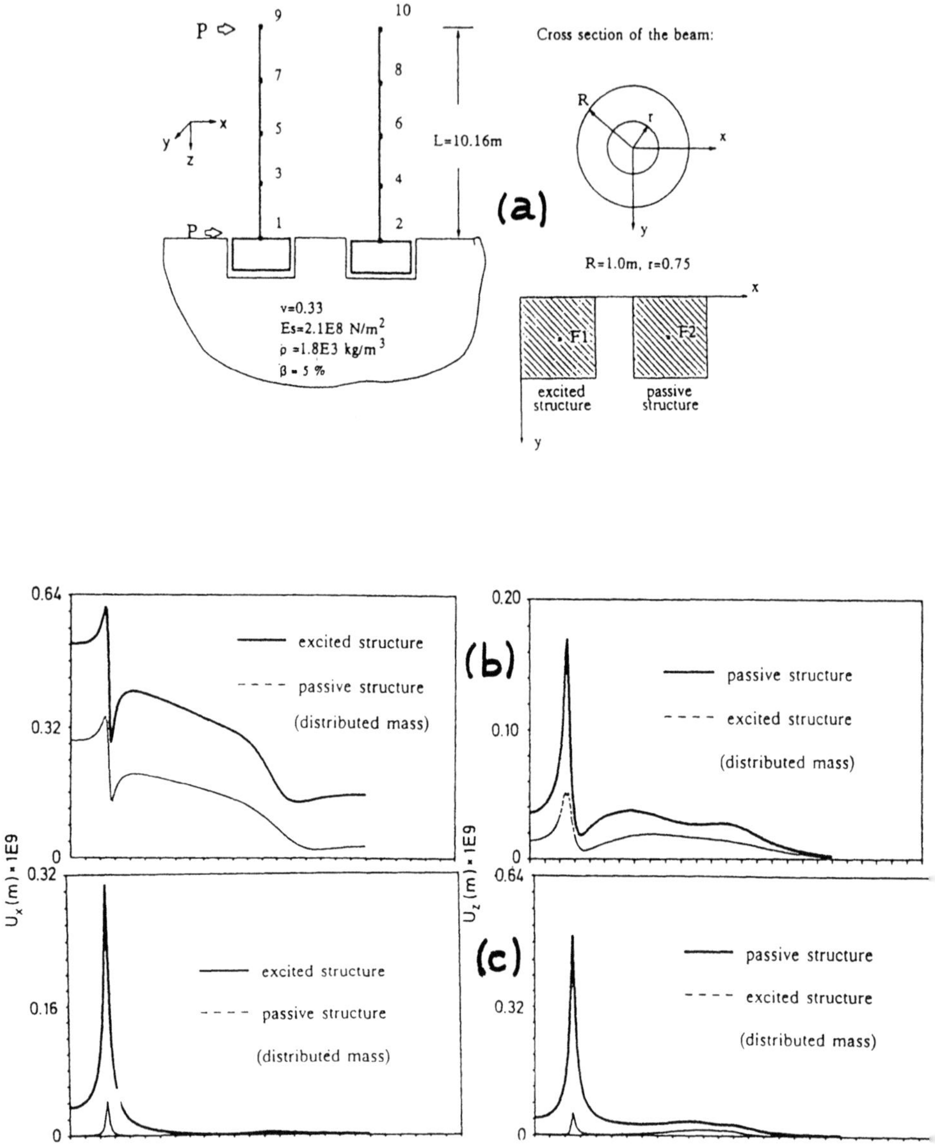

Figure 7. Response of two adjacent elastic, chimneys supported by rigid embedded foundations to a harmonic horizontal load: (a) system geometry; (b) load at foundation base; (c) load at chimney top.

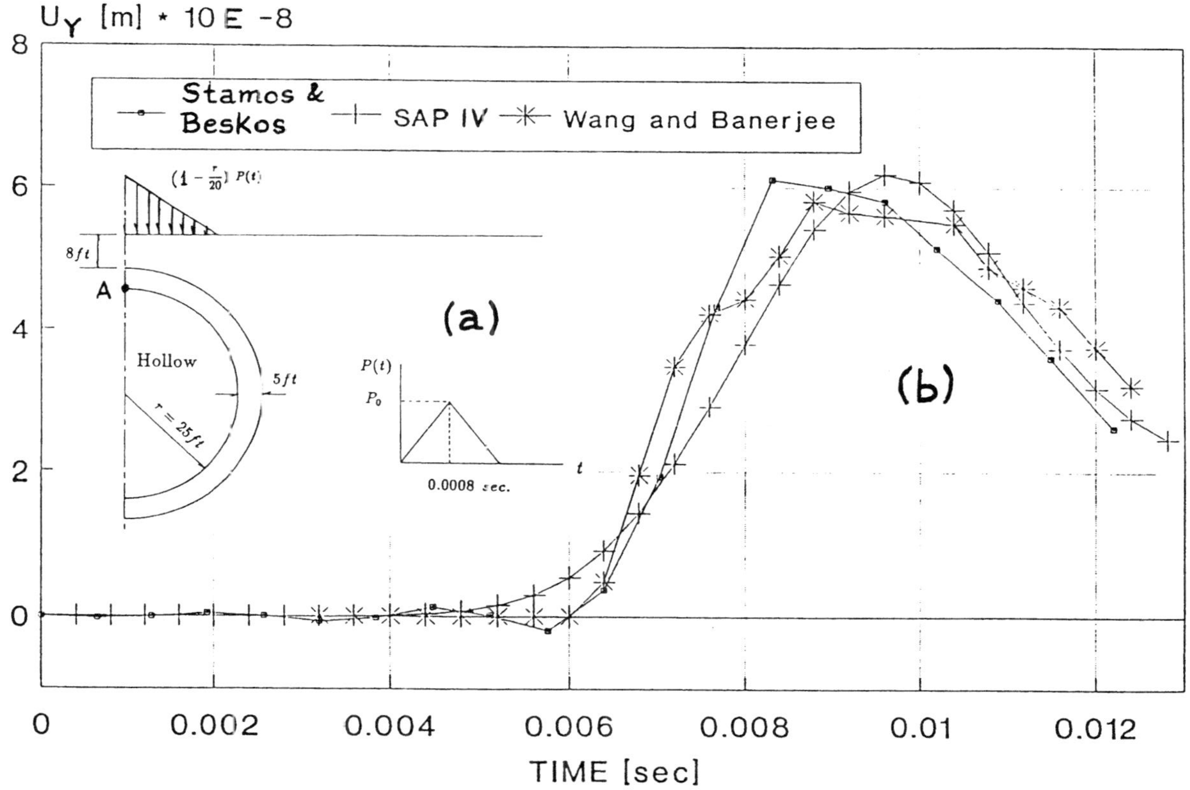

Figure 8. Undergrour.d shperical shell under ground surface dynamic load: (a) geometry of structure; (b) vertical deflection history of point A of the shell.

and Banerjee (1990) and the axisymmetric FEM capability of SAP IV (Bathe et al 1973). In Wang and Banerjee (1990) 24 line quadratic elements were used for the shell, 48 elements for the free soil surface and 8 enclosing elements to surround the area of interest. The finite element modelling involved 248 4-node isoparametric quadrilateral elements for the shell and 372 elements for the soil and the FEM response was obtained by numerical step-by-step time integration (Wilson's θ) in conjunction with a consistent mass option (Stamos and Beskos 1992 a). The results show a satisfactory agreement.

5.6. EXAMPLE 6

Consider a spherical cavity in the infinite elastic space (i.e., at a very big depth from ground surface), which is lined by a shell of thickness $h = 0.01\ \alpha$ and inner radius α, as shown in Fig. 9a. The material constants of the soil are E_m, $v_m = 0.25$ and ϱ_m and those of the shell $E_s = 2.5\ E_m$, $v_s = 0.20$ and $\varrho_s = 1.56\ \varrho_m$. A plane compressional transient wave of the form $u^f = (V_o/c_1)(c_1t\text{-}x\text{-}\alpha)\ H\ (c_1t\text{-}x\text{-}\alpha)$ propagates along the x direction in the soil medium and is diffracted by the shell as shown in Fig. 9 a. In the above expression for the free field motion, V_o is the velocity jump, H is the Heaviside operator and $-\alpha$ defines the point on the x axis where the wave front is located at time $t = 0$. Thus the wave reaches the shell at $t = 0$ and lasts for a time t_o so that $c_1t_o = 10\alpha$.

Stamos and Beskos (1992 a) analysed this problem by the direct BEM in conjunction with Laplace transform and the full-space Green's function. The discretization of 1/4 of every surface (interior and exterior) of the shell involved 6 quadratic quadrilateral boundary elements, while the solution involved 20 values of the transform parameter s. Fig. 9 b portrays the displacement histories along the x axis of points F, T and E of the shell as obtained by the above method. In the same figure the numerical results of Mathews and Geers (1987) are also shown for comparison. The very good agreement between the two results is apparent.

6. Conclusions

A combination of tutorial and state-of-the-art aspects on the application of the conventional, direct, frequency or time domain BEM to linear dynamic soil-structure interaction problems has been presented.

The dynamic response of foundations and above ground structures as well as underground structures to externally applied loads or incident seismic waves of arbitrary spatial and temporal variation have been discussed and illustrated by representative numerical examples.

The general conclusion is that the BEM alone, or coupled with the FEM, is ideally suited for the above mentioned problems and represents an accurate and efficient solution tool.

It is expected that, hardware improvements and theoretical advances in nonlinear BEM analysis, will enable the analyst to solve in the near future larger order problems involving material and/or geometric nonlinearities.

Acknowledgements

The author is grateful to the Greek General Secretariat for Research and Technology

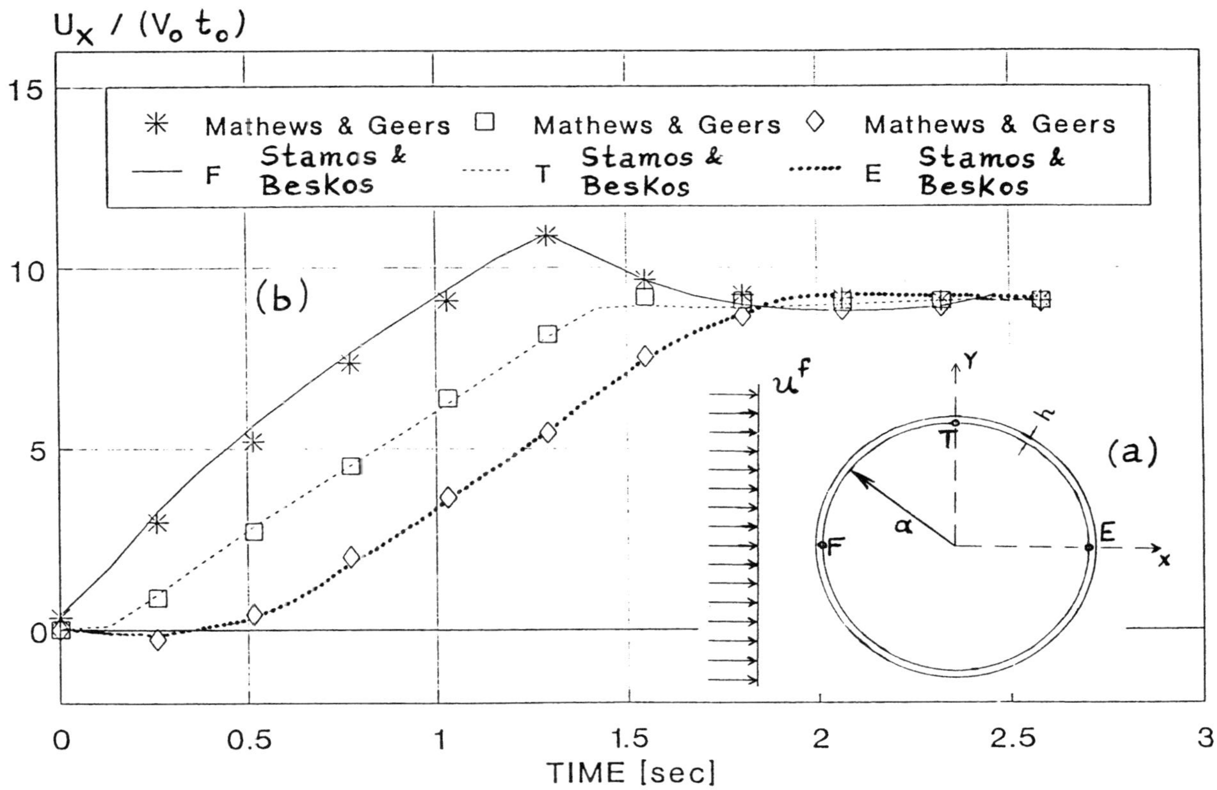

Figure 9. Underground spherical shell under transient compressional wave: (a) geometry of structure; (b) horizontal deflection history of points F, T and E of the shell.

(grant 7821/19-5-88) and to the Greek-German Bilateral Scientific Cooperation Program for supporting his work on the use of the BEM to wave propagation and soil-structure interaction. Thanks are also due to Mrs. Helen Alexandridis-Balis for her excellent typing of this work.

7. References

Abascal, R. and Dominguez, J. (1984 a) 'Dynamic behavior of strip footings on non-homogeneous viscoelastic soils', in D.E. Beskos, T. Krauthammer and I. Vardoulakis (eds.), Dynamic Soil-Structure Interaction, A.A. Balkema, Rotterdam, pp. 25-35.

Abascal, R. and Dominguez, J. (1984 b) 'Dynamic compliances of strip footings on non-homogeneous soils', in C.A. Brebbia (ed.), Boundary Elements VI, Springer-Verlag, Berlin, pp. 7.15-7.26.

Abascal, R. and Dominguez, J. (1985) 'Dynamic response of embedded strip foundations subject to obliquely incident waves' in C.A. Brebbia and G. Maier (eds.), Boundary Elements VII, Springer-Verlag, Berlin, pp. 6.63-6.69.

Abascal, R. and Dominguez, J. (1986) 'Vibrations of footings on zoned viscoelastic soils' Journal of Engineering Mechanics of the ASCE 112, 433-447.

Achenbach, J.D. (1973) Wave Propagation in Elastic Solids, North-Holland, Amsterdam.

Ahmad, S. and Banerjee, P.K. (1988) 'Time-domain transient elastodynamic analysis of 3-D solids by BEM', International Journal for Numerical Methods in Engineering 26, 1709-1728.

Ahmad, S., Israil, A.S.M. and Chen, K. (1988) 'Comparison of dynamic stiffness of rigid square and rectangular foundations by DBEM and IBEM' International Journal for Numerical and Analytical Methods in Geomechanics 12, 657-678.

Ahmad, S. and Manolis, G.D. (1987) 'Dynamic analysis of 3-D structures by a transformed boundary element method, Computational Mechanics 2, 185-196.

Alarcon, E. and Dominguez, J. (1989) 'Impedance of foundations using the boundary integral equation methods', in R.P. Shaw et al (eds.), Innovative Numerical Analysis for the Engineering Sciences, University Press of Virginia, Charlottesville, pp. 473-483.

Alarcon, E., Dominguez, J. and del Cano, F. (1980) 'Dynamic stiffness of foundations' in C.A. Brebbia (ed.), New Developments in Boundary Element Methods, Springer-Verlag, Berlin, pp. 264-280.

Antes, H. (1990) 'Dynamic interaction analysis in wave propagation problems by a time-dependent boundary element method', in G. Kuhn and H. Mang (eds.), Discretization Methods in Structural Mechanics, Springer-Verlag, Berlin, pp. 105-114.

Antes, H. and von Estorff, O. (1989) 'Dynamic response analysis of rigid foundations and of elastic structures by boundary element procedures', Soil Dynamics and Earthquake Engineering 8, 68-74.

Aubry, D. and Clouteau, D. (1991) 'Seismic site effect and soil-structure interaction' in W.B. Krätzig et al (eds.), Structural Dynamics, A.A. Balkema, Rotterdam, pp. 731-738.

Auersch, L. and Schmid, G. (1990) 'A simple boundary element formulation and its applications to wavefield excited soil-structure interaction', Earthquake Engineering and Structural Dynamics 19, 931-947.

Banerjee, P.K., Ahmad, S. and Manolis, G.D. (1987) 'Advanced elastodynamic analysis, in D.E. Beskos' (ed.), Boundary Element Methods in Mechanics, Elsevier Science Publishers, Amsterdam, pp. 257-284.

Bathe, K.J., Wilson, E.L. and Peterson, F.E. (1973) SAP IV, A Structural Analysis Program for Static and Dynamic Response of Linear Systems, Report No. EERC 73-11, University of California, Berkeley, California, USA.

Beskos, D.E. (1987) 'Boundary element methods in dynamic analysis', Applied Mechanics Reviews 40, 1-23.

Beskos, D.E. (1988) 'Boundary element methods in geomechanics', in C.A. Brebbia (ed.), Boundary Elements X, Vol. 4, Springer-Verlag, Berlin, pp. 3-28.

Beskos, D.E. (1992) 'Wave propagation through ground', in G.D. Manolis and T.G. Davies (eds.), Boundary Element Techniques in Geomechanics, Computational Mechanics Publications, Southampton, chapter 10.

Bielak, J., MacCamy, R.C., McGhee, D.S. and Barry, A. (1991) 'Unified symmetric BEM-FEM for site effects on ground motion -SH waves', Journal of Engineering Mechanics of the ASCE 117, 2265-2285.

Chouw, N., Le, R. and Schmid, G. (1991 a) 'Vibration transmitting behavior of the soil', in W.B. Krätzig et al (eds.), Structural Dynamics, pp. 701-708.

Chouw, N., Le, R. and Schmid, G. (1991 b) 'Propagationof vibration in a soil layer over bedrock', Engineering Analysis with Boundary Elements 8, 125-131.

Chouw, N. and Schmid, G. (1991) 'Influence of the geometrical effects of the soil on structure-soil-structure interaction', in W.B. Krätzig et al (eds.), Structural Dynamics, A.A. Balkema, Rotterdam, pp. 795-802.

Dominguez, J. (1978 a) Dynamic Stiffness of Rectangular Foundations, Report No R78-20, Department of Civil Engineering, Massachusetts Institute of Technology, Cambridge.

Dominguez, J. (1978 b) Response of Embedded Foundations to Travelling Waves, Report No R78-24, Department of Civil Engineering, Massachusetts Institute of Technology, Cambridge.

Dominguez, J. and Abascal, R. (1987) 'Dynamics of foundations', in C.A. Brebbia (ed.), Topics in Boundary Element Research, Vol. 4, Springer-Verlag, Berlin, pp. 27-75.

Dominguez, J. and Abascal, R. (1989) 'Seismic response of strip footings on zoned viscoelastic soils', Journal of Engineering Mechanics of the ASCE 115, 913-934.

Dominguez, J. and Meise, T. (1991) 'On the use of the BEM for wave propagation in infinite domains', Engineering Analysis with Boundary Elements 8, 132-138.

Emperador, J.M. and Dominguez, J. (1989) 'Dynamic response of axisymmetric embedded foundations', Earthquake Engineering and Structural Dynamics 18, 1105-1117.

Eringen, A.C. and Suhubi, E.S. (1975) Elastodynamics, Vol. II, Linear Theory, Academic Press, New York.

von Estorff, O. and Antes, H. (1991) 'Dynamic response of tunnel structures', in W.B. Krätzig et al (eds.), Structural Dynamics, A.A. Balkema, Rotterdam, pp. 803-810.

von Estorff, O. and Kausel, E. (1989) 'Coupling of boundary and finite elements for soil-structure interaction problems' Earthquake Engineering and Structural Dynamics 18, 1065-1075.

von Estorff, O. and Prabucki, M.J. (1990) 'Dynamic response in the time domain by coupled boundary and finite elements, Computational Mechanics 6, 35-46.

von Estorff, O. and Schmid, G. (1984) 'Application of the boundary element method to the analysis of the vibration behavior of strip foundations on a soil layer', in D.E. Beskos, T. Krauthammer and I. Vardoulakis (eds.), A.A. Balkema, Rotterdam, pp. 11-17.

von Estorff, O., Stamos, A.A., Beskos, D.E. and Antes, H. (1991) 'Dynamic interaction effects in underground traffic systems', Engineering Analysis with Boundary Elements

8, 167-175.

Gaitanaros, A.P. and Karabalis, D.L. (1988) 'Dynamic analysis of 3-D flexible embedde foundations by a frequency domain BEM-FEM', Earthquake Engineering and Structura Dynamics 16, 653-674.

Gallego, R. and Dominguez, J. (1990) 'A unified formulation of two existing time-domai boundary element approaches', Communications in Applied Numerical Methods 6, 17 25.

Gaul, L., Klein, P. and Plenge, M. (1988) 'Continuum-boundary element and experimenta models of soil-foundations interaction', in G. Swoboda (ed.), Numerical Methods i Geomechanics Innsbruck 1988, A.A. Balkema, Rotterdam, pp. 1649-1661.

Gaul, L., Schanz, M. and Fielder, C. (1991) 'Viscoelastic formulations of BEM in time an frequency domain', in S.N. Atluri, D.E. Beskos, R. Jones and G. Yagawa (eds.) Computational Mechanics '91, ICES Publications, Atlanta, pp. 304-310.

Gomez Lera, M.S., Alarcon, E. and Dominguez, J. (1985) 'On the use of a 3-D fundamenta solution for axisymmetric steady-state dynamic problems', in C.A. Brebbia and G. Maie (eds.), Boundary Elements VII, Springer-Verlag, Berlin, pp. 6.127-6.136.

Gonsalves, I.R., Shippy, D.J. and Rizzo, F.J. (1990) 'Direct boundary integral equation for elastodynamics in 3-D half-spaces', Computational Mechanics 6, 279-292.

Huh, Y. and Schmid, G. (1984) 'Application of boundary elements to soil-structur interaction problems', Engineering Analysis 1, 170-173.

Huh, Y. and Schmid, G. (1986) 'Soil-structure interaction in seismic environment', in D.L Karabalis (ed.), Recent Applications in Computational Mechanics, American Societ of Civil Engineers, New York, pp. 186-200.

Israil, A.S.M. and Ahmad, S. (1989) 'Dynamic vertical compliance of strip foundations i layered soils', Earthquake Engineering and Structural Dynamics 18, 933-950.

Israil, A.S.M. and Banerjee, P.K. (1990) 'Advanced time domain formulation of BEM fo two-dimensional transient elastodynamics', International Journal for Numerical Method in Engineering 29, 1421-1440.

Karabalis, D.L. (1988) 'Dynamic soil-structure interaction by BEM', in C.A. Brebbia (ed.) Boundary Elements X, Vol. 4, Springer-Verlag, Berlin, pp. 351-374.

Karabalis, D.L. (1991) 'A simplified 3-D time domain BEM for dynamic soil-structur interaction problems', Engineering Analysis with Boundary Elements 8, 139-145.

Karabalis, D.L. and Beskos, D.E. (1984) 'Dynamic response of 3-D rigid surfac foundations by time domain boundary element method', Earthquake Engineering an Structural Dynamics 12, 73-93.

Karabalis, D.L. and Beskos, D.E. (1985) 'Dynamic response of 3-D flexible foundations by time domain BEM and FEM', Soil Dynamics and Earthquake Engineering 4, 91-101.

Karabalis, D.L. and Beskos, D.E. (1986) 'Dynamic analysis of 3-D rigid embedde foundations by time domain boundary element method', Computer Methods in Applie Mechanics and Engineering 56, 91-120.

Karabalis, D.L. and Beskos, D.E. (1987 a) 'Dynamic soil-structure interaction', in D.E Beskos (ed.), Boundary Element Methods in Mechanics, Elsevier Science Publishers Amsterdam, pp. 499-562.

Karabalis, D.L. and Beskos, D.E. (1987 b) 'Three-dimensional soil-structure interaction by boundary element methods', in C.A. Brebbia (ed.), Topics in Boundary Elemen Research, Vol. 4, Springer-Verlag, Berlin, pp. 1-26.

Karabalis, D.L. and Beskos, D.E. (1990) 'Letter to the editor on Time-domain transien elastodynamic analysis of 3-D solids by BEM', International Journal for Numerical Methods in Engineering 29, 211-215.

Karabalis, D.L. and Huang, C.F.D. (1989) 'Kinematic and inertial soil-structure interaction by time domain BEM', in C.A. Brebbia and J.J. Connor (eds.), Advances in Boundary Elements, Vol. 3, Computational Mechanics Publications, Southampton, pp. 359-374.

Karabalis, D.L. and Mohammadi, M. (1991) 'Foundation - soil-foundation dynamics using a 3-D frequency domain BEM', in C.A. Brebbia and G.S. Gipson (eds.), Boundary Elements XIII, Computational Mechanics Publications, Southampton, pp. 447-456.

Kitahara, M., Ando, T. and Muranishi, Y. (1987) 'On the effect of viscosity for the dynamic strain of underground structures', in M. Tanaka and Q.H. Du (eds.), Theory and Applications of Boundary Element Methods, Pergamon Press, Oxford, pp. 75-84.

Kitamura, Y. and Sakurai, S. (1979) 'Dynamic stiffness for rectangular rigid foundations on a semi-infinite elastic medium', International Journal for Numerical and Analytical Methods in Geomechanics 3, 159-171.

Kobayashi, S. (1985) 'Fundamentals of boundary integral equation methods in elastodynamics', in C.A. Brebbia (ed.), Topics in Boundary Element Research, Vol. 2, Springer-Verlag, Berlin, pp. 1-54.

Kobayashi, S. (1987) "Elastodynamics', in D.E. Beskos (ed.), Boundary Element Methods in Mechanics, Elsevier Science Publishers, Amsterdam, pp. 191-255.

Kobayashi, S. (1985) 'Applications of boundary integral equation method to geomechanics', in T. Kawamoto and Y. Ichikawa (eds.), Numerical Methods in Geomechanics Nagoya 1985, A.A. Balkema, Rotterdam, pp. 83-92.

Kobayashi, S. and Mori, K. (1986) 'Three-dimensional dynamic analysis of soil-structure interactions by boundary integral equation-finite element combined method', in R.P. Shaw et al (eds.), Innovative Numerical Methods in Engineering, Springer-Verlag, Berlin, pp. 613-618.

Kobayashi, S. and Nishimura, N. (1982) ' Transient stress analysis of tunnels and caverns of arbitrary shape due to travelling waves', in P.K. Banerjee and R.P. Shaw (eds.), Developments in Boundary Element Methods-2, Applied Science Publishers, London, pp. 177-210.

Kobayashi, S. and Nishimura, N. (1983) 'Analysis of dynamic soil-structure interactions by boundary integral equation method', in P. Lascaux (ed.), Numerical Methods in Engineering, Pluralis, Paris, pp. 353-362.

Kobayashi, S., Nishimura, N. and Mori, K. (1986) 'Applications of boundary element-finite element combined method to three-dimensional viscoelastodynamic problems', in Q.H. Du (ed.), Boundary Elements, Pergamon Press, Oxford, pp. 67-74.

Luco, J.E. (1982) ' Linear soil-structure interaction: a review', in S.K. Datta (ed.), Earthquake Ground Motion and its Effects on Structures, American Society of Mechanical Engineers, New York, pp. 41-57.

Luco, J.E. and Wong, H.L. (1990) 'Forced vibration of Lotung containment model: theory and observations', Journal of Engineering Mechanics of the ASCE 116, 845-861.

Lu, Y.Y., Belytschko, T. and Liu, W.K. (1991) 'A variationally coupled FE-BE method for elasticity and fracture mechanics', Computer Methods in Applied Mechanics and Engineering 85, 21-37.

Mang, H.A., Torzicky, P. and Chen, Z.Y. (1989) 'On the mechanical inconsistency of symmetrization of unsymmetric coupling matrices for BEFEM discretization of solids, Computational Mechanics 4, 301-308.

Manolis, G.D. (1983) 'Dynamic behavior of underground structures', Shock and Vibration Digest 15 (11), 7-18.

Manolis, G.D. and Beskos, D.E. (1983) 'Dynamic response of lined tunnels by an isoparametric boundary element method', Computer Methods in Applied Mechanics and Engineering 36, 291-307.

Manolis, G.D. and Beskos, D.E. (1988) 'Boundary Element Methods in Elastodynamics, Unwin Hyman (sold to Chapman and Hall), London.

Manolis, G.D. and Davies, T.G. (1992) Boundary Element Techniques in Geomechanics, Computational Mechanics Publications, Southampton.

Mathews, I.C. and Geers, T.L. (1987) 'A doubly asymptotic nonreflecting boundary for ground-shock analysis', Journal of Applied Mechanics of the ASME 54, 489-497.

Mohammadi, M. and Karabalis, D.L. (1990) '3-D soil-structure interaction analysis by BEM: comparison studies and computational aspects', Soil Dynamics and Earthquake Engineering 9, 96-108.

Nishimura, N. and Kobayashi, S. (1989) 'Advanced substructured analysis of elastodynamic wave propagation problems, in P.K. Banerjee and R.B. Wilson (eds.), Developments in Boundary Element Methods-5, Elsevier Applied Science, London, pp.157-186.

Ottenstreuer, M. (1982) 'Frequency dependent dynamic response of footings', in A.S. Cakmak, A.M. Abdel-Ghaffar and C.A. Brebbia (eds.), Soil Dynamics and Earthquake Engineering, A.A. Balkema, Rotterdam, pp. 799-809.

Ottenstreuer, M. and Schmid, G. (1981) 'Boundary elements applied to soil-foundation interaction, in C.A. Brebbia (ed.), Boundary Element Methods, Springer-Verlag, Berlin, pp. 293-309.

Pao, Y.H. and Mow, C.C. (1973) Diffraction of Elastic Waves and Dynamic Stress Concentrations, Crane Russak, New York.

Rajapakse, R.K.N.D. and Wang, Y. (1991) 'Elastodynamic Green's functions of an orthotropic half plane', Journal of Engineering Mechanics of the ASCE 117, 588-604.

Reng, Z.X., Wiberg, N.E. and Tullberg, O. (1986) 'Dynamic soil-structure interaction problem including viscous damping', in Q.H. Du (ed.), Boundary Elements, Pergamon Press, Oxford, 507-520.

Sato, T., Kawase, H. and Yoshida, K. (1983) 'Dynamic response analysis of rigid foundations subjected to seismic waves by boundary element method', in C.A. Brebbia, T. Futagami and M. Tanaka (eds.), Boundary Elements, Springer-Verlag, Berlin, pp. 765-774.

Savidis, S.A. and Richter, T. (1979) 'Dynamic response of elastic plates on the surface of the half-space', International Journal for Numerical and Analytical Methods in Geomechanics 3, 245-254.

Schmid, G., Huh, Y. and von Estorff, O. (1984) 'Evaluation of compliance functions of soil foundations systems by boundary element method', in Proccedings of 8th World Conference on Earthquake Engineering, San Fransisco, CA, July 1984, Prentice Hall, Englewood Cliffs, N.J., pp. 1041-1047.

Spyrakos, C.C. and Antes, H. (1986) 'Time domain boundary element method approaches in elastodynamics: a comparatice study', Computers and Structures 24, 529-535.

Spyrakos, C.C. and Beskos, D.E. (1986 a) 'Dynamic response of rigid strip foundations by time domain boundary element method', International Journal for Numerical Methods in Engineering 23, 1547-1565.

Spyrakos, C.C. and Beskos, D.E. (1986 b) 'Dynamic response of flexible strip foundations

by boundary and finite elements', Soil Dynamics and Earthquake Engineering 5, 84-96.
Spyrakos, C.C., Patel, P.N. and Beskos, D.E. (1986) 'Dynamic analysis of flexible embedded foundations:plane strain case', in G.A. Keramidas and C.A. Brebbia (eds.), Computational Methods and Experimental Measurements, Springer-Verlag, Berlin, pp. 615-624.
Stamos, A.A. and Beskos, D.E. (1992 a) 'Dynamic soil-structure interaction in 3-D underground structures', in Proccedings of BEM-14, Seville, Spain,November 3-6, 1992.
Stamos, A.A. and Beskos, D.E. (1992 b) 'Three-dimensional wave diffraction by long lined tunnels in half-space', in Proceedings of the 1st National Congress on Computational Mechanics, Athens, Greece, September 3-4, 1992.
Stamos, A.A., von Estorff, O., Antes, H. and Beskos, D.E. (1992) 'Vibration isolation in road-tunnel traffic systems', Engineering Analysis with Boundary Elements, to be submitted.
Tanaka, M. (1985) 'Time-space boundary element methods for viscoelasticity', in C.A. Brebbia (ed.), Variational Methods in Engineering, Springer-Verlag, Berlin, pp. 8.3-8.16.
Tassoulas, J.L. (1989) 'Dynamic soil-structure interaction', in D.E. Beskos (ed.), Boundary Element Methods in Structural Analysis, American Society of Civil Engineers, New York, pp. 273-308.
Timoshenko, S.P. and Goodier, J.N. (1970) Theory of Elasticity, 3rd Edition, McGraw-Hill Book Company, New York.
Tong, J. and Kuribayashi, E. (1988) 'Dynamic responses of flexible rectangular foundations embedded into a viscoelastic half-space to obliquely incident waves', Soils and Foundations of the Japanese Society of Soil Mechanics and Foundation Engineering 28, 56-64.
Tullberg, O. and Wiberg, N.E. (1985) 'Boundary element solutions for some problems in soil dynamics', in T. Kawamoto and Y. Ichikawa (eds.), Numerical Methods in Geomechanics Nagoya 1985, A.A. Balkema, Rotterdam, pp. 157-164.
Vardoulakis, I.G., Beskos, D.E., Leung, K.L., Dasgupta, B. and Sterling, R.L. (1987) 'Computation of vibration levels in underground space', International Journal of Rock Mechanics and Mining Sciences & Geomechanics Abstracts 24, 291-298.
Wang, H.C. and Banrjee, P.K. (1990) 'Axisymmetric transient elastodynamic analysis by boundary element method', International Journal of Solids and Structures 26, 401-415.
Wang, S. and Schmid, G. (1992) 'Dynamic structure-soil-structure interaction by FEM and BEM', Computational Mechanics 9, 347-357.
Whittaker, W.L. and Christiano, P. (1982 a) 'Dynamic response of plate on elastic half-space', Journal of Engineering Mechanics of the ASCE 108, 133-154.
Whittaker, W.L. and Christiano, P. (1982 b) 'Response of a plate on elastic half-space to harmonic waves', Earthquake Engineering and Structural Dynamics 10, 255-266.
Wolf, J.P. (1985) Dynamic Soil-Structure Interaction, Prentice Hall, Englewood Cliffs, N.J.
Wolf, J.P. (1988) Soil-Structure Interaction Analysis in Time Domain, Prentice Hall, Englewood Cliffs, N.J.
Wolf, J.P. and Darbre, G.R. (1984 a) 'Dynamic stiffness matrix of soil by the boundary element method: conceptual aspects', Earthquake Engineering and Structural Dynamics 12, 385-400.
Wolf, J.P. and Darbre, G.R. (1984 b) 'Dynamic stiffness matrix of soil by the boundary element method: embedded foundation', Earthquake Engineering and Structural Dynamics 12, 401-416.

Zhang, L.P. and Chopra, A.K. (1991) 'Three-dimensional analysis of spatially varying ground motions around a uniform canyon in a homogeneous half-space', Earthquake Engineering and Structural Dynamics 20, 911-926.

A STRUCTURAL ENGINEER'S VIEW OF SOIL-STRUCTURE-INTERACTION

RAY W. CLOUGH
Nishkian Professor Emeritus
Department of Civil Engineering
University of California
Berkeley, CA 94720

ABSTRACT. In this paper, the sequence of significant developments in the methods of analysis of dynamic earthquake interaction between structures and their foundation media is outlined. The development is divided into four phases, starting with a rigid foundation medium where there is no interaction with the structure and ending with a description of the frequency domain response analysis of a structure on a continuum foundation system. The purpose of the paper is to explain the interaction analyses in the context of typical structural dynamics analysis procedures.

1. INTRODUCTION

In earthquake engineering practice, it is well-recognized now that the foundation material on which a building is constructed may interact dynamically with the structure during its response to earthquake excitation -- to the extent that the maximum stresses and deflections in the system are modified significantly from the values that would have been developed if it were on a rigid foundation. However, forty years ago when the methods of analysis of structural response to earthquake motions were just beginning to be developed, such interaction effects were considered to be of little consequence, and hence were ignored. A research study by R. G. Merritt and G. W. Housner published in the Bulletin of the Seismological Society of America (BSSA) in October 1954 did much to encourage this lack of interest in soil-structure interaction (SSI). Their paper, entitled "Effect of Foundation Compliance on Earthquake Stresses in a Multistory Building", described the effect that variations in the ratio of building stiffness to foundation stiffness might have on the amplitude of response to earthquake excitation. They found that changes of this compliance ratio within the range considered practically useful for multistory building construction caused a maximum change of only 10 percent in the fundamental period of system vibration, and concluded that the base shear force due to earthquake excitation in a typical tall building is hardly affected by foundation flexibility

P. Gülkan and R. W. Clough (eds.), Developments in Dynamic Soil-Structure Interaction, 91–109.

even for the most flexible foundation that might be considered suitable for such a building.

When I read the paper shortly after it was written, it convinced me that I need not take account of foundation flexibility in tall building dynamic response analyses. Moreover, this attitude apparently was prevalent throughout the world at that time, as evidenced by the fact that no papers on building-foundation interaction were presented at either the First or Second World Conference on Earthquake Engineering (WCEE), held in San Francisco in 1956 and in Tokyo in 1960, respectively. The first mention of SSI at the World Conferences appeared in a single paper in the Proceedings of 3WCEE held in New Zealand in 1965; however, it wasn't until eight years later at the 5WCEE held in Rome in 1973 that any significant amount of SSI research was reported.

Interest in SSI effects on the response of structures to earthquakes has continued to expand dramatically during the nearly two decades since the Rome Conference, largely with impetus from the developing nuclear power industry. In defense of Merritt and Housner, however, we should note that the massive and rigid nuclear power station structures founded on the relatively soft soils usually available adjacent to water sources, which were the subject of much of this SSI research, are not at all similar to the tall light-weight buildings they considered. Hence these more recent developments do not negate the conclusions presented by Merritt and Housner, they merely extend greatly the range of parameters under consideration.

As a structural engineer dealing with the analysis of structural response to earthquakes for more than forty years, but only peripherally concerned with SSI, I have had a unique observation point from which to follow the development of the methods of SSI analysis. It is the purpose of this paper to describe the development of these methods from that point of view, with the hope that it will be a useful supplement to the description given by the other authors at this meeting, most of whom have been directly concerned with such developments.

For convenience, I have divided the sequence of this analytical development into four phases. Noting at the outset that significant earthquake response analyses could not be performed until the advent of digital computers, Phase I of the account begins with reference to one of the earliest earthquake response analyses of a tall building done by computer [1]. The building considered is the Alexander Building in San Francisco, a fifteen story steel frame structure. It was an obvious choice for the analysis because its vibration mode shapes and frequencies already had been evaluated analytically by Robison and Rinne[2], and the vibration frequencies had been measured in the field by Blume and Hesselmeyer [3]. As was true in all but the simplest of examples, linear elastic response was assumed and the mode-superposition method of analysis was adopted to avoid over-taxing the capacity of the computers then available.

The essential assumption regarding the building performance made in all of the Phase I analyses was that it is supported on a rigid foundation, and that the motion applied to the building's base is a previously recorded earthquake acceleration history. Thus it was implicitly assumed that SSI does not affect the building response. In contrast, for the Phase II development of analytical procedures, it was assumed that a body of flexible soil is interposed between the base of the building and the rigid foundation rock; hence a limited amount of SSI is included in the response. However, the earthquake motion applied to the base of the soil layer by the foundation rock is an accelerogram that had been recorded previously by a strong-motion seismograph located at the soil surface.

The obvious deficiency of the Phase II analyses is that the motions actually occurring at the base of the soil layer cannot be the same as those recorded at its free surface. Accordingly, in the Phase III stage of development a correction was made for this deficiency, either by calculation of a modified base rock motion by deconvolution of the free-field surface record or by employing a formulation of the analysis procedure that applies the recorded accelerogram as a free-field input. Because both the deconvolved rigid base input and the free-field input are applied to the same mathematical model consisting of a soil layer interposed between the building foundation and the rigid rock base, and because the same free-field motion is assumed to exist at the surface of the soil layer, both of these analysis procedures lead to the same results.

These Phase III analysis procedures still have a major deficiency, however, in that the soil layer is assumed to be supported on a rigid rock base. As a consequence of this assumption, all energy transmitted into the soil, due both to the earthquake motion applied at the base and to the vibratory response of the building above, must be reflected back to the building and no energy can be lost by radiation into the underlying rock. The Phase IV development corrects for this deficiency by introducing an elastic medium under the soil layer. Typically this medium is assumed to extend to infinity, so any energy passing into it from above is not reflected but instead radiates away. Because this energy does not contribute further to the vibratory response of the building, its loss by radiation may constitute a significant damping mechanism on the earthquake response behavior.

The formulation of the structural response equations applicable to each of these Phases in the development of earthquake analysis procedures is discussed in separate sections of this paper. For convenience in visualization, the structure under consideration is taken to be a planar multistory frame, but the analysis is equally applicable to any type of structure -- planar or three-dimensional.

2. PHASE I - STRUCTURE ON RIGID SUPPORTS

In the early days of earthquake response analysis of structures, soil-structure interaction was ignored as was noted earlier and the supports of the structure were assumed to move directly in accordance with the prescribed earthquake motions. At first, all supports were assumed to be attached to a single rigid block of the foundation material and only a single component of motion was considered to act in this block. (The analysis described in Reference 1 is an example based on these assumptions.) However, as time went by the complexity of the structural systems increased and it became important to take account of earthquake motions acting in two or three rectangular coordinate axes. Subsequently it was noted that input excitation through a single rigid base block is realistic only if the structure is quite small. If it is large relative to the typical wave length of the earthquake motions propagating through the earth, it is necessary to allow for different base motions to be applied to the widely spaced supports.

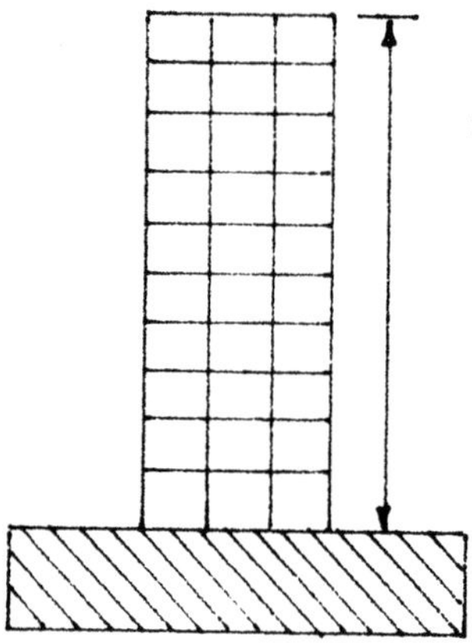

Figure 1. Structure with rigid base block input

The equations of earthquake response motion for a structure with only a single rigid base block input are developed here with reference to the system depicted in Fig. 1. The earthquake input is denoted $\underline{v}_g(t)$, and in principle this might include six components -- translations along and rotations about three rectangular coordinate axes. However, in actuality it is inconceivable that a base block of any significant size would be subjected to appreciable rigid body rotations; accordingly it will be assumed that $\underline{v}_g(t)$ is limited to translations along one, two, or three global coordinate axes. The response motions of the frame nodal points are designated $\underline{v}^t(t)$, where the superscript denotes that these are total motions; these may include as many as three components of rotation, as well as the corresponding translations at each node.

The equations of motion express the dynamic equilibrium of the forces acting in all degrees of freedom -- the resisting forces of inertia $\underline{f}_I(t)$, of damping $\underline{f}_D(t)$, and of elasticity $\underline{f}_S(t)$, as well as the externally applied nodal loads $\underline{p}(t)$ which act in the opposing sense. Thus the equilibrium equation is expressed by

$$\underline{f}_I(t) + \underline{f}_D(t) + \underline{f}_S(t) = \underline{p}(t) \tag{1}$$

But describing the resisting forces in terms of the corresponding physical property matrices, and partitioning the forces and motions as well as the property matrices into sets associated with the structure and with the base block, this becomes

$$\begin{bmatrix} m & m_g \\ m_g^T & m_{gg} \end{bmatrix} \begin{Bmatrix} \ddot{v}^t(t) \\ \ddot{v}_g(t) \end{Bmatrix} + \begin{bmatrix} c & c_g \\ c_g^T & c_{gg} \end{bmatrix} \begin{Bmatrix} \dot{v}^t(t) \\ \dot{v}_g(t) \end{Bmatrix} + \begin{bmatrix} k & k_g \\ k_g^T & k_{gg} \end{bmatrix} \begin{Bmatrix} v^t(t) \\ v_g(t) \end{Bmatrix} = \begin{Bmatrix} 0 \\ p_g(t) \end{Bmatrix} \qquad (2)$$

in which it will be noted that no external forces are applied directly to the structure in a typical earthquake excitation situation.

Ncw the equilibrium expressed by these equations of motion is important only for the forces acting in the structure because the base block forces $\underline{p}_g(t)$ serve to equilibrate the forces at the support points; accordingly only the upper partition of Eq. (2) need be considered at this time. Also, it is convenient to transfer to the right side of the equation all forces induced by coupling with the base block motions $\underline{v}_g(t)$, and on this basis the equilibrium equation becomes

$$\underline{m}\ddot{\underline{v}}^t(t) + \underline{c}\dot{\underline{v}}^t(t) + \underline{k}\underline{v}^t(t) = -\underline{m}_g\ddot{\underline{v}}_g(t) - \underline{c}_g\dot{\underline{v}}_g(t) - \underline{k}_g\underline{v}_g(t) \qquad (3)$$

in which $\underline{m}$, $\underline{c}$, and $\underline{k}$ are the structure mass, damping, and stiffness matrices, respectively, and the coupling effects between the base block and the structure are denoted by corresponding property matrices with the subscript "g".

Equation (3) is a valid expression of dynamic equilibrium for the structure, but it is not convenient for general use because the earthquake excitation is expressed in terms of the base displacements and velocities in addition to the acceleration values typically provided by a strong motion earthquake record. To derive a simplified version of the response equation, the total structure displacements, $\underline{v}^t(t)$, are expressed as the sum of a "quasistatic" component $\underline{v}^s(t)$ which would result from static application of the support motions plus a "dynamic" (or relative) component $\underline{v}(t)$ that may exist when the support motions vanish, thus

$$\underline{v}^t(t) = \underline{v}^s(t) + \underline{v}(t) \qquad (4)$$

These quasistatic displacements may be expressed conveniently by a matrix of displacement influence coefficients $\underline{r}$ in which each column vector lists the displacement of each structure degree of freedom due to a unit value of a single static support displacement, thus

$$\underline{v}^s(t) = \underline{r}\,\underline{v}_g(t) \qquad (5)$$

For the rigid body structure motions that result from the specified translation displacements of the base block $\underline{v}_g(t)$ in the situation shown in Fig. 1, each of these vectors consists of a "one" for each nodal displacement component in the direction of the base translation and "zeros" for all other nodal components.

Substituting Eqs. (4) and (5) and their time derivatives into Eq. (3), and transferring all terms associated with the base block motions $\underline{v}_g(t)$ to the right hand side leads to

$$\underline{m}\ddot{\underline{v}}(t) + \underline{c}\dot{\underline{v}}(t) + \underline{k}\underline{v}(t) = -(\underline{m}\underline{r} + \underline{m}_g)\ddot{\underline{v}}_g(t) - (\underline{c}\underline{r} + \underline{c}_g)\dot{\underline{v}}_g(t) - (\underline{k}\underline{r} + \underline{k}_g)\underline{v}_g(t) \quad (6)$$

which still appears to require consideration of the base block velocity and displacement. However, if an expression for the quasistatic displacement transformation $\underline{r}$ is derived by consideration of static equilibrium, a great simplification may be obtained. Thus if the static equilibrium equation is extracted from Eq. (6) by setting to zero all time derivatives, and noting that the dynamic displacements $\underline{v}(t)$ also vanish in the static condition, the equation becomes

$$\underline{0} = (\underline{k}\underline{r} + \underline{k}_g)\underline{v}_g(t) \quad (7)$$

from which the quasistatic displacement transformation matrix is seen to be given by

$$\underline{r} = -\underline{k}^{-1}\underline{k}_g \quad (8)$$

Equation (7) shows that the last term on the right hand side of Eq. (6) vanishes. Similarly, the preceding term must vanish if the system has stiffness-proportional damping; moreover, even in cases where the damping is not entirely stiffness-proportional, the damping forces are very small in comparison with those of stiffness or inertia and can be neglected. Hence the equation of motion may be written finally as

$$\underline{m}\ddot{\underline{v}}(t) + \underline{c}\dot{\underline{v}}(t) + \underline{k}\underline{v}(t) = -(\underline{m}\underline{r} + \underline{m}_g)\ddot{\underline{v}}_g(t) \quad (9)$$

and it is apparent that the effective earthquake loading term on the right hand side of this equation is a function simply of the earthquake acceleration history $\ddot{\underline{v}}_g(t)$.

Equation (9) is applicable to a structure having its supports localized on a single rigid foundation block; but as was mentioned earlier, for a structure of great horizontal extent such as a dam or a long span bridge it is apparent that different earthquake motions might be acting at the various support points. Consequently, it is necessary to generalize the meaning of the seismic input expression, as follows

$$\ddot{\underline{v}}_g(t) = \begin{Bmatrix} \ddot{v}_{g1}(t) \\ \ddot{v}_{g2}(t) \\ \ddot{v}_{g3}(t) \\ \cdot \\ \cdot \\ \cdot \end{Bmatrix} \qquad (10)$$

in which the subscripts identify the earthquake motions acting at a set of independent rigid support blocks. Figure 2 depicts such a system.

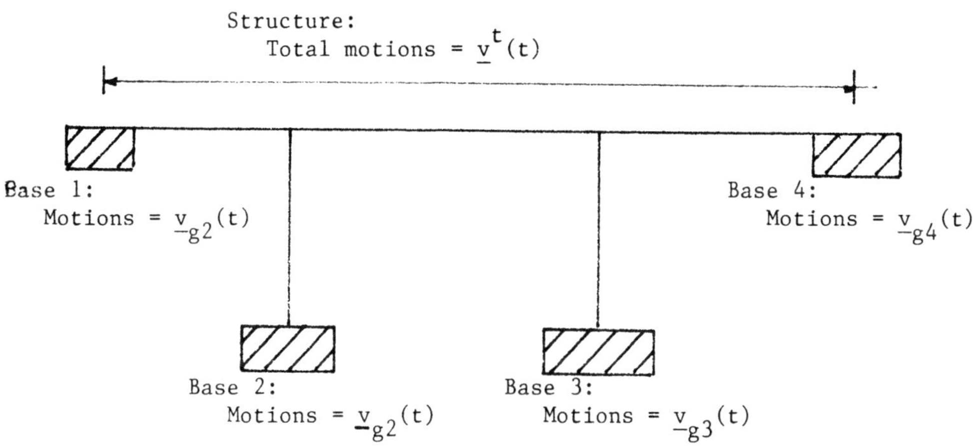

Figure 2. Structure with multiple rigid base supports

Note that each of the support motions indicated in this figure may include as many as three components of translation; rotations also could be included if desired, but rotations of large base blocks are considered to be of negligible importance, as mentioned earlier

The basic difference between this case and that shown in Fig. 1 is that any input motion component given in Fig. 2 may induce deformations and stresses in the structure rather than only the simple rigid body motions that were described previously. Fortunately, the quasistatic displacement transformation matrix that was derived previously, $\underline{r} = -\underline{k}^{-1}\underline{k}_g$, which represents the solution of a statically indeterminate analysis problem, is valid for the multiple support block case. In fact this expression even could be used to derive the rigid body kinematics of a system on a single support block, if desired.

Consequently the dynamic nodal displacements for the system of Fig. 2 are given by the solution of Eq. 9 just as they were in the single support block case. However because deformations and stresses of this structure may result from the quasistatic displacements as well as from the dynamic motions, the member stresses must be obtained by applying the stress transformations to the total nodal displacements,

$\underline{v}^t(t)$, rather than just to the dynamic component.

3. PHASE II - STRUCTURE ON A BODY OF SOIL

Although it was convenient to neglect foundation flexibility in the formulation of the Phase I analyses, it was apparent in most of the sites under consideration that the foundation material was far from rigid and that its deformations might contribute appreciably to the earthquake response, especially if the structure were very stiff and massive. Thus in formulating the Phase II type analyses, this flexibility was modeled by interposing a body of soil between the base of the structure and the "rigid" foundation rock. In the analysis of buildings of the type shown in Fig. 1, the flexible soil layer was modeled by a mesh of 2D (or 3D) linear elastic solid finite elements added at the base of the structure. This mesh was then supported by the rigid base through which the earthquake motions $\underline{v}_g(t)$ were applied to the system, as depicted in Fig. 3. It should be noted that Eq. (9) represents the response equation for this combined system just as well as it does for the case of Fig. 1 in which there is no soil layer because the soil is treated simply as part of the structural system.

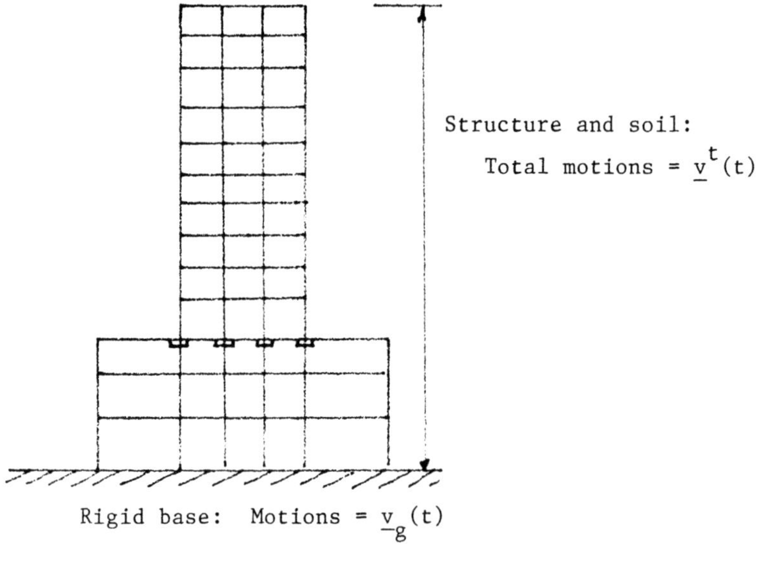

Figure 3. Soil layer between structure and rigid base

This type of combined model usually was formulated by the standard finite element "assembly" procedure, with the result that the system stiffness and mass matrices incorporated the appropriate physical properties of the soil layer. In developing the model of the building, "substructure" techniques that were pioneered in the aviation engineering industry often were employed[4]. These were especially effective in modeling tall buildings in which a group of members might be used to represent a single story segment, or even a several story segment of the building, and this substructure might be duplicated several times in assembling the complete structure. Use of substructures (or "super-elements" as they often were called) in this way was a great convenience in building up the mathematical model, particularly if the members of each material type could be grouped together in separate substructures. Typically the soil layer at the base of the building was treated as a separate substructure having mass density and modulus of elasticity values differing greatly from those of the building substructures; of particular significance was the fact that the damping ratio of the soil might differ from those of the building materials by an order of magnitude.

When the substructure matrices were assembled to form the complete system mass and stiffness matrices, these were then used in a standard eigenproblem analysis to calculate the vibration mode shapes and frequencies, and of course these results might differ significantly from those obtained for the building supported directly on the rigid base. But the modal response equations derived from these mode shapes were still uncoupled with regard to the mass and stiffness forces because of the modal orthogonality properties of the combined system. However, if a modal damping matrix for the system were constructed similarly by assembly of mass - and stiffness-proportional damping matrices for the substructures, and if the damping ratios of the substructures were not all the same, the resulting modal response equations were found to be coupled by modal damping forces because the combined system damping was not proportional. In other words, the ratio of the damping forces to the inertia or stiffness forces in the various substructures was not constant.

This result was not surprising; it could be anticipated that the damped vibration mode shapes for a system having much greater damping in the soil substructure than in the building would differ from the undamped mode shapes which would pertain if the system were proportionally damped. However, the serious consequence of non-proportional damping is that the standard mode-superposition analysis procedure is not applicable due to the damping coupling of the modal response equations. It is possible to avoid this difficulty by solving the damped eigenproblem and thereby obtaining a set of complex damped mode shapes with orthogonality properties that lead to a set of uncoupled modal response equations. However, the computational effort required to calculate the complex modal properties and to use them in a mode-superposition analysis is so great that a typical structural engineer prefers to develop the coupled modal equations using the undamped mode shapes, and then to solve a truncated set of these coupled equations in the time domain using a step-by-step procedure as described in Reference [5].

In closing this discussion of the Phase II development, it may be of interest to mention that the formulation of a combined finite element model of a structure and its foundation rock long has been customary in the seismic analysis of concrete dams. For such systems it is evident that flexibility of the foundation may have a major influence on the stresses developed in the very rigid dam, and it is standard practice to model the foundation rock with a mesh of isoparametric solid elements having the effective modulus of elasticity measured for that material. However, a question that was immediately apparent in developing such a model was how large a volume of rock should be included. In evaluation of static stresses in arch dams it was standard practice (based on physical model tests) to use a foundation block that extended in all directions from the dam-rock interface a distance equal to the dam height. However, there was some question that this criterion was equally applicable to the dynamic response mechanism, so a series of vibration analyses was made using successively increased sizes of foundation blocks. As the size increased, it soon became apparent that the dominant vibration modes were those of the foundation rock, with little participation or stressing of the dam. For this reason, in the seismic analysis of concrete dams it has become customary to include the contribution of the foundation rock to the system stiffness matrix, but to neglect any foundation contribution to the mass matrix. This neglect of the foundation mass in the dynamic analysis has an additional advantage in that the earthquake motions at the base of the foundation block are the same as its free-field surface motions because there is no wave propagation in this massless medium.

4. PHASE III - FREE-FIELD INPUT FORMULATION

Although the use of a massless foundation medium avoids the need for a Phase III type analysis to deal with the difference between the earthquake motions at the base of the foundation rock and its free-field surface motions, it is evident that the system behavior will be better represented by a model that takes account of the mass as well as the stiffness properties of the foundation. The first approach to dealing with this problem was developed by geotechnical engineers. They calculated what the motion at depth must have been to be consistent with the observed free-field motions, using a frequency domain analysis procedure called "deconvolution". Such an analysis is quite simple for a soil layer with horizontally uniform properties; but it also can be performed for any more complicated soil system that is specified to be lying between the rigid base and the bottom surface of the building as defined by a 2D or 3D finite element mesh. An early deconvolution analysis of a system with general configuration was the calculation of the earthquake motions at the base of the Pacoima Dam-foundation rock finite element model starting with the earthquake accelerogram recorded on the ridge adjacent to the crest of the dam, as reported in Reference [6].

To a structural engineer, a more attractive way of dealing with this unknown base input problem is to formulate the earthquake response equations directly in terms of the free-field motions rather than the base excitation. In this type of formulation, we consider again the combined system portrayed in Fig. 3 and note again that $\underline{v}^t(t)$ represents the total motions of all degrees of freedom, both soil and structure; so the response of the combined system still can be expressed by Eq. (3), which is repeated here and designated Eq. (A) for convenience in this derivation

$$\underline{m}\ddot{\underline{v}}^t(t) + \underline{c}\dot{\underline{v}}^t(t) + \underline{k}\underline{v}^t(t) = -\underline{m}_g\ddot{\underline{v}}_g(t) - \underline{c}_g\dot{\underline{v}}_g(t) - \underline{k}_g\underline{v}_g(t) \qquad (A)$$

Now in order to develop the free-field equations, it is necessary to express the system property matrices as the combination of contributions associated with each of the two substructures, using a tilde to identify the soil matrices and an over-bar for the structure matrices, thus

$$\begin{array}{ccccc} & & \underline{\text{Soil}} & & \underline{\text{Structure}} \\ \underline{m} & = & \underline{\tilde{m}} & + & \underline{\bar{m}} \\ \underline{c} & = & \underline{\tilde{c}} & + & \underline{\bar{c}} \\ \underline{k} & = & \underline{\tilde{k}} & + & \underline{\bar{k}} \end{array} \qquad (A\text{-}1)$$

It should be noted here that the dimension of each of these matrices corresponds to the number of degrees of freedom of the combined system, but that the nonzero property coefficients relate only to the degrees of freedom of each substructure. Additionally, in the total motion of the degrees of freedom it is necessary to distinguish the free-field motion that would exist if no structure were present [denoted by a tilde, $\underline{\tilde{v}}(t)$] from the additional motion that results when the structure is included in the system [denoted by an over-bar, $\bar{v}^t(t)$], thus

$$\underline{v}^t(t) = \underset{\text{free field}}{\underline{\tilde{v}}(t)} + \underset{\text{added}}{\underline{\bar{v}}^t(t)} \qquad (A\text{-}2)$$

Substituting Eqs. (A-1) and (A-2) into Eq. (A) leads to

$$\left[\underline{\bar{m}} + \underline{\tilde{m}}\right]\left\{\ddot{\underline{\tilde{v}}}(t) + \ddot{\underline{\bar{v}}}^t(t)\right\} + \left[\underline{\bar{c}} + \underline{\tilde{c}}\right]\left\{\dot{\underline{\tilde{v}}}(t) + \dot{\underline{\bar{v}}}^t(t)\right\} + \left[\underline{\bar{k}} + \underline{\tilde{k}}\right]\left\{\underline{\tilde{v}}(t) + \underline{\bar{v}}^t(t)\right\}$$
$$= -\underline{m}_g\ddot{\underline{v}}_g(t) - \underline{c}_g\dot{\underline{v}}_g(t) - \underline{k}_g\underline{v}_g(t) \qquad (B)$$

Now the free-field response equation may be extracted from Eq.(B) by considering only the free-field property matrices and the corresponding motion terms on the left hand side (i.e., by omitting the structure properties and the added motion it causes) thus obtaining

$$\underline{\tilde{m}}\ddot{\underline{\tilde{v}}}(t) + \underline{\tilde{c}}\dot{\underline{\tilde{v}}}(t) + \underline{\tilde{k}}\underline{\tilde{v}}(t) = -\underline{m}_g\ddot{\underline{v}}_g(t) - \underline{c}_g\dot{\underline{v}}_g(t) - \underline{k}_g\underline{v}_g(t) \qquad (C)$$

It will be noted that the effective input force on the right side of Eq. (C) is the same as that in Eq. (A), because the soil system has the same soil coupling coefficients and the same base rock motions as apply to the combined system. For this reason, the right side of Eq. (B) may be replaced by the left side of Eq. (C). Multiplying out the products in Eq. (B), making the substitution from Eq. (C), cancelling terms, and transferring the remaining free-field motion terms to the right side leads to the following Eq. D:

$$\underline{m}\ddot{\bar{\underline{v}}}(t) + \underline{c}\dot{\bar{\underline{v}}}(t) + \underline{k}\bar{\underline{v}}(t) = -\bar{\underline{m}}\ddot{\check{\underline{v}}}(t) - \bar{\underline{c}}\dot{\check{\underline{v}}}(t) - \bar{\underline{k}}\check{\underline{v}}(t) \qquad (D)$$

in which the property terms on the left side represent the combined system while those on the right refer to the added structure only.

This is one formulation of the response equation for the entire system expressed in terms of the free-field excitation. However, it can be simplified to a form where the input involves only the free-field acceleration by following a procedure similar to that employed in reducing Eq. (3) to Eq. (9); that is, by expressing the total added motions $\underline{v}^t(t)$ as follows:

$$\bar{\underline{v}}^t(t) = \bar{\underline{v}}(t) + \bar{\underline{v}}^s(t) \qquad (D\text{-}1)$$

where $\underline{v}^s(t)$ is the quasistatic motion due to the static free-field displacements and $\underline{v}(t)$ is the dynamic added motion. The quasistatic contribution is expressed conveniently by a quasistatic free-field displacement transformation matrix $\tilde{\underline{r}}$ as follows:

$$\bar{\underline{v}}^s = \hat{\underline{r}}\,\check{\underline{v}} \qquad (D\text{-}2)$$

where each column of $\tilde{\underline{r}}$ lists the displacements associated with a unit value of a single free-field static displacement component. Introducing Eqs. (D-1) and (D-2) in Eq. (D) and transferring the free-field input terms to the right side leads to

$$\underline{m}\ddot{\bar{\underline{v}}}(t) + \underline{c}\dot{\bar{\underline{v}}}(t) + \underline{k}\bar{\underline{v}}(t) = -[\underline{m}\hat{\underline{r}} + \bar{\underline{m}}]\ddot{\check{\underline{v}}}(t) - [\underline{c}\hat{\underline{r}} + \bar{\underline{c}}]\dot{\check{\underline{v}}}(t) - [\underline{k}\hat{\underline{r}} + \bar{\underline{k}}]\check{\underline{v}}(t) \qquad (D\text{-}3)$$

which is completely analogous to Eq. (6). But as was seen with Eq. (6), this may be simplified by consideration of the static case in which both the time derivatives and the dynamic added motion vanish. Thus the static result is

$$\underline{0} = [\underline{k}\hat{\underline{r}} + \bar{\underline{k}}]\check{\underline{v}} \qquad (D\text{-}4)$$

from which the transformation matrix is given by

$$\underline{\tilde{r}} = -\underline{k}^{-1}\underline{\bar{k}} \tag{D-5}$$

Equation (D-4) shows that the last term on the right side of Eq. (D-3) vanishes, and following the reasoning that led to the derivation of Eq. 9 it is assumed that the second term may be neglected as well; so the final form of the free-field input response equation becomes

$$\underline{m}\ddot{\underline{\bar{v}}}(t) + \underline{c}\dot{\underline{\bar{v}}}(t) + \underline{k}\underline{\bar{v}}(t) = -[\underline{m}\underline{\tilde{r}} + \underline{\bar{m}}]\ddot{\underline{\tilde{v}}}(t) \tag{11}$$

In contrast with Eq. (9), which was developed for the rigid base input, it is evident that the dynamic response given directly from this free-field input analysis is the added motion resulting from changing the system from the free-field configuration by the addition of the structure.

Thus the total motion $\underline{v}^t$ now consists of the original free-field motion $\underline{\tilde{v}}$ (which pertains only to the free-field portion of the system) plus this dynamic response term $\underline{\bar{v}}$. In addition, a quasistatic contribution $\underline{\bar{v}}^s$ is developed in the entire system due to the free-field motions of the soil interacting with the added structure. Hence, after introducing the quasistatic displacement transformation of Eq. (D-2), the result is

$$\underline{\bar{v}}^t = \underline{\tilde{v}} + \underline{\bar{v}} + \underline{\tilde{r}}\underline{\tilde{v}} \tag{12}$$

in which the quasistatic displacement transformation $\underline{\tilde{r}}$ is given by Eq. (D-5). The major change in Eq. (11) from Eq. (9), other than in the response quantity, is in the effective load term on the right hand side. This now depends on the mass of the added structure $\underline{\bar{m}}$ rather than the base coupling mass $\underline{m}_g$, and of course on the free-field quasistatic transformation $\underline{\tilde{r}}$ rather than on the rigid base transformation $\underline{r}$.

Further consideration of the submatrices making up the final term in Eq. (11) will show that it is only the degrees of freedom at the interface between the free-field soil and the added structure that contribute to the quasistatic response component. Hence, if these interface free-field motions are denoted by $\underline{\tilde{v}}_f$, and if the corresponding columns of the quasistatic transformation are indicated by $\underline{\tilde{r}}_f$ while those of the added structure mass matrix are given by $\underline{\bar{m}}_f$, the resulting simplified version of the free-field input response equation becomes

$$\underline{m}\ddot{\underline{\bar{v}}}(t) + \underline{c}\dot{\underline{\bar{v}}}(t) + \underline{k}\underline{\bar{v}}(t) = -[\underline{m}\underline{\tilde{r}}_f + \underline{\bar{m}}_f]\ddot{\underline{\tilde{v}}}_f(t) \tag{13}$$

It should be noted again that results given by this free-field formulation are identical to those that would be obtained from a rigid base solution in which the base input is determined by a deconvolution analysis; thus they are valid only if the actual soil system is supported on a rigid base so as to correspond with this mathematical model.

5. PHASE IV - SUBSTRUCTURE SSI FORMULATION

Although substructure formulations of the analysis of complicated structural systems had been used extensively by structural engineers long before the importance of the dynamic SSI mechanism had been recognized by earthquake engineers, the concept of substructure analysis of earthquake response now tends to be associated mainly with SSI problems. By treating the soil system and the structure it supports as two separate substructures, the analysis procedure is improved in two important ways as compared with a direct analysis: (1) the formulation of the free-field input equations is simplified, and (2) the soil system can be represented as a continuum rather than as a discretized finite element model and thus need not include a rigid base rock constraint. An early example of a paper describing the substructure method of SSI analysis is listed here as Reference [7].

For this discussion, the simplest form of continuum soil model as depicted by Fig. 4 will be considered first. This system is equivalent to that of Fig. 3 except that a rigid foundation block is interposed between

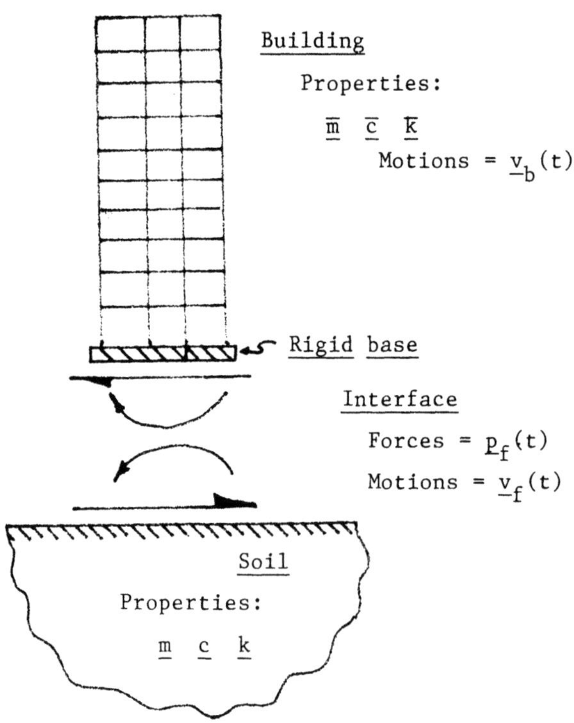

Figure 4. Substructure model of soil-structure interaction

the building and the elastic half-space. This limits the displacements $\underline{v}_f(t)$ developed at the base of the building as well as at the surface of the soil to the rigid body base block motions. These are shown here as a rotation and a lateral translation for a planar model, but in principle all six components of rigid body motion could be included for analysis of a 3D building model.

The free-field formulation of the equation of motion, Eq. (13), may be written for the building substructure shown in Fig. 4, as follows:

$$\underline{m}\ddot{\bar{\underline{v}}}(t) + \underline{c}\dot{\bar{\underline{v}}}(t) + \underline{k}\bar{\underline{v}}(t) = -[\underline{m}\tilde{\underline{r}}_f + \bar{\underline{m}}_f]\ddot{\underline{v}}_f(t) - \begin{Bmatrix} 0 \\ p_f(t) \end{Bmatrix} \quad (14)$$

where it should be noted that $\bar{\underline{v}}$ represents the dynamic motions of the building relative to its base block, and that it includes all degrees of freedom of the system. Also, $\ddot{\underline{v}}_f(t)$ represents the free-field accelerations of the soil at the interface (which are constrained to match the rigid body base motions) and the last term on the right side of Eq. (14) represents the corresponding set of interacting forces applied by the soil to the building which, of course, are non-zero only at the building base.

Before the response of the combined soil-structure system may be calculated, the building substructure of Eq. (14) must be assembled with the soil substructure, but because the soil has frequency dependent properties it is convenient to carry out the response analysis in the frequency domain. Accordingly, Eq. (14) is converted to the frequency domain by taking its Fourier transform, with the following result:

$$[-\bar{\omega}^2\underline{m} + (1+i\eta)\underline{k}]\underline{V}(i\bar{\omega}) = -[\underline{m}\tilde{\underline{r}}_f + \underline{m}_f]\ddot{\underline{V}}(i\bar{\omega}) - \begin{Bmatrix} 0 \\ P(i\bar{\omega}) \end{Bmatrix} \quad (15)$$

It will be noted that capital letters are used here to denote the frequency domain equivalents of displacement quantities represented in the time domain by lower case letters in Eq. (14). Note also that the viscous damping coefficient, c, in Eq. (14) has been replaced by the hysteretic damping coefficient, η, because the hysteretic damping is frequency independent. When the system is operating at resonance, the two types of damping have exactly the same effect if $\eta = c/m\omega$.

The quantity in brackets at the left side of Eq. (15) is called the building impedance, and is denoted here by

$$\underline{I}_b(i\bar{\omega}) \equiv [-\bar{\omega}^2\underline{m} + (1+i\eta)\underline{k}] \quad (15a)$$

so that Eq. (15) becomes

$$\underline{I}_b(i\bar{\omega})\underline{V}(i\bar{\omega}) = -[\underline{m}\tilde{\underline{r}}_f + \bar{\underline{m}}_f]\ddot{\underline{V}}_f(i\bar{\omega}) - \begin{Bmatrix} 0 \\ P(i\bar{\omega}) \end{Bmatrix} \quad (15b)$$

The corresponding frequency domain equation for the soil substructure may be written as follows:

$$\underline{I}_f(i\bar{\omega})\,\underline{V}(i\bar{\omega}) = \begin{Bmatrix} 0 \\ P(i\bar{\omega}) \end{Bmatrix} \tag{16}$$

in which $\underline{I}_f(i\bar{\omega})$ represents the impedance of the half space soil region. The frequency dependent coefficients of this matrix have been evaluated for a variety of foundation conditions during years of SSI research.

Now because the same degrees of freedom are used in the equations of motion for each of the substructures, they can be combined by direct stiffness assembly, leading to:

$$\left[\underline{I}_b(i\bar{\omega}) + \underline{I}_f(i\bar{\omega})\right]\underline{V}(i\bar{\omega}) = -\left[\underline{m}\,\underline{\tilde{r}}_f + \underline{\bar{m}}_f\right]\ddot{\underline{V}}_f(i\bar{\omega}) \tag{17}$$

in which the opposing interaction forces $\underline{P}(i\bar{\omega})$ have cancelled out. The final frequency domain expression for the dynamic response $\underline{V}(i\bar{\omega})$ is obtained by solving Eq. (17); in symbolic form the solution may be expressed as follows

$$\underline{V}(i\bar{\omega}) = -\left[\underline{I}_b(i\bar{\omega}) + \underline{I}_f(i\bar{\omega})\right]^{-1}\left[\underline{m}\,\underline{\tilde{r}}_f + \underline{\bar{m}}_f\right]\ddot{\underline{V}}_f(i\bar{\omega}) \tag{18}$$

The expression for the time history of response then is given by the Inverse Fourier Transform (IFT) of Eq. (18), i.e.,

$$\underline{v}(t) = \mathrm{IFT}\left[\underline{V}(i\bar{\omega})\right] \tag{19}$$

Of course, to carry out this analysis, the simultaneous equation set of Eq. (18) must be solved for a large number of discrete frequencies $\bar{\omega}$, hence it is a major computational task. Hcwever, a large part of the SSI effort in recent years has been directed toward improving the efficiency of frequency domain analysis with the result that it is now considered no more formidable than the corresponding time domain analysis and yet it offers great advantages in representing the dynamic behavior of continuum soil systems.

The obvious limitation of the SSI system shown in Fig. 4 is the building base located at the interface with the foundation medium. Although this base is assumed to be rigid in the analysis discussed above, such rigidity can be approached only in special circumstances. In many cases, the flexibility of the building base may be comparable with that of the soil, and the mathematical model should characterize this flexibility in a realistic way, such as by the finite element method. To maintain displacement compatibility with such a building model, then, the soil model also should be subject to the finite element displacement patterns. Reference 7 describes representation of the soil in this manner, deriving a matrix of impedance coefficients for a half space subjected to finite element compatible surface displacements.

Many extensions of such ideas have now been developed, including the concept of hybrid modeling in which a soil substructure formed as an assemblage of finite elements is used to connect the structure to the soil continuum. However, detailed discussion of such developments is beyond the scope of this brief survey.

6. CONCLUDING REMARKS

6.1. This brief survey has outlined a structural engineer's impression of developments in techniques for dealing with the interaction between a structure and its foundation medium in the field of earthquake structural response analysis. The account starts with the era when such interaction was considered to be of negligible importance, and the prescribed earthquake excitation was applied directly to the structural supports. Then the continually increasing complexity of the analysis procedures is described, and it is clear that this chronology reflects not only the improving understanding of the interaction mechanism but also the increasing computational capabilities provided by the rapidly expanding computer hardware and software industries. However, in the area of earthquake engineering these improved capabilities sometimes have had a detrimental effect in emphasizing greater refinement of the mathematical model in preference to seeking better insight into the dynamic behavior. In actual fact, the quest for extreme refinement in the mathematical model often is a waste of effort, in view of the simplifying assumptions that have to be made about the system geometry, material properties, and dynamic response mechanism; also the mass of numerical data derived from a refined model may tend to obscure the true nature of the response behavior.

6.2 In most cases the purpose of the response analysis is to obtain an indication of the earthquake safety of an engineered system, and this requires: (1) a reasonably reliable model of the system, and (2) an adequate estimate of the earthquake excitation to which it may be subjected. At best, the ground motion estimate is only a crude approximation of the maximum credible earthquake (MCE) for that location; consequently the mathematical model need provide no more than a reasonable approximation of the expected structural response behavior.

6.3 Although the uncertainties of the seismic input estimate undoubtedly are much greater than those anticipated in the mathematical model, it is desirable to validate the model by physical experiments when possible because of the many assumptions that may have been introduced in its formulation. In the case of a system subject to SSI, such experiments generally can be made only in the field at full scale because of the difficulty of producing a reduced scale model of the foundation medium. However the usual method of performing such tests, involving a harmonic shaker attached to the structure, generally provide only very limited information on the SSI mechanism because very little energy is transmitted into the foundation (although they may be useful in validating the mathematical model of the structure). For this reason it is likely that SSI test data must be obtained primarily from recordings of response to actual earthquakes, and hence the instrumentation of a wide range of SSI systems is one of today's most urgent requirements in the field of earthquake engineering. Only when records have been obtained of the seismic input and of the response of the interacting structure can the validity of the SSI analysis procedures be demonstrated.

6.4 In the seismic safety analysis of a structure, the basic question under consideration is whether its performance will be "adequate" when it is subjected to a MCE. The definition of adequate performance is not well established, but clearly collapse and its related life hazard must be avoided. In most cases, a level of performance considerably less than collapse will be taken as the acceptable response limit, but it is important to note that even this limit usually is well above the elastic deformation limit of the structure. On the other hand, most of the SSI analysis procedures are valid only within the linear elastic range. Obviously the elastic limit deformation could be taken as a highly conservative upper limit of acceptability, but well-designed structures can safely accommodate forces and displacements that greatly exceed this limit. For this reason, except in the case of brittle structures which are not appropriate for construction in active seismic regions, the question of seismic safety is not answered by analysis, but rather by design. The key to structural safety during earthquakes is ductility, that is, the ability to maintain significant resistance to deformation when the deformations have exceeded the limits of linear elasticity. Hence the system must be well tied together so as to maintain continuity, and the displacements calculated by a linear elastic analysis have little significance.

7. REFERENCES

[1] Clough, Ray W. "On the Importance of Higher Modes of Vibration in the Earthquake Response of a Tall Building", Bulletin, Seismological Society of America, Vol. 45, 1955, pp. 289-301.

[2] Robison, E. C. and Blume, J. E. "Vibration Characteristics and Earthquake Forces for a Fifteen-Story Building", unpublished MS.

[3] Blume, J. A. and Hesselmeyer, H. L. "Reconciliation of the Computed and Observed Periods of Vibration of a Fifteen Story Building", unpublished Stanford University thesis, 1934.

[4] Przemieniecki, J. S. "Matrix Structural Analysis of Substructures", J. Amer. Inst. Aeron., Astron., Vol. 1, (1963) pp. 138-147.

[5] Clough, R. W. and Mojtahedi, S. "Earthquake Response Analysis Considering Non-Proportional Damping", Earthquake Engineering and Structural Dynamics, Vol. 4, No. 5, 1976, pp. 489-496.

[6] Reimer, R. B., Clough, R. W. and Raphael, J. M. "Evaluation of the Pacoima Dam Accelerogram", Proceedings, 5WCEE, Rome, Italy, 1974, pp. 2328-2337.

[7] Gutierrez, J. A. and Chopra, A. K. "A Substructure Method for the Earthquake Analysis of Structures including Structure-Soil Interaction", Earthquake Engineering and Structural Dynamics, Vol. 6, No. 1, 1978, pp. 51-70.

ABSORBING BOUNDARY CONDITIONS IN SOIL-STRUCTURE INTERACTION ANALYSES

Y. MENGI
Middle East Technical University,
Department of Engineering Sciences,
Ankara, Turkey

A.K. TANRIKULU
Cukurova University,
Department of Civil Engineering,
Adana, Turkey

ABSTRACT. The use of finite element method in soil-structure interaction analysis dictates that the infinite soil medium be truncated along a certain boundary, called artificial boundary, separating the soil-structure system into near and far fields. In conjunction with the analysis of near field, some artificial boundaries capable of transmitting waves from near to far field are proposed in literature. In the present study, first these artificial boundary conditions are reviewed, and then they are assessed by using some chosen sample problems. The comparisons are presented in both time and frequency spaces.

1. Introduction

There are two basic methods which are being used in soil-structure interaction analysis: complete method (CM) and substructure method (SM) [1,2]. Complete method has three main steps: (a) description of seismic environment, (b) free field analysis, (c) interaction analysis. On the other hand, the main steps of the substructure method are: (a) description of seismic environment, (b) free field analysis, (c) determination of impedance functions, (d) analysis of superstructure. Depending upon the type of foundation and irregularities which may exist in soil system, the free field analysis may also involve scattering analysis [2].

The steps (b) and (c) of both CM and SM involve infinite soil media. The analyses in these steps are usually carried out by finite element methods (FEM) [2,3]. The use of FEM for this purpose dictates that the infinite medium be truncated along certain boundaries (called artificial boundaries) and thus be reduced to a finite region (near field). In order to have meaningful results, the artificial boundaries, which actually do not exist, should be capable to transmit waves from near to far field without reflections, or at least the wave reflections back into near field should be minimized. To this end various boundary conditions (bc's) for artificial boundaries are proposed [1,16], which are named as absorbing or artificial or transmitting bc's in literature. The artificial bc's may be also interpreted as the constitutive equations for the interaction forces between near and far fields; thus their performance in soil-structure interaction analysis depends on how correctly they describe these forces.

P. Gülkan and R. W. Clough (eds.), Developments in Dynamic Soil-Structure Interaction, 111–146.

The artificial bc's can be classified under the following titles: (a) viscous, (b) superposition, (c) paraxial, (d) extrapolation, (e) horizontal wave bc's. Viscous and paraxial bc's can be formulated both in time and frequency domains whereas (superposition, extrapolation) and horizontal wave bc's can be used only in time and frequency spaces respectively. The viscous bc's is the most commonly used bc's in practice as it has a simple form suitable for finite element formulation and nonlinear analysis. Even though horizontal wave (HW) bc's are frequency dependent, their use in soil-structure interaction analysis has become popular in recent years and are installed in some general purpose computer softwares, such as, FLUSH [3]. This is because the performance of HW bc's is much better than the others for vertical artificial boundaries (which can be placed very close to the structure when they are used [3,16]) and because they can be used in approximate nonlinear analysis when combined with equivalent linear method [3].

In the present study, first, the artificial bc's are reviewed and their limitations are discussed. Then, the performances of various artificial bc's are compared by using some chosen sample problems. The comparisons are presented in both time and frequency spaces. The work is organized as follows. In Chapter 2, we review basic equations of wave propagation relevant to the derivation of artificial bc's. In Chapter 3, we derive, at some length, the artificial bc's and discuss their limitations implied by the derivations. Finally, in Chapter 4, we assess the performance of various artificial bc's by employing some example problems.

2. Some Definitions in Wave Mechanics

In this section, some aspects of wave propagation relevant to the establishment of absorbing boundary conditions (bc's) being used in soil-structure interaction analysis will be reviewed.

2.1 HARMONIC WAVE

A harmonic wave propagating in a medium is described by the equation

$$\begin{aligned} u &= A e^{i(\omega t - \boldsymbol{k}.\boldsymbol{x})} \\ &= A e^{i(\omega t - k_j x_j)} \end{aligned} \tag{1}$$

where

- u : disturbance which propagates
- A : amplitude of the disturbance
- ω : angular frequency
- t : time
- $\boldsymbol{k}$: wave number vector
- $\boldsymbol{x}$: position vector
- i : imaginary number.

It will be assumed that Eq.(1) is written in an x_i Cartesian coordinate system. In the second line of Eq.(1), the summation convention is used and it is assumed that latin subcripts have the range from 1 to 3.

If all of the components of the wave number vector are real, then one has

$$\boldsymbol{k} = k\,\boldsymbol{n} \tag{2}$$

where k is wave number defined by

$$k = |\boldsymbol{k}| \tag{3}$$

and $\boldsymbol{n}$ is unit vector describing the direction in which the harmonic wave propagates. For the wave velocity c and wave length λ of the harmonic wave one has the relations

$$\begin{aligned} c &= \frac{\omega}{k} \\ \lambda &= \frac{2\pi}{k} \end{aligned} \tag{4}$$

Satisfaction of the field equations of the medium dictates that an equation of the form

$$F(\boldsymbol{k},\omega)=0 \tag{5}$$

is to be satisfied. For a given ω, Eq.(5) describes surfaces in wave number space, which will be called wave number surfaces (WNS). A point on WNS in $\boldsymbol{k}$ space has the distance $k=\omega/c$ from the origin.

When the components of the wave number vector $\boldsymbol{k}$ are real, the harmonic wave equation, Eq.(1), can be written in terms of apparent (phase) wave velocities as

$$u=Ae^{i\omega(t-\frac{x_j}{c_j})} \tag{6}$$

where

$$c_j=\frac{\omega}{k_j} \tag{7}$$

designates the apparent wave velocity in x_j direction. It should be noted that the summation is implied over the index j in Eq.(6).

2.2. HARMONIC WAVES IN A LINEAR ELASTIC MEDIUM

Displacement equations of motion for a linear anisotropic elastic solid are, in a Cartesian coordinate system,

$$c_{ijkl}\partial_{lj}u_k + f_i = \rho\, \ddot{u}_i \tag{8}$$

where

- c_{ijkl} : elastic coefficients
- f_i : body force per unit volume
- u_i : displacement components
- ρ : mass density

and ∂_i and dot designate respectively the partial derivatives with respect to x_i and time. In Eq.(8), the indicial notation is used.

The harmonic wave

$$u_i=A_i e^{i(\omega t-k_j x_j)} \tag{9}$$

will satisfy Eq.(8) in the absence of body forces if

$$(c_{ijkl}k_l k_j - \delta_{ik}\rho\omega^2)A_k=0 \tag{10}$$

where δ_{ik} is Kronecker's delta. To have nontrivial solution in Eq.(10), one should have

$$F(\boldsymbol{k},\omega)=\det(c_{ijkl}k_l k_j - \delta_{ik}\rho\omega^2)=0 \tag{11}$$

which is the characteristic equation for harmonic waves propagating in a linear anisotropic elastic medium. For a given ω, Eq.(11) describes three sheets of closed surfaces (WNS) in $\boldsymbol{k}$ space, which are symmetric with respect to the origin of $\boldsymbol{k}$ coordinate system.

If the material is isotropic, one has

$$c_{ijkl} = \lambda\delta_{ij}\delta_{kl} + \mu(\delta_{ik}\delta_{jl} + \delta_{il}\delta_{jk}) \tag{12}$$

where λ and μ are Lame's constants, for which Eq.(11) takes the form

$$F(\boldsymbol{k},\omega) = (k_j k_j - \frac{\omega^2}{c_p^2})\ (k_l k_l - \frac{\omega^2}{c_s^2})^2 = 0 \tag{13}$$

Here, $c_p = ((2\mu+\lambda)/\rho)^{1/2}$ is dilatational and $c_s = (\mu/\rho)^{1/2}$ is shear wave velocity. Eq.(13) shows that the waves propagating in an isotropic elastic medium are composed of one wave propagating with the velocity c_p and two waves with the velocity c_s, which are called respectively dilatational (P) and shear (SV or SH) waves. From Eq.(10) it follows that the amplitude vector of the dilatational wave is parallel to the wave propagation direction, and those of two shear waves are perpendicular to it and orthogonal to each other.

Eq.(13) shows that WNS's for isotropic elastic media are composed of two concentric spheres whose center is at the origin of $\boldsymbol{k}$ space, and whose radii are (ω/c_s) and (ω/c_p) respectively (see Fig.1).

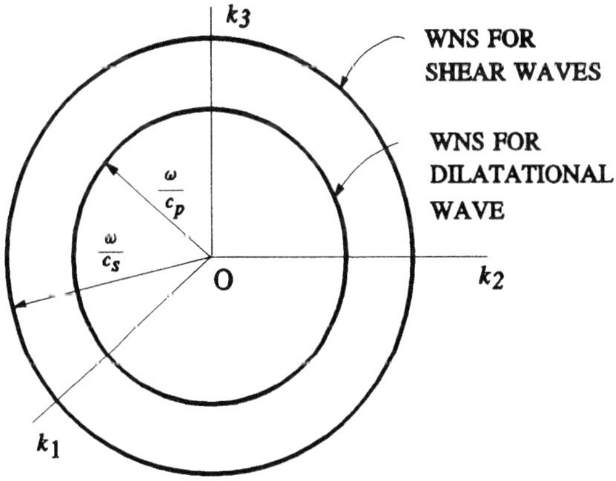

Figure 1 Wave number surfaces for isotropic elastic media

2.3. REFLECTION AND REFRACTION OF HARMONIC WAVES IN ISOTROPIC ELASTIC MEDIA

Let us consider two half spaces 1 and 2 which are bonded to each other and made of two different isotropic elastic materials having the properties (λ,μ,ρ) and $(\bar{\lambda},\bar{\mu},\bar{\rho})$ respectively (see Fig.2). The medium is referred to an x_i Cartesian coordinate system in which the interface of two half spaces coincides with the $(x_1$-$x_3)$ plane and the x_2 axis directed downwards.

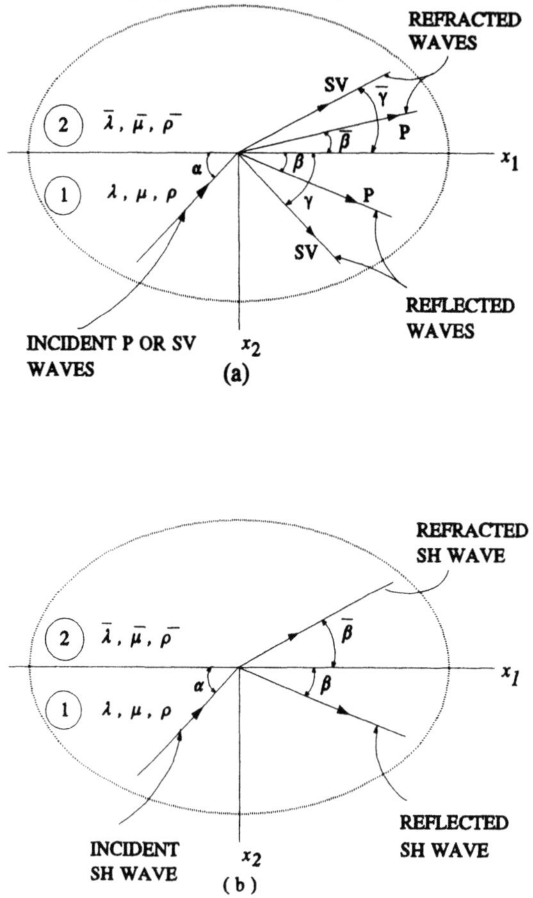

Figure 2 Reflection and refraction of (a) P or SV wave (b) SH wave

Let us consider reflection and refraction of an incident wave which propagates in x_1-x_2 plane. In agreement with the terminology being used in literature, the two shear waves having the displacement vector in x_1-x_2 plane and perpendicular to it will be called SV and SH waves respectively.

Satisfaction of continuity conditions at the interface for all x_1 requires that

$$k_1 : \textit{be the same for incident, reflected and refracted waves} \tag{14}$$

In view of Eq.(7), this condition can also be written in terms of the apparent wave velocity component c_1 as

$$c_1 : \textit{be the same for incident, reflected and refracted waves} \tag{15}$$

From the satisfaction of interface conditions it follows that the incident P or SV wave results in two reflected waves which are P and SV waves, and two refracted waves which are again P and SV waves. It may be shown that when the incident angle α (see Fig.2) becomes less than certain critical values, some of the reflected and refracted waves are converted into surface waves [17]. On the other hand, the incident SH wave yields reflected and refracted SH waves only (see Fig.2).

From the condition, Eq.(14) or (15), it follows that

$$\frac{c}{\cos\alpha} = \frac{c_p}{\cos\beta} = \frac{c_s}{\cos\gamma} = \frac{\bar{c}_p}{\cos\bar{\beta}} = \frac{\bar{c}_s}{\cos\bar{\gamma}} \tag{16}$$

for incident P or SV wave and

$$\frac{c_s}{\cos\alpha} = \frac{c_s}{\cos\beta} = \frac{\bar{c}_s}{\cos\bar{\beta}} \tag{17}$$

for incident SH wave, which are called Snell's Law. In Eq.(16), $c=c_s$ for incident SV wave and $c=c_p$ for incident P wave. In Eqs.(16) and (17), (c_s, c_p) and ($\bar{c}_s, \bar{c}_p$) are the wave velocities in the half spaces 1 and 2 respectively. For the definitions of the angles appearing in Eqs.(16) and (17), see Fig.2.

The amplitudes of reflected and refracted waves in terms of that of the incident wave can be determined by using interface conditions.

2.4. ONE DIMENSIONAL WAVE PROPAGATION IN AN INFINITE ISOTROPIC ELASTIC MEDIUM

A plane wave propagating in an arbitrary direction x in an infinite isotropic elastic medium can be written in space-time domain as

$$u = \mathrm{f}(t - \frac{x}{c}) + \mathrm{g}(t + \frac{x}{c}) \tag{18}$$

where f and g are some functions, u is displacement; $c=c_p$ and $u//x$ for dilatational waves, and $c=c_s$ and $u \perp x$ for shear waves. The function f in Eq.(18) represents the wave propagating in the positive direction of x axis (outgoing wave) and the function g designates the wave propagating in the negative direction of x axis (incoming wave).

Eq.(18) represents waves propagating without dispersion (i.e., without changes in the wave profiles). This is because the medium is elastic, and because the wave is a plane wave and the medium is infinite (i.e., no boundary effects exist).

2.5. CHARACTERISTIC PLANE FOR ONE DIMENSIONAL WAVES

Let us consider plane elastic waves propagating in an x direction. Further let us assume that the medium is bounded by two planes which are perpendicular to x axis at $x=0$ and $x=L$. In this case, the wave pattern will be composed of outgoing and incoming waves. These waves can be described more conveniently by using the (x-t) characteristic plane shown in Fig.3. In this figure, the lines $x\text{-}ct=const.$ and $x+ct=const.$ represent the characteristic lines along which outgoing and incoming waves propagate respectively. Reflections of waves are governed by the boundary conditions at the boundaries $x \doteq 0$ and $x=L$, i.e., along the lines B1 and B2 in Fig.3. When the boundary is free, the displacement is reflected with the same sign and the stress with the opposite sign; when the boundary is fixed, the opposite situation holds.

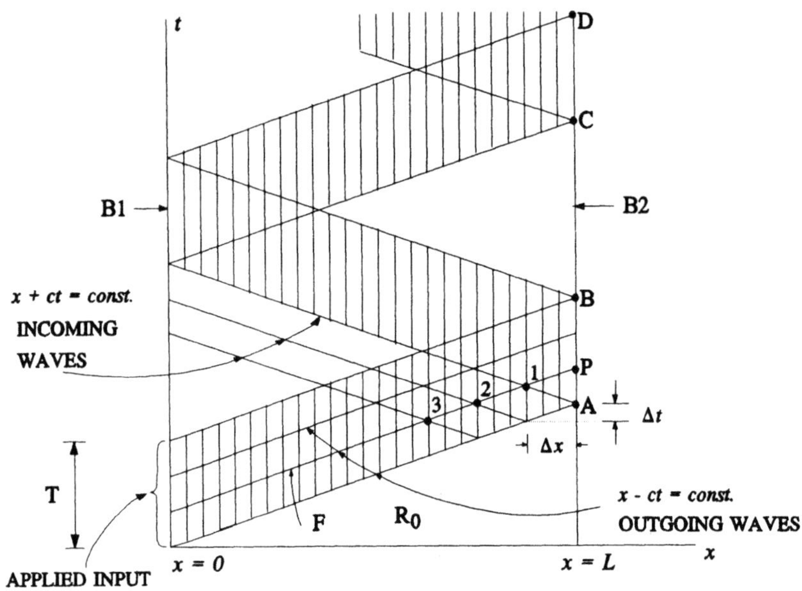

Figure 3 Characteristic plane for one dimensional wave propagation

Now, let us assume that the waves are generated by an input of the duration T, which is applied to the left boundary $x=0$. Further assume that the medium is initially at rest. In this case, the region R_0 in Fig.3 is undisturbed. The shaded regions represent incident and reflected waves. Along the line B2, the line segment AB describes the first reflection; and line segment CD and others, higher order reflections at the right boundary $x=L$.

3. Absorbing Boundary Conditions

3.1. GENERAL CONSIDERATIONS

Let us consider a system shown in Fig.4 which is composed of a structure and an elastic half space representing soil, and which is under the influence of some dynamic loads. To perform soil-structure interaction analysis for this system by finite elements, we truncate the half space along a certain surface (L), called artificial boundary, which divides the system into two parts, namely, near field (A) and far field (B), and retain the near field in the analysis. To eliminate reflections at the artificial boundary, some special conditions must be used on L. These conditions are termed as artificial or absorbing or transmitting boundary conditions.

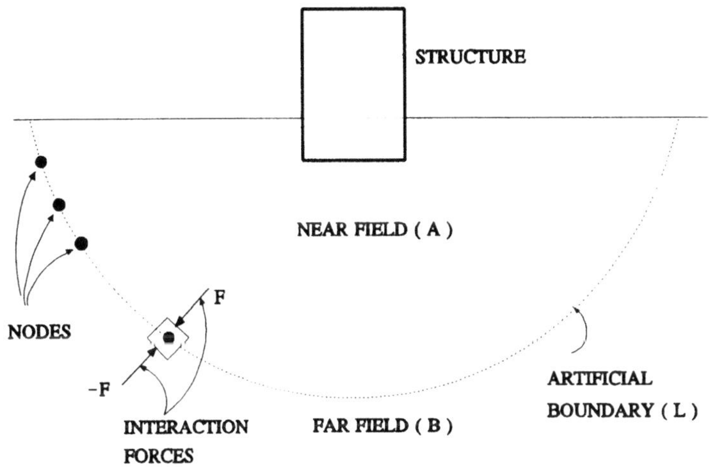

Figure 4 Description of artificial boundary

Within the context of finite element formulation, we now proceed establishing the general form of the conditions to be imposed on the artificial boundary L. To this end we consider a certain node on L.-$\boldsymbol{F}$ and $\boldsymbol{F}$ in Fig.4 designate respectively interaction forces acting on far and near fields. The use of finite element formulation dictates that the costitutive equation for the interaction forces acting on near field can be written approximately in frequency space as

$$-\boldsymbol{F}(\omega)=\boldsymbol{S}(\omega)\ \boldsymbol{u}(\omega)$$

or

$$\boldsymbol{F}(\omega)+\boldsymbol{S}(\omega)\,\boldsymbol{u}(\omega)=\mathbf{0} \tag{19}$$

where $\boldsymbol{F}$ and $\boldsymbol{u}$ are column vectors containing force and displacement components of all the nodes on L and $\boldsymbol{S}$ is a square matrix containing impedance functions of the far field B. Eq.(19) is valid when the dynamic source is contained in the near field A. In the presence of free field motion, Eq.(19) should be modified as

$$F(\omega) - F^f(\omega) + S(\omega)\,(u(\omega) - u^f(\omega)) = 0 \tag{20}$$

where F^f and u^f are free field forces and displacements at the nodes of L. The impedance matrix can be split into real and imaginary parts as

$$S(\omega) = K(\omega) + i\omega C(\omega) \tag{21}$$

in which the imaginary part C describes energy dissipation due to radiation of waves from near to far field.

Eq.(19) can be interpreted as the artificial boundary condition to be imposed on the boundary L of near field; but, its use in the analysis in this form is not feasible since it involves the determination of the impedance matrix S for the infinite far field, which we wish to eliminate. It is further worth noting that the condition in Eq.(19) is not local in both time and space.

Before we close this section we comment on some properties of the impedance matrix. Generally, S does not vanish as $\omega \to \infty$ [1]. In fact, K and C parts of S in Eq.(21) approach some finite values, say, K_i and C_i, as $\omega \to \infty$. In view of this observation, the impedance matrix S can be expressed as the sum of the irregular part S_i and regular part S_r:

$$S(\omega) = S_i(\omega) + S_r(\omega) \tag{22}$$

where

$$\begin{aligned} S_i(\omega) &= K_i + i\omega C_i \\ S_r(\omega) &= (K(\omega) - K_i) + i\omega(C(\omega) - C_i) \end{aligned} \tag{23}$$

From the last two equations we see that K and C parts of S_r vanish as $\omega \to \infty$. Using these definitions, we can now write Eq.(19) in time domain as

$$F(t) + K_i u(t) + C_i \dot{u}(t) + \int_0^t S_r(t-\tau) u(\tau) \mathrm{d}\tau = 0 \tag{24}$$

where, not to complicate the notation, the symbols which are used for the quantities in frequency space are kept the same in time domain, and the kernel $S_r(t)$ is the inverse Fourier transform of $S_r(\omega)$.

From Eqs.(22) and (23) it follows that when the response involves high frequencies S can be approximated as

$$S(\omega) \cong i\omega C_i \tag{25}$$

which leads to, in view of Eq.(19),

$$F(\omega) + i\omega C_i u(\omega) = 0 \tag{26}$$

or in time domain

$$F(t) + C_i \dot{u}(t) = 0 \tag{27}$$

Eq.(26) or (27) represents the asymptotic form of artificial boundary conditions valid when $\omega \to \infty$. These boundary conditions imply that when ω is large the energy in near field radiates to far field through some coupled dashpots located at the nodes of the artificial boundary L (see Fig.4).

3.2. VISCOUS BC'S

With the object of deriving viscous bc's we first consider a plane wave propagating in an x direction in an isotropic elastic medium (see Fig.5).Without loss of generality we may assume that the x axis lies in (x_1-x_2) plane. Let $\boldsymbol{n}$ and s be unit vectors (in (x_1-x_2) plane) in x direction and perpendicular to it respectively. Further, we consider a plane artificial boundary L which is perpendicular to x axis and located at $x=L$. Upon the strike of the plane wave at the artificial boundary, we will have both outgoing (incident) and incoming (reflected) waves as described in Eq.(18). Our object is to eliminate the reflected wave g in Eq.(18) after the plane wave strikes the artificial boundary. To this end we require that the contribution to the motion at artificial boundary comes only from the outgoing plane wave f in Eq.(18), i.e.,

$$u = \mathrm{f}(t - \frac{x}{c}) \tag{28}$$

where $c=c_p$ and $u//x$ for P waves, and $c=c_s$ and and $u \perp x$ for shear waves. Eq.(28) implies that the spatial and time derivatives of the displacement u are related to each other by

$$\partial_x u + \frac{1}{c}\dot{u} = 0 \tag{29}$$

on the artificial boundary L.

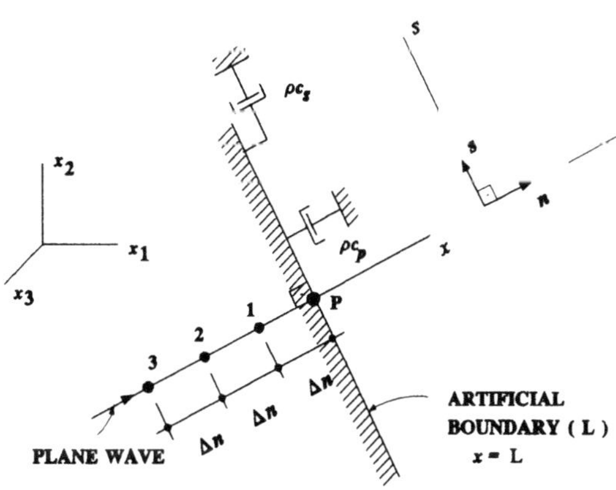

Figure 5 Description of viscous bc's

Using constitutive equations of isotropic elastic materials, Eq.(29) can be written for P and SV waves respectively as

$$\begin{aligned} \tau_{nn} + \rho\, c_p\, \dot{u}_n &= 0 \\ \tau_{ns} + \rho\, c_s\, \dot{u}_s &= 0 \end{aligned} \tag{30}$$

where τ_{nn} and τ_{ns} are normal and tangential components of the traction vector t^n acting on L, and u_n and u_s are normal and tangential displacement components (see Fig.5). The second equation in Eqs.(30) is also valid for SH waves provided that τ_{ns} and u_s appearing in that equation are to be understood as the out-of-plane shear stress and displacement components.

The conditions in Eqs.(30) are called viscous bc's. They are local in space and time, and have suitable forms for finite element and nonlinear analyses. Viscous bc's can be imposed on artificial boundary by putting viscous dampers in normal and tangential directions of L as indicated in Fig.5. The coefficients of dampers in normal and tangential directions are respectively (ρc_p) and (ρc_s).

Derivation of Eqs.(30) implies that viscous bc's are exact only for plane waves impinging the artificial boundary at right angles. They constitute approximate absorbing bc's for inclined body waves. The approximation improves as the incident angle approaches 90° and, in view of the findings established in the previous section, as the frequency gets larger.

J.Lysmer and R.L.Kuhlemeyer in [12] generalized viscous bc's, Eqs.(30), as

$$\begin{aligned} \tau_{nn} + a\rho\, c_p\, \dot{u}_n &= 0 \\ \tau_{ns} + b\rho\, c_s\, \dot{u}_s &= 0 \end{aligned} \tag{31}$$

by introducting the nondimensional constants a and b and studied the influence of these constants on the performance of viscous bc's. To this end, they investigated the reflection of P and SV waves at the artificial boundary L by using viscous bc's in Eqs.(31) (see Fig.6). In what follows, some details of that study are presented.

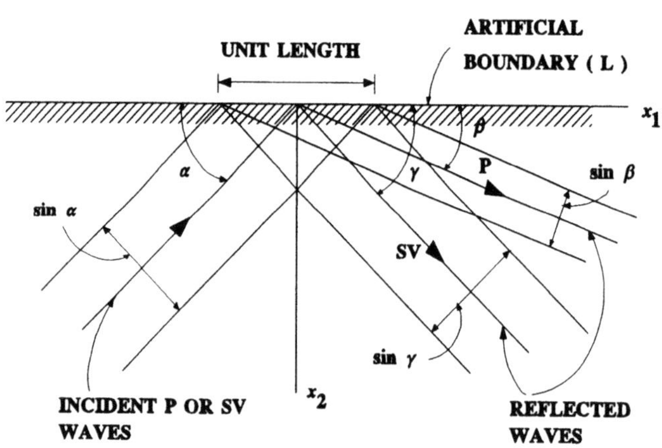

Figure 6 Reflection of P or SV wave at an artificial boundary

Let the amplitudes of the incident wave and reflected P and SV waves in Fig.6 be A, B and C respectively. The incident angle α and the angles of reflected waves are related by Snell's law, Eq.(16) (for the definitions of these angles see Fig.6). The ratios of the amplitudes of reflected waves to that of the incident wave, i.e., (B/A) and (C/A), are governed by the viscous bc's, Eqs.(31). Let $\dot{E}_i$ and $\dot{E}_r$ be average energy rates defined per unit length of the artficial boundary L and associated with the incident and reflected waves respectively. After some manipulations, for the ratio of the energy rate of reflected waves to that of the incident wave one can obtain

$$\frac{\dot{E}_r}{\dot{E}_i} = \frac{c_p}{c}\left(\frac{B}{A}\right)^2 \frac{\sin\beta}{\sin\alpha} + \frac{c_s}{c}\left(\frac{C}{A}\right)^2 \frac{\sin\gamma}{\sin\alpha} \tag{32}$$

where $c=c_p$ if the incident wave is P wave and $c=c_s$ if it is SV wave.

In [12], the variation of the energy ratio $(\dot{E}_r/\dot{E}_i)$ with the incident angle α is studied for various values of the constants a and b, and it is found that the use of $a=b=1$ is suitable for practical purposes. These values of a and b correspond to keeping the viscous bc's in their original form, Eqs.(30).

Fig.7 shows the variation of the energy ratio $(\dot{E}_r/\dot{E}_i)$ with the incident angle α for incident P wave. The results are obtained by using a Poisson's ratio of (1/3) and original viscous bc's Eqs.(30). The effective energy ratio appearing in the figures is defined by

$$\frac{\dot{E}_r}{\dot{E}_i}\sin\alpha$$

which is a better measure for the reflected energy (compared to energy ratio) since it takes into account the amount of energy arriving at the boundary. Fig.7 shows that viscous bc's perform well for a wide range of incident angles and its performance , as anticipated, gets better as the incident angle approaches 90°.

White, Valliappan and Lee in [13] restudied the generalized viscous bc's, Eqs.(31), and established the values of the constants a and b by minimizing the reflected energy. In that work they presented a table which gives the optimal values of a and b as functions of Poisson's ratio. However, they found that the use of the optimal values of a and b in Eqs.(31) does not bring major improvement to the performance of viscous bc's compared to simply taking $a=b=1$.

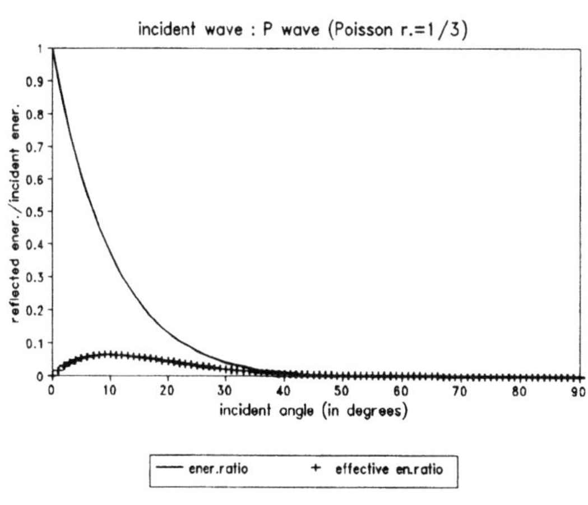

Figure 7 Energy ratio for incident P wave reflecting at viscous boundary

3.3. PARAXIAL BC'S

Paraxial bc's are aimed at eliminating the reflected waves at an artificial boundary with a higher order approximation compared viscous bc's [4,14]. In what follows, we discuss paraxial bc's first for out-of-plane, then for inplane motions of isotropic elastic media.

3.3.1. *Out-of-Plane Motion (SH Waves).* In discussing this case, we refer to Fig.8 in which we assume that the out-of-plane motion is generated by inclined SH waves propagating in x_1-x_2 plane. The artificial boundary (L) is chosen to coincide with the plane $x_2=0$. The reflection of a component of SH waves is shown in the figure. It may be noted that the incident angle α varies from one component to the other of SH waves.

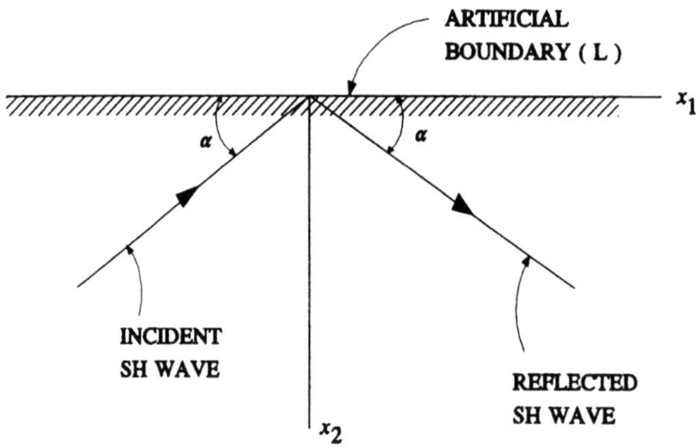

Figure 8 Reflection of SH wave at an artificial boundary

The motion generated by a SH wave component has the form

$$u = A e^{i(\omega t - k_1 x_1 + k_2 x_2)} \tag{33}$$

where u is out-of-plane displacement component, (i.e., $u \perp (x_1$-$x_2))$, $k_1 \geq 0$, and $k_2 > 0$ for the incoming SH wave and $k_2 < 0$ for the reflected SH wave. The wave numbers k_1, k_2 and frequency ω in Eq.(33) are related through the characteristic equation, the second factor in Eq.(13), i.e.,

$$k_1^2 + k_2^2 - \frac{\omega^2}{c_s^2} = 0$$

which can be solved for k_2 as

$$k_2 = \mp \sqrt{\frac{\omega^2}{c_s^2} - k_1^2} \tag{34}$$

Since our object is to transmit SH waves through the artificial boundary without reflections, we choose plus sign in Eq.(34), i.e.,

$$k_2 = + \frac{\omega}{c_s} \sqrt{1 - (\frac{k_1 c_s}{\omega})^2} \tag{35}$$

In view of Eqs.(2) and (7), for the factor $(k_1 c_s/\omega)$ appearing in this equation one can write

$$\frac{k_1 c_s}{\omega} = \frac{c_s}{c_1} = \cos\alpha \tag{36}$$

which shows that the second factor in the square root in Eq.(35) is smaller than one.

At this point, we assume that the deviation of the incident angle α from the right angle is small and we expand the square root in Eq.(35) in terms of $(k_1 c_s/\omega)$ using Taylor's formula, which gives

$$k_2 = \frac{\omega}{c_s} \left[1 - \frac{1}{2} \left(\frac{k_1 c_s}{\omega} \right)^2 - \dots \right] \tag{37}$$

By retaining various number of terms in Eq.(37), various order paraxial bc's can be derived. We will designate the paraxial bc's which correspond to keeping one and two terms in Eq.(37) by P1 and P2 respectively. When the correspondence rules

$$\begin{aligned} i\omega &\rightarrow \partial_t \\ ik_2 &\rightarrow \partial_2 \\ -ik_1 &\rightarrow \partial_1 \ , \end{aligned} \tag{38}$$

which can be obtained from Eq.(33), are used, P1 and P2 bc's can be obtained from Eq.(37) in space-time domain as

$$\begin{aligned} \text{P1}: \quad & \partial_2 u = \frac{1}{c_s} \dot{u} \\ \text{P2}: \quad & \partial_2 \dot{u} = \frac{1}{c_s} \ddot{u} - \frac{c_s}{2} \partial_{11} u \end{aligned} \tag{39}$$

When these bc's are generalized for an artificial boundary point having the outer normal axis n and tangential axis s (see Fig.5), they become

$$\text{P1}: \quad \partial_n u = -\frac{1}{c_s}\dot{u}$$
$$\text{P2}: \quad \partial_n \dot{u} = -\frac{1}{c_s}\ddot{u} + \frac{c_s}{2}\partial_{ss} u \tag{40}$$

We observe that P1 bc is identical with the viscous bc in Eq.(29). In view of Eqs.(36) and (37) we further see that paraxial bc's in Eqs.(40) are exact for SH wave components impinging the artificial boundary at right angles.

The wave number surfaces (WNS) as approximated by the paraxial bc's, Eqs.(39), are compared in Fig.9 with the exact WNS. This figure indicates that as the order of paraxial bc's increases the range of the incident angles α for which the approximation is valid also increases.

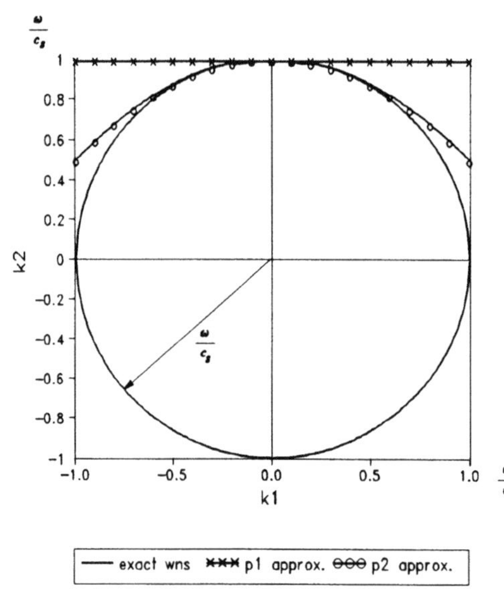

Figure 9 Approximations by paraxial bc's for out-of-plane motions

3.3.2 *Inplane motion (P+SV waves)*. We assume that the inplane motion is generated by P and SV waves as described in Fig.6. The artificial boundary plane (L) is defined by $x_2=0$.

In this case, the reflected P and SV waves are coupled. They decouple only when $\alpha=90°$. For this reason, the procedure which was used for out-of-plane case can not be employed for inplane case. In what follows the procedure adopted in [4] in establishing paraxial bc's for plane strain case will be reviewed.

For the inplane displacement components u_1 and u_2, one has an equation of the form of Eq.(33), where $k_1 \geq 0$, and we require that $k_2 > 0$ so that the artificial boundary L transmits waves without reflections. u_1 and u_2 should satisfy the displacement equations of motion, which in wave number-frequency space can be written in matrix form as

$$\left[\left(\frac{k_1}{\omega}\right)^2 \boldsymbol{A}_1 + \left(\frac{k_2}{\omega}\right)^2 \boldsymbol{A}_2 - \left(\frac{k_1}{\omega}\right)\left(\frac{k_2}{\omega}\right) \boldsymbol{A}_3 - \boldsymbol{I}\right] \boldsymbol{u}^* = \boldsymbol{0} \tag{41}$$

where $\boldsymbol{I}$ is (2×2) identity matrix, $\boldsymbol{u}=(u_1,u_2)^{\mathrm{T}}$, $\boldsymbol{u}^*$ is a vector describing Fourier amplitudes of u_1 and u_2 and

$$\boldsymbol{A}_1 = \begin{bmatrix} c_p^2 & 0 \\ 0 & c_s^2 \end{bmatrix}, \quad \boldsymbol{A}_2 = \begin{bmatrix} c_s^2 & 0 \\ 0 & c_p^2 \end{bmatrix}, \quad \boldsymbol{A}_3 = \left(c_p^2 - c_s^2\right) \begin{bmatrix} 0 & 1 \\ 1 & 0 \end{bmatrix}$$

In view of inspection of paraxial bc's obtained for out-of-plane case, Eqs.(39), we write paraxial bc's for inplane case in the form

$$\begin{aligned} &\mathrm{P1}: \quad \partial_2 \boldsymbol{u} = \boldsymbol{B}_0\, \dot{\boldsymbol{u}} \\ &\mathrm{P2}: \quad \partial_2 \dot{\boldsymbol{u}} = \boldsymbol{C}_0\, \ddot{\boldsymbol{u}} + \boldsymbol{C}_1\, \partial_{11} \boldsymbol{u} + \boldsymbol{C}_2\, \partial_1 \dot{\boldsymbol{u}} \end{aligned} \tag{42}$$

which, using Eq.(33), can be written in Fourier space as

$$\begin{aligned} &\mathrm{P1}: \quad \left(\frac{k_2}{\omega}\right) \boldsymbol{u}^* = \boldsymbol{B}_0\, \boldsymbol{u}^* \\ &\mathrm{P2}: \quad \left(\frac{k_2}{\omega}\right) \boldsymbol{u}^* = \boldsymbol{C}_0\, \boldsymbol{u}^* + \left(\frac{k_1}{\omega}\right)^2 \boldsymbol{C}_1\, \boldsymbol{u}^* - \left(\frac{k_1}{\omega}\right) \boldsymbol{C}_2\, \boldsymbol{u}^* \end{aligned} \tag{43}$$

where $\boldsymbol{B}_0$, $\boldsymbol{C}_0$, $\boldsymbol{C}_1$, $\boldsymbol{C}_2$ are some (2×2) matrices which are to be determined. To this end, we require that paraxial bc's, Eqs.(43), satisfy equations of motion, Eq.(41), which leads to, after some manipulations,

$$\begin{aligned} &\mathrm{P1}: \quad \boldsymbol{A}_2\, \boldsymbol{B}_0^2 - \boldsymbol{I} = \boldsymbol{0} \\ &\mathrm{P2}: \quad \boldsymbol{A}_2\, \boldsymbol{C}_0^2 - \boldsymbol{I} = \boldsymbol{0} \\ &\qquad\quad \boldsymbol{A}_2 \left(\boldsymbol{C}_0\, \boldsymbol{C}_2 + \boldsymbol{C}_2\, \boldsymbol{C}_0\right) + \boldsymbol{A}_3\, \boldsymbol{C}_0 = \boldsymbol{0} \\ &\qquad\quad \boldsymbol{A}_1 + \boldsymbol{A}_2 \left(\boldsymbol{C}_0\, \boldsymbol{C}_1 + \boldsymbol{C}_1\, \boldsymbol{C}_0 + \boldsymbol{C}_2^2\right) + \boldsymbol{A}_3\, \boldsymbol{C}_2 = \boldsymbol{0} \end{aligned}$$

from which $\boldsymbol{B}_0$ and $\boldsymbol{C}_i$ matrices can be determined as

$$B_0 = C_0 = \begin{bmatrix} \frac{1}{c_s} & 0 \\ 0 & \frac{1}{c_p} \end{bmatrix}$$

$$C_2 = (c_s - c_p) \begin{bmatrix} 0 & \frac{1}{c_s} \\ \frac{1}{c_p} & 0 \end{bmatrix} \tag{44}$$

$$C_1 = \frac{1}{2} \begin{bmatrix} (c_s - 2c_p) & 0 \\ 0 & (c_p - 2c_s) \end{bmatrix}$$

It may be noted that the matrix C_2 in Eqs.(42) represents the coupling between dilatational and shear waves.

The approximations implied by paraxial bc's, Eqs.(42), with regard to their transmitting inplane waves through artificial boundaries are described in Fig.10 in wave number space for the Poisson ratio $\nu = 1/3$.

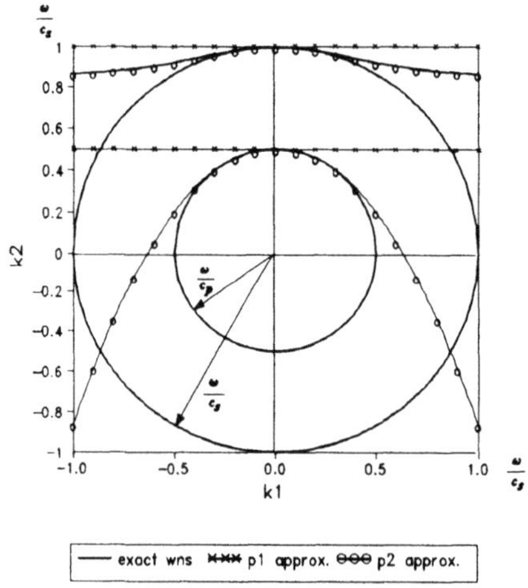

Figure 10 Approximations by paraxial bc's for inplane motions (Poisson's ratio = 1/3)

Paraxial bc's in Eqs.(42) can be generalized for an artificial boundary point having the outer normal axis n and tangential axis s (see Fig.5) as

$$\begin{aligned} \mathrm{P1}: &\quad \partial_n \boldsymbol{u} = -\boldsymbol{B}_0 \dot{\boldsymbol{u}} \\ \mathrm{P2}: &\quad \partial_n \dot{\boldsymbol{u}} = -\boldsymbol{C}_0 \ddot{\boldsymbol{u}} - \boldsymbol{C}_1 \partial_{ss} \boldsymbol{u} - \boldsymbol{C}_2 \partial_s \dot{\boldsymbol{u}} \end{aligned} \tag{45}$$

where $\boldsymbol{u}$ is a vector composed of normal and tangential displacement components, i.e., $\boldsymbol{u}=(u_n,u_s)^T$.

3.4. EXTRAPOLATION ALGORITHM (EA) FOR ARTIFICIAL BOUNDARIES

First, we explain this algorithm for one dimensional case, and for that we refer to Fig.3 which shows the characteristic plane for one dimensional waves propagating in x direction. Suppose that we obtain the solution numerically by using the time increment Δt and spatial increment Δx which are related by $\Delta x = c\ \Delta t$, where c is wave velocity.Further, suppose that the right boundary $x=L$ (B2) is artificial boundary. The outgoing waves coming to this boundary propagate along the characteristic lines $x\text{-}ct=$ *const*. We consider one member of this characteristic family which intersects the artificial boundary B2 at the point P (this line is designated by F in Fig.3). To transmit the wave through the point P without reflections, we should require that the contribution to the motion at P comes only from the points located along the characteristic line F. In EA [1], we approximate the motion at P by extrapolating the values at the points 1, 2, 3, etc. of the line F; if these points are equally spaced and the polynomial approximation is used, this leads to

$$u(\mathrm{P}) = \sum_{j=1}^{n} (-1)^{j+1}\ C_j^n\ u^j \tag{46}$$

where C_j^n are Binomial coefficients defined by

$$C_j^n = \frac{n!}{(n-j)!\ j!}\ , \tag{47}$$

n designates the number of points used in EA and u^j is the displacement value at the point j, i.e.,

$$u^j = u(x - j\Delta x,\ t - j\Delta t) \tag{48}$$

In the last equation, $(x,\ t)$ are the coordinates of the boundary point P in the $(x\text{-}t)$ plane (see Fig.3). The EA formula, Eq.(46), for $n=1,2,3$ yields

$$\begin{aligned} n=1: &\quad u(\mathrm{P}) = u^1 \\ n=2: &\quad u(\mathrm{P}) = 2u^1 - u^2 \\ n=3: &\quad u(\mathrm{P}) = 3u^1 - 3u^2 + u^3 \end{aligned} \tag{49}$$

We now generalize EA for three dimensional case. Before we do that, it may be noted that the wave velocities in characteristic space correspond to asymptotic phase (apparent) velocities associated with infinite wave number or frequency. These asymptotic velocities are equal to either shear or Rayleigh or Stoneley wave velocity depending upon geometric properties of the medium [18]. It may be further noted that the velocities of Rayleigh and Stoneley waves are close to that of shear wave velocity. In view of these observations, the use of c_s is suggested in EA for three dimensional case [1].

We now turn our attention to writing the generalized EA. We consider a point P on the artificial boundary L having the normal axis n and tangential axis s (see Fig.5). We extrapolate the motion at P by using the shear velocity c_s and the points 1,2,3, etc. located along the normal axis n. If the points are equally spaced and the time increment Δt is constant, then EA formulas for three dimensional case can be written, in view of Eq.(46), as

$$\boldsymbol{u}(\mathrm{P},t) = \sum_{j=1}^{n} (-1)^{j+1} \, C_j^n \, \boldsymbol{u}^j \tag{50}$$

where $\boldsymbol{u}$ is displacement vector and $\boldsymbol{u}^j$ is defined by

$$\boldsymbol{u}^j = \boldsymbol{u}\,(\boldsymbol{x}^j\,, t - j\,\Delta t) \tag{51}$$

in which $\boldsymbol{x}^j$ describes the position of the point j. It may be noted that the spatial increment Δn along the normal axis n (see Fig.5) is related to the time increment Δt by $\Delta n = c_s\ \Delta t$.

3.5. SUPERPOSITION BC'S

We first discuss superposition bc's for one dimensional case. To this end we refer to Fig.3 which shows the characteristic plane for the waves propagating in x direction. We suppose that the right boundary $x=L$ (B2) is an artificial boundary. Further, we suppose that we obtain the solution numerically using the increments Δx and Δt which are related by $\Delta x = c\ \Delta t$.

Consider a particular point on B2, say, P in Fig.3. From the discussions presented in the previous chapter, we know that the displacement wave reflects with the same sign when B2 is free and with the opposite sign when B2 is fixed. Thus, obtaining two solutions at each time step, one using fixed bc's and the other using free bc's on B2, and averaging them will cancel approximately the reflected waves at B2. We say "approximately" because we assume that we obtain the solution numerically. Since we do the averaging at every time step, the reflections are eliminated as they occur [1], which, in turn, prevents higher order reflections. It may be noted that as the solutions for two types of bc's (at each time step) will be the same in interior region, the computations can be shortened by introducing a boundary zone adjacent to the artificial boundary $x=L$ and obtaining the two different solutions only in that zone.

Superposition bc's is extended to multi dimensional case by W.D. Smith [10]. To explain the numerical implementation of that study we refer to Fig.5 and consider an artificial boundary point P with the normal axis n and two orthogonal tangential axes s_1 and s_2. The use of three dimensional superposition bc's involves, at each time step, the computation of two solutions using bc's

$$\begin{aligned} &\textit{for solution } 1: \\ &u_n = 0\,,\ \tau_{ns_1} = 0\,,\ \tau_{ns_2} = 0 \\ &\textit{for solution } 2: \\ &u_{s_1} = 0\,,\ u_{s_2} = 0\,,\ \tau_{nn} = 0 \end{aligned} \tag{52}$$

at the artificial boundary points and averaging them. In Eqs.(52), (u_n, u_{s_1}, u_{s_2}) and (τ_{nn}, τ_{ns_1}, τ_{ns_2}) designate respectively the displacement and traction vector components in n, s_1 and s_2 directions. In [10] it is pointed out that superposition bc's work not only for body but also for surface waves.

3.6. HORIZONTAL WAVE (HW) BC'S SUITABLE FOR FINITE ELEMENT ANALYSIS

3.6.1. *General Considerations*. We first explain the reasons for using HW bc's in soil-structure interaction analysis. To this end, we consider the system in Fig.11, which consists of a structure and an elastic slab of depth H representing the soil. The slab rests on a rigid bedrock. We assume that the response of the system is governed mainly by a motion generated by waves propagating in horizontal direction. We will call these waves generalized surface waves in the sense that the medium in which they propagate is not semi-infinite, instead, it is a slab of finite depth, implying that the wave motion will be effected by the reflections at both upper and lower faces of the slab. Suppose that we wish to analyze the soil-structure system by using the finite element method (FEM). For that, we truncate the system along the vertical planes L and R located respectively on the left and right sides of the structure (see Fig.11) and perform the analysis in the near field which consists of the structure and the soil region between L and R. With regard to locations of L and R we have two options. In the first option, we keep L and R sufficiently away from the structure so that the analysis in the vicinity of the structure are not effected much by the types of bc's used on L and R; but, in this case, the size of the finite element mesh would be very large, thus the computational load would be extremely high. The second option involves keeping L and R close to the structure; but, in this case we use some special bc's on L and R capable of transmitting waves propagating in horizontal direction. The use of these special bc's (which we call HW bc's) in soil-structure analysis would reduce the size of the near field, and thus, the computational load considerably.

In what follows, we present the derivation of HW bc's for out-of-plane motion, and as the procedure is similar, we refer the reader to the reference 15 for HW bc's for inplane motion. These two bc's will be termed respectively as generalized Love and Rayleigh wave bc's in this study.

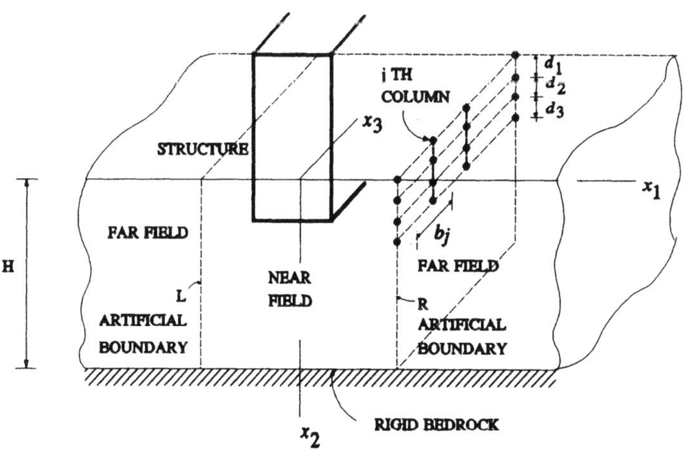

Figure 11 Description of artificial boundaries for surface waves

3.6.2. *Generalized Love Wave bc's*. Love wave bc's were first established by J. Lysmer and G. Waas in [6] in a form suitable for finite element analysis. Here, we derive these bc's by using shorter and somewhat different formulation than the one employed in [6]. For the derivation, we refer to Fig.12 which shows the free field system (FFS) for the soil-structure interaction problem described in Fig.11. FFS is referred to the cartesian coordinate system shown in the figure in which x_1 axis is oriented in horizontal direction and x_2 axis is directed downwards. For the generalized Love waves propagating in x_1 direction, one has

$$u = u^* \,(x_2, k_1, \omega)\, e^{i(\omega t - k_1 x_1)} \tag{53}$$

where u is the out-of-plane displacement component, i.e., $u \perp (x_1\text{-}x_2)$ and u^* designates the Fourier amplitude of u. Not to complicate the notation, if there is no ambiguity, we will drop asterisk in u^* and designate it simply by u.

In view of Eq.(53), the equation governing u in Fourier space becomes

$$\partial_{22} u - k_1^2 u + \frac{\omega^2}{c_s^2} u = 0 \tag{54}$$

and the shear stress τ_{13}, τ_{23} (see Fig.13) are related to u by

$$\begin{aligned} \tau_{13} &= -ik_1 \mu u \\ \tau_{23} &= \mu \partial_2 u \end{aligned} \tag{55}$$

As the Love wave bc's remain the same for both right and left boundaries, R and L, below we present the derivation only for the right boundary R. To this end, we subdivide the slab in

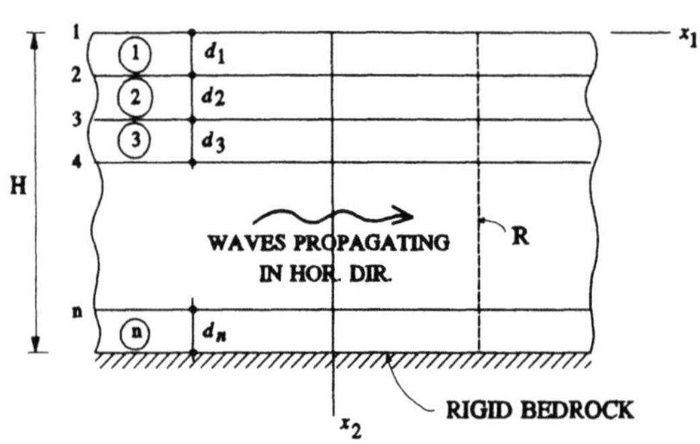

Figure 12 Surface waves in free field

Fig.12 into thin layers so that the planes used in the layering pass through the nodes at R, which are a part of the mesh employed in near filed. The number of thin layers is designated by n in the figure. We now consider a typical thin layer of the thickness d shown in Fig.13. The forces and displacements at the top face T and lower face B of the thin layer (in x_3 direction) are designated respectively by (F_1, u_1) and (F_2, u_2) in the figure. The element forces

F_1 and F_2 are defined per unit area of the faces T and B, and thus they are related to the shear stress component τ_{23} by

$$\begin{aligned} F_1 &= -\tau_{23} \quad on \quad T \\ F_2 &= +\tau_{23} \quad on \quad B \end{aligned} \tag{56}$$

The element equations for the thin layer relating F_i to u_i can be obtained by integrating Eq.(54) and using Eqs.(55) and (56). They can be written in matrix form as [1]

$$\underbrace{\begin{bmatrix} F_1 \\ F_2 \end{bmatrix}}_{\boldsymbol{F}} = \underbrace{\frac{k_2 \mu}{\sin k_2 d} \begin{bmatrix} \cos k_2 d & -1 \\ -1 & \cos k_2 d \end{bmatrix}}_{\boldsymbol{D}} \underbrace{\begin{bmatrix} u_1 \\ u_2 \end{bmatrix}}_{\boldsymbol{u}} \tag{57}$$

where k_2 is the wave number for x_2 direction, which, in view of Eq.(13), satisfies

$$k_1^2 + k_2^2 - \frac{\omega^2}{c_s^2} = 0 \tag{58}$$

and the matrix $\boldsymbol{D}$ in Eq.(57) represents the dynamic stiffness matrix for the thin layer.

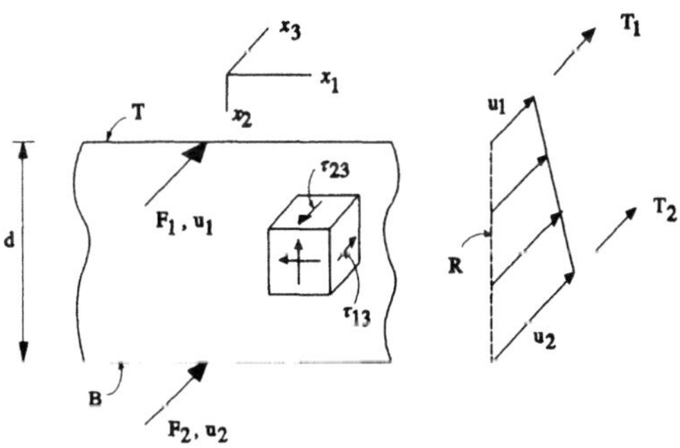

Figure 13 A typical layer element

At this point, since the layers are assumed to be thin, we expand sin and cos terms in Eq.(57) in terms of their arguments using Taylor's formula and retain first two terms in the expansions. This, in view of Eq.(58), leads to, the following approximation for $\boldsymbol{D}$:

$$\boldsymbol{D} \cong \boldsymbol{K}_e - \omega^2 \boldsymbol{M}_e \tag{59}$$

with

$$\boldsymbol{K}_e = \frac{\mu}{d}\begin{bmatrix} 1+\dfrac{(k_1 d)^2}{3} & -1+\dfrac{(k_1 d)^2}{6} \\ -1+\dfrac{(k_1 d)^2}{6} & 1+\dfrac{(k_1 d)^2}{3} \end{bmatrix} \tag{60}$$

$$\boldsymbol{M}_e = \rho\, d \begin{bmatrix} \dfrac{1}{3} & \dfrac{1}{6} \\ \dfrac{1}{6} & \dfrac{1}{3} \end{bmatrix} \tag{61}$$

In Eq.(59), $\boldsymbol{M}_e$ represents consistent element mass matrix. $\boldsymbol{K}_e$ can be interpreted as generalized element stiffness matrix. We call it generalized since it contains the wave number k_1.

Governing equations for the layered system in Fig.12 can be obtained by assembling the element equations in Eq.(57) and by using Eqs.(60) and (61). When the rigid base condition at $x_2=H$ is taken into account, this yields a system of equations of the form

$$(\boldsymbol{A} + k_1^2 \boldsymbol{B} - \omega^2 \boldsymbol{M})\, \boldsymbol{u} = \boldsymbol{0} \tag{62}$$

where $\boldsymbol{u}=(u_1\ u_2\ \dots\ u_n)^T$ is n dimensional displacement vector, and $\boldsymbol{M}$ (system mass matrix), $\boldsymbol{A}$ and $\boldsymbol{B}$ are some constant $(n \times n)$ matrices. From Eqs.(60) and (61) it follows that $\boldsymbol{A}$, $\boldsymbol{B}$ and $\boldsymbol{M}$ are all symmetric and tridiagonal. The right hand side of Eq.(62) is zero because no applied forces exist in the layered system.

For a given ω, Eq.(62) represents an eigenvalue problem of the form

$$(\boldsymbol{C} - \lambda \boldsymbol{B})\, \boldsymbol{u} = \boldsymbol{0} \tag{63}$$

in which $\lambda=k_1^2$ is eigenvalue and $\boldsymbol{C}$ is defined by

$$\boldsymbol{C} = \omega^2 \boldsymbol{M} - \boldsymbol{A} \tag{64}$$

Since $\boldsymbol{C}$ and $\boldsymbol{B}$ matrices in Eq.(63) are real and symmetric, it follows that eigenvalues would be all real and eigenvectors would be orthogonal with respect to $\boldsymbol{C}$ and $\boldsymbol{B}$ matrices. Let the eigenvectors of Eq.(63) be $(\boldsymbol{V}_1, \dots, \boldsymbol{V}_n)$ and be normalized with respect to $\boldsymbol{B}$ matrix. Then, the general solution of Eq.(63) can be expressed in terms of eigenvectors as

$$\boldsymbol{u} = \sum_{i=1}^{n} \alpha_i \boldsymbol{V}_i = \boldsymbol{V}\boldsymbol{\alpha} \tag{65}$$

where $\boldsymbol{V}$ is the modal matrix whose columns are composed of eigenvectors and $\boldsymbol{\alpha}$ is modal participation vector which can be related to $\boldsymbol{u}$, in view of orthogonality conditions, by

$$\alpha = V^T B\, u \tag{66}$$

We now turn our attention to deriving transmitting bc's for the artificial boundary R. We note that the interaction forces acting on this boundary are caused by the shear stress component τ_{13}. For a typical thin layer shown in Fig.13, these interaction forces are designated by T_1 and T_2 (whose positive directions are taken to be in x_3 direction). At this point we assume that the displacement varies linearly over the thickness d of the thin layer. Thus, in view of the first of Eqs.(55), the distribution of τ_{13} would also be linear over d. When we lump these stresses at the upper and lower levels of the layer using virtual work principle, for the interaction forces T_1 and T_2 we can write, using the first of Eqs.(55),

$$\begin{bmatrix} T_1 \\ T_2 \end{bmatrix} = -ik_1 \mu d \begin{bmatrix} \frac{1}{3} & \frac{1}{6} \\ \frac{1}{6} & \frac{1}{3} \end{bmatrix} \begin{bmatrix} u_1 \\ u_2 \end{bmatrix} \tag{67}$$

If we assemble these equations for the layered shown in Fig.12 and take into account the rigid base conditions at $x_2=H$, we get

$$T = -ik_1\, B\, u \tag{68}$$

where $T=(T_1, \ldots , T_n)^T$ describe the interaction forces exerted by far field at the nodes of near field on R, $u=(u_1, \ldots , u_n)^T$ and B is an $(n\times n)$ matrix which was defined previously in Eq.(62).

The wave number k_1 in Eq.(68) is related to the eigenvalue of the eigenvalue problem in Eq.(63) by

$$k_1 = \mp \sqrt{\lambda} \tag{69}$$

In order to have the artificial boundary R to transmit the waves from near to far field without reflections, we should require that k_1 be either positive if it is real or have negative imaginary part if it is imaginary. In view of Eq.(69), this yields

$$\begin{aligned} k_1 &= +\sqrt{\lambda} && \textit{when}\ \ \lambda > 0 \\ k_1 &= -i\sqrt{|\lambda|} && \textit{when}\ \ \lambda < 0 \end{aligned} \tag{70}$$

Let k_1^j $(j=1, \ldots, n)$ be the value of k_1 corresponding to the eigenvalue λ_j, which is determined using Eqs.(70). Let V_j be the mode shape (eigen) vector associated with the eigenvalue λ_j. Then, Eq.(68) can be written for the jth as

$$T^j = -ik_1^j\, \alpha_j\, B\, V_j \tag{71}$$

where T^j and α_j designate respectively the interaction force contribution and modal participation factor of the jth mode. When Eq.(71) is combined for all of the modes, one gets

$$\boldsymbol{T} = -\boldsymbol{B}\ \boldsymbol{V}\ \boldsymbol{H}\ \boldsymbol{\alpha} \tag{72}$$

where $\boldsymbol{H}$ is a diagonal matrix of the form

$$\boldsymbol{H} = diag\,(ik_1^j) \quad (j=1, \ldots, n) \tag{73}$$

Elimination of α in Eq.(72) by using Eq.(66) finally gives

$$\boldsymbol{T} + \boldsymbol{Q}\ \boldsymbol{u} = \boldsymbol{0} \tag{74}$$

where $\boldsymbol{Q}$ is a symmetric matrix defined by

$$\boldsymbol{Q} = \boldsymbol{B}\ \boldsymbol{V}\ \boldsymbol{H}\ \boldsymbol{V}^T\ \boldsymbol{B} \tag{75}$$

Eq.(74) together with Eq.(75) constitutes the generalized Love wave bc's for the artificial boundary R. The interaction forces $\boldsymbol{T}$ in Eq.(74) are defined per unit length in x_3 direction (see Fig.11). As mentioned previously, Love wave bc's for the left artificial boundary L are again given by Eq.(74) with Eq.(75).

In the presence of free field motion, the artificial bc's, Eq.(74), should be modified, in view of Eq.(20), as

$$\boldsymbol{T} - \boldsymbol{T}^f + \boldsymbol{Q}\,(\boldsymbol{u} - \boldsymbol{u}^f) = \boldsymbol{0} \tag{76}$$

where $\boldsymbol{T}^f$ and $\boldsymbol{u}^f$ are respectively free field forces and displacements on the boundaries L and R, $\boldsymbol{u}$ denotes total displacements.

The bc's in Eq.(76) are derived for generalized Love waves. Using a similar procedure, an equation having the same form as that of Eq.(76) can be obtained for the bc's associated with generalized Rayleigh waves [15]. In the discussions which will follow, the matrix $\boldsymbol{Q}$ in Eq.(75) will be designated by $\boldsymbol{S}$ for Rayleigh wave bc's.

3.6.3. *Finite Element Equations for Near Field.* For the sake of completeness, we now present the finite element equations for the near field of the system shown in Fig.11. Using Eq.(76) they can be written in frequency domain as, in the absence of material damping,

$$(-\omega^2 \boldsymbol{M} + \boldsymbol{E} + \boldsymbol{K})\,\boldsymbol{u} = \boldsymbol{F} \tag{77}$$

where $\boldsymbol{K}$ and $\boldsymbol{M}$ are system stiffness and mass matrices (of the near field), and $\boldsymbol{E}$ designates the contribution coming to the coefficient matrix from the artificial bc's on L and R, which can be defined by

$$\boldsymbol{E} = \sum_{j=1}^{m} b_j\,(\boldsymbol{Q}_j + \boldsymbol{S}_j) \tag{78}$$

where $\boldsymbol{Q}$ and $\boldsymbol{S}$ are the matrices appearing respectively in Love and Rayleigh wave bc's. Each term in the summation in Eq.(78) represents the contribution coming from the nodes lying on a different vertical column on L or R (see Fig.11); b_j denotes the average width for the jth vertical column; and m designates the total number of vertical columns on L and R. The force vector $\boldsymbol{F}$ in Eq.(77) is defined by

$$\boldsymbol{F} = \boldsymbol{F}^f + \boldsymbol{E}\ \boldsymbol{u}^f$$

where $\boldsymbol{F}^f$ describes the forces acting at the nodes of L and R caused by the free field motion.

In [3,16] it is pointed out that the use of Love and Rayleigh wave bc's, instead of viscous bc's, on the vertical boundaries L and R enables one to put these boundaries very close to the structure, thus to reduce the size of the near field and the computational load.

3.6.4. *The Types of Absorbing bc's Used In FLUSH.* Fig.14 shows the 2-D (plane strain) model used in FLUSH [3] to simulate 3-D effects in soil-structure analysis. "b" in the figure designates the width of the slice of the 2-D model considered in the analysis. The model is equipped with two types bc's:

a) Shear wave viscous bc's are used at the nodes of the front face (F) and back face (B) of the plane strain slice (see Fig.14). These bc's simulate the radiation of energy in x_3 direction (i.e., in the direction perpendicular to the plane of 2-D model) from the slice to the surrounding medium; thus, they approximately account for 3-D effects.

b) Generalized Rayleigh wave bc's are used at the nodes of the lateral vertical boundaries L and R. These bc's describe the energy radiation in inplane horizontal (x_1) direction.

In FLUSH, it is assumed that the model rests on a rigid base (bedrock). However, if desired, the flexibility of the half-space underneath the model can be taken into account by using viscous bc's at the base [16]. These viscous bc's describe the energy radiation from the model to the underlying half-space.

In [16], the use of either generalized Rayleigh wave or viscous bc's is suggested for the lateral vertical boundaries L and R. However, as stated in the previous section, the size of the finite element model associated with Rayleigh bc's is considerably smaller than that corresponding to viscous bc's. This explains the reason for the preference made in FLUSH for the type of absorbing bc's on L and R.

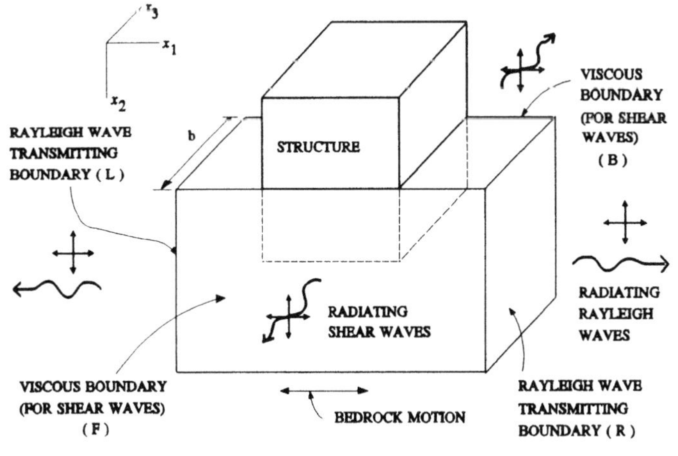

Figure 14 The soil-structure model used in FLUSH

4. Comparison of various artificial bc's

In this chapter, with the object of comparing the performance of various bc's and studying the influence of the location of artificial boundaries, we study some sample problems.

4.1. HALF SPACE PROBLEM

Here, we consider an elastic half space which is referred to an x_i Cartesian coordinate system in which the (x_1-x_3) plane coincides with the top surface of the half space (see Fig.15). The half space is excited in out-of-plane direction by subjecting it to shear stresses on the top surface. The spatial and time variations of the applied stress are trapezoidal having the maximum intensity s_o. A plane horizontal artificial boundary is considered at the distance d from the top surface, i.e., at $x_2=d$.

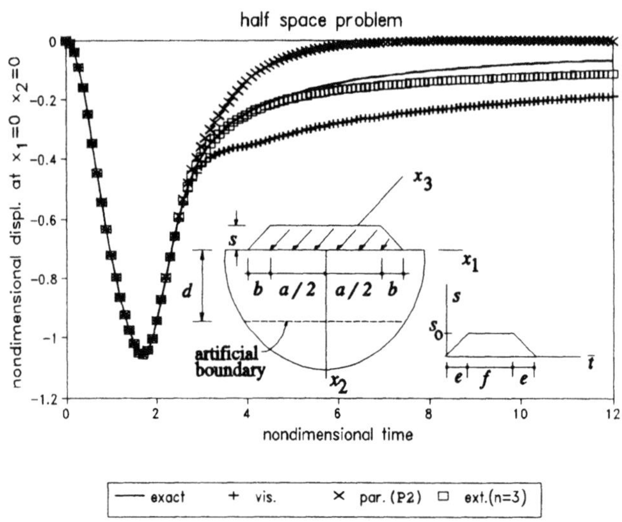

Figure 15 Comparison of various artificial bc's for the displacement at $x_1=0$, $x_2=0$ (half space problem, $\bar{d}=1$)

This problem is solved by a special technique proposed in [19], which involves the use of Fourier transform for the coordinate x_1 and the method of characteristics. The results obtained for various types of artificial bc's are presented in Figs.(15-17). The nondimensional quantities appearing in the figures are defined by

$$\bar{u} = \frac{\mu}{s_o a} u \quad ; \quad \bar{\tau}_{13} = \frac{\tau_{13}}{s_o} \quad ; \quad \bar{t} = \frac{a}{c_s} \quad ; \quad \bar{d} = \frac{d}{a} \tag{79}$$

where bar designates the nondimensional value of a quantity and a is defined in Fig.15. The results are obtained by using the spatial increments $\Delta x_2=0.05a$ and $\Delta x_1=0.025a$ and

$$\frac{b}{a} = 0.5 \quad ; \quad f = 1 \quad ; \quad e = 0.5 \tag{80}$$

It may be noted that the time increment Δt is related to Δx_2 by $\Delta t = \Delta x_2 / c_s$ and that the parameters a, b, e and f appearing in Eqs.(80) define the spatial and time variations of the applied shear stress (see Fig.15).

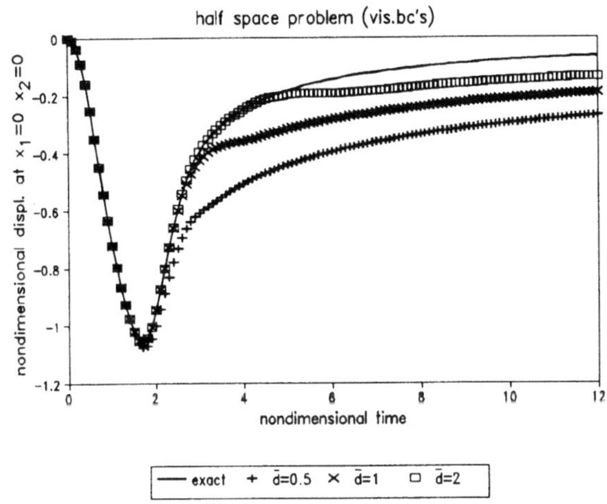

Figure 16 The influence of the location of artificial boundary on the response for viscous bc's (half space problem)

The results in Fig.15 are obtained by putting the artificial boundary at the depth $d = a$ and by using various artificial bc's. Form the comparison of the curves in the figure showing time variation of the displacement at the top surface, we see that extrapolation algorithm with $n = 3$ gives the best result, but the results obtained by viscous and paraxial bc's are also acceptable.

Figs.16 and 17 pertain to the comparisons of the results obtained by putting the artificial boundary at various distances from the source. The study of the figures indicates that the curves corresponding to viscous and paraxial bc's approach the exact one from different sides as the depth of the artificial boundary increases. The figures may also suggest that the artificial boundary may function well when it is put at a depth which is approximately equal to the width of the applied strip load.

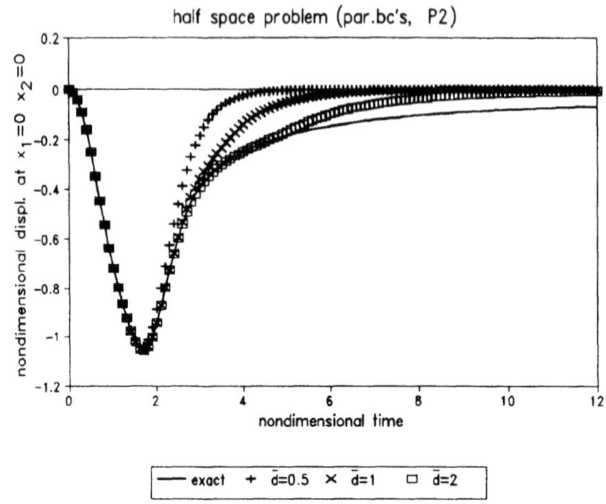

Figure 17 The influence of the location of artificial boundary on the response for paraxial bc's (P2) (half space problem)

4.2. LAYER PROBLEM

This problem involves an elastic slab of depth H resting on a rigid bedrock (see Fig.18). The slab is excited by out-of-plane shear stresses acting on the top surface of the slab. Since the coordinate system, and the spatial and time variations of the applied shear stresses are the same as those used for the half space problem, they will not be repeated here. Two vertical artificial boundary planes are put symmetrically on both sides of the applied strip load at the distance d from the $(x_2\text{-}x_3)$ plane.

The governing equations are integrated again by using the technique which was used for the half space problem. The results are presented in Fig.18. The nondimensional quantities appearing in the figure are defined by

$$\bar{u} = \frac{\mu}{s_o H} u \quad ; \quad \bar{t} = \frac{H}{c_s} \quad ; \quad \bar{d} = \frac{d}{H} \tag{81}$$

The results are obtained by using the following values of the load parameter (for the definitions of these parameters, see Fig.18):

$$\frac{a}{H} = 0.4 \quad ; \quad \frac{b}{H} = 0.05 \quad ; \quad f = 1 \quad ; \quad e = 0.5 \tag{82}$$

In numerical computations, the increments $\Delta x_1 = 0.025H$, $\Delta x_2 = 0.05H$ and $\Delta t = \Delta x_2 / c_s$ are used. The distance of the artificial boundaries from the $(x_2\text{-}x_3)$ plane is chosen as $d = 2H$.

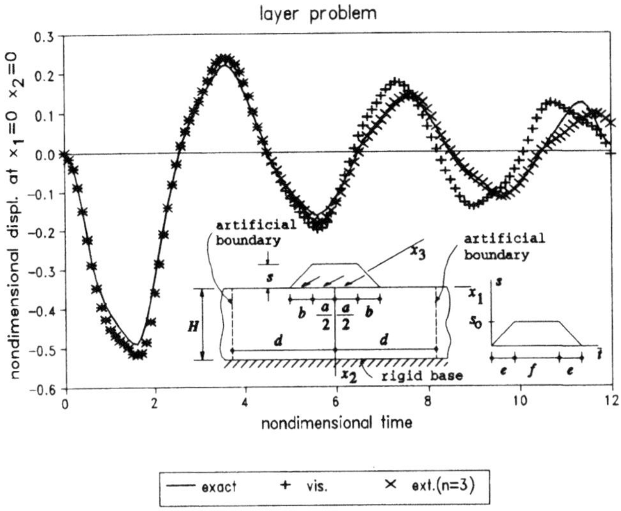

Figure 18 Comparison of various artificial bc's for the displacement at $x_1=0$, $x_2=0$ (layer problem, $\bar{d}=2$)

In Fig.18, the exact time variation of the displacement at the top surface is compared with the approximate one which is obtained by using viscous and extrapolation bc's. In the figure, the influence of energy radiation (due to waves propagating in horizontal direction) is apparent, which causes a decay in the amplitudes of the oscillations as time increases. From the figure we see that extrapolation algorithm with $n=3$ functions better than viscous bc's and that both of the bc's perform well in transmitting the waves through artificial boundaries.

4.3. COMPLIANCE AND INPUT MOTION FUNCTIONS FOR A RIGID STRIP FOUNDATION

Here, we study the performance of viscous bc's in the determination of compliance and input motion functions for rigid foundations. For this purpose we consider a rigid strip foundation of width $2b$ welded to an elastic half space (see Fig.19). The system is referred to a Cartesian x_i coordinate system in which (x_1-x_3) plane coincides with the top surface of the half space, x_3 axis lies along the mid-line of the strip foundation and x_2 axis is directed downwards. The forces acting on the top of the foundation in x_1 and x_2 directions are designated respectively by F_1 and F_2 in the figure, and the moment about x_3 axis, by M_3 (in Fourier space).

Seismic environment is described by inclined plane P and SV waves propagating in (x_1-x_2) plane as shown in Fig.19. The inclination angle α of the incident wave (IW) is related to those (β,γ) of reflected waves by

$$\frac{c}{\cos\alpha} = \frac{c_p}{\cos\beta} = \frac{c_s}{\cos\gamma} \tag{83}$$

where $c=c_p$ when IW is P wave and $c=c_s$ when IW is SV wave. For the surface motion, in the absence of rigid foundation, one can write

$$u^g = U^g e^{i\omega(t - \frac{x_1}{c_1})} \tag{84}$$

where u^g is the displacement distribution on the top surface, c_1 is the apparent (phase) wave velocity in x_1 direction defined by

$$c_1 = \frac{c}{\cos\alpha} \tag{85}$$

$U^g = (U_1{}^g,\ U_2{}^g)^T$ in Eq.(84) are the Fourier amplitudes of the free surface motion. These amplitudes can be related to the amplitude of IW so that free traction bc's on the top surface of the half space are satisfied.

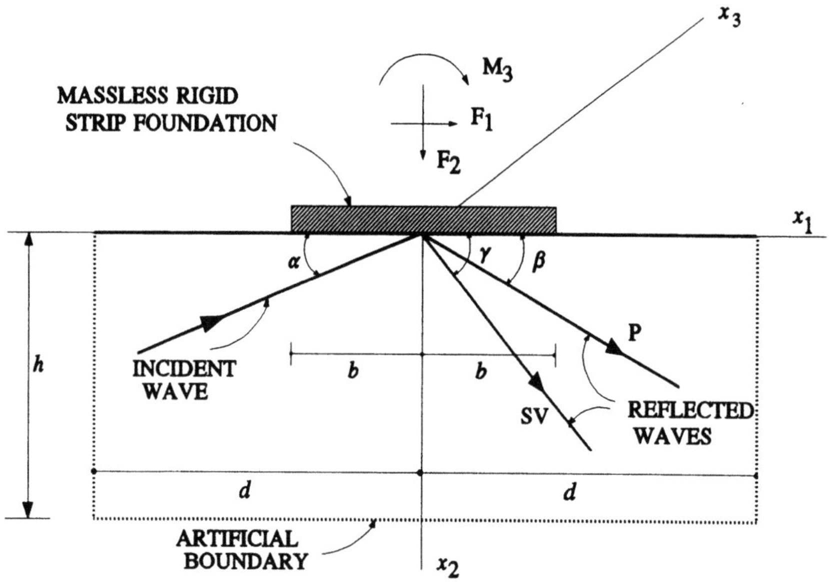

Figure 19 Strip foundation under the influence of body waves

Through elastodynamic analysis, for the interaction forces $F = (F_1, F_2, M_3)^T$ one can obtain the equation

$$F = K\left(U - S\,U^g\right) \tag{86}$$

in frequency space. Here, $U = (U_1, U_2, \alpha_3)^T$ designate translations and rotation of the rigid strip foundation, K is the impedance matrix having the form

$$K = \begin{bmatrix} K_{11} & 0 & K_{13} \\ 0 & K_{22} & 0 \\ K_{13} & 0 & K_{33} \end{bmatrix} \tag{87}$$

and S is the input motion matrix of the form

$$S = \begin{bmatrix} S_{11} & S_{12} \\ S_{21} & S_{22} \\ S_{31} & S_{32} \end{bmatrix} \tag{88}$$

Eq.(86) relates the interaction forces F to the translations and rotation of the foundation and to the free surface displacements. It may be noted that the compliance matrix C of the foundation is related to its impedance matrix by

$$C = K^{-1}$$

From Eq.(86) we see that the matrix S, in the absence of interaction forces, determines the translations and rotation of the rigid foundation in terms of input displacement amplitudes by

$$U = S\, U^g \tag{89}$$

The elements of the matrices K and S can be established by solving some elastodynamic problems subject to some mixed boundary conditions.

We now present some numerical examples. In the first example, we determine the vertical compliance $C_{22}=K_{22}^{-1}$ by using BEM and viscous bc's along the artificial boundary indicated in Fig.19 and compare it with the exact results taken from [22]. In obtaining the results the Poisson ratio is chosen to be 0.25 and the depth of the horizontal artificial boundary is taken to be $h=2b$. The distance d of the vertical artificial boundaries from (x_2-x_3) plane is kept varied. The results are shown for the real part of vertical compliance function in Fig.20. The nondimensional quantities appearing in the figure are defined by

$$\bar{\omega} = \frac{\omega b}{c_s} \;\; ; \;\; \bar{d} = \frac{d}{b} \;\; ; \;\; \bar{C}_{22} = \pi \mu\, C_{22} \tag{90}$$

From Fig.20 we see that the results improve as the frequency increases, which is consistent with the fact that viscous bc's are valid for high frequencies (see Chapter 3, the section on viscous bc's). The figure also shows that as the distance of the vertical artificial boundaries from the source becomes larger, the results get somewhat better.

In the second example, we determine the input motion coefficient S_{22} by using BEM and viscous bc's along the artificial boundaries indicated in Fig.19 and compare the results with the exact ones. The comparison is made for two different values of the incident angle α, namely, for $\alpha=0°$, 45°. The results, which are obtained for Poisson's ratio$=0.25$, $h/b=2$, $d/b=4$, are presented in Figs.21 and 22 for the real and imaginary parts of S_{22} respectively. The nondimensional frequency in the figures is again defined in Eq.(90). The figures show that the comparison is satisfactory except for lower frequencies for which viscous bc's are not valid.

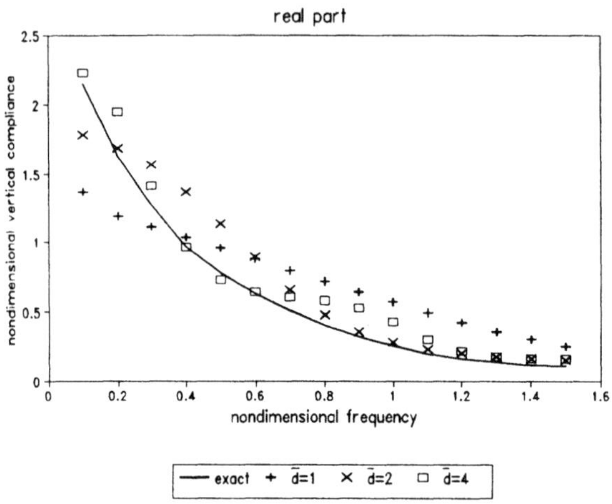

Figure 20 Real part of vertical compliance function for rigid foundation (Poisson's ratio= 0.25)

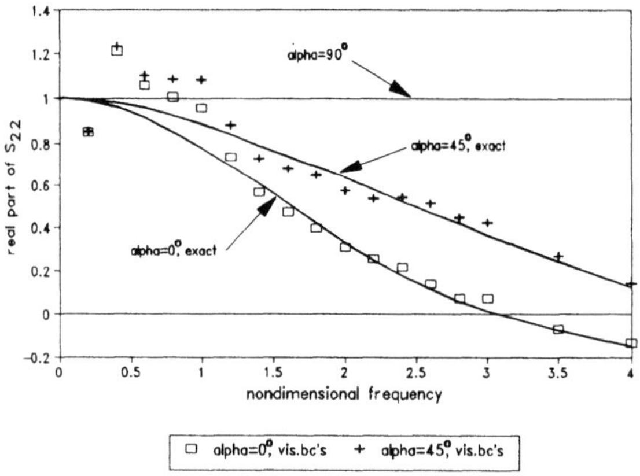

Figure 21 Real part of the input motion function S_{22} (Poisson's ratio= 0.25)

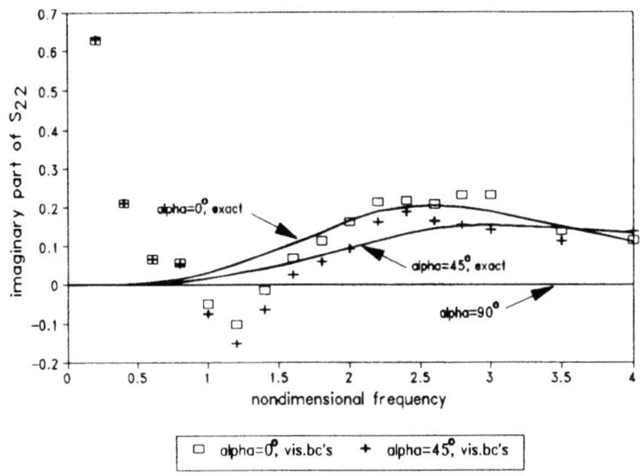

Figure 22 Imaginary part of the input motion function S_{22} (Poisson's ratio= 0.25)

5. References

1. J. P. Wolf, *Soil-Structure-Interaction Analysis in Time Domain*, Printice-Hall, Englewood Cliffs, New Jersey, 1988.
2. J. Lysmer, *Analytical Procedures in Soil Dynamics*, Report of Earthquake Engineering Research Center, UCB, Report No. UCB/EERC-78/29, 1978.
3. J. Lysmer, T. Udaka, C. Tsai and H. B. Seed, *FLUSH*, Report of Earthquake Engineering Research Center, UCB, Report No. UCB/EERC-75/30, 1975.
4. R. Clayton and B. Engquist, "Absorbing Boundary Conditions for Acoustic and Elastic Wave Equations", *Bulletin of the Seismological Society of America*, **67**, 1529-1540 (1977).
5. M. Cohen and P. C. Jennings, "Silent Boundary Methods for Transient Analysis", in *Computational Methods for Transient Analysis*, T.Belytschko and T.J.R.Hughes, eds.
6. J. Lysmer and G. Waas, "Shear Waves in Plane Infinite Structures", *Journal of the Engineering Mechanics Division, ASCE*, **98**, 85-105 (1972).
7. R. R. Kunar and J. Marti, "A Nonreflecting Boundary for Explicit Calculations", *Computational Methods for Infinite Domain Media-Structure Interaction, Applied Mechanics Division, ASME*, **46**, 183-204 (1981).
8. Z. P. Liao and H. L. Wong, "A Transmitting Boundary for the Numerical Simulation of Elastic Wave Propagation", *Soil Dynamics and Earthquake Engineering*, **3**, 174-183 (1984).
9. J. M. Roesset and M. M. Ettouney, "Transmitting Boundaries: A Comparison", *International Journal for Numerical and Analytical Methods in Geomechanics*, **1**, 151-176 (1977).
10. W. D. Smith, "A Nonreflecting Plane Boundary for Wave Propagating Problems", *Journal of Computational Physics*, **15**, 492-503 (1986).

11. J. P. Wolf, "A Comparison of Time-Domain Transmitting Boundaries", *Earthquake Engineering and Structural Dynamics*, **14**, 655-673 (1986).

12. J. Lysmer and R. L. Kuhlemeyer, "Finite Dynamic Model for Infinite Media", *Journal of the Engineering Mechanics Division, ASCE*, **95** , 859-877 (1969).

13. W. White, S. Valliappan and I. K. Lee, "Unified Boundary for Finite Dynamic Models", *Journal of the Engineering Mechanics Division, ASCE*, **103**, 949-964 (1977).

14. B. Engquist and A. Majda, "Absorbing Boundary Conditions for the Numerical Solution of Waves", *Mathematics of Computations*, **31**, 629-651 (1977).

15. G. Waas and J. Lysmer, "Vibrations of Footings Embedded in Layered Media", *Proceedings of the Water Experimental Station, Symposium on the Applications of the Finite Element Method in Geotechnical Engineering*, U. S. Army Engineers, Water Experimental Station, Vicksburg, Miss., 1972.

16. A. G. Masso, J. Lysmer, J. C. Chen and H. B. Seed, *Soil-Structure-Interaction in Different Seismic Environments*, Report of Earthquake Engineering Research Center, UCB, Report No. UCB/EERC-79/18, 1979.

17. H. D. McNiven and Y. Mengi, "Critical Angles Associated with the Reflection-Refraction of Elastic Waves at an Interface", *The Journal of the Acoustical Society of America*, **44**, 1658-1663 (1968).

18. Y. Mengi and H. D. McNiven, "Asymptotic Phase Velocities of Axisymmetric Waves in Composite Rods", *The Journal of the Acoustical Society of America*, **42**, 66-72 (1967).

19. Y. Mengi and A. K. Tanrikulu, "A Numerical Technique for Two-Dimensional Transient Wave Propagation Analyses", *Communications in Applied Numerical Methods*, **6**, 623-632 (1990).

20. H. L. Wong and J. E. Luco, "Dynamic Response of Rectangular Foundations to Obliquely Incident Seismic Waves", *Earthquake Engineering and Structural Dynamics*, **6**, 3-16 (1978).

21. C. A. Brebbia and J. Dominguez, *Boundary Elements, An Introductory Course*, McGraw-Hill, New York, 1989.

22. J. E. Luco and R. A. Westmann, "Dynamic Response of A Rigid Footing Bonded to An Elastic Half Space", *Journal of Applied Mechanics*, **39 E**, 527-534 (1972).

SOIL-STRUCTURE INTERACTION IN LARGE STRUCTURES SUBJECTED TO INCOHERENT GROUND MOTIONS

M. NOVAK and O. RAMADAN
Faculty of Engineering Science
The University of Western Ontario
London, Ontario N6A 5B9
Canada

ABSTRACT. Characteristics, description and effects on structural response of spatial incoherence of seismic ground motions are treated considering structure-foundation interaction. Governing equations of the motion are formulated and solved either in terms of random vibration or by means of simulated motions and FFT. Examples involving a concrete gravity dam and buried pipelines are given.

1. Introduction

The performance and safety of earthquake-resisting structures can be enhanced by improving the understanding and representation of earthquake ground motions. One aspect of these motions, relevant for the analysis of extended structures such as large dams, pipelines, tunnels and long bridges, is spatial variability. The spatial variability of ground motion may be caused by either well defined seismic waves which travel across the structure (wave passage effect) or by waves that are not fully correlated (synchronized) in space (spatial incoherence). While the wave passage effect can be specified deterministically, spatial incoherence, which results from various random factors at the source and along the wave paths, calls for stochastic description.

The effects of wave passage on the response of structures were recognized rather early (Newmark 1969). However, the study of the spatial incoherence of seismic ground motion started relatively recently but the body of information on the subject is increasing rapidly. Novak and Hindy (1979) and Hindy and Novak (1980) investigated the response of buried pipelines; Luco and Wong (1986) and Luco and Mita (1987) analyzed the response of massless footings on elastic halfspace; response of gravity dams was examined by Novak and Suen (1987), who limited their investigation to vertical vibration, and by Ramadan and Novak (1991a, 1992a,b) who considered horizontal response; idealized superstructures were investigated by Mita and Luco (1986), Veletsos and Prasad (1989) and Veletsos and Tang (1989); finally, Berrah and Kausel (1989) proposed a modified response spectrum method together with a modal combination rule for both discrete and continuous systems.

All this research has shown that both the wave passage effect and spatial incoherence of ground motions can have significant effects on structural response. The response can be

P. Gülkan and R. W. Clough (eds.), Developments in Dynamic Soil-Structure Interaction, 147–165.

reduced or amplified and some types of response may occur that remain completely unforeseen if spatial variability is not accounted for.

This paper outlines some of the approaches which may be used for the evaluation of the seismic response of structures accounting for structure-foundation interaction and wave passage effect as well as spatial incoherence of the ground motion. Both deterministic and stochastic representation of ground motions are described, response analysis and internal forces evaluation are outlined and a few examples are given.

2. Representation of Spatially Correlated Seismic Ground Motions

Spatially correlated seismic ground motions are considered here in two ways: as a random process and as a deterministic process. In the former approach, the ground motions are described in terms of the auto- and cross-spectra of the process. In the deterministic case, the motions are represented by their time histories. These time histories may be actual motions recorded during past earthquakes or artificially simulated ones. The latter are more common due to a paucity of data recorded at large scale seismographic arrays. The selection of a specific approach for a particular application depends on the problem to be solved, the resources available, and the form of the solution required. For example, the random approach is most suitable for the response analysis of linear systems while the deterministic representation is necessary for systems with strong nonlinearities. The deterministic approach can also be more illustrative and is still preferred by most practising engineers.

2.1 GROUND MOTION AS A RANDOM PROCESS

For the analysis of the seismic response of structures allowing for randomness of ground shaking in both space and time, complex cross-spectral density functions (cross-spectra) of the ground motions have to be specified for any two stations along the structure-foundation interface. For simplicity, the random field of ground motions is assumed to be homogeneous. This assumption is supported by the analysis of the accelerograms recorded by the SMART-1 seismographic array in Lotung, Taiwan (Harichandran & Vanmarcke 1986). The cross-spectrum of the ground motion between any two stations i and j, $S_{ij}(\vec{r}_i,\vec{r}_j,\omega)$, can then be expressed approximately in terms of the local invariant auto-spectrum, $S(\omega)$, and the real-valued square root of the coherence function (root coherence), $R(\vec{r}_i,\vec{r}_j,\omega)$, where $\vec{r}_i$, $\vec{r}_j$ are position vectors and ω is circular frequency (in rad/sec). For a truly homogeneous process, $R(\vec{r}_i,\vec{r}_j,\omega)$ does not depend on the positions of the two stations $\vec{r}_i$ and $\vec{r}_j$ but rather on the separation vector $\vec{r} = \vec{r}_i - \vec{r}_j$ or $\vec{r} = \vec{r}_j - \vec{r}_i$. The dependence of the coherency function, $R(\vec{r},\omega)$, on the separation vector $\vec{r}$ makes R direction-dependent (anisotropic). The coherency function is direction-dependent because different separation directions are associated with different wave paths. If the separation range of interest is much smaller than the epicentral distance, however, the direction-dependence of $R(\vec{r},\omega)$ becomes insignificant. In this case, the coherency function depends on the true separation $r = |\vec{r}|$ and the cross-spectrum of the ground motion can be described as

$$S(r,\omega) = S(\omega)R(r,\omega)\exp(i\frac{\omega r}{v_a}) \tag{1}$$

In Eq. 1, $R(r,\omega)$ is the spatial coherency of the ground motion caused by various random factors at the source and along the wave paths, while $\exp(i\,\omega r/v_a)$ expresses the effect of a plane travelling wave; v_a is the apparent wave propagation velocity in the direction of the separation vector $\vec{r}$ and is assumed to be constant at least within the dimensions of the structure considered.

2.1.1 *Local Auto-Spectrum of Ground Accelerations.* A number of mathematical models were suggested for the auto-spectrum of the ground acceleration including white noise, filtered white noise and the impulse process (see, for example, Liu 1970). While the auto-spectra of both the filtered white noise and the impulse process may be adequate for some applications, the white noise representation, while mathematically simple, does not provide any similarity to earthquakes. The acceleration auto-spectrum assumed here is of the modified Kanai-Tajimi type (Clough and Penzien 1975) given by

$$S_a(\omega) = S_o \frac{[1+4\zeta_g^2(\frac{\omega}{\omega_g})^2]}{[1-(\frac{\omega}{\omega_g})^2]^2 + 4\zeta_g^2(\frac{\omega}{\omega_g})^2} \frac{(\frac{\omega}{\omega_f})^4}{[1-(\frac{\omega}{\omega_f})^2]^2 + 4\zeta_f^2(\frac{\omega}{\omega_f})^2} \tag{2}$$

in which S_o is the white-noise bedrock acceleration spectrum, ω_g, ω_f are resonant frequencies and ζ_g, ζ_f are damping ratios. The first filter was chosen based on an extensive time series analysis of strong motion records while the second filter is used to eliminate the undesirable singularities in the velocity and displacement spectra which occur at $\omega=0$. For convenience, the parameters of the two filters may be expressed as related, e.g. $\omega_f=0.1\,\omega_g$; $\zeta_f=\zeta_g$. By a suitable choice of these ratios various spectra can be formulated to describe a variety of ground motions featuring different frequency contents (Figure 1). The power spectrum of the local ground displacement is $S_d(\omega)=S_a(\omega)/\omega^4$.

2.1.2 *Spatial Coherency of Seismic Ground Motion.* To establish a realistic model for the coherency function, records from specially designed arrays of strong-motion seismographs are needed. The SMART-1 seismographic array was the first large-scale digital array to become operational. The analysis of the data provided by this array revealed useful information about the ground motion coherency. The data were analyzed by a number of researchers (Harichandran and Vanmarcke 1986 and others) who proposed a few mathematical models for the coherency function.

The first study to incorporate the coherency function in evaluating the seismic response of structures appears to be the paper on buried pipelines published by Novak and Hindy (1979). The study was done before the SMART-1 array data became available. Therefore, based on previous studies on wave scattering in inhomogeneous media, and with respect to the properties of other random phenomena such as turbulence of natural wind, the coherency function was assumed in a simple exponential form, i.e.

$$R(r,\omega) = \exp[-c(\frac{\omega r}{V})^\gamma] \tag{3}$$

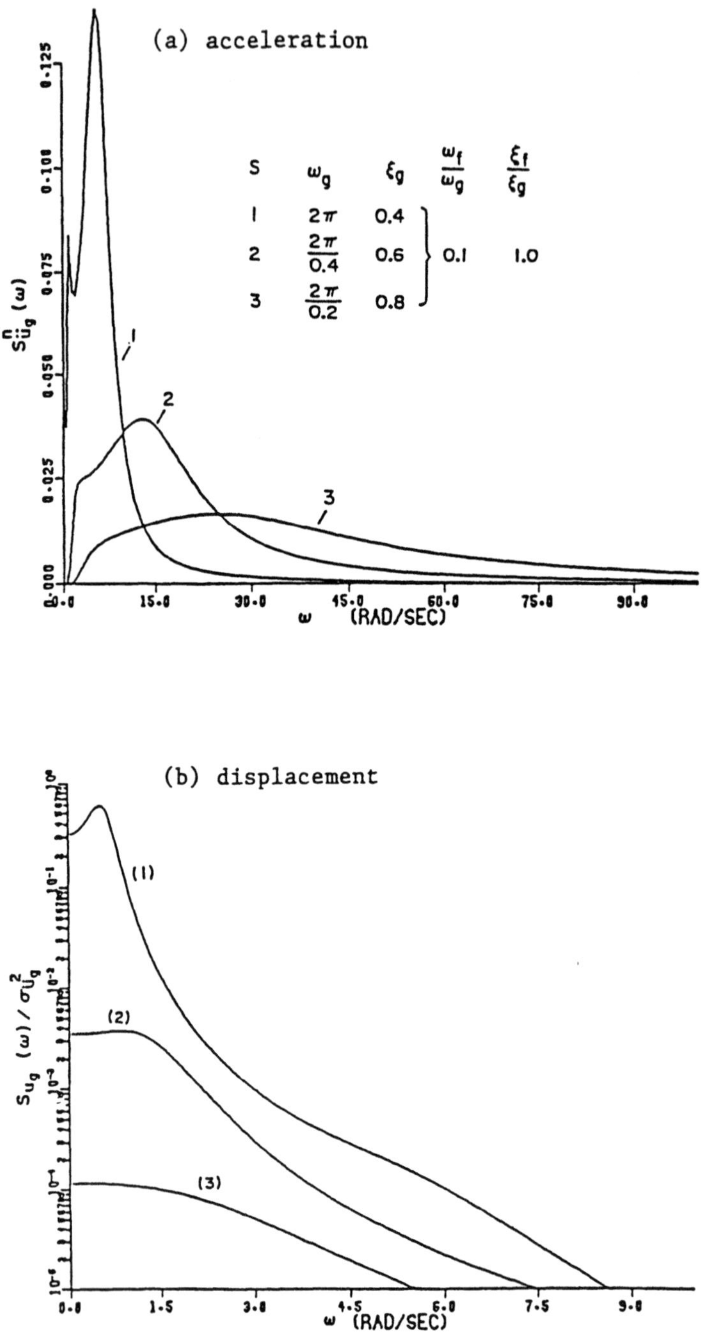

Figure 1 Spectra of Ground Acceleration and Displacement Normalized to Have Unit Acceleration Variance.

In Eq. 3, c is an exponential decay factor which depends on the focal depth, fault size and the degree of irregularities and soil layering along the wave path; γ is a power constant depending on the fault mechanism, epicentral distance and other factors; and V is a suitable wave velocity which is introduced to make the coherency dependent on both the frequency (wave length) and separation and thus on the nondimensional frequency parameter (ωr/V). The coherency model given by Eq. 3, while involving two independent constants only, was found to be in good agreement with the five-constant model suggested by Harichandran and Vanmarcke (1986) (Novak 1987). In Figure 2 the model is also compared with the eight-constant anisotropic model described in Hao (1991) for radial and tangential separations, respectively. Again, the agreement is quite adequate for engineering applications. Therefore, the model by Eq. 3 with γ=1.0 is adopted in the rest of this paper. Upward propagating shear waves are assumed and therefore, the exponential part of Eq. 1 is omitted.

The degree of the coherency by Eq. 3 is described by the constant c or the ratio c/V. However, this parameter does not completely describe the degree of the spatial coherence of the strong motion part of the random process as the latter also depends on the power distribution with frequency (i.e. auto-spectrum). For a specific value of the ratio c/V, the ground motion would be less correlated if the energy contained in the process were associated with high frequencies than if they were associated with low frequencies. By integrating the correlation coefficient between any two stations with respect to their separation, r, over the range 0 to ∞, the correlation length of the ground acceleration can be defined as (Hindy and Novak 1980)

$$L_a = \int_0^{\infty} \int_0^{\infty} R(r,\omega) S_a^n(\omega) \, d\omega \, dr \tag{4}$$

and for the ground displacement

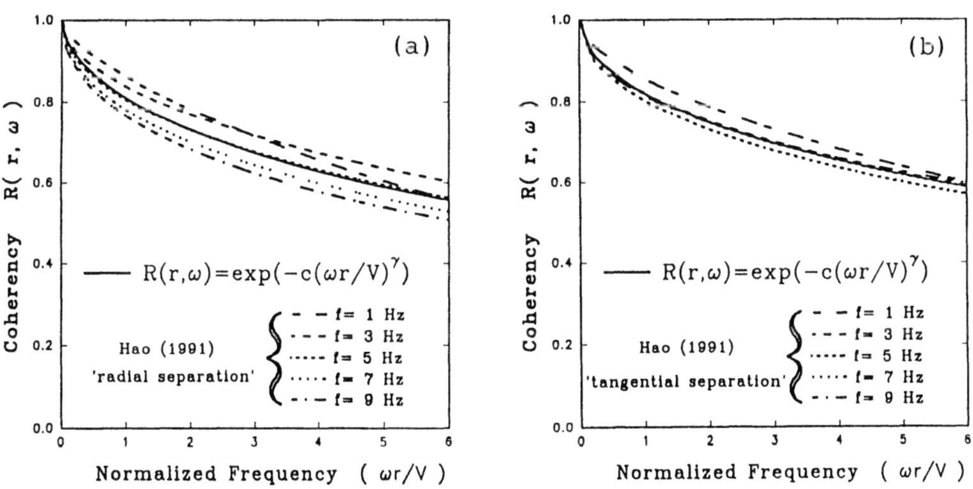

Figure 2 Fitting the Coherency Function by Eq. 3 to That in Hao (1991): (a) Radial Separations; (b) Tangential Separations

$$L_d = \int_0^\infty \int_0^\infty R(r,\omega) S_d^n(\omega)\, d\omega\, dr \tag{5}$$

In Eqs. 4 and 5, $S_a^n(\omega) = S_a(\omega)/\sigma_a^2$ is the normalized acceleration spectrum, $S_d^n(\omega) = S_d(\omega)/\sigma_d^2 = S_a(\omega)/(\omega^4\sigma_d^2)$ is the normalized displacement spectrum, and σ_a^2, σ_d^2 are the variances of the ground acceleration and displacement, respectively. Both $S_a(\omega)$ and $S_d(\omega)$ are defined for positive frequencies only, i.e. they are one-sided. The correlation lengths L_a and L_d may also be expressed in terms of the frequency-dependent correlation length, ℓ_ω, in the form

$$L_a = \int_0^\infty \ell_\omega S_a^n(\omega)\, d\omega \tag{6}$$

$$L_d = \int_0^\infty \ell_\omega S_d^n(\omega)\, d\omega \tag{7}$$

where

$$\ell_\omega = \int_0^\infty R(r,\omega)\, dr \tag{8}$$

The comparison of these correlation lengths with the size of a particular structure indicates whether the effects of the ground motion spatial incoherence could be significant or not.

2.2 SIMULATION OF SPATIALLY INCOHERENT GROUND MOTIONS

To evaluate the seismic response of a structure including the effects of ground motion incoherency in a deterministic way, time histories of the ground motion at all nodes on the structure-foundation interface are needed. Due to a paucity in recorded ground motions at specific sites, artificially generated, simulated motions are usually used. While the simulation techniques, which consider ground motion variations in time alone, have received considerable attention (e.g. Chang et al. 1982 & Polhemus and Cakmak 1981), only a few procedures which include both the temporal and the spatial variations were published (Sinozuka and Deodatis 1988, Spanos and Mignolet 1990, Ramadan and Novak 1991a). Spatially incoherent ground motions are simulated here using the technique due to Ramadan and Novak (1991a). This technique is fully consistent with the random approach methodology outlined above and generates motions which match both a target auto-spectrum and a target coherency function. The technique is general and can simulate motions in one-, two- and three-dimensional domains. It is appropriate for both isotropic and anisotropic coherency structures.

Consider the realizations of ground displacements treated as spatially variable homogeneous, stationary random processes with a one-sided, local displacement power spectrum $S_d(\omega)$ and a square-root coherence function, or coherency, $R(r,\omega)$. For the case of co-linear stations assumed here, such motions can be generated by

$$u_s(t) = \sum_{k=0}^{N} \sum_{i=1}^{M} a_k \Big\{ \sqrt{S_d(\omega_{ik})\Delta\omega}\, \cos[\omega_{ik}(t-\tau_s)+\phi_{ik}+\pi ks/L_i\,] + \sqrt{S_d(\omega'_{ik})\Delta\omega}\, \cos[\omega'_{ik}(t-\tau_s)+\phi'_{ik}-\pi ks/L_i]\Big\} \tag{9}$$

In Eq. 9, s is the coordinate with respect to an arbitrary reference point, ϕ_{ik}, ϕ_{ik}' are two statistically independent random variables uniformly distributed between 0 and 2π, t is time and $\tau_s = s/v_a$ is the time lag due to surface waves where v_a is the apparent wave velocity measured along the s-axis; a_k, k=0,1,...N and L_i, i=1,2,...M are respectively amplitude modifiers and correlation lengths dependent on the coherency function; finally, $\Delta\omega$ is the frequency step, and ω_{ik}, ω_{ik}' are two different circular frequencies given by

$$\omega_{ik} = i(\Delta\omega) - k(\Delta\omega)/N \tag{10a}$$

$$\omega'_{ik} = i(\Delta\omega) - (k - \frac{1}{2})\Delta\omega/N \tag{10b}$$

$$\Delta\omega = \omega_{max}/M \tag{10c}$$

where ω_{max} is a mathematical, or physical, cut-off frequency and M is the number of frequency bands. If the coherency function given by Eq. 3 is assumed ($\gamma=1$), the correlation lengths and amplitude modifiers become

$$L_i = \alpha\, V/[c(i - \frac{1}{2})\Delta\omega] \; , \quad i=1,2,...M,$$

$$a_0^2 = \frac{1}{\alpha}(1 - e^{-\alpha}), \text{ and}$$

$$a_k^2 = 2[1 - (-1)^k e^{-\alpha}]/(\alpha[1 + (\pi k/\alpha)^2]) \quad k=1,2,...N$$

in which α is a constant which controls the degree of accuracy. When $\alpha \geq 3$ and $N \geq 5$, the auto-spectrum and coherency function of the simulated motions compare well with the target values. For more details about the technique in its general form, refer to Ramadan and Novak (1991a).

For nonlinear systems and in many practical applications, the response is usually evaluated by integrating the governing equations of motion in the time domain. In such cases, time histories of the ground velocity and acceleration are also required. Ground velocities, $\dot{u}_s(t)$, and accelerations, $\ddot{u}_s(t)$, compatible with the ground displacements $u_s(t)$, may be generated as the first and second time derivatives of Eq. 9, i.e. $\dot{u}_s(t) = \partial u_s(t)/\partial t$ and $\ddot{u}_s(t) = \partial^2 u_s(t)/\partial t^2$.

To convert the stationary motions generated in the way outlined to nonstationary ones, a suitable envelope function is superimposed.

As an example, ground motions are simulated at 11 equidistant nodes, s=0,85.3,170.6,...853 m using Eqs. 2, 3 and 9. The spectrum parameters are $S_o=0.01$, $\omega_g=5\pi$, and $\zeta_g=0.6$ while the other constants assume the values of M=100, $\omega_{max}=60$ rad/s, $\alpha=3$, N=10, and $v_a \rightarrow \infty$ in Eq. 1 so that no travelling waves are considered. Figures 3a and b show the variations of the simulated motions with distance for the coherency parameter c/V=0.001 and 0.002, respectively. For these ground motions the correlation lengths, defined by Eq. 7, assume the values of 3259 and 1630 m, respectively. Figure 3 suggests that a tall chimney or tower with plan dimensions in the order of 25 m would not be affected by the spatial variability, or incoherence, of the ground motion; but a 500 m long bridge, pipeline or dam would be

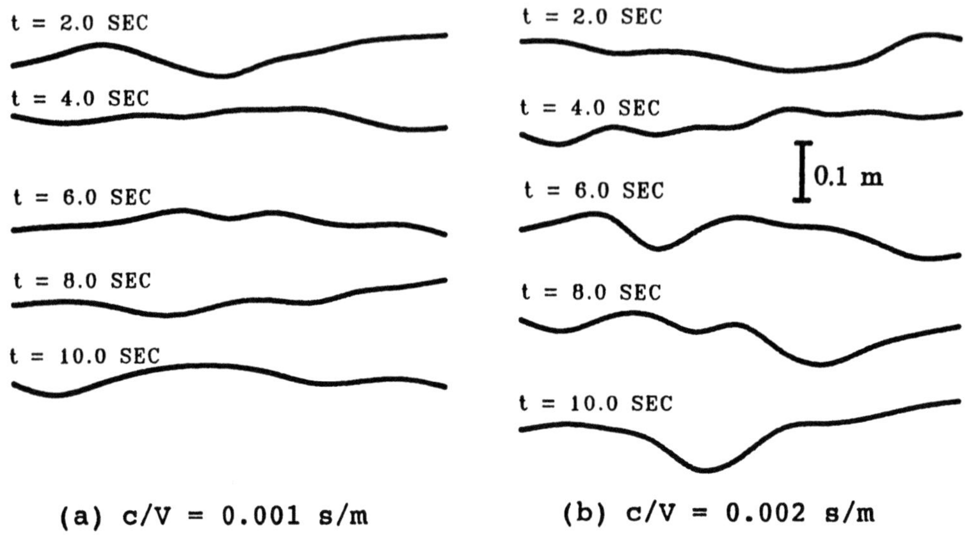

Figure 3 Simulated Ground Motion at Five Time Instants: (a) c/V=0.001; (b) c/V=0.002 s/m.

significantly affected by this variability. Also, the effects of the motions shown in Figure 3b for the smaller correlation length L_d=1630 m would be more severe than those of the motions from Figure 3a with L_d=3259 m. Variations of the seismic ground displacements over larger distances are given in Figure 4 where the motions over a 5 km x 5 km area are displayed; all spectrum and coherency parameters are the same as those of the motions in Figure 3a. The ground accelerations compatible with the motions in Figure 3a are displayed for the first six stations in Figure 5. Nonstationary characteristics were introduced by superimposing a time-envelope in the shape of a half-sine wave. Variations in the ground accelerations from one node to another are clearly visible in this figure. The simulated motions' auto-spectrum, cross-spectrum and coherency function for a separation of 50 m and c/V=0.001 are compared with the target functions in Figure 6. The agreement is good and quite adequate for engineering applications.

3. Governing Equations of Motion and Their Solution

The formulation of the governing equations of motion depends on the structural system of interest. Assume a continuously supported system such as a dam or a buried pipeline discretized into a number of nodes. Denote the ground motion vector, $\{y_g\}$, the vector of relative motions between the ground and the structure, $\{y\}$, and the vector of the absolute motions of the structure $\{Y\}=\{y\}+\{y_g\}$. Realizing that the structural stiffness is related to the absolute motions $\{Y\}$ while the foundation resistance derives from the relative motions $\{y\}$, the equations of motion for the structure-foundation system are

$$[\,m\,]\{\ddot{Y}\}+[K_s]\{Y\}+[K_f]\{y\} = \{0\} \tag{11}$$

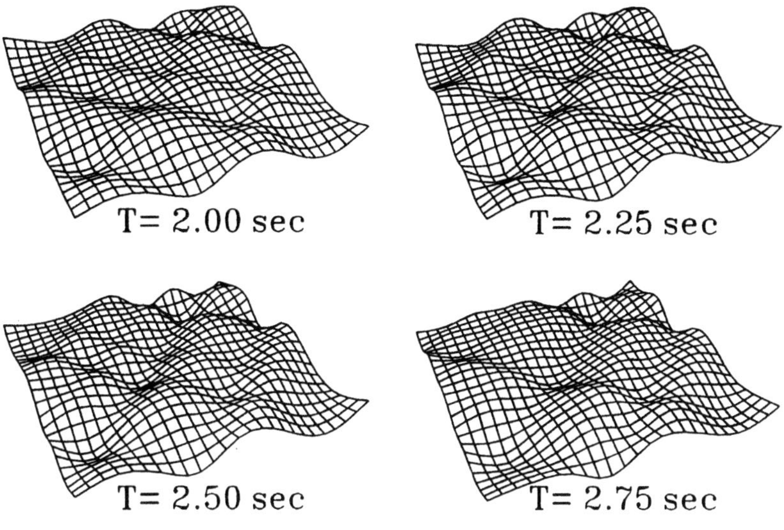

Figure 4 Incoherent Ground Motions Simulated Over a 5 km x 5 km Area, Shown at Different Instances (c/V=0.001)

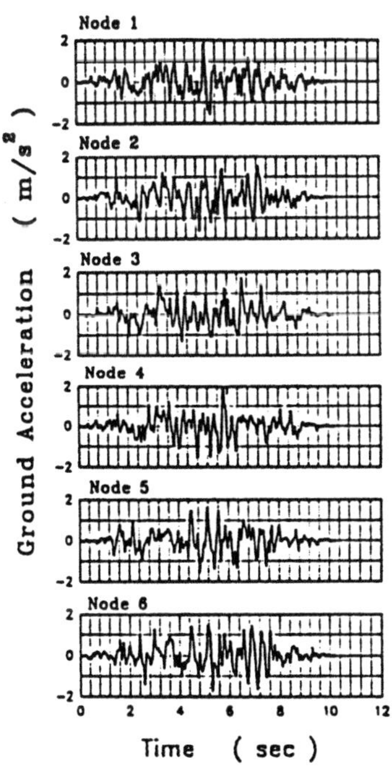

Figure 5 Simulated Ground Accelerations (c/V=0.001)

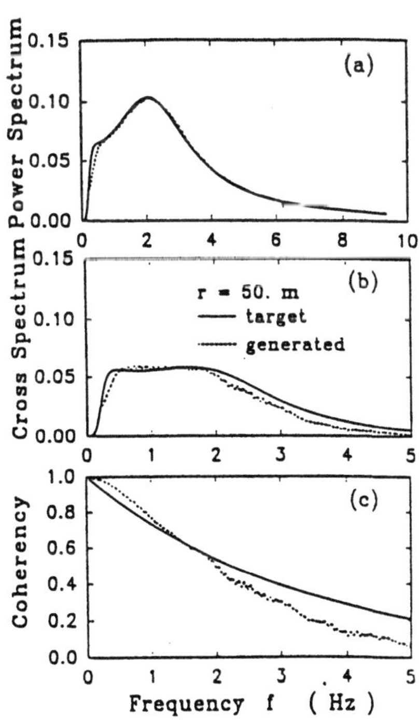

Figure 6 Target and Generated Spectra and Coherency Functions

in which [m] is the mass matrix, $[K_s]$ is the complex structural stiffness matrix and $[K_f]$ = $[K_f(\omega)]$ is the complex foundation stiffness matrix depending on frequency. (Effects of structure-fluid interaction may be included in the matrix $[K_s]$.) Denoting the total stiffness matrix

$$[K_t] = [K_s] + [K_f] \tag{12}$$

the equations of motion are in terms of the relative motion

$$[\,m\,]\{\ddot{y}\}+[K_t]\{y\} = -[\,m\,]\{\ddot{y}_g\} - [K_s]\{y_g\} \tag{13}$$

or in terms of the absolute motions, preferred here because of the subsequent stress calculation,

$$[\,m\,]\{\ddot{Y}\} + [K_t]\{Y\} = \{K_f\}\{y_g\} \tag{14}$$

In Eqs. 11 to 14, the displacement vectors contain all active degrees of freedom. For a three-dimensional problem, six degrees of freedom (three translations and three rotations) may be considered at each node. However, the ground motion vector, $\{y_g\}$, has non-zero elements associated only with the active translations. This formulation may also be used for multiply supported systems which consist of large numbers of short spans so that no additional nodes need be considered between the supports. For large spans or multi-storey structures, the formulation is somewhat different and involves a decomposition of the structural response into a pseudo-static component and a dynamic component (Clough and Penzien 1975). The suitable procedure for solving Eq. 14 depends on whether the ground motion is described as random or deterministic.

3.1 RESPONSE TO RANDOM GROUND MOTION

The response to random ground shaking can be evaluated using either a direct analysis or modal superposition. The response of many soil-structure systems can be treated as stationary because the damping derived from soil-structure interaction is large. If it is not, the corrections for nonstationarity can be made as described in Clough and Penzien (1975).

The direct random vibration analysis bypasses the free vibration analysis and readily accounts for the variation in foundation stiffness with frequency. After some manipulations, the spectral matrix of the response of the system described by Eq. 14 is obtained and from the individual spectra the variances of the absolute displacement at stations i follow as

$$\overline{Y_i^2} = \mathrm{Re} \int_0^{\infty} S_d(\omega) < H_i(\omega)> [K_f(\omega)][R(r,\omega)][K_f^*(\omega)]^T < H_i^*(\omega) >^T d\omega \tag{15}$$

in which $< H_i(\omega) >$ is the ith row of the transfer matrix $[H(\omega)]$ and the asterisk denotes a complex conjugate. From the variance, the peak response and all other response statistics are obtained as usual.

For the modal random vibration analysis, the modes of free vibration, natural frequencies and modal damping ratios have to be established first. This task is complicated by the frequency dependency of the foundation stiffness matrix but the knowledge of the free vibration characteristics is useful. Because many vibration modes participate in the response, the modal response analysis loses the main advantage it has in standard structures. More details on the random vibration analysis can be found in Novak and Suen (1987).

3.2 RESPONSE TO DETERMINISTIC GROUND MOTION

The dynamic response of structures subjected to deterministic dynamic loading is usually evaluated by numerical integration of the equations of motion in the time domain. However, the application of this technique to structure-foundation systems is complicated by the frequency-dependence of the foundation stiffness. Alternatively, to solve Eq. 14 and allow for the frequency-dependence of the foundation stiffness, the input ground motions are first transferred to the frequency domain using the Fast Fourier Transform (FFT), then the response is calculated in the frequency domain, and finally, the response is transferred back to the time domain through an inverse FFT operation. This procedure is known as the complex response analysis and provides the response in both the frequency and time domains.

For continuous structures with distributed mass and simple geometry and boundary conditions, system discretization may not be necessary and the differential equation may be integrated directly. This may lead to a closed form representation for the system natural frequencies and vibration modes. An example of continuous structures are buried pipelines investigated by Novak and Hindy (1979) and Hindy and Novak (1980).

3.3 INTERNAL FORCES AND STRESSES

When the absolute displacements vector is evaluated deterministically for a certain structure, internal forces and stresses can be easily calculated using the theories of structural analysis. If the response is evaluated stochastically, however, the internal forces are described by their spectral matrices. The integral of these spectra over frequency yields the internal forces variances from which the root-mean-squares are obtained. The maximum (peak) values are evaluated using a suitable peak factor. For more details see Hindy and Novak (1980) and Novak and Suen (1987).

4. Examples of Structural Response to Spatially Correlated Ground Motions

In this section, some of the studies conducted on the response of structures to spatially correlated ground motions are summarized. Buried pipelines and concrete gravity dams are discussed in some detail and the response of a few other systems is briefly mentioned.

4.1 BURIED PIPELINES

Buried pipelines are used to transport various substances such as oil, water and gas over great distances. Both travelling waves and spatially correlated random ground motions have significant effects on buried pipelines. The type and magnitude of the pipe strains and stresses depend on the type of the seismic wave and its angle of incidence, θ. Maximum axial strain occurs at $\theta=90°$ for P-waves and at $\theta=45°$ for S-waves while the maximum bending strain corresponds to $\theta=50$-$60°$ for P-waves and $\theta=90°$ for S-waves, respectively. Pipe-soil interaction reduces both the axial and bending stresses and the stress reduction increases for softer soils. The axial and lateral responses of a pipeline to random ground motion feature axial and bending stresses that are almost constant along the pipe length except near its ends. The stresses due to ground motion incoherence can be considerable but the actual stress level depends on the actual degree of ground motion incoherence and other factors. For more details on the buried pipelines response, see Hindy and Novak (1980) and other literature.

4.2 CONCRETE GRAVITY DAMS

The analysis of dams for seismic effects is in practice usually conducted using a two-dimensional finite element model of its slice or monolith. This model provides the basic information but ignores the interaction between adjacent monoliths and does not account for the possible variations of seismic ground motions along the dam longitudinal axis. To indicate the effects of ground motion spatial incoherence in both the vertical and lateral directions some results are presented for an 853 m long dam with a cross-section similar to that of the Koyna Dam of Western India. Under the effects of vertical incoherent ground motions, the dam bends in the vertical plane passing through its longitudinal axis. When subjected to horizontal incoherent motions, the dam bends in the horizontal plane and undergoes torsional vibration as well. In both cases, the dam length to height ratio and the length to width ratio are assumed large enough so that its response can be evaluated in terms of the beam theory. Dam-foundation interaction is included and the through-rock interaction (or coupling) of foundation elements is also considered. Dam-reservoir interaction is accounted for in the evaluation of the dam lateral response.

The root-mean-square (rms) of the dam stresses and response generated by the vertical component of the ground motion are displayed in Figure 7. The ground motion is modelled as a random process using Eqs. 1, 2 and 3 with $c/V=0.0008$; $\omega_g=10\omega_f=2\pi$ and $\zeta_g=\zeta_f=0.4$ for case (a) and $\omega_g=10\omega_f=10\pi$ and $\zeta_g=\zeta_f=0.8$ for case (b). The rms ground acceleration is 0.61 m/s^2 in both cases. The distribution of the dam response to incoherent ground motions is similar to that of the buried pipelines and the resulting stresses are quite significant. The stresses are much smaller in case (b) because the response to the incoherent ground motions depends on ground displacement rather than acceleration and the rms of the ground displacement for case (a) is about 30 times higher than that for case (b). The effect of dam-foundation interaction is shown in Figure 8 which displays the spectrum of the dam motion normalized by the spectrum of the input ground motion. This interaction amplifies the low frequency components of the ground motion but suppresses the high frequency components. The effect of foundation stiffness on dam stresses is profound. As the foundation stiffness increases, the dam is forced to follow the ground motion and the stresses increase. More data on the dam response to the ground vertical motion can be found in Novak and Suen (1987).

The dam response to the horizontal component of spatially correlated (SC) ground motion is investigated using the deterministic ground motions generated by Eq. 9 and plotted in Figures 3a and 5. The analysis follows Section 3.2. For five time instants, Figures 9 and 10

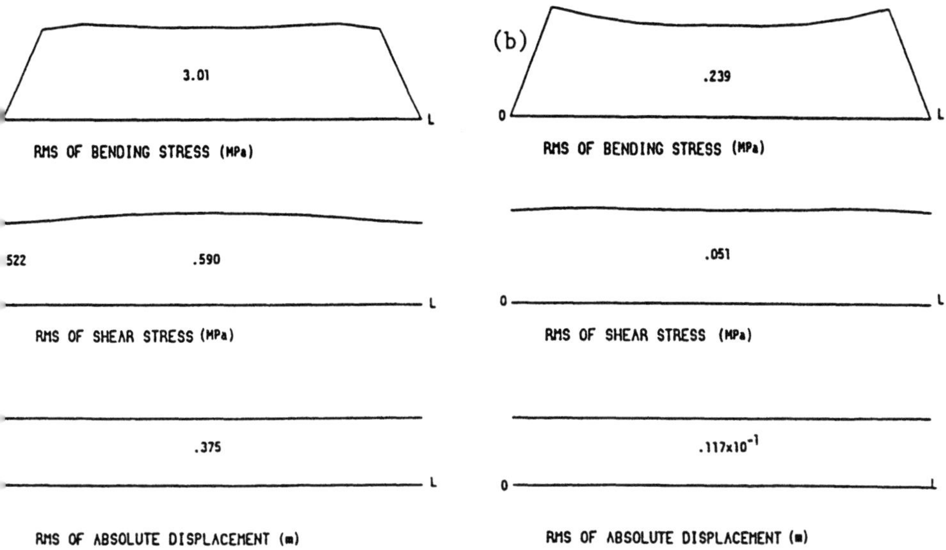

Figure 7 Dam Vertical Response and Stresses for (a) Narrow Band Spectrum (1) and (b) Broad Band Spectrum (3).

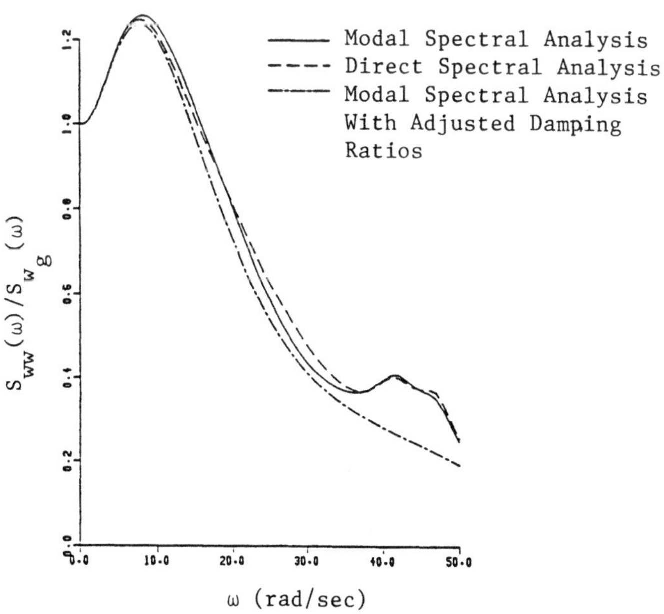

Figure 8 Spectrum of Absolute Dam Displacement Normalized by Ground Motion Spectrum (Vertical Response).

show the dam horizontal response and the corresponding distribution of the bending moments along the dam. For comparison, the dam response to fully correlated (FC) ground motions is also shown in these figures. While the dam vibrates almost as a rigid body under the fully correlated ground motions, it bends and twists significantly under the spatially incoherent motions. The maximum bending stress and the maximum shear stress resulting from dam bending and torsion are 8.76 and 1.46 MPa, respectively. The ratio of the dam motion to the ground motion at the dam midpoint is displayed in Figure 11. Comparing Figure 8 to Figure 11, it is apparent that the damping generated by the foundation in the vertical direction is considerably higher than that produced in the horizontal direction. Figure 11 also shows that there are more higher vibration modes present in the dam response to the SC motions than in the response to the FC motions. The effect of foundation rock stiffness on dam response is displayed in Figure 12 where the Fourier spectrum of the dam horizontal response at the dam midpoint is plotted for different values of foundation shear wave velocity, V_s. This spectrum is normalized by that of the ground motion and was calculated for the fully correlated input ground motions. As the foundation stiffness increases, the system natural frequencies increase but the foundation damping decreases. The sudden drop in the dam response, appearing in Figures 11 and 12 at ω=25.13 rad/s, is due to the dam-reservoir interaction; this effect is discussed later herein.

As the foundation tributary areas are so close, the through-foundation coupling (or interaction) of these areas was included in the analysis. This increases the computational costs dramatically. To investigate the importance of this effect, the dam response was also evaluated assuming a Winkler type foundation (i.e. without coupling). The results are compared to the response obtained with the foundation coupling included in Figure 13, in which the Fourier amplitudes of dam bending and torsional moments at its midpoint are shown. When the through-foundation coupling is ignored, the errors in the dam critical bending and shear stresses are -8% and +5%, respectively. This suggests that the effect of the through-foundation coupling of tributary elements on the dam stresses is not very significant.

The results presented above correspond to a monolithic dam. In practice, concrete gravity dams are usually built in separate blocks (monoliths) connected by monolith vertical joints. The presence of these joints affects the dam response. The effect of these joints can be included in an approximate way as described in Ramadan and Novak (1992b). Figure 14 shows the effect of the joints on bending and torsional responses at the dam midpoint. The presence of joints results in considerable reduction in dam bending at all frequencies. The torsional moment decreases at lower frequencies but increases at higher frequencies. This indicates that the presence of monolith joints reduces the dam response in the horizontal plane but increases the dam response in the plane of its cross-section. For the example dam, the presence of the monolith joints reduces the maximum bending stress to 1.66 MPa but increases the maximum shear stress to 1.91 MPa.

The effect of dam-reservoir interaction may be inferred from Figure 15 where the dam midpoint response is displayed for both empty and full reservoirs. Figure 15a presents the Fourier spectra of the dam horizontal displacement normalized by those of the ground motion, while Figure 15b displays the Fourier spectra of dam bending moments. Both figures show that reservoir interaction results in an increase in dam response for frequencies lower than the reservoir fundamental resonance (25.13 rad/s in this example), followed by a response reduction at higher frequencies. As the vibration frequency approaches the reservoir second resonance (75.39 rad/s, not shown in Figure 15), a slight increase in dam response is

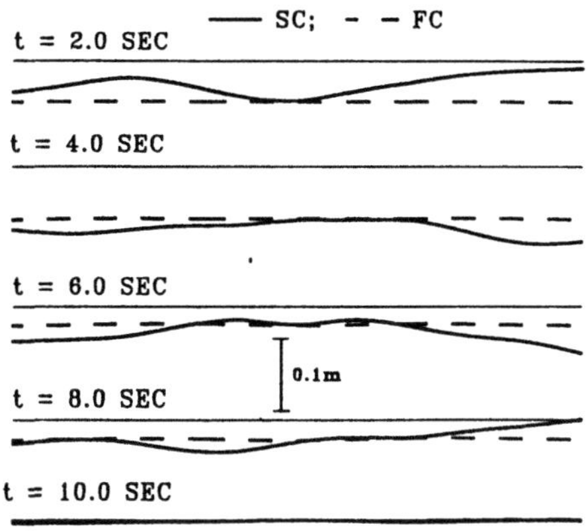

Figure 9 Lateral Response Along Dam Axis

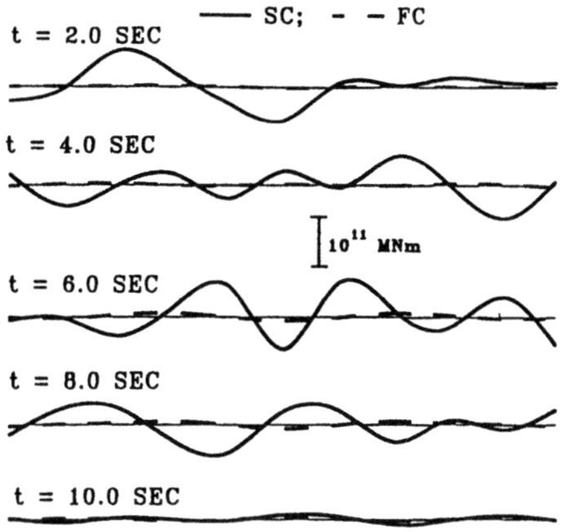

Figure 10 Bending Moment Distribution Along Dam Axis

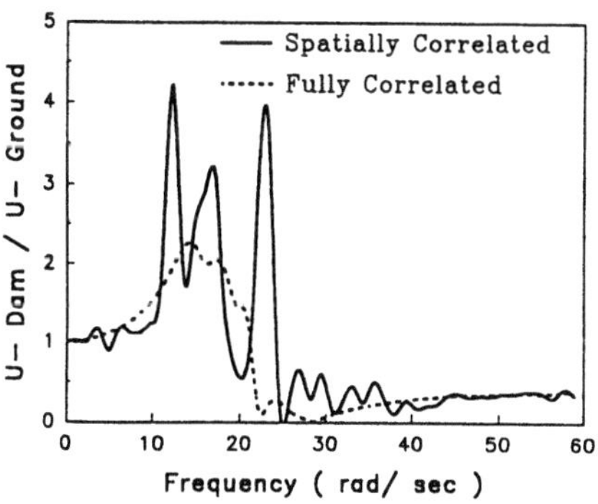

Figure 11 Fourier Spectrum of Dam Motion Normalized by Ground Motion Spectrum at Midpoint.

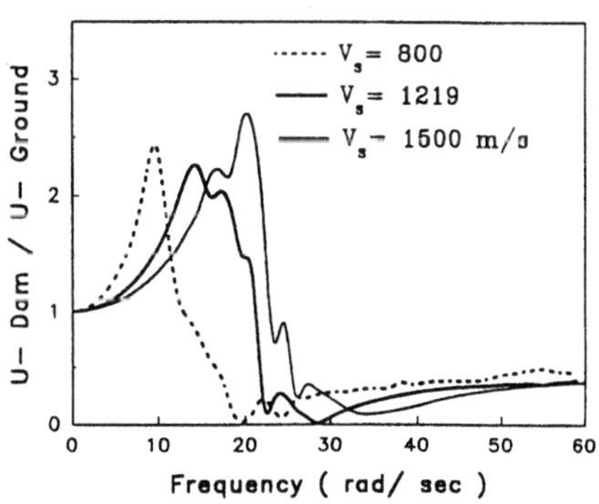

Figure 12 Effect of Foundation Stiffness on Dam Response to Fully Correlated Motions.

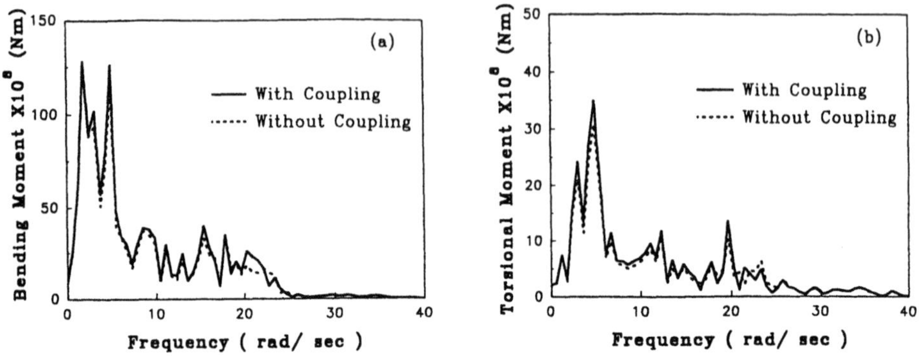

Figure 13 Effect of Through-Foundation Coupling on Fourier Spectra of Dam Response: (a) Bending Moment; (b) Torsional Moment.

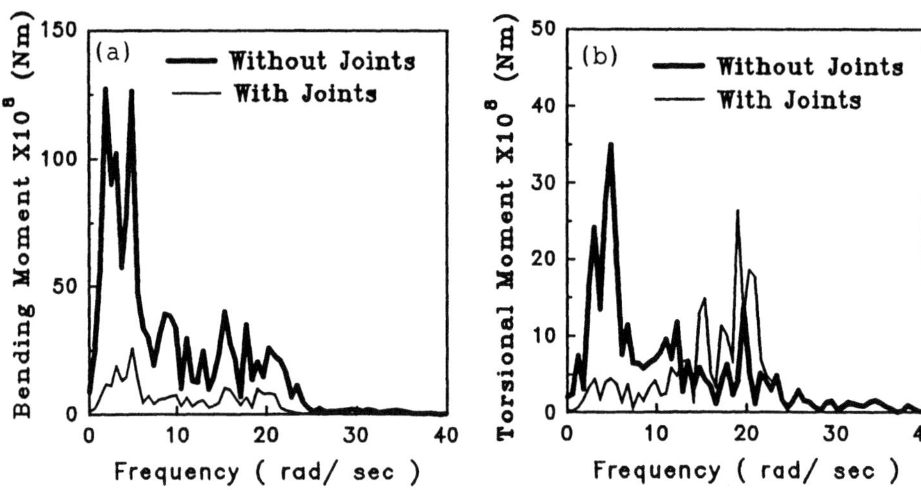

Figure 14 Effect of Monolith Joints on Fourier Spectra of Dam Response at Midpoint: (a) Bending Moment; (b) Torsional Moment.

also indicated.

Finally, the dam response to the spatially incoherent ground motions is compared with its response to a deterministic (fully correlated) wave travelling along the dam longitudinal axis in Figure 16. The dam internal forces caused by the deterministic travelling waves are not similar to those due to incoherent ground motions and are much smaller. The travelling waves generate peak moments that are only 20 per cent of the bending moments and 50 per cent of the torsional moments caused by the incoherent motions. More details on the

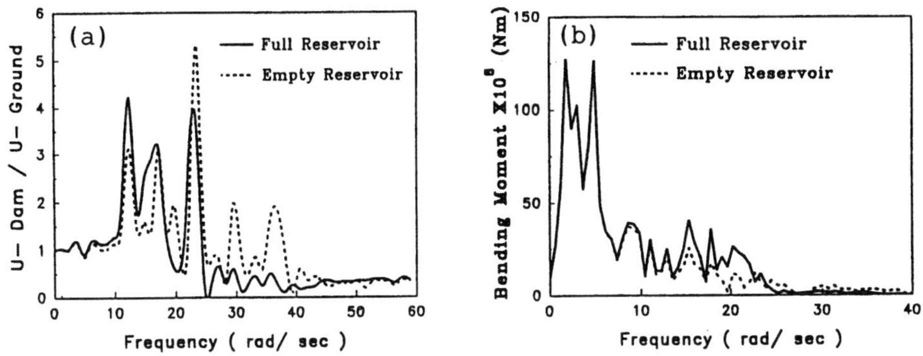

Figure 15 Effect of Dam-Reservoir Interaction on Fourier Spectra of Dam Response: (a) Horizontal Translation; (b) Bending Moment.

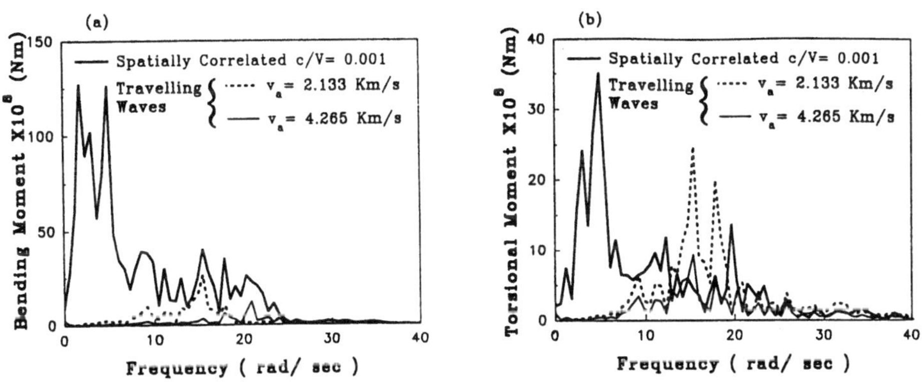

Figure 16 Dam Response to Spatially Correlated Motions and Fully Correlated Travelling Waves.

dam horizontal response are given in Ramadan and Novak (1992a,b).

It should be emphasized that the dam stresses calculated here remain completely unforeseen when the spatial incoherence of ground motion is not considered. These stresses are particularly significant because they should be added to the critical stresses evaluated using the conventional two-dimensional model.

4.3 OTHER CASES

Other structures under incoherent ground motions were also investigated. Luco and Wong (1986) studied the response of a massless square footing bonded to a uniform viscoelastic halfspace while Luco and Mita (1987) considered a circular footing. The analysis was conducted using a stationary random vibration approach and all three components of the ground motion were considered and assumed correlated. They concluded that ground motion incoherency reduces the translational responses of rigid footings and produces a rotational response (rocking and torsion) even for perfectly symmetric cases. Their study also suggested that for small degrees of ground motion incoherence the effects of travelling waves are qualitatively and quantitatively similar to those of the spatial incoherence. The results of the rigid footing analysis were later used to analyze simplified models of multi-storey and containment buildings with two planes of symmetry (Mita and Luco 1986). It was found that ground motion incoherence reduces the high frequency components of the translational response on the axis of symmetry of the superstructure but leads to marked torsional and rocking responses. These effects are quite similar to those of turbulent wind.

Veletsos and Prasad (1989) analyzed the random response of a two-degree-of-freedom structure resting on the surface of a viscoelastic halfspace. The restraining action of the supporting medium was approximated by a series of mutually independent springs of the Winkler type. This simplification reduced the complexity of the problem considerably and lead, in some cases, to closed form solutions. The conclusions are similar to those reported by Luco and Mita (1987) but the separate effects of inertial and kinematic interaction were also presented. Veletsos and Tang (1989) proposed an approximate deterministic approach for evaluating the structural response to incoherent ground motions. In this approach, the ground motion is described by an acceleration time history and a spatial coherency function.

5. Conclusions

Deterministic and stochastic procedures for the response analysis of structures to spatially incoherent seismic ground motions which account for structure-foundation interaction were discussed. Numerical examples suggest that both wave passage and spatial incoherence of ground motions may have significant effects on seismic response of structures. These effects may be beneficial or negative depending on the frequency contents of the input ground motion and the degree of spatial variability, type and size of the structure, the relative stiffnesses of the structure and its foundation, and other factors. The effects are more critical for extensive structures because tall structures suffer from the variations of the ground motion in the time alone while extensive structures are affected by the ground motion variation in space as well.

6. References

Berrah, M. and Kausel, E. (1989). "Modified response spectrum model for the design of structures subjected to spatially varying seismic excitations," Report No. R90-2, MIT Department of Civil Engineering, Cambridge, Massachusetts.

Chang, M.K., Kwiatkowski, J.W. and Robert, F.N. (1982). "ARMA models for earthquake ground motions," Earthq. Engrg. Struc. Dyn., 10, pp. 651-662.

Clough, R.W. and Penzien, J. (1975). Dynamics of structures. McGraw-Hill, Inc., New York.

Hao, Hong (1991). "Response of multiply supported rigid plate to spatially correlated seismic excitations," Earthq. Engrg. Struc. Dyn., 20, pp. 821-835.

Harichandran, R.S. and Vanmarcke, E.H. (1986). "Stochastic variation of earthquake ground motion in space and time," J. Eng. Mech., ASCE, Feb., 154-174.

Hindy, A. and Novak, M. (1980). "Response of pipelines to random ground motion," J. Eng. Mech., ASCE, EM2, pp. 339-360.
Liu, S.C. (1970). "Synthesis of stochastic representations of ground motions," Bell System Technical Journal, 49(4), pp. 521-541.
Luco, J.E. and Wong, H.L. (1986). "Response of a rigid foundation to a spatially random ground motion," Earthq. Engrg. Struc. Dyn., 14, pp. 891-908.
Luco, J.E. and Mita, A. (1987). "Response of a circular foundation to a spatially random ground motion," J. Eng. Mech., ASCE, 113(1), pp. 1-15.
Mita, A. and Luco, J.E. (1986). "Response of structures to a spatially random ground motion," Proc. Third U.S. National Conf. on Earthquake Engineering, Charleston, S.C., pp. 907-918.
Newmark, N.M. (1969). "Torsion in symmetric buildings," Proc. 4WCEE, Santiago, Chile, pp. A3-19-A3-32.
Novak, M. (1987). Discussion, J. Engrg. Mech., ASCE, Vol. 113, No. 8, pp. 1267-1270.
Novak, M. and Hindy, A. (1979). "Seismic response of buried pipelines," 3rd Canadian Conf. on Earthq. Engrg., Montreal, 1, pp. 177-203.
Novak, M. and Suen, E. (1987). "Dam-foundation interaction under spatially correlated random ground motion," In Development in Geotechnical Engineering (A.S. Cakmak, Ed.), Elsevier, pp. 25-39.
Polhemus, N.W. and Cakmak, A.S. (1981). "Simulation of earthquake ground motions using autoregressive moving average (ARMA) models," Earthq. Engrg. Struc. Dyn., 9, pp. 343-354.
Ramadan, O. and Novak, M. (1991a). "Strains in long dams due to spatially correlated random ground motions," In P.D. Spanos and C.A. Prebbia (eds.), Computational stochastic mechanics, CMP and Elsevier, pp. 583-594.
Ramadan, O. and Novak, M. (1991b). "Synthesizing spatially incoherent random ground motions for earthquake response analysis," Report No. GEOT-14-91, Fac. Eng. Sc., U.W.O., London, Canada.
Ramadan, O. and Novak, M. (1992a). "Dam response to incoherent ground motions," Proc. of the 10WCEE, Madrid, Spain, July 19-25.
Ramadan, O. and Novak, M. (1992b). "Dam-reservoir-foundation interaction under incoherent seismic excitations," Proc. of the 45th Canadian Geotechnical Society Conference, Toronto, Canada, Oct. 25-28.
Shinozuka, M. and Deodatis, G. (1988). "Stochastic wave models for simulation of seismic ground motion," Proc. of the Workshop on Spatial Variation of Earthquake Ground Motion, Dunwalke, Princeton University, Nov. 7-9.
Spanos, P.D. and Mignolet, M.P. (1990). "Simulation of stationary random processes: two-stage MA to ARMA approach," J. Engrg. Mech., ASCE, 116(3), pp. 620-641.
Veletsos, A.S. and Tang, Y. (1989). "A deterministic assessment of effects of ground motion incoherence," Report No. NCEER-89-0037, National Center for Earthquake Engineering Research, State University of New York at Buffalo.
Veletsos, A.S. and Prasad, A.M. (1989). "Seismic interaction of structures and soils: stochastic approach," J. Struc. Eng., ASCE, 115(4), pp. 935-956.

SOIL-STRUCTURE INTERACTION ANALYSIS USING MULTIPLE INPUT MOTIONS

JOSEPH PENZIEN
International Civil Engineering Consultants, Inc.
Berkeley, California, U.S.A.

CHENG-HSING CHEN; YANG-JYE LEE
National Taiwan University
Taipei, Taiwan, R.O.C.

HAI-NAN HSIAU
Taiwan Power Company
Taipei, Taiwan, R.O.C.

IKUO KATAYAMA
Tokyo Electric Power Services Company
Tokyo, Japan

ABSTRACT: A soil-structure interaction analysis procedure using hybrid modelling is presented which takes into account the effects of wave scattering, spatial variations of seismic input motions, and the nonlinear response of soils. The effectiveness and efficiency of the procedure are investigated through correlation studies using field-model test results.

1. Introduction

The HASSI computer program for solving three-dimensional soil-structure interaction problems makes use of hybrid modelling and the substructure method of analysis. The dynamic properties of the structure and a finite region of soil in the near-field surrounding its foundation are modelled using finite elements and the corresponding properties of the soil in the far-field are modelled by complex frequency-dependent impedance functions along an hemispherical near-field/far-field interface (Gupta, et al. 1982). The program treats soil nonlinearities by the equivalent linearization procedure.

Through previous correlation studies using field-model test results, it has been shown that this program can effectively predict the dynamic response of a soil-structure system under either forced vibration or earthquake excitation (Katayama, et al. 1987; Katayama, et al. 1989; Chen, et al. 1990). Because the field models were of small-scale, the variations of the free-field ground motions over their base dimensions were negligible; thus, uniform motions could be used as inputs in these correlation studies. Likewise, for the same reason, wave scattering effects were negligible, allowing them to be ignored.

However, for large-scale embedded structures, having base dimensions of the same order of magnitude as the predominant wave lengths in the free-field ground motions, both multiple inputs

P. Gülkan and R. W. Clough (eds.), Developments in Dynamic Soil-Structure Interaction, 167–178.

and wave scattering effects must be included to obtain realistic soil-structure interaction results.

To provide the capability of treating these cases, the HASSI program has recently been improved by implementing specialized structural substructuring and superposition procedures for including both multiple free-field ground-motion inputs and wave scattering effects in a single formulation (Lin 1980; Penzien, et al. 1989). The analytical techniques used in this improved program (HASSI-8) allow these features to be treated in an effective and efficient manner (Katayama, et al. 1991).

While results from the above-mentioned correlation studies have already been published, the analytical procedures implemented into the HASSI-8 computer program have not been published. Therefore, it is the purpose of this paper to present these procedures and to provide selected results from the correlation studies.

2. Soil-Structure System

For an axisymmetric soil-structure system, the near-field hybrid model can be represented by finite elements as shown in Figure 1.

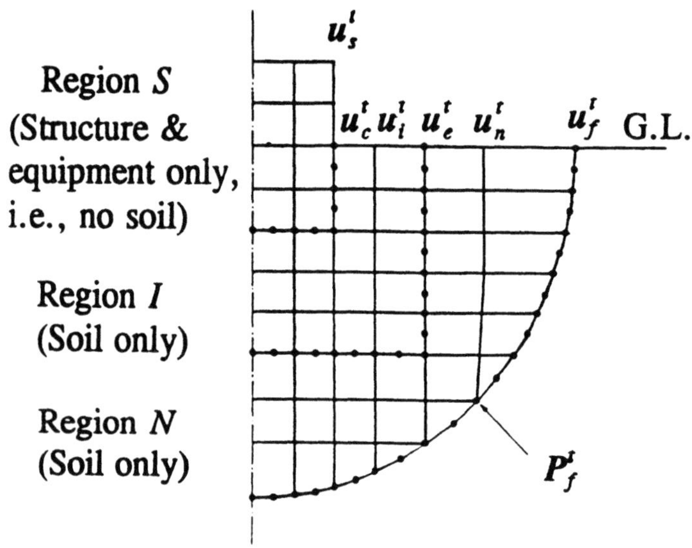

Figure 1. Complete near-field system

To take account of the strain-dependent soil nonlinearities, it has been subdivided into subregions *S*, *I*, and *N*, representing the region of the structure, the region of the soil subjected to significant soil-structure interaction effects, and the region of the soil experiencing negligible soil-structure interaction effects, respectively. The hemispherical interface between the near- and far-fields is subjected to an interaction force vector P_f^t resulting from the combined motion of the near-field/far-field system. The total-displacement time-histories of all nodes shown in Figure 1 include the effects of free-field soil motions, soil-structure interaction, and wave scattering. The controlling equations of these motions are

$$\begin{bmatrix} S^S_{ss} & S^S_{sc} & 0 & 0 & 0 & 0 \\ S^S_{cs} & S^{SI}_{cc} & S^I_{ci} & S^I_{ce} & 0 & 0 \\ 0 & S^I_{ic} & S^t_{ii} & S^f_{ie} & 0 & 0 \\ 0 & S^I_{ec} & S^t_{ei} & S^{IN}_{ee} & S^N_{en} & S^N_{ef} \\ 0 & 0 & 0 & S^N_{ne} & S^N_{nn} & S^N_{nf} \\ 0 & 0 & 0 & S^N_{fe} & S^N_{fn} & S^N_{ff} \end{bmatrix} \begin{Bmatrix} u^t_s \\ u^t_c \\ u^t_i \\ u^t_e \\ u^t_n \\ u^t_f \end{Bmatrix} = \begin{Bmatrix} 0 \\ 0 \\ 0 \\ 0 \\ 0 \\ P^t_f \end{Bmatrix} \tag{1}$$

in which the S matrices on the left-hand side are the assembled impedance matrices corresponding to each region mentioned above, including contributions from mass, stiffness, and complex-stiffness damping.

3. Auxiliary Substructure

Consider now the soil system under free-field conditions without any surface cavity present as shown in Figure 2.

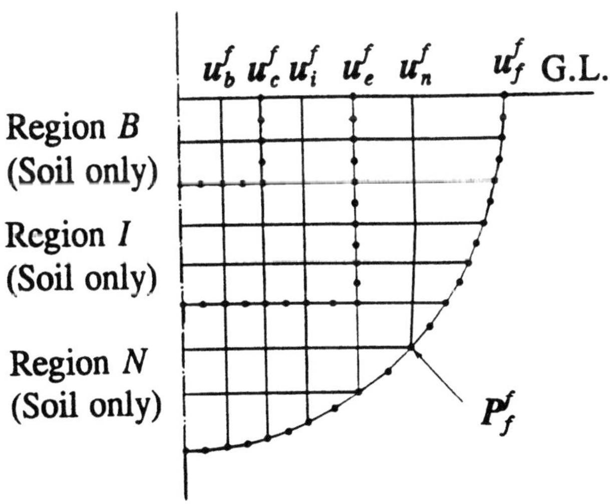

Figure 2. Free-field system

The soils are divided into regions B, I, N, and F corresponding to the same soil-structure regions shown in Figure 1. The nodal displacement time-histories u^f_b, u^f_c, u^f_i, u^f_e, u^f_n, and u^f_f are the free-field soil motions which can be evaluated by hypothesizing types of seismic waves present and specifying free-field control motions at the surface axis-of-symmetry location. In engineering, it is common practice to assume vertically propagating shear and compression waves in horizontally-layered soils for this purpose.

Isolating region N in Figure 2, it will be subjected to free-field excitation forces P_f^f in the local degrees of freedom (DOF) at the hemispherical boundary and external forces P_e^f applied in the DOF at the I/N interface location. The equations of motion for this isolated region under free-field conditions can be written as

$$\begin{bmatrix} S_{ee}^N & S_{en}^N & S_{ef}^N \\ S_{ne}^N & S_{nn}^N & S_{nf}^N \\ S_{fe}^N & S_{fn}^N & S_{ff}^N \end{bmatrix} \begin{Bmatrix} u_e^f \\ u_n^f \\ u_f^f \end{Bmatrix} = \begin{Bmatrix} P_e^f \\ 0 \\ P_f^f \end{Bmatrix} \tag{2}$$

Consider once again isolated region N, but this time, suppose the nodal displacement time-histories in all DOF are the free-field soil motions except those in the DOF at boundary "e" where they are constrained to zero as indicated in Figure 3.

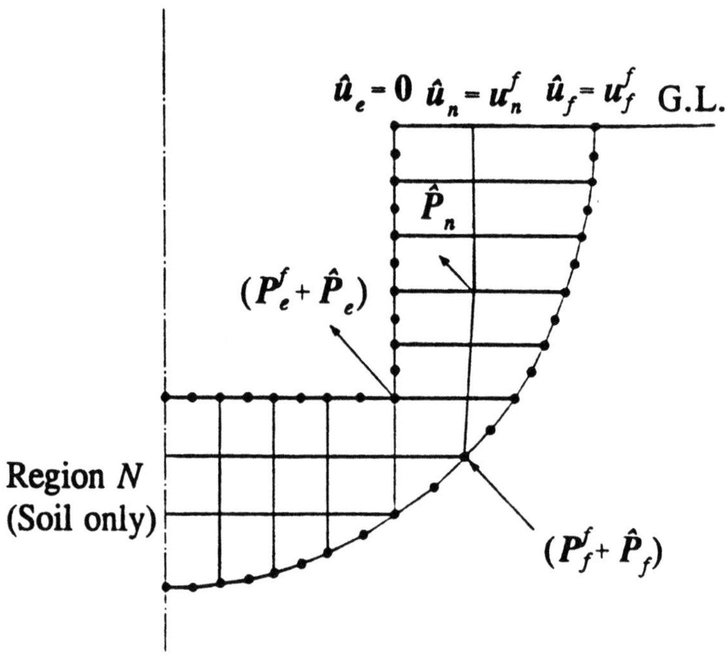

Figure 3. External forces applied to Region N to produce free-field motions in all DOF, except in the DOF of boundary "e" which have zero motions

Under this condition, additional sets of external forces $\hat{P}_e$, $\hat{P}_n$, and $\hat{P}_f$ must be imposed on the nodes at boundary "e", within region N, and at boundary "f", respectively, as given by Eqs. (2) upon substituting $\langle -u_e^f \;\; 0 \;\; 0 \rangle^T$ and $\langle \hat{P}_e \;\; \hat{P}_n \;\; \hat{P}_f \rangle^T$ for the displacement and load vectors, respectively, i.e., as given by

$$\begin{bmatrix} S_{ee}^N & S_{en}^N & S_{ef}^N \\ S_{ne}^N & S_{nn}^N & S_{nf}^N \\ S_{fe}^N & S_{fn}^N & S_{ff}^N \end{bmatrix} \begin{Bmatrix} -u_e^f \\ 0 \\ 0 \end{Bmatrix} = \begin{Bmatrix} \hat{P}_e \\ \hat{P}_n \\ \hat{P}_f \end{Bmatrix} \tag{3}$$

Adding Eqs. (2) and (3) and substituting Eq. (3) for vector $\langle \hat{P}_e \;\; \hat{P}_n \;\; \hat{P}_f \rangle^T$ in the resulting equation, one obtains the equations of motion for the auxiliary substructure shown in Figure 3 in the form

$$\begin{bmatrix} S_{ee}^N & S_{en}^N & S_{ef}^N \\ S_{ne}^N & S_{nn}^N & S_{nf}^N \\ S_{fe}^N & S_{fn}^N & S_{ff}^N \end{bmatrix} \begin{Bmatrix} 0 \\ u_n^f \\ u_f^f \end{Bmatrix} = \begin{Bmatrix} 0 \\ 0 \\ P_f^f \end{Bmatrix} + \begin{bmatrix} 0 & S_{en}^N & S_{ef}^N \\ -S_{ne}^N & 0 & 0 \\ -S_{fe}^N & 0 & 0 \end{bmatrix} \begin{Bmatrix} u_e^f \\ u_n^f \\ u_f^f \end{Bmatrix} \tag{4}$$

4. Analytical Model of HASSI-8

By subtracting Eqs. (4) representing the auxiliary substructure (Figure 3) from Eqs. (1) representing the original soil-structural system (Figure 1), a new set of equations are obtained representing the soil/structure system shown in Figure 4.

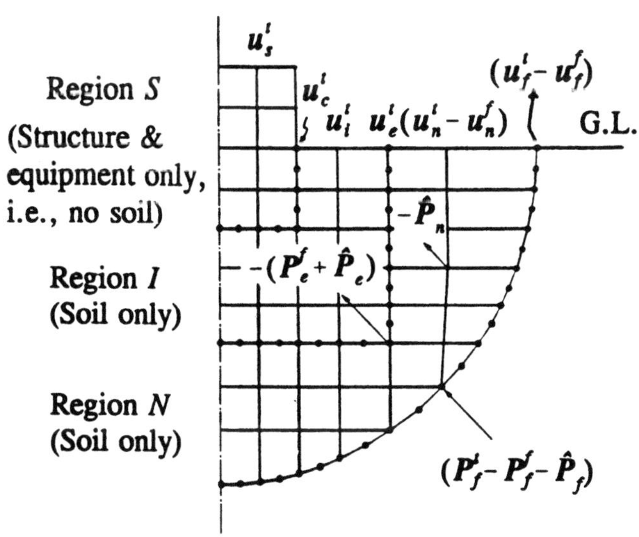

Figure 4. External forces applied to the complete near-field region to produce relative motions in the "n" and "f" DOF of Region N, but total motions in all other DOF

Noting that

$$P_f^t - P_f^f = -G_{ff}(u_f^t - u_f^f) \tag{5}$$

where G_{ff} is the far-field impedance matrix as developed by Gupta, et al. (1982), this new set of equations takes the form

$$\begin{bmatrix} S_{ss}^S & S_{sc}^S & 0 & 0 & 0 & 0 \\ S_{cs}^S & S_{cc}^{SI} & S_{ci}^I & S_{ce}^I & 0 & 0 \\ 0 & S_{ic}^I & S_{ii}^I & S_{ie}^I & 0 & 0 \\ 0 & S_{ec}^I & S_{ei}^I & S_{ee}^{IN} & S_{en}^N & S_{ef}^N \\ 0 & 0 & 0 & S_{ne}^N & S_{nn}^N & S_{nf}^N \\ 0 & 0 & 0 & S_{fe}^N & S_{fn}^N & (S_{ff}^N + G_{ff}) \end{bmatrix} \begin{Bmatrix} u_s^t \\ u_c^t \\ u_i^t \\ u_e^t \\ (u_n^t - u_n^f) \\ (u_f^t - u_f^f) \end{Bmatrix} = \begin{bmatrix} 0 & 0 & 0 \\ 0 & 0 & 0 \\ 0 & 0 & 0 \\ 0 & -S_{en}^N & -S_{ef}^N \\ S_{ne}^N & 0 & 0 \\ S_{fe}^N & 0 & 0 \end{bmatrix} \begin{Bmatrix} u_e^f \\ u_n^f \\ u_f^f \end{Bmatrix} \tag{6}$$

Once the free-field soil displacements u_e^f, u_n^f, and u_f^f are known, this equation can be solved giving the total displacements in the DOF of "*s*", "*c*", "*i*", and "*e*", and the relative displacements in the DOF of "*n*" and "*f*". The general formulation leading to this equation has considered the spatial variations of the free-field ground motions and has included the effects of both SSI and wave scattering.

To adjust for soil nonlinearities, the HASSI-8 program uses the iteration procedure of the equivalent linearization method (Katayama, et al. 1989; Chen, et al. 1990). In the site response analysis to determine the free-field ground motions u_e^f, u_n^f, and u_f^f, this iteration procedure is first applied in calculating the strain-compatible shear modulus and damping ratio for each soil element in the near-field. These values are then used as initial soil properties in the soil-structure interaction analysis which follows. Using the shear-strain levels obtained in this analysis, the soil properties in Region *I* of the near-field, where the effects of soil-structure interaction are significant, can be further modified. The soil properties in Regions *N* and *F* may be assumed to remain constant however, during the SSI analysis phase, due to the negligible influence of soil-structure interaction in these regions. This assumption allows, using condensation procedures, the generation of impedances acting in the "*c*" and "*i*" boundary DOF of Region *I* which will fully account for its interaction with the neighboring regions; thus, greatly simplifying the analytical formulation of the complete soil-structure system and similarly reducing computation effort during the iteration process. The HASSI-8 analysis can be performed in a rather small computer using a relatively small amount of operating time.

5. Correlations with Field Test Data

To verify the HASSI-8 computer program capability, correlation studies were carried out using field-test results obtained for the EPRI/TPC containment model located in Lotung, Taiwan (Tang 1987; Chen, et al. 1990) and the TEPSCO containment model located at its Kazusaminato site in Japan (Katayama, et al. 1987). These reinforced concrete models were built for the specific purpose of investigating the effects of seismic soil-structure interaction.

5.1. EPRI/TPC CONTAINMENT MODEL

The EPRI/TPC 1/4-scale model consists of a hollow cylinder of thickness 30 cm supporting a thick roof slab and resting on a basemat. It has an outside diameter of 10.5 m and a total height of 15 m with 4.7 m embedded in the ground and it is supported on very deep alluvial deposits consisting of relatively loose silty sands and sandy silts.

The dynamic response of the complete soil-structure system during the May 20, 1986 Hualien earthquake will now be examined. To conduct this analysis, the HASSI-8 model was constructed in accordance with the above formulations as shown in Figure 5.

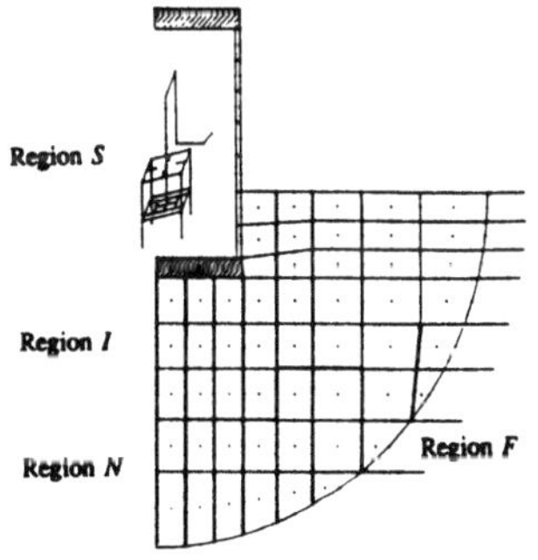

Figure 5. Hybrid model of 1/4-scale containment model at Lotung

Both the roof and basemat of the containment structure were modelled as rigid blocks and the internal steam generator and its supporting frame were modelled by beam elements. The hollow cylinder and the soil in the near-field were modelled by variable-node axisymmetric solid elements. For the seismic analysis, three-components of acceleration recorded 50 m away from the model were specified as the control motions for calculating the strain-compatible free-field soil motions.

Using the vertically propagating shear-wave and compression-wave models to evaluate the three-component free-field ground motions, the spatially varying input motions were determined and the complete soil-structure interaction analysis was completed using 3 iterations to obtain strain-compatible responses of the soils in Region *I*. The predicted time-history responses correlated very well with the corresponding measured responses. Due to limited space herein, comparisons of corresponding responses will be shown only for those motions recorded on the basemat and on the roof and they will be made relative to their acceleration response spectra.

The acceleration response spectra for the predicted (*P*) and recorded (*R*) east-west (E-W), north-south (N-S), and vertical (V) motions at the top east-side location on the basemat, using 5% damping, are shown in Figure 6. The agreement of corresponding spectra for the predicted and recorded motions are quite good, except those representing the N-S motions in the higher frequency range. This discrepancy is due primarily to the fact that the field foundation conditions are not fully symmetrical as assumed in the HASSI-8 modelling. Clearly the axisymmetrical analytical model simulates the E-W behavior considerably better than the N-S behavior.

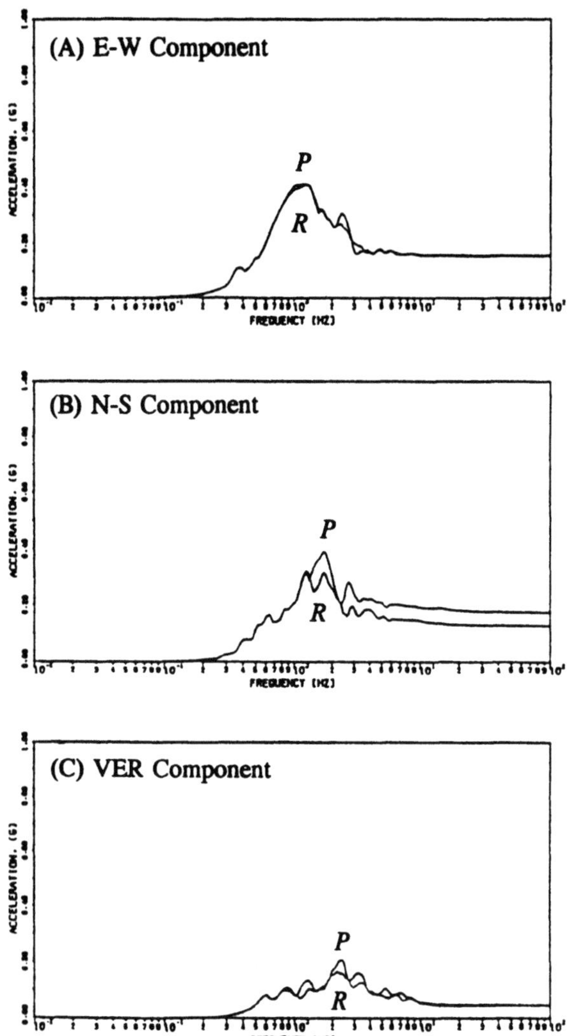

Figure 6. Comparison of the acceleration response spectra for the predicted (*P*) and recorded (*R*) motions at the top east-edge location on the basemat using 5% damping

Figure 7 shows acceleration response spectra similar to those of Figure 6, except that the motions represented are those at the top east-edge location of the roof slab. Again, the agreement of corresponding spectra for the predicted (*P*) and recorded (*R*) motions are quite good except in the approximate frequency range 2-6 Hz. The discrepancy in this range, which is quite large for the N-S motions, is due to the lack of full-symmetry conditions in the actual field foundation as mentioned earlier.

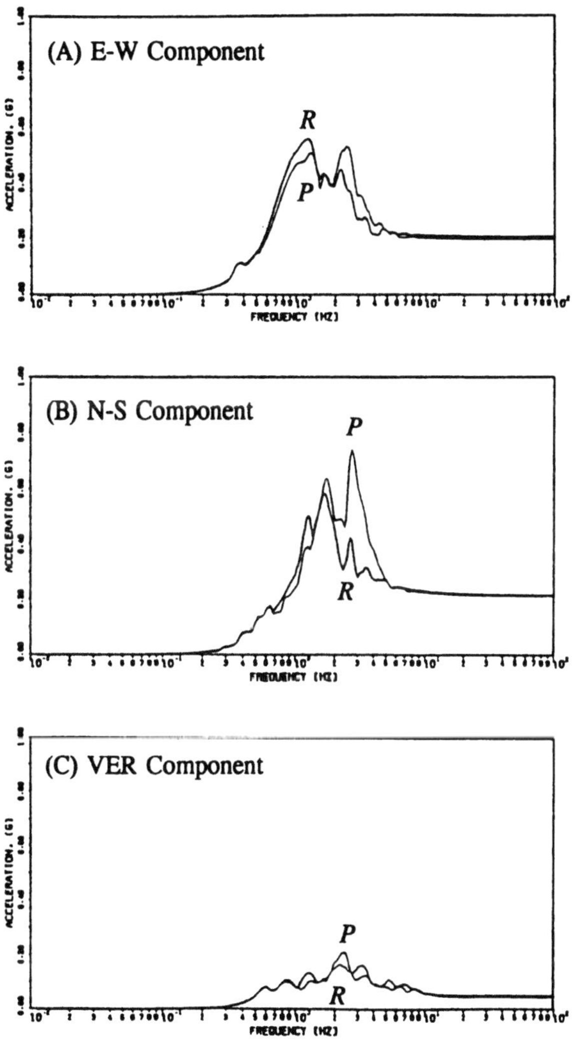

Figure 7. Comparison of the acceleration response spectra for the predicted and recorded motions at the top east-edge location on the roof slab using 5% damping

The HASSI-8 computing time required to obtain the time-history seismic response of the complete soil-structure system shown in Figure 5 was about 30 hours for a 3-iteration solution using the Micro Vax II computer. This time is much less than that required by other commonly used methods for conducting three-dimensional nonlinear soil-structure interaction analyses.

5.2. TEPSCO CONTAINMENT MODEL

The TEPSCO containment model consists of a hollow cylinder supporting a thick roof slab and resting on a basemat as shown in Figure 8.

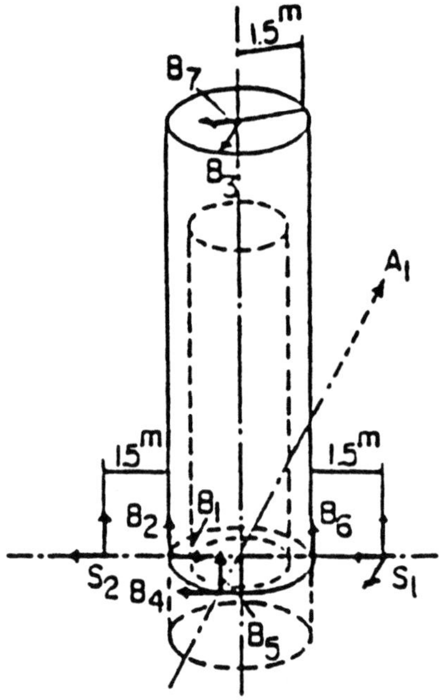

Figure 8. TEPSCO containment model showing seismograph locations

It has an outside diameter of 3.0 m and a total height of 10 m with 1.5 m embedded in the ground. It is supported on a medium coarse sand layer having a 300 m/sec shear-wave velocity at the small strain level. The dynamic response of the complete soil/structure system during the 1987 Off Bousou Peninsula earthquake will now be examined.

For correlation purposes, the HASSI-8 model of this system was constructed in a manner similar to that described previously for the Lotung system (Figure 5). Using strain-dependent soil properties obtained by laboratory tests and three orthogonal components of surface motion measured at station No. S_1, located 1.5 m away from the outer-boundary of the model (see Figure 8) as the control free-field input motion, the dynamic response of the complete soil/structure system was obtained.

The predicted (*P*) time-histories of acceleration in the N-S and E-W directions at the center of the top surface of the roof slab are shown by the dashed curves in Figures 9(a) and 9(b), respectively, where they can be compared with their corresponding recorded (*R*) motions shown by the solid curves.

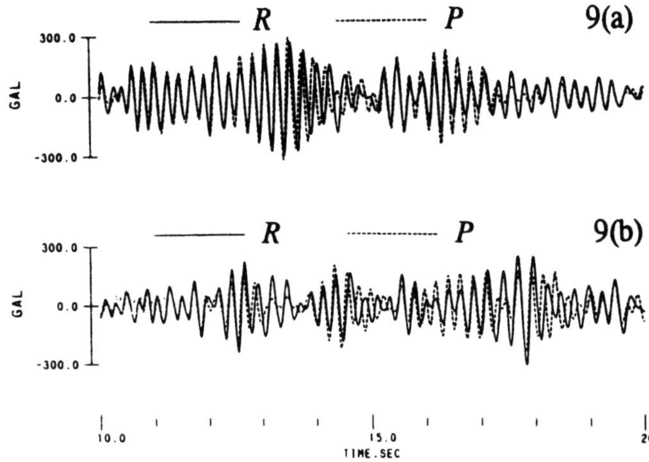

Figure 9. Time-histories of acceleration in the N-S and E-W directions on top surface of roof slab

The 5%-damping acceleration response spectra for these motions are shown in Figures 10(a) and 10(b). The agreement between corresponding time-histories in Figure 9 and their acceleration response spectra in Figure 10 are generally very good. The CPU and I/O times to carry out the above correlation studies were 30 and 60 minutes respectively, on FACOM-M/38R.

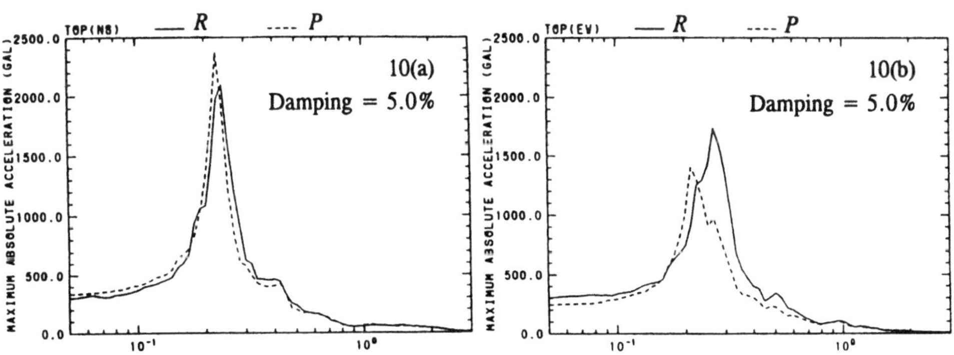

Figure 10. Acceleration response spectra (5% damping) for N-S and E-W motions on top surface of roof slab

5.3. WAVE SCATTERING AND MULTIPLE GROUND-MOTION INPUTS

Prior to conducting all of the above-described HASSI-8 correlation studies, identical studies were carried out using the HASSI-7 computer program. These two computer programs differ only in the added features of the HASSI-8 program to include the effects of wave scattering and multiple ground-motion inputs. Upon comparing corresponding results generated by these two programs, only small differences are noticeable. This close comparison is to be expected as the base dimensions of both the EPRI/TPC and TEPSCO field-test models are too small to significantly induce these effects. The operations in the HASSI-8 program which treat them were, however, found to be functioning in accordance with the analytical formulations previously given. Therefore, the authors are confident such effects, which are significant for large-scale structures, can be evaluated with reasonable accuracy using the HASSI-8 program.

6. Conclusions

The analytical procedures of the HASSI-8 computer program uses spatially varying ground motions at some distance from the soil-foundation interface to solve a three-dimensional soil-structure interaction problem and they include the effects of wave scattering, multiple input motions, and strain-dependency of soil properties. As demonstrated herein, the formulations and solution techniques employed make this program very effective and efficient.

7. Acknowledgement

The computer code HASSI-8 was developed under the cooperative support of the Tokyo Electric Power Services Company (TEPSCO) and the Taiwan Power Company (TPC). The authors express their sincere thanks and appreciation to TEPSCO and TPC for their financial support and assistance, and to the Electric Power Research Institute of the U.S.A., TPC and TEPSCO for providing the field-test data referred to herein.

8. REFERENCES

Chen, C. H., Lee, Y. J., Jean, W. Y., Katayama, I., and Penzien, J. (1990) 'Correlation of predicted seismic response using hybrid modelling with EPRI/TPC Lotung experimental data', *Earthquake Engineering & Structural Dynamics* 19, 993-1024.

Gupta, S., Penzien, J., Lin, T. W., and Yeh, C. S. (1982) 'Three-dimensional hybrid modelling of soil-structure interaction', *Earthquake Engineering & Structural Dynamics* 10, 69-87.

Katayama, I., Ohsumi, T., Kubo, Y., Nakase, H., and Penzien, J. (1987) 'Experimental validation of soil-structure interaction analysis', *Proc.* of US-Korea Joint Seminar on Critical Engineering Systems, Seoul, Korea, 1, 495-506.

Katayama, I., Chen, C. H., Lee, Y. J., Jean, W. Y., and Penzien, J. (1989) 'Near-field soil-structure interaction analysis using nonlinear hybrid modelling', SMiRT-10, Anaheim, California, K05/2, 127-132.

Katayama, I., Tsuzuki, T., and Penzien, J. (1991) 'Wave scattering effect in soil-structure interaction', SMiRT-11, Tokyo, Japan, K07/3, 153-158.

Lin, T. W. (1980) 'Dynamic response of embedded structure subjected to horizontal shear waves', Ph.D. thesis, National Taiwan University, Taipei, Taiwan, R.O.C.

Penzien, J., Chen, C. H., Lee, Y. J., and Jean, W. Y. (1989) 'Computer program HASSI-8, hybrid analysis of soil-structure interaction', Report prepared by Eastern International Engineers, Inc. (Penzien, J. et al.) for Tokyo Electric Power Services Co., Ltd.

Tang, H. T. (1987) 'Large scale soil-structure interaction', EPRI Report NP-5513-SR, Palo Alto, California, U.S.A.

INFLUENCE OF DIFFERENT SOIL MODELING CRITERIA IN SSI ANALYSIS

P. GÜLKAN
Department of Civil Engineering
Middle East Technical University
06531 Ankara
Turkey

ABSTRACT. This paper is in the form of two nearly independent parts. In the first, an overview of the mechanistic aspects of modeling soil structure systems and solution techniques are presented in textbook fashion. The second part deals with a particular application in which blind predictions were made for the measured earthquake response of a 1/4-scale nuclear containment structure built on the SMART site near Lotung, Taiwan.

1. Introduction

The mathematical relationships describing the stress-strain behavior of a small amount of matter in a structural model is called the "material model" or the "constitutive model". The formulation of the constitutive model is probably the most important step in obtaining the solution of practical engineering problems; this importance is even greater for geomechanical systems comprising representations of structural and geological components. For geomaterials a representative formulation is often critical because, depending on strain levels, highly nonlinear behavior can be obtained. For this reason a whole branch of geomechanical engineering has been developed whose task is to formulate and to calibrate geotechnical models [e.g. Dungar and Studer (1986), Chen and Baladi (1985), Desai (1984), Desai and Gallagher (1984), Wittke (1979), Gudehus (1977)]. For most structural components the representation of material constitutive relations can be done with greater confidence.

For the purpose of making a broad overview of the subject some general concepts are introduced in the next three sections. This treatment is necessarily textbook-like, and follows the order contained in Dungar (1986). The details are discussed in a selective way because the objective here is to demonstrate that the difficulties encountered in practice concerning soil modeling can be overcome if the entity being modeled is large enough, and the boundaries and loading circumstances are defined within certain limits. It may then be possible to treat soil deposits as linear elastic, or nonlinear elastic, or perfectly plastic for the purpose of engineering stress analysis.

P. Gülkan and R. W. Clough (eds.), Developments in Dynamic Soil-Structure Interaction, 179–217.

2. Elastic, Piecewise Elastic and Elastic-Plastic Models

Geomaterials, like many other types of materials, exhibit nonlinear behavior in the form of brittle failure or plastic behavior as shown in Figure 1. Whereas much of current research in geomechanics is oriented towards the formulation of increasingly more involved models for soil- and rock-like materials, it is also true that many problems exist which permit a linear elastic model to be assumed either as a first approximation, or even as the definitive model itself, provided the elastic properties have been properly adjusted taking into account the expected ranges of strain.

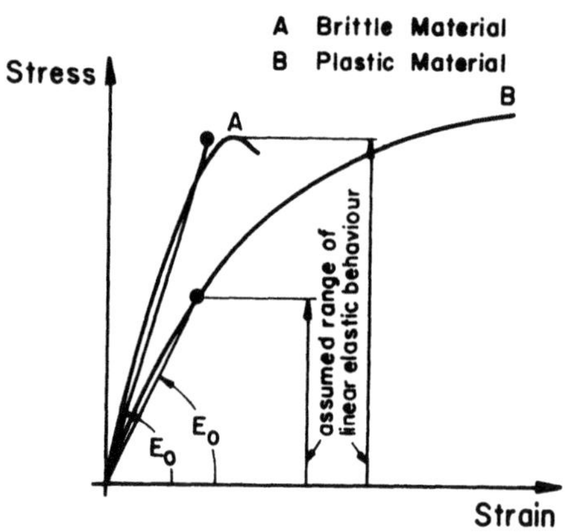

Figure 1. Stress-strain characteristics of brittle and plastic material [after Dungar (1986)]

The finite element method is used in current engineering practice for the routine solution of many different types of problems, and many different elements have been developed for two- or three-dimensional analysis. With this facility at hand, it is also possible to assume linear elastic behavior dependent on strain levels - the piecewise elastic method of analysis as described in Figure 2. This approach basically involves making adjustments in the elastic properties which are illustrated by a single proportionality constant in Figure 2, with the load being applied incrementally over an arbitrarily defined pseudo-time axis.

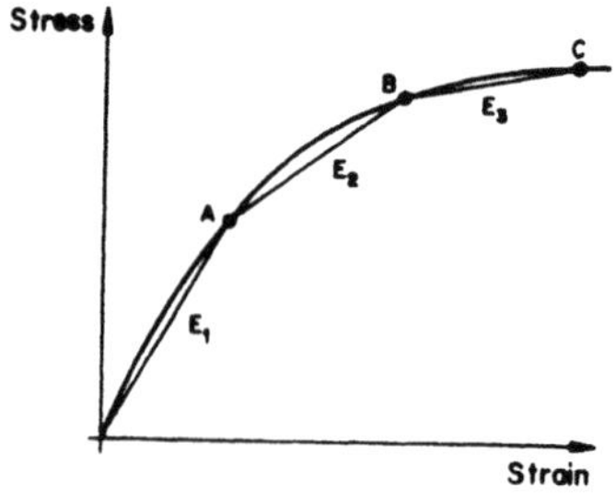

Figure 2. Piecewise elastic representation of a plastic material

In the case of dynamic loading, similar piecewise linear stresses may be pursued on a step-by-step basis throughout the duration of the externally defined effect. In this case a large number of adjustments must be made in the elastic properties with provision made for reversed loading paths. The solution strategy in conducting such analyses may be modified by identifying equivalent elastic parameters which correspond to the range of average strain conditions attained during the response phase. The equivalent linearization method has found much popularity in the response analysis of soil-structure systems, and we will devote more attention to it in the sections that follow.

The basic concept involved in elasto-plastic models is illustrated for the case of uniaxial stress-strain in Figure 3. Here, elastic response occurs until a given level of stress denoted by point A is reached. The importance of point A lies in the fact that once stress level exceeds the elastic limit $\sigma_y = \sigma_a$, any increment of stress is accompanied by an irrecoverable increment of plastic strain. The curve ABC defines the yield curve, and if at point B with the corresponding point $\sigma_y = \sigma_a$ an unloading occurs, then the elastic component of the total strain, ϵ, is recovered, whereas the plastic component, ϵ^p, is retained. On reloading linear elastic behavior is obtained until the stress level reaches σ_y again. So for this simple example we may write

$$\sigma_y = f(\epsilon^p) \tag{1}$$

Equation (1) is called the hardening law. When σ_y increases with increasing ϵ^p, the material is said to strain harden; when it remains constant then it is called perfectly plastic, and when it decreases it is said to strain soften.

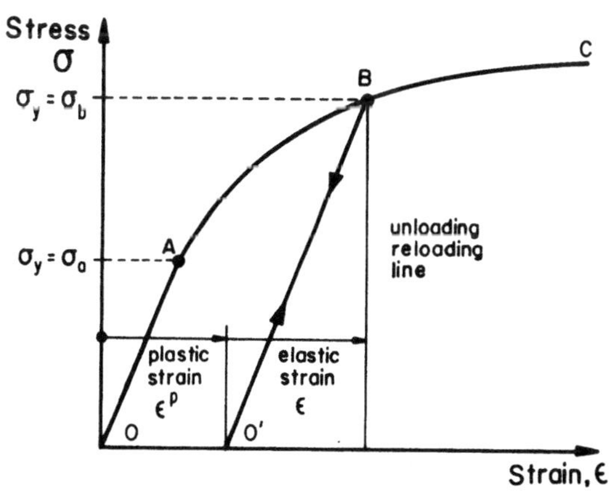

Figure 3. Yield stress, plastic strain and repeated loading for a plastic material

Depending on strain levels and the loading "history" geomaterials exhibit combinations of all of these modes of behavior. In a strict sense, Equation (1) must be written for all components of stress and strain for generality.

2.1. INVARIANTS OF STRESS AND STRAIN

Uniaxial stress-strain serves only as a conceptual tool in reference to the actual three-dimensional state of stress valid for the majority of problems encountered in practice. For any material model to be of use in a general solution code it must contain provisions for representing stress-strain relationships in a multiple stress state. If we ignore the planes of weakness or discontinuity which exist in most geomaterial deposits, then models of multiaxial behavior can be based on the principal components of stress and strain which are independent of the coordinate axes.

In geotechnical engineering stresses are commonly expressed in terms of the effective stress which is the total stress, that is the stress caused by the loads acting on the boundaries of a finite volume of material minus the pore pressure. Using the sign convention that positive stresses are tensile, and collecting all stresses into a column vector

$$\boldsymbol{\sigma} = \boldsymbol{\sigma}' - \mathbf{p} \tag{2}$$

where $\boldsymbol{\sigma}$ and $\boldsymbol{\sigma}'$ are the vectors of total and effective stress, and $\mathbf{p}$ the pore pressure vector

$$\mathbf{p} = \{p_o \;\; p_o \;\; p_o \;\; 0 \;\; 0 \;\; 0\}^T \tag{3}$$

where

$$p_o = \text{pore water pressure}$$

One way of expressing the stress invariants p', q, and Θ is illustrated in Figure 4. Accordingly,

$$p' = -\frac{1}{3}(\sigma'_1 + \sigma'_2 + \sigma'_3) \tag{4}$$

$$q = \frac{1}{\sqrt{2}}[(\sigma'_1 - \sigma'_2)^2 + (\sigma'_2 - \sigma'_3)^2 + (\sigma'_3 - \sigma'_1)^2]^{1/2} \tag{5}$$

$$\theta = \tan^{-1}\left[\frac{\sigma'_1 - 2\sigma'_2 + \sigma'_3}{\sqrt{3}(\sigma'_1 - \sigma'_3)}\right] \tag{6}$$

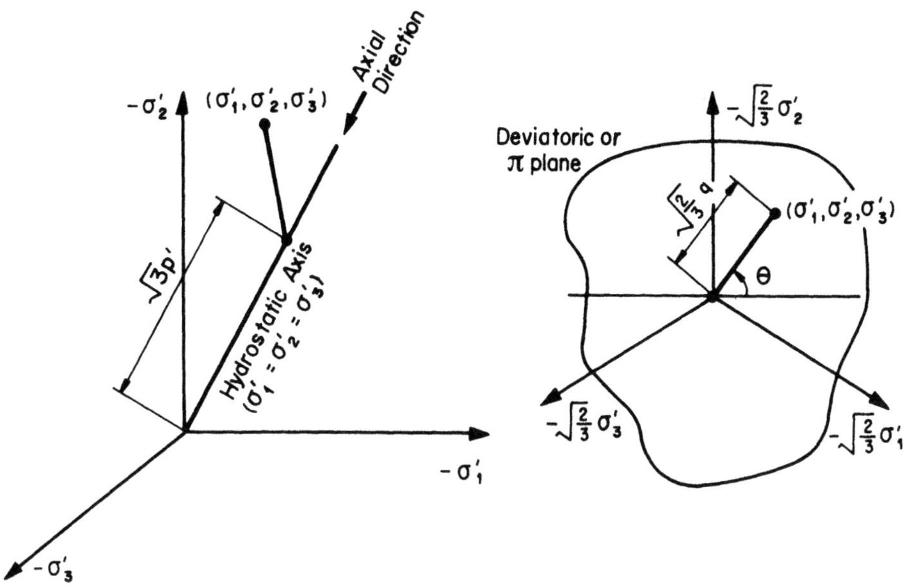

Figure 4. Stress invariants p', q and θ; the deviatoric plane is normal to the hydrostatic axis

The plane normal to the hydrostatic axis is called the deviatoric (or π) plane. The distance from the origin to this plane is $\sqrt{3}\,p'$, and the distance from the hydrostatic axis to any point within this plane is given by

$$\sqrt{2}\, q / 3 \tag{7}$$

The three quantities given by Equations (4) - (6) therefore uniquely define a stress state σ'_1, σ'_2, and σ'_3. In the special case of the triaxial test [see, e.g. Prater and Studer (1986)] we obtain

$$p' = -\frac{1}{3}(\sigma'_1 + 2\sigma'_3) \tag{4a}$$

$$q = \sigma_3 - \sigma_1 \tag{5a}$$

The principal strain increments can also be used in plasticity models, and these may be resolved into volumetric and deviatoric components within the octahedral plane.

3. Some Elasto-Plastic Models

As preliminary considerations to the development of elastoplastic models which will be presented in the sections to follow, brief statements will be made of some well known yield criteria for multistress conditions. This discussion is of course less than exhaustive. For this purpose the yield curve shown in Figure 3 may be extended into a surface expressed in terms of the effective stress vector σ' and the plastic strain ϵ^p

$$f(\sigma',\kappa) = 0 \tag{7}$$

where κ is the hardening parameter. In Figure 5 a series of functions similar to Equation (7) each corresponding to different plastic strains ϵ^p_i (i = 1,2 . . n) is illustrated. A point $p(\sigma')$ is at a normalized distance given by Equation (7) from the yield surface. This distance is zero when the point lies on the surface, and the function is negative when the point lies inside the surface.

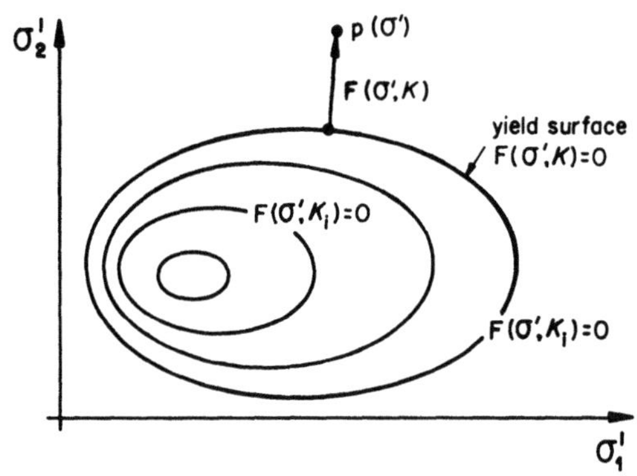

Figure 5. Yield surface in two dimensional space

3.1. VON MISES CRITERION

With reference to the isotropic material assumption, in which case the yield surface can be expressed in terms of the stress invariant, a simple yield surface consisting of a cylinder of radius r whose axis is the hydrostatic axis may be defined:

$$F = q - r = 0 \tag{8}$$

This surface is described in Figure 6. When results from a triaxial test are employed then

$$F = q - 2c = 0 \tag{9}$$

where c is the cohesion, a positive number.

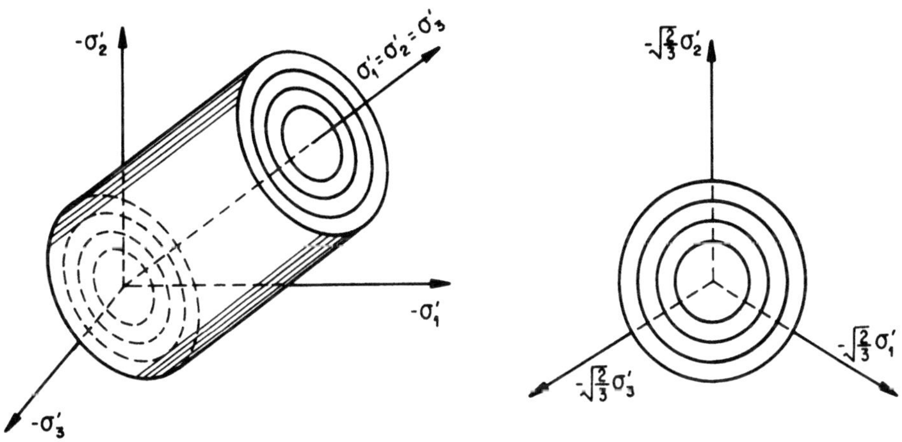

Figure 6. The von Mises yield criterion

3.2. TRESCA CRITERION

In triaxial compression conditions Equation (9) corresponds to the maximum shear stress criterion postulated by Tresca. In the general case we can express this as

$$F = (\sigma_3 - \sigma_1) - 2c = 0 \tag{10}$$

In Figure 7 we picture Equation (10) in the deviatoric plane along with the von Mises criterion.

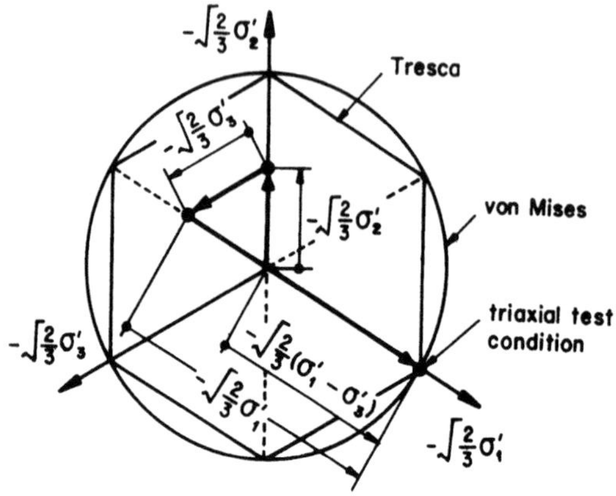

Figure 7. Von Mises and Tresca criteria in the deviatoric plane

3.3. MOHR-COULOMB CRITERION

Under drained conditions geomaterials under shear exhibit sensitivity to normal pressure. For triaxial test conditions a close approximation to the actual strength can be obtained as described in Figure 8.

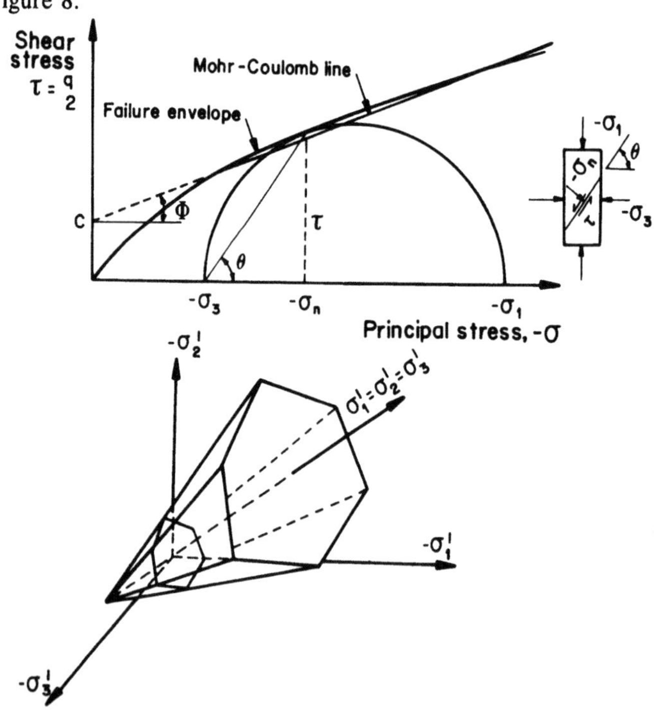

Figure 8. Construction of Mohr-Coulomb line

$$\tau = c - \sigma_n \tan \phi \tag{11}$$

where τ is the maximum shear stress on the failure plane, σ_n the corresponding normal stress (negative for compression) and ϕ the friction angle. The corresponding failure surface can be generalized as

$$F = -\sigma_1(1 - \sin \phi) + \sigma_3(1 + \sin \phi) - 2c = 0 \tag{12}$$

Note that Equation (12) reduces to Equation (10) for $\phi = 0$. In Figure 9 all three of these criteria are shown on the deviatoric plane. Further refinements in the way these criteria are defined are possible; many of these represent efforts to eliminate the sharp corners of the failure surfaces in Figures 7 or 9 which lead to numerical difficulties.

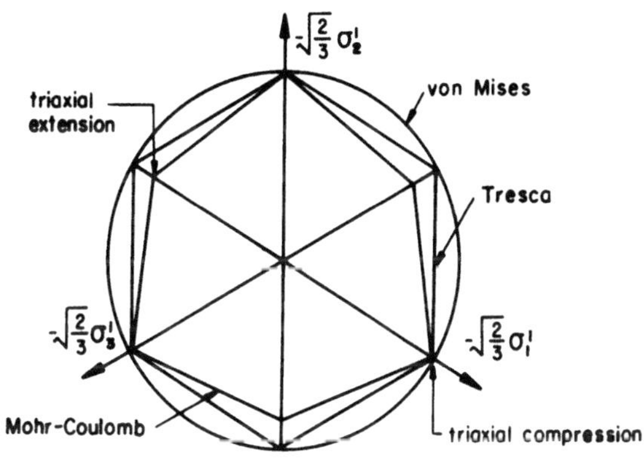

Figure 9. Comparison of yield criteria

3.4. DRUCKER-PRAGER CRITERION

The von Mises and Tresca hypotheses described above are both independent of the hydrostatic pressure p_o, and are experimentally verified to be valid for metal-like materials. A modification of the von Mises criterion proposed by Drucker can be formulated as

$$3 \alpha \sigma_m + \frac{1}{\sqrt{2}} \sqrt{(\sigma'^2_1 + \sigma'^2_2 + \sigma'^2_3)} - k = 0 \tag{13}$$

where α and k are given constants which can be expressed in terms of the constants ϕ and c of the Mohr-Coulomb hypothesis as follows:

$$\alpha = \frac{2 \sin\phi}{\sqrt{3}\,(3 - \sin\phi)} \tag{14}$$

$$k = \frac{6\, c \cos\phi}{\sqrt{3}\,(3 - \sin\phi)} \tag{15}$$

and

$$\alpha_m = \frac{1}{3}\,(\sigma_1 + \sigma_2 + \sigma_3) \tag{16}$$

4. Linear Analysis Procedures

Much of current analysis techniques rely on the finite element method, so it will be necessary to cover some well-known procedures with reference to detailed derivation of element properties of a particular element known as the isoparametric quadrilateral or Q8 element [Weaver and Johnston (1984)]. With the exception of the stiffness properties of the element used in the analysis all linear structural analysis procedures share the following analysis steps:

1. Decide on complexity of analysis required for problem at hand
2. Select appropriate dimensions of mathematical model
3. Specify different material zones of model
4. Generate mesh
5. Define loads and other external effects
6. Assemble structural stiffness matrix
7. Solve for displacements for each load case
8. Use displacements to work stresses for each load case

Currently available software have made many of these steps to appear to be invisible to the user, and can be executed with great ease.

4.1. ELEMENT STIFFNESS FORMULATION

The 8-noded isoparametric element Q8 is shown in its physical and mapping space formats in

Figure 10. Nodal displacements consist of x- and y-displacements of each node. Thus

$$\mathbf{q} = \{u_1\ v_1\ u_2\ v_2 \ .\ .\ .\ u_8\ v_8\}^T \tag{17}$$

$$= \{q_1\ q_2\ .\ .\ \qquad q_{16}\}^T$$

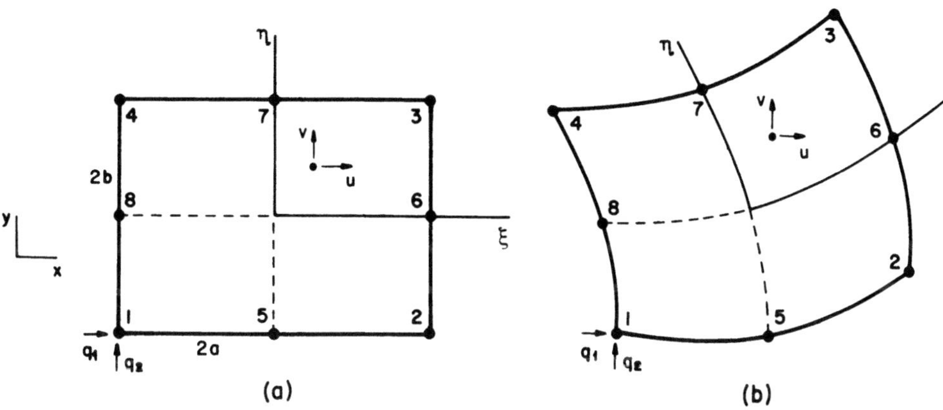

Figure 10. Element Q8

The following form for the displacement functions may be assumed

$$u = c_1 + c_2\xi + c_3\eta + c_4\xi^2 + c_5\xi\eta + c_6\eta^2 + c_7\xi^2\eta + c_8\xi\eta^2 \tag{18a}$$

$$v = c_9 + c_{10}\xi + c_{11}\eta + c_{12}\xi^2 + c_{13}\xi\eta + c_{14}\eta^2 + c_{15}\xi^2\eta + c_{16}\xi\eta^2 \tag{18b}$$

Evaluating the generic displacements expressed in Equation (18) at the nodes leads to the following form

$$u = \Sigma\, f_i\, u_i \qquad i = 1,2,\ldots,8 \tag{19a}$$

$$v = \Sigma\, f_i\, v_i \qquad i = 1,2,\ldots,8 \tag{19b}$$

where

$$f_i = \frac{1}{4}(1 + \xi_0)(1 + \eta_0)(\xi_0 + \eta_0 - 1) \qquad i = 1,2,3,4 \tag{20a}$$

$$f_i = \frac{1}{2}(1 - \xi^2)(1 + \eta_0) \qquad i = 5,6 \tag{20b}$$

$$f_i = \frac{1}{2}(1 + \xi_0)(1 - \eta^2) \qquad i = 7,8 \tag{20c}$$

with

$$\xi_o = \xi_i\, \xi \tag{21a}$$

$$\eta_o = \eta_i\, \eta \tag{21b}$$

For the 8 nodes in Figure 10 the ξ_i and η_i values are listed in Table 1.

TABLE 1. Nodal coordinates for element Q8

i	1	2	3	4	5	6	7	8
ξ_i	-1	1	1	-1	0	1	0	-1
η_i	-1	-1	1	1	-1	0	1	0

The stiffness relation will be obtained via the virtual work formulation, so we need to obtain the expressions for the strains within the element. The generic displacements in Equation (19) are expressible via the interpolation functions given in Equation (20), so that we must differentiate them according to the chain rule. With a comma preceding a subscripted variable denoting partial differentiation with respect to that variable we can obtain the expressions in Table 2 for this purpose.

TABLE 2. Shape function derivatives

i	$f_{i,\xi}$	$f_{i,\eta}$
1	$\frac{1}{4}(2\xi + \eta)(1 - \eta)$	$\frac{1}{4}(1 - \xi)(2\eta + \xi)$
2	$\frac{1}{4}(2\xi - \eta)(1 - \eta)$	$\frac{1}{4}(1 + \xi)(2\eta + \xi)$
3	$\frac{1}{4}(2\xi + \eta)(1 + \eta)$	$\frac{1}{4}(1 + \xi)(2\eta + \xi)$
4	$\frac{1}{4}(2\xi - \eta)(1 + \eta)$	$\frac{1}{4}(1 - \xi)(2\eta - \xi)$
5	$-\xi(1 - \eta)$	$-\frac{1}{2}(1 - \xi^2)$
6	$\frac{1}{2}(1 - \eta^2)$	$-(1 + \xi)\eta$
7	$-\xi(1 + \eta)$	$\frac{1}{2}(1 - \xi^2)$
8	$-\frac{1}{2}(1 - \eta^2)$	$-(1 - \xi)\eta$

$$\epsilon_x = \frac{\partial u}{\partial x} \tag{22a}$$

$$\epsilon_y = \frac{\partial v}{\partial y} \tag{22b}$$

$$\gamma_{xy} = \frac{\partial u}{\partial y} + \frac{\partial v}{\partial x} \tag{22c}$$

If we take the geometric interpolation functions to be the same as the displacement shape functions given by Equation (20), this will mean that the local coordinates ξ and η will become curvilinear, and the sides of the element will become quadratic. With

$$\begin{aligned} x &= \Sigma\, f_i\, x_i \qquad\qquad i = 1,2,\ldots,8 \\ y &= \Sigma\, f_i\, y_i \qquad\qquad i = 1,2,\ldots,8 \end{aligned} \tag{23}$$

we seek a way of formulating Equation (22). Note that expressions like

$$\frac{\partial f}{\partial x} = \frac{\partial f}{\partial \xi}\frac{\partial \xi}{\partial x} + \frac{\partial f}{\partial \eta}\frac{\partial \eta}{\partial x} \tag{24}$$

can not be evaluated directly because we can not solve for ξ and η in terms of x and y. Taking the reverse approach

$$\frac{\partial f}{\partial \xi} = \frac{\partial f}{\partial x}\frac{\partial x}{\partial \xi} + \frac{\partial f}{\partial y}\frac{\partial y}{\partial \xi} \tag{25}$$

and

$$\frac{\partial f}{\partial \eta} = \frac{\partial f}{\partial x}\frac{\partial x}{\partial \eta} + \frac{\partial f}{\partial y}\frac{\partial y}{\partial \eta} \tag{26}$$

or

$$\begin{Bmatrix} \dfrac{\partial f}{\partial \xi} \\ \dfrac{\partial f}{\partial \eta} \end{Bmatrix} = \begin{bmatrix} \dfrac{\partial x}{\partial \xi} & \dfrac{\partial y}{\partial \xi} \\ \dfrac{\partial x}{\partial \eta} & \dfrac{\partial y}{\partial \eta} \end{bmatrix} \begin{Bmatrix} \dfrac{\partial f}{\partial x} \\ \dfrac{\partial f}{\partial y} \end{Bmatrix} \tag{27}$$

$$= \begin{bmatrix} x_{,\xi} & y_{,\xi} \\ x_{,\eta} & y_{,\eta} \end{bmatrix} \begin{Bmatrix} \dfrac{\partial f}{\partial x} \\ \dfrac{\partial f}{\partial y} \end{Bmatrix}$$

$$= \begin{bmatrix} \Sigma\, f_{i,\xi}\, x_i & \Sigma\, f_{i,\xi}\, y_i \\ \Sigma\, f_{i,\eta}\, x_i & \Sigma\, f_{i,\eta}\, y_i \end{bmatrix} \begin{Bmatrix} \dfrac{\partial f}{\partial x} \\ \dfrac{\partial f}{\partial y} \end{Bmatrix} \tag{28}$$

The first entity of size 2x2 on the right side of Equation (28) is the Jacobian matrix **J** which can be derived explicitly given the expressions in Table 2. We can then write

$$\begin{Bmatrix} \dfrac{\partial f}{\partial x} \\ \dfrac{\partial f}{\partial y} \end{Bmatrix} = \mathbf{J}^{-1} \begin{Bmatrix} \dfrac{\partial f}{\partial \xi} \\ \dfrac{\partial f}{\partial \eta} \end{Bmatrix} \tag{29}$$

The Jacobian matrix in Equation (28) can be expressed as a product

$$\mathbf{J} = \mathbf{D}_L \, \mathbf{C}_N \tag{30}$$

where

$$\mathbf{D}_L = \begin{bmatrix} f_{1,\xi} & f_{2,\xi} & \cdots\cdots & f_{8,\xi} \\ f_{1,\eta} & f_{2,\eta} & \cdots\cdots & f_{8,\eta} \end{bmatrix} \tag{31}$$

and

$$\mathbf{C}_N = \begin{bmatrix} x_1 & y_1 \\ x_2 & y_2 \\ & \\ . & . \\ . & . \\ x_8 & y_8 \end{bmatrix} \tag{32}$$

Note that

$$\mathbf{J}^{-1} = \frac{1}{|\,\mathbf{J}\,|} \begin{bmatrix} J_{22} & -J_{12} \\ -J_{21} & J_{11} \end{bmatrix} \qquad = (\mathbf{D}_L \quad \mathbf{C}_N)^{-1} \tag{33}$$

Expressions containing **J** can not be integrated in closed form because they appear in denominator positions, so they must be evaluated numerically.

The strain-displacement relationships for the element are

$$\epsilon_i = \mathbf{B}_i \;\; \mathbf{q}_i \qquad\qquad i = 1,2,\ldots,8 \tag{34}$$

From Equation (22)

$$\mathbf{B}_i = \begin{bmatrix} f_{i,x} & 0 \\ 0 & f_{i,y} \\ f_{i,y} & f_{i,x} \end{bmatrix} \tag{35}$$

The submatrix in Equation (35) can be written with the help of Equations (33), (29) and the expressions in Table 2. Expanding Equation (34) to include all components

$$\epsilon = \mathbf{B}\,\mathbf{q} \tag{34a}$$

For isotropic materials under plane strain conditions the following relations exist between stresses and strains

$$\sigma = \mathbf{D}\,\epsilon \tag{36}$$

where

$$\mathbf{D} = \frac{E}{(1 - \nu^2)} \begin{bmatrix} 1 & \nu & 0 \\ \nu & 1 & 0 \\ 0 & 0 & (1 - \nu)/2 \end{bmatrix}$$

Then

$$\sigma = \mathbf{D}\,\mathbf{B}\,\mathbf{q} \tag{37}$$

Equations (34a) and (37) serve to express the virtual strain energy corresponding to virtual displacements $\delta\mathbf{q}$ at the nodes

$$\delta\bar{U} = \int_{vol} \delta\mathbf{q}^T\,\mathbf{B}^T\,\mathbf{D}\,\mathbf{B}\,d(vol)\,\mathbf{q} \tag{38}$$

The corresponding external virtual work is

$$\delta \bar{\mathbf{w}}_e = \delta \mathbf{q}^T \mathbf{f}_{el} \tag{39}$$

Equating the last two expressions, and noting that arbitrary virtual displacements can be factored out we obtain the element stiffness matrix $\mathbf{f}_{el}$

$$\mathbf{f}_{el} = \int_{vol} \mathbf{B}^T \mathbf{D} \mathbf{B} \, d(vol) \, \mathbf{q} \tag{40}$$

$$= \mathbf{k}_{el} \, \mathbf{q} \tag{40a}$$

In a domain with constant thickness t defined by the local coordinates ξ and η we can define the stiffness matrix of the element as

$$\mathbf{k}_{el} = t \int \int \mathbf{B}^T(\xi,\eta) \, \mathbf{D} \, \mathbf{B}(\xi,\eta) \; | \, \mathbf{J}(\xi,\eta) \, | \; d\xi \, d\eta \tag{41}$$

or in a numerical scheme

$$\mathbf{k}_{el} = t \, \Sigma\Sigma \, R_j R_k \, \mathbf{B}^T(\xi_j,\eta_k) \, \mathbf{D} \, \mathbf{B}(\xi_j,\eta_k) \; | \, \mathbf{J}(\xi_j,\eta_k) \, | \tag{42}$$

In Equation (42) the matrix **B** and **J** are functions of the coordinates (ξ_j,η_k) of the integration points, and R_j and R_k are the corresponding weighting factors. Equivalent nodal loads due to body forces (b_x b_y) are

$$\mathbf{f}_b = t \int \int \mathbf{f}^T(\xi,\eta) \, \mathbf{b}(\xi,\eta) \; | \, \mathbf{J}(\xi,\eta) \, | \; d\xi \, d\eta \tag{43}$$

Adding the element stiffness matrices and force vectors on an element-by-element basis we obtain the equilibrium equations for the entire structure

$$\mathbf{f}_{sys} = \mathbf{k}_{sys} \, \mathbf{q}_{sys} \tag{44}$$

4.2. DYNAMIC ANALYSIS OF LINEAR DAMPED SYSTEMS

The shape functions f_i used in deriving the stiffness equations can be used in formulating the mass matrix of vibrating systems in a consistent way. Taking the time differentials of Equation (19)

$$\begin{Bmatrix} \ddot{u} \\ \ddot{v} \end{Bmatrix} = \Sigma\, f_i \begin{Bmatrix} \ddot{u}_i \\ \ddot{v}_i \end{Bmatrix} \tag{45}$$

the corresponding body forces become

$$\begin{Bmatrix} f_{bx} \\ f_{by} \end{Bmatrix} = \rho \begin{Bmatrix} \ddot{u} \\ \ddot{v} \end{Bmatrix} \tag{46}$$

where ρ is the mass density of the material. The element mass matrix $\mathbf{m}_{el}$ denotes the relationship between element nodal forces and element nodal accelerations as

$$\mathbf{f}_{in} = \mathbf{m}_{el}\, \ddot{\mathbf{q}}_i \tag{47}$$

Making use of Equations (39) and (43) we obtain

$$\mathbf{m}_{el} = \int \rho\, \mathbf{f}^T\, \mathbf{f}\, d(vol) \tag{48}$$

where the row vector $\mathbf{f}$ lists the 8 interpolation functions given by Equation (20). Adding the mass matrix components for non-suppressed degrees of freedom we obtain for the entire system

$$\mathbf{m}\, \ddot{\mathbf{q}} + \mathbf{c}\, \dot{\mathbf{q}} + \mathbf{k}_{sys}\, \mathbf{q} = \mathbf{F(t)} \tag{49}$$

Equation (49) considers damping effects only within the vibrating system, that is when energy losses are confined to the domain in motion, and are the form of viscous or material damping. In soil-structure analysis this assumption is not tenable because the effects of wave propagation away from the region under consideration are usually more important. With a judicious formulation of the system parameters many problems can be solved within what are called engineering bounds of error.

Equation (49) can be solved in a a number of ways [Clough and Penzien (1975), Bathe and Wilson (1976)]. If the modal decomposition method is used then for each mode ϕ_i considered in the solution an equation of the following form is obtained if the external disturbance is in the form of a ground acceleration $a_g(t)$:

$$\ddot{y}_i + 2\beta_i\omega_i\dot{y}_i + \omega_i^2 y_i = - \frac{L_i}{m_i} a_g(t) \tag{50}$$

In Equation (50) y_i is the time dependent displacement in the i'th normal mode defined uniquely by its shape vector $\boldsymbol{\phi}_i$, frequency ω_i and damping ratio ξ_i, with

$$L_i = \mathbf{m}^T \boldsymbol{\phi}_i \mathbf{I} \tag{51}$$

$$m_i = \mathbf{m}^T \boldsymbol{\phi}_i \mathbf{m} \tag{52}$$

where the vector **I** contains the direction cosines between the degrees of freedom of the system and the direction in which the ground motion occurs.

4.3. DYNAMIC ANALYSIS OF NONLINEAR DAMPED SYSTEMS

The dynamic response of geomaterial structures often involves significant nonlinear response of these materials. In engineering practice such systems can be analyzed either by equivalent linear methods, or by more rigorous approaches. A complete treatment of nonlinear dynamic analysis is not within the scope of this overview, but texts exist for a full coverage [e.g. Bathe (1982)].

Equivalent linearization forms the basis of a general solution technique in mechanics, and is partially based on the observation that hysteretic material damping is frequency independent in the range of frequencies of interest in earthquakes. For linear analysis of soil-structure systems, the nonlinear dynamic soil behavior is approximated by an equivalent secant modulus and an equivalent viscous damping [Seed and Idriss (1970), Seed et al. (1986)]. An attractive feature of this approach is that once equivalent linear properties have been established, then the advantages offered by frequency response methods can be exploited [Lysmer et al. (1975)]. Soil-structure interaction analyses are conducted in two stages, one for establishing the equivalent properties, and the second for the calculation of the system response on the basis of these properties.

While the equivalent linear response analysis has the advantage that established procedures of linear dynamics can be implemented, it gives no clue as to the progressive deterioration of initial characteristics. Also, nearly all solutions are for vertically propagating shear waves in horizontally layered 2-D soil system.

In the nonlinear formulation the equations of equilibrium take the form

$$\mathbf{m}\,\ddot{\mathbf{q}} + \mathbf{c}\,\dot{\mathbf{q}} + \mathbf{p}(t) = \mathbf{F}(t) \tag{53}$$

In this equation $\mathbf{p}(t)$ denotes the nodal forces which are equivalent at time t to the element stresses. These forces are no longer equal to $\mathbf{k}_{sys}$ $\mathbf{q}$, **but must** be determined from the contribution of all elements. For implicit integration schemes **the** tangent stiffness matrix $\mathbf{k}_T$ is also required. This is derived by adding the tangent stiffness matrices of the individual elements.

A nonlinear dynamic analysis requires a time-step marching procedure. When at time t, $\mathbf{q}(t)$, $\mathbf{q}(t)$, and $\mathbf{q}(t)$ are known so that Equation (53) is satisfied then the same quantities must be determined at $t + \Delta t$ using exactly the same equation of equilibrium. At $t = 0$ the initial displacement and velocity vectors need to be given. In the implicit Newmark method Equation (53) is written at time $t + \Delta t$

$$\mathbf{m}\,\ddot{\mathbf{q}}(t + \Delta t) + \mathbf{c}\,\dot{\mathbf{q}}(t + \Delta t) + \mathbf{p}(t + \Delta t) = \mathbf{F}(t + \Delta t) \tag{53a}$$

Denoting

$$\Delta\ddot{\mathbf{q}} = \ddot{\mathbf{q}}(t + \Delta t) - \ddot{\mathbf{q}}(t) \tag{54}$$

the velocity and displacement at $(t + \Delta t)$ can be expressed as

$$\dot{\mathbf{q}}(t + \Delta t) = \dot{\mathbf{q}}(t) + \Delta t\,\ddot{\mathbf{q}}(t) + \delta\,\Delta t\,\Delta\ddot{\mathbf{q}} \tag{55}$$

$$\mathbf{q}(t + \Delta t) = \mathbf{q}(t) + \Delta t\,\dot{\mathbf{q}}(t) + \tfrac{1}{2}(\Delta t)^2\,\ddot{\mathbf{q}}(t) + \alpha(\Delta t)^2\,\Delta\ddot{\mathbf{q}} \tag{56}$$

Choosing $\delta = 0.5$ and $\alpha = 0.25$ will yield an unconditionally stable algorithm. Let us denote by $\mathbf{q}^{(i)}$, $\mathbf{q}^{(i)}$ and $\mathbf{q}^{(i)}$ the current values at the i'th iteration for $\mathbf{q}(t + \Delta t)$, etc. Then

$$\mathbf{q}^{(i)} = \mathbf{q}^{(i-1)} + \Delta\mathbf{q}^{(i)} \tag{57}$$

$$\dot{\mathbf{q}}^{(i)} = \dot{\mathbf{q}}^{(i-1)} + 2\,\Delta\mathbf{q}^{(i)} / \Delta t \tag{58}$$

$$\ddot{\mathbf{q}}^{(i)} = \ddot{\mathbf{q}}^{(i-1)} + 4\,\Delta\mathbf{q}^{(i)} / (\Delta t)^2 \tag{59}$$

Here $\Delta\mathbf{q}^{(i)}$ is the incremental displacement calculated during the same increment. For the internal joint forces we can write

$$\mathbf{p}^{(i)} = \mathbf{p}^{(i-1)} + \mathbf{k}_T\,\Delta\mathbf{q}^{(i)} \tag{60}$$

where $\mathbf{k}_T$ is the tangent stiffness matrix for the displacement configuration $\mathbf{q}^{(i-1)}$. Substituting Equations (57) - (60) into Equation (53a), a linear system of equations expressed by

$$\mathbf{k}^* \, \Delta\mathbf{q}^{(i)} = \mathbf{R}^* \tag{61}$$

is obtained where

$$\mathbf{k}^* = \mathbf{k}_T(\mathbf{q}^{(i-1)}) + 4\,\mathbf{m} \,/\, (\Delta t)^2 + 2\,\mathbf{c} \,/\, \Delta t \tag{62}$$

$$\mathbf{R}^* = \mathbf{F}(t + \Delta t) - \mathbf{p}(\mathbf{q}^{(i-1)}) - \mathbf{m}\,\ddot{\mathbf{q}}^{(i-1)} - \mathbf{c}\,\dot{\mathbf{q}}^{(i-1)} \tag{63}$$

Equation (61) must be solved for $\Delta\mathbf{q}^{(i)}$ for each iteration at each time step. $\mathbf{k}^*$ is called the expanded tangent stiffness matrix and $\mathbf{R}^*$ the expanded residual force vector.

Use of plasticity models in Equation (53) affords a more rigorous way of expressing the tangent stiffness properties of the elements, and both loading and unloading "histories" can be accommodated by tracking the displacement paths of the elements [Anderheggen et al. (1985)].

5. Correlation of Predicted Dynamic Response with Measurements: Lotung Experiment

5.1. BACKGROUND INFORMATION

The Standard Review Plan of the United States Nuclear Regulatory Commission (USNRC) requires that for modeling soil-structure interaction (SSI) both the half space and the finite boundaries methods be applied and the enveloped results be used in the design of Category I structures. This requirement implies the recognition that assumptions and approximations made in each method may introduce many uncertainties into the results. There is therefore a need to calibrate these methodologies in a realistic way. Mindful of this need, NRC sponsored a research project on the validation of SSI methodologies in cooperation with the Electric Power Research Institute (EPRI). Argonne National Laboratory (ANL) coordinated the project on behalf of NRC.

The data base for this validation was collected on a 1/4-scale containment model in Lotung, Taiwan, where earthquakes occur frequently. The test structure was constructed and instrumented by EPRI in cooperation with the Taiwan Power Company (TPC). It is situated in the area of the Strong Motion Array Taiwan, No. 1 (SMART-1) which has been sponsored by the US National Science Foundation (NSF), and is maintained by the Institute of Earth Sciences of Academia Sinica in Taiwan [Bolt (1982), Graves and Philippacopoulos (1986), Tang (1986), Chen et al. (1990)]. A number of strong motion events have already been captured on recordings on or near the containment model.

In addition to earthquake monitoring forced vibration tests were performed to obtain the dynamic characteristics of the soil-structure system. These experiments were performed with the help of eccentric mass vibrators to provide steady-state sinusoidal excitation.

As consultants in civil safety matters to the Swiss Federal Office of Energy, Nuclear Safety Department (HSK), the Swiss firm of Basler & Hofmann (B&H) has been involved over the past 20 years in the review of the seismic design and requalification of nuclear installations in Switzerland. During the course of this involvement B&H has made extensive use of their own SSI computer codes, and when the firm became aware of the USNRC research project on the validation of SSI methodologies, it asked to participate in the activities through HSK. This led to a cooperative agreement between USNRC and HSK in which the former would supply the input data, and HSK would furnish through B&H the blind predictions for the response to forced vibration tests and the earthquake input motions [Basler & Hofmann (1987a), (1987b), (1987c)]. In this part of the paper a selected overview will be given for part of this work.

5.2. THE SOIL-CONTAINMENT MODEL

The location of the containment model within the SMART-1 array site in Lotung, Taiwan, is shown in Figure 11. An extensive program of subsurface geological and geotechnical investigations was intituted so that a precise modeling of the soil system could be done. The results from the geotechnical investigations were released to the researchers in the form of "data packages" which included the following components:

1. Borehole data supplemented by laboratory tests (generally to max. 60 m with one hole to 150 m; SPT values, water level, soil index tests, direct shear, unconfined compression, unconsolidated undrained triaxial compression, resonant column, cyclic triaxial, specific gravity, monotonic/cyclic stress-strain)

2. Seismic refraction

3. Up-hole, cross-hole seismic survey

The seismic cross-hole and up-hole experiments showed that the average shear wave velocity profile at the site began from a minimum of 100 m/s at the surface, attaining a value of 300 m/s at 60 m depth. The compression wave velocity was nearly constant at about 1500 m/s which is typical of saturated formations. The soil profile which formed the basis of many subsequent minor adjustements is shown in Figure 12. Basically, this included three principal layers. The first layer of silty fine sands with some silts and clays of low plasticity extended from the surface to a depth of 15 m. The second layer which contained mostly silty fine sands with gravel inclusions extended from 15 to 35 m depth. Below this depth the third major soil layer consisted largely of silts and silty clays.

The ground surface and subsurface features were such that level ground and horizontal layering could be assumed for the site. The rock level was at a depth of some 400 m below the surface so for practical purposes a point of termination for the soil depth had to be made; depending on the degree of refinement for the soils model being used, this depth ranged between 40 - 60 m. A considerable amount of judgment was exercised in fine tuning the properties of a given version of the model.

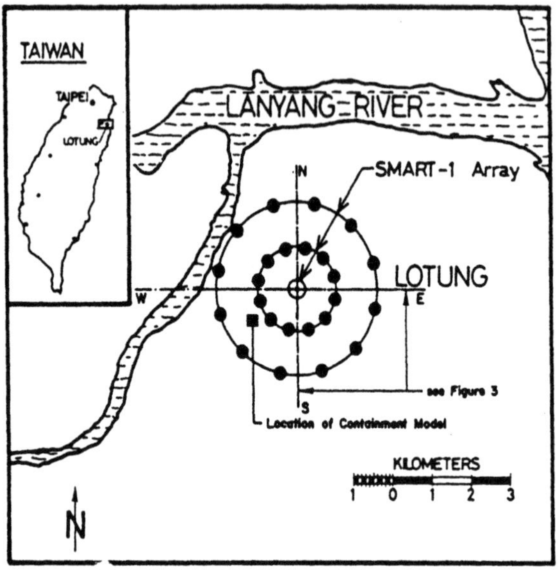

Figure 11. Location of containment model within the SMART-1 array in Lotung, Taiwan

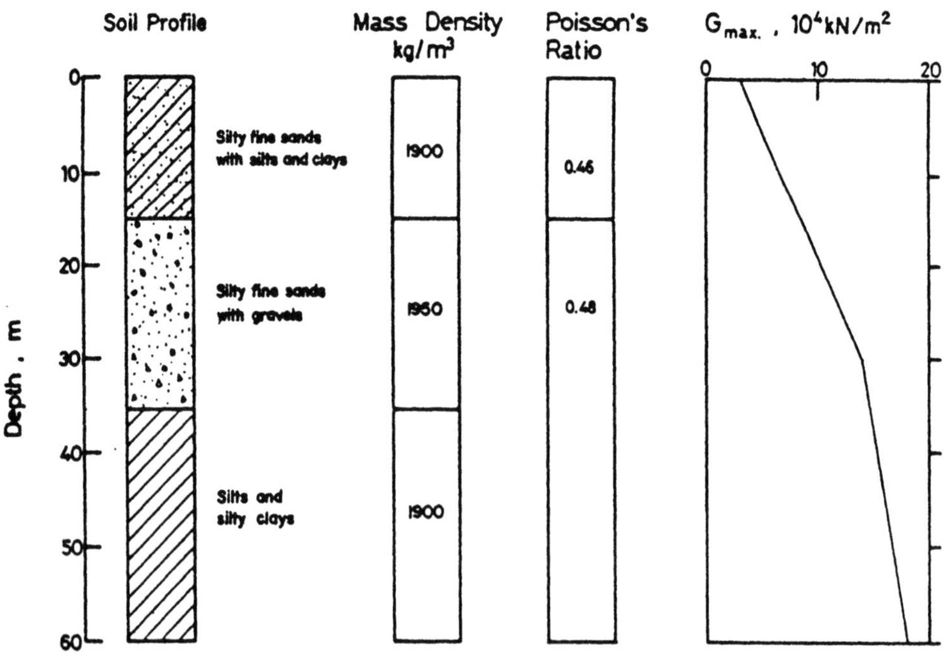

Figure 12. Basic soil model

A cross-section of the 1/4-scale model in Lotung is shown in Figure 13. The model consisted of a reinforced concrete cylinder with a heavy roof stiffened with one-way girders and a heavy basemat. It was more than 15 m tall, with an outside radius of 5.25 m. Of the overall height, about 1/4 was embedded in the surrounding soil. Except for a mockup steam generator placed asymmetrically within the structure the model was essentially empty during the period to which reference is now made. The total mass of the containment model was approximately 800 t.

During forced vibration tests the response of the model was caused by eccentric mass shakers placed at the edge of the roof, with the line of the resultant force directed either radially or tangentially. The shaker was operated at a discrete number of frequencies at which the force amplitude varied harmonically, and the maximum amplitude was a function of the square of the frequency. The response of the containment model to forced vibrations was monitored at eleven points by a total of 20 uniaxial acceleration transducers some of which are identified in Figure 13.

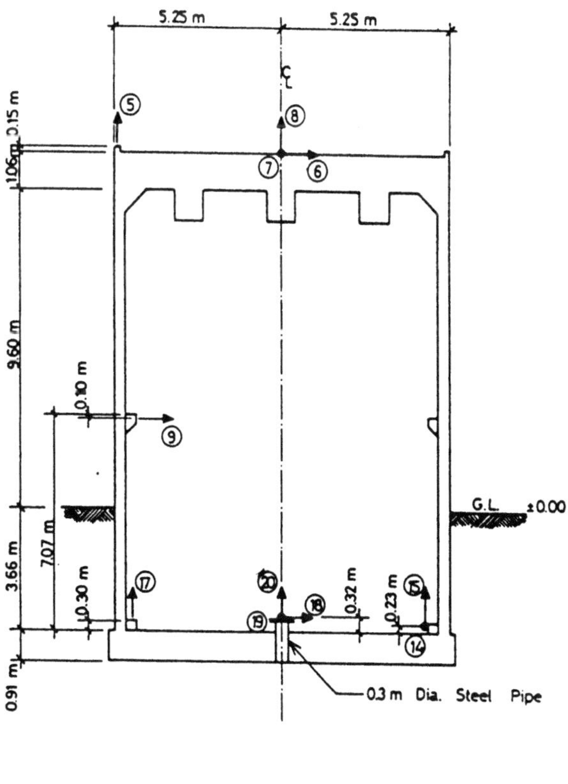

Figure 13. Cross-section of containment model

The earthquake response of the model and the soil surrounding it was monitored by 33 triaxial strong motion accelerometers and 12 pressure gages. Within the model itself there were 8 strong motion accelerometers of type FBA-13. Four instruments were located inside the containment on the circumference of the basemat, one at each cardinal point. The remaining four were arranged in the same way at an elevation of 8.1 m above ground level on the inside of the containment wall. Two additional instruments on the steam generator mockup are not considered in this treatment because no predictions were made for them.

The positions of the 15 surface strong motion accelerometers and the 8 downhole instruments are shown in Figure 14 in conjunction with the finite element meshes used either for horizontal or for vertical ground motion inputs. Not all instrument locations were included in either of the analyses.

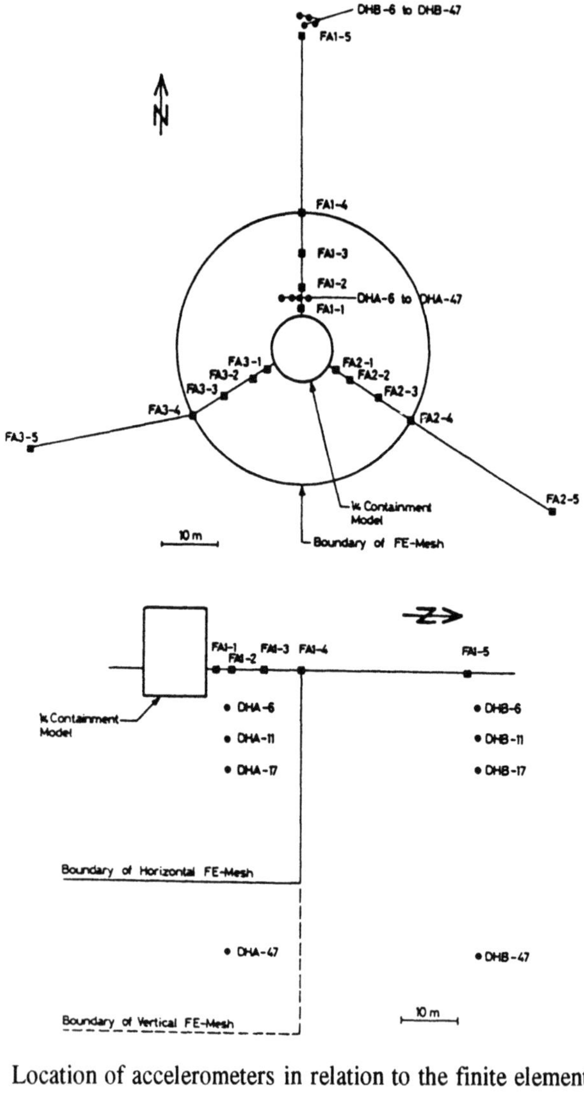

Figure 14. Location of accelerometers in relation to the finite element mesh

5.3. SSI METHODOLOGY

A modified version of the computer code BHLUSH/BHBOUND [Basler & Hofmann (1986)] coded for the analysis of axisymmetric soil-structure systems under arbitrary earthquake or forced vibration input was used in the analyses. The forced vibration input can be specified at any point in the structure while earthquake input is assumed to occur only at the rigid base. This requires that for SSI analysis, any earthquake motion recorded near the surface be deconvolved down to that formation.

The non-axisymmetric nature of the earthquake motion is considered in the analysis by expanding load and displacement vectors into Fourier series. In this formulation the discretized soil-structure system is defined by nodal rings which can displace and deform horizontally and vertically.

Radiation damping is an important phenomenon to be considered in SSI amalysis. In the code this effect is accounted for by means of transmitting boundaries. The radial extension of the finite element mesh used in any analysis is necessarily limited, so the energy escaping from the vicinity of the structure would be reflected from the vertical sides of the model unless these are equipped with numerical devices called transmitting boundaries. The frequency dependent stiffness matrices and load vectors representing the effects of soil layers extending to infinity at the sides of the model are computed by means of BHBOUND which requires as input the thickness and properties of the outermost soil column. The output from BHBOUND is then used by BHLUSH in the SSI analysis. In the vertical direction this energy absorbing boundary does not exist, so the depth of the soil-structure model must be chosen deep enough to avoid unwanted reflections from the base.

Frequency independent material damping is considered by means of complex elastic constants whose strain dependence is accounted for by means of a built-in library of material curves. Primary nonlinearity effects of the free-field response may be considered in terms of the free field solution. Secondary nonlinearity arising from the interaction of the structure with the soil in its immediate vicinity is considered in an iterative way.

5.4. SELECTED RESULTS

The Lotung exercise was a learn-as-you-go-along one. The information packages concerning both the forced vibration tests and the earthquake recording as well as the background soils exploration results were released in a step-by-step fashion so that as soon as one milestone was reached in terms of the data available up to that point, one learned what the actual measurements indicated. This permitted an evolving refinement of the models, and one could identify possible shortcomings of the previous ones on the basis of now-quantified mismatches at hand, and do any adjustments so that, hopefully, the next time around the agreement would be better. For example, two successive models referred to as B_1 and C_h are shown in Figure 15. The first was finalized on the basis of the initial batch of soils data, and used in the prediction of the steady-state tests. When measurements during forced vibrations were released so that a better model could be elaborated for the earthquake response (for which only the input was supplied), one could go back, and with the help of some hindsight, identify the deficiencies of B_1 and build C_h. The improvement in the accuracy of the predictions displayed in Figure 16 is noteworthy.

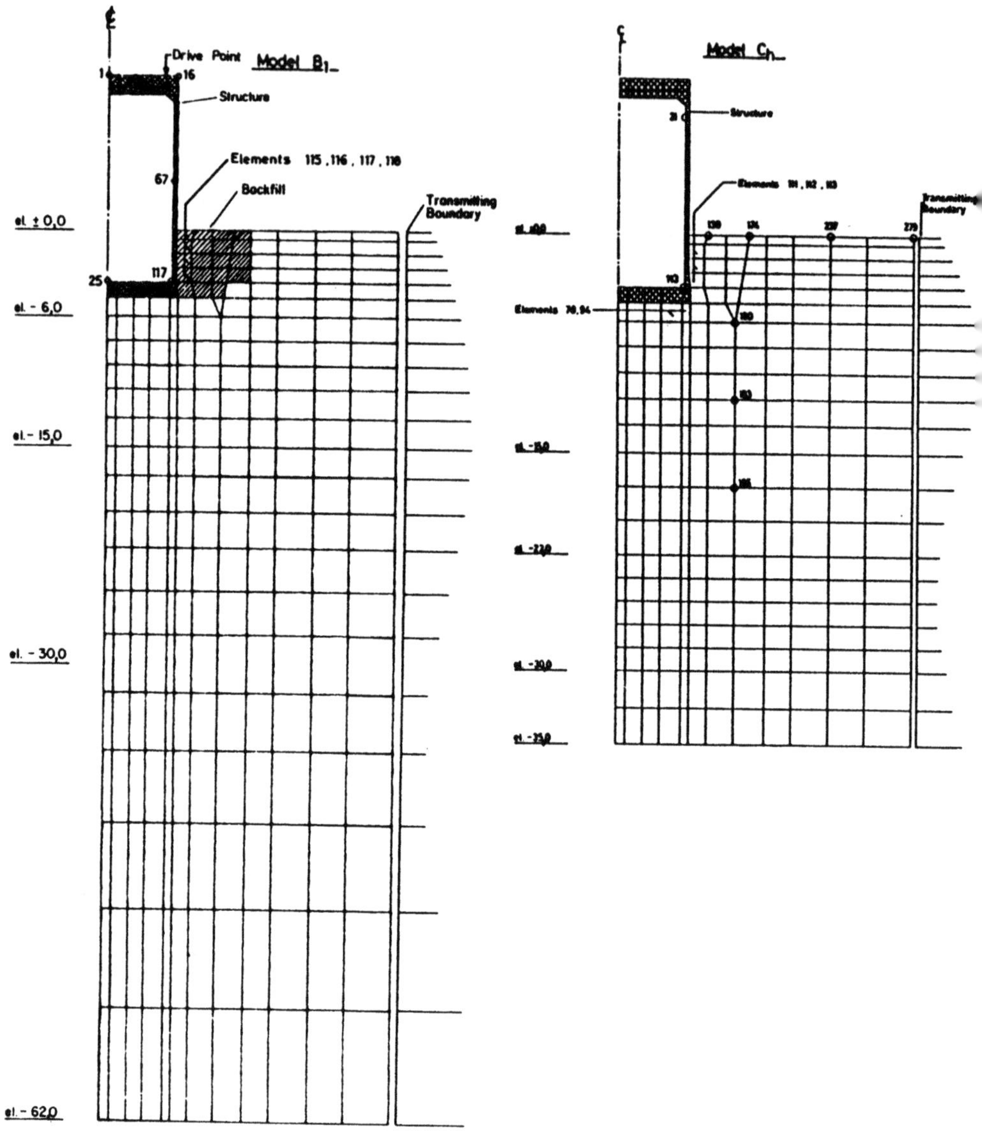

Figure 15. Finite element meshes for models B_1 and C_h

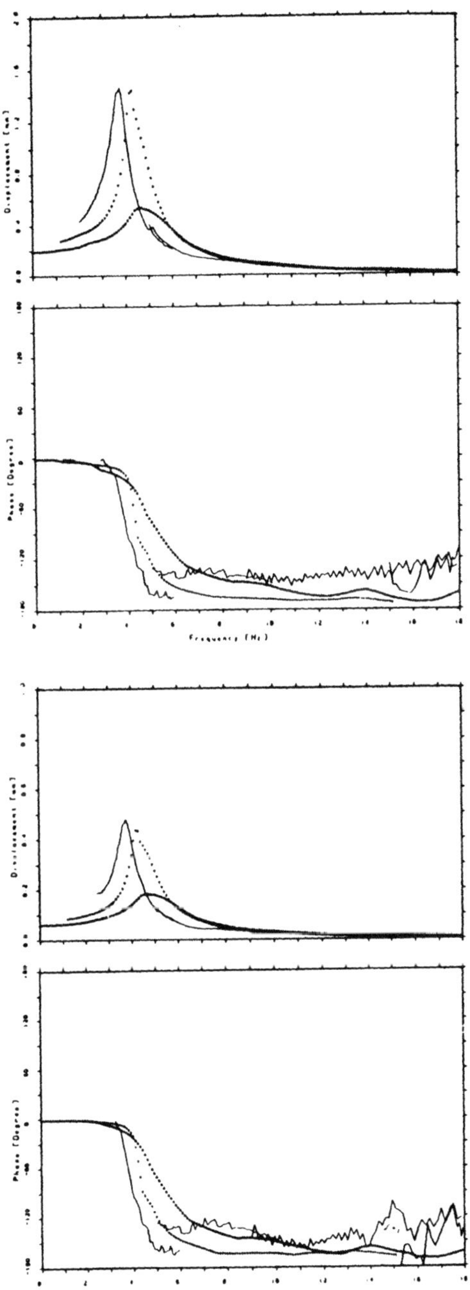

Horizontal

Vertical

Figure 16. Measured and calculated horizontal and vertical displacements at point 16 (see Figure 15). (-----): measured, (ΔΔΔΔΔ): model B_1, (+ + + + +): model C_h

The earthquake whose response was predicted occurred on 20 May 1986 at a point 66 km south of Lotung. The point at which the basic input motion was recorded is indicated in Figure 14 as Station FA1-5. The three components of the ground acceleration at this point are shown in Figure 17, and the corresponding acceleration response spectra for 5 percent damping in Figure 18.

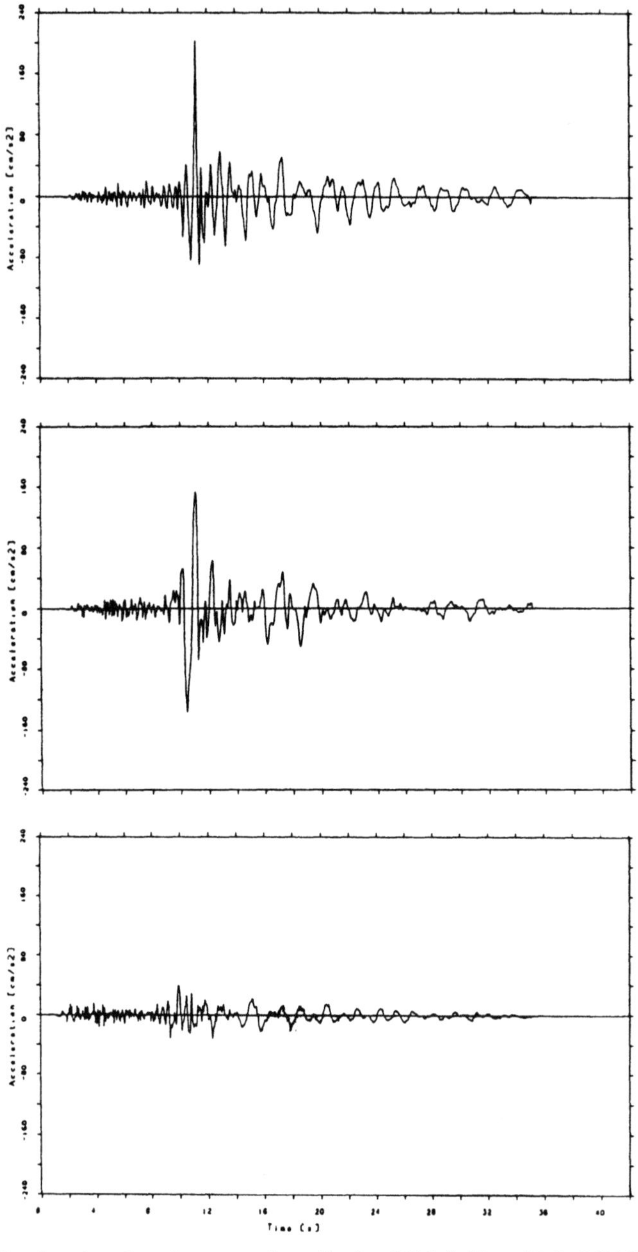

Figure 17. Acceleration-time records at Station FA1-5. Top: N-S; Middle: E-W; Bottom: V

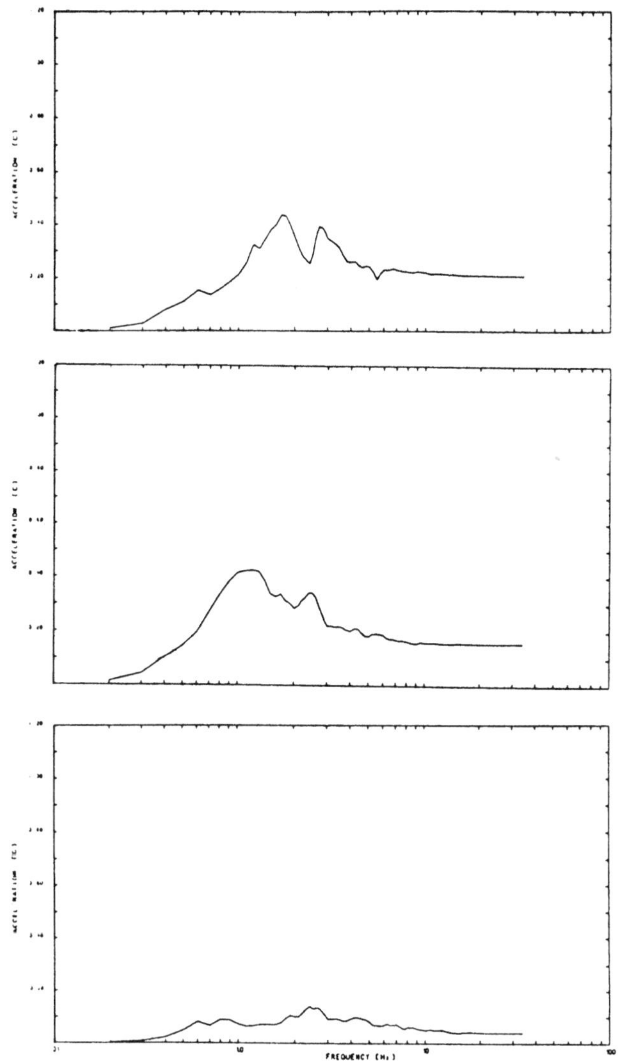

Figure 18. Acceleration spectra for motions in Figure 17

The original digitized record contained more than 7100 ordinates at a sampling rate of 200 Hz, and required a considerable computational effort. A forward FFT was applied to this record, and then it was backtransformed at 25 Hz. This way, all important frequency components up to this frequency cut-off were preserved.

5.4.1. *Free-Field Analysis.* The objectives of this phase of the analysis were as follows: (a) Find the strain-compatible dynamic soil properties in the free-field, so that the effects of nonlinear soil behavior could be considered in the SSI analysis phase, (b) calculate the variation of the earthquake motion with depth in the free-field, and (c) deconvolve the given acceleration records from the surface to the level of the rigid base.

At first model subset C_h was used for deconvolving the E-W component, and then the N-S component which led to lower strains (and therefore higher equivalent shear modulus values and lower equivalent damping ratios). Reasoning that under simultaneous vertical and horizontal ground motions, the more degraded soil properties would control the free-field response, the E-W results displayed in Figure 19 were used in further calculations. For free-field analysis in the vertical direction calculations were conducted to a depth of 35 m, and then these were extrapolated to 60 m.

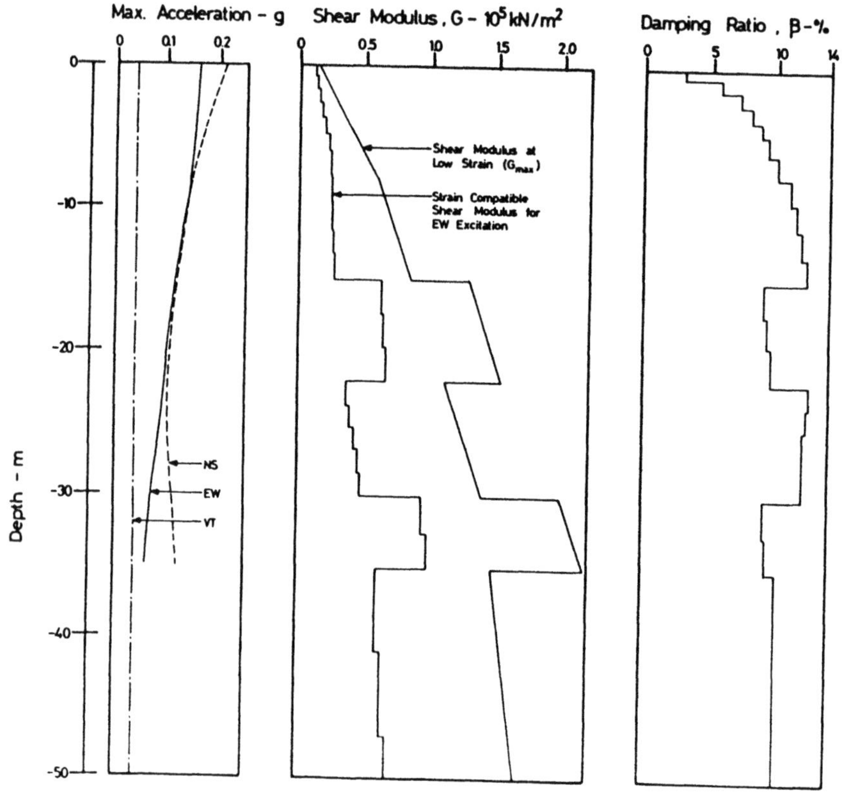

Figure 19. Summary results of free-field analysis

5.4.2. *Earthquake Response SSI Analysis.* This set of calculations was conducted separately for each of the three components specified at station FA1-5. An initial solution of the soil-structure system represented by the soil model subset C_h, with strain compatible dynamic soil properties displayed in Figure 20, was obtained. The strains in the soil domain accounting also for the interaction effects of the structure were calculated to a depth of 22 m. Then with new shear modulus and damping ratios, an iterated new solution was sought which resulted in the final response of the soil-structure system caused by the E-W motion.

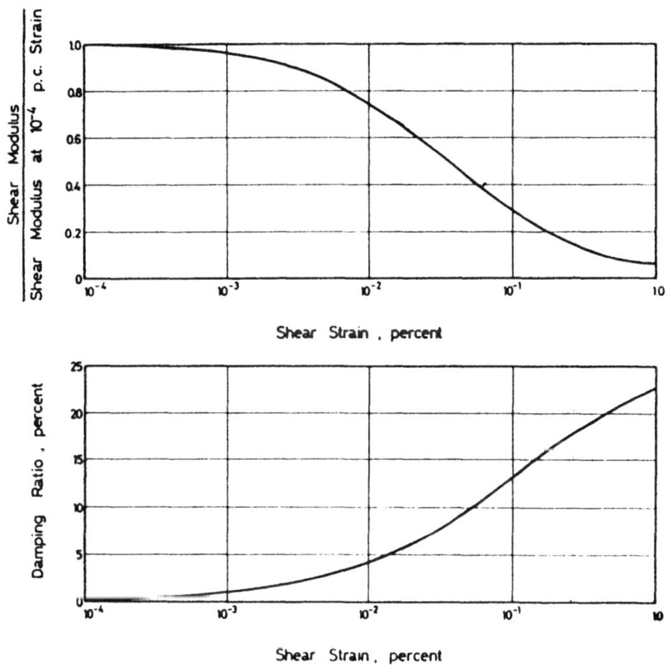

Figure 20. Strain compatible shear rigidity and damping in the soil

The procedure described above was repeated for the N-S and V components, and the calculated response was transformed to the recordings made at points of interest. It would be to belabor the point if calculated versus measured time records were to be presented for all coordinates in the soil-structure system model. In the interest of selectivity we present the acceleration spectra (for 5 percent damping) for calculated and measured responses at the following points:

Accelerometer Station FA1-1 (see Figure 14 for location) displayed in Figure 21

Inner ledge at top of model (point 21 in Figure 15) displayed in Figure 22

Inside corner of bottom slab (point 113 in Figure 15) displayed in Figure 23

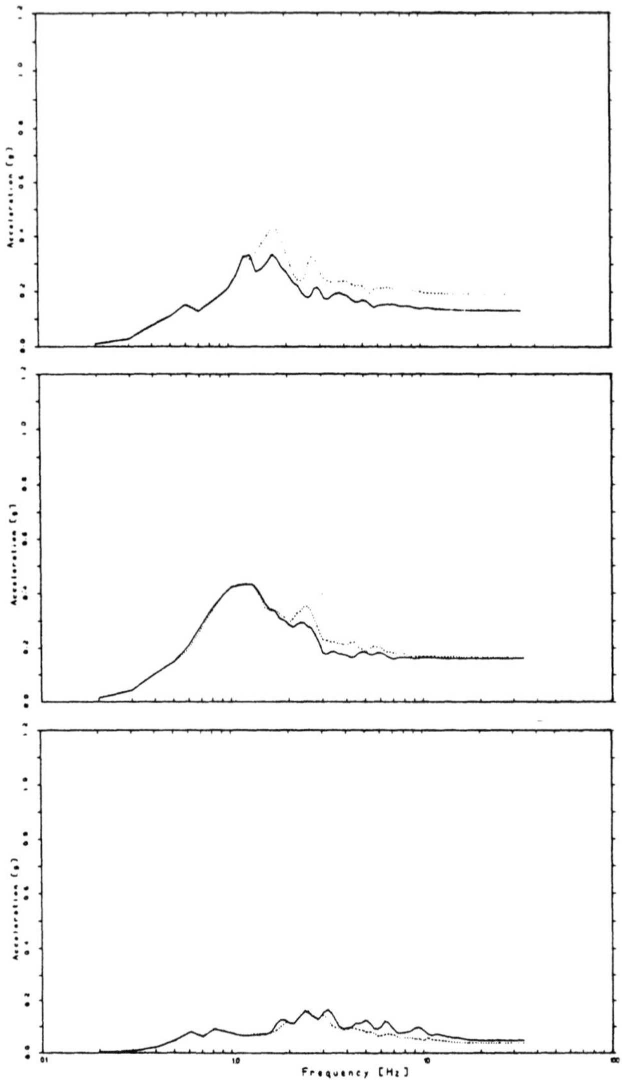

Figure 21. Measured (-----) and calculated (- - - -) acceleration spectra at Station FA1-1

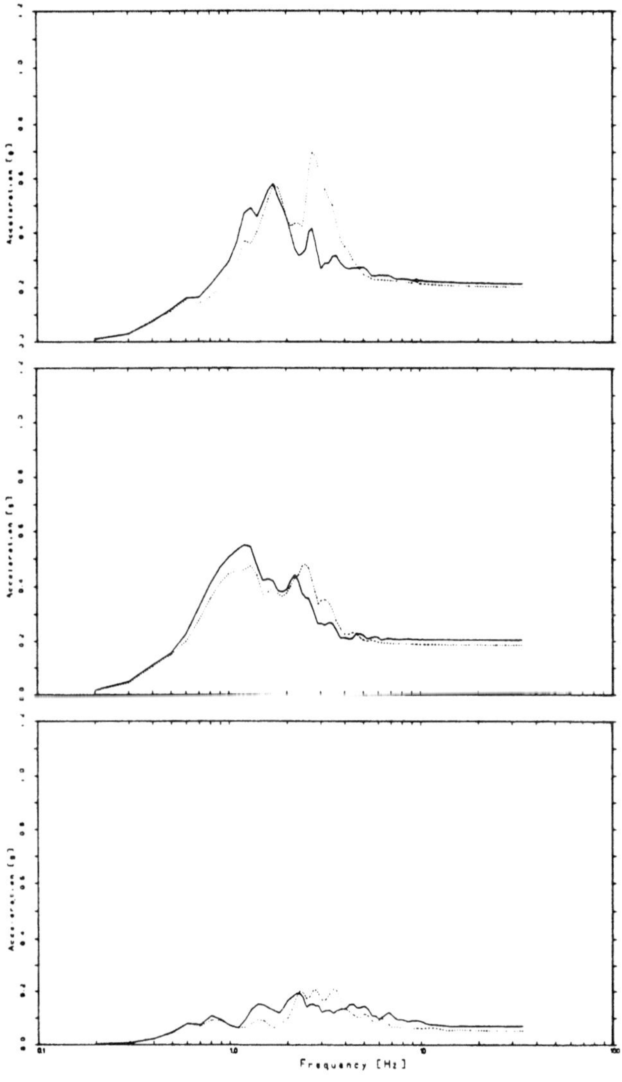

Figure 22. Measured (-----) and calculated (- - - -) acceleration spectra at point 21

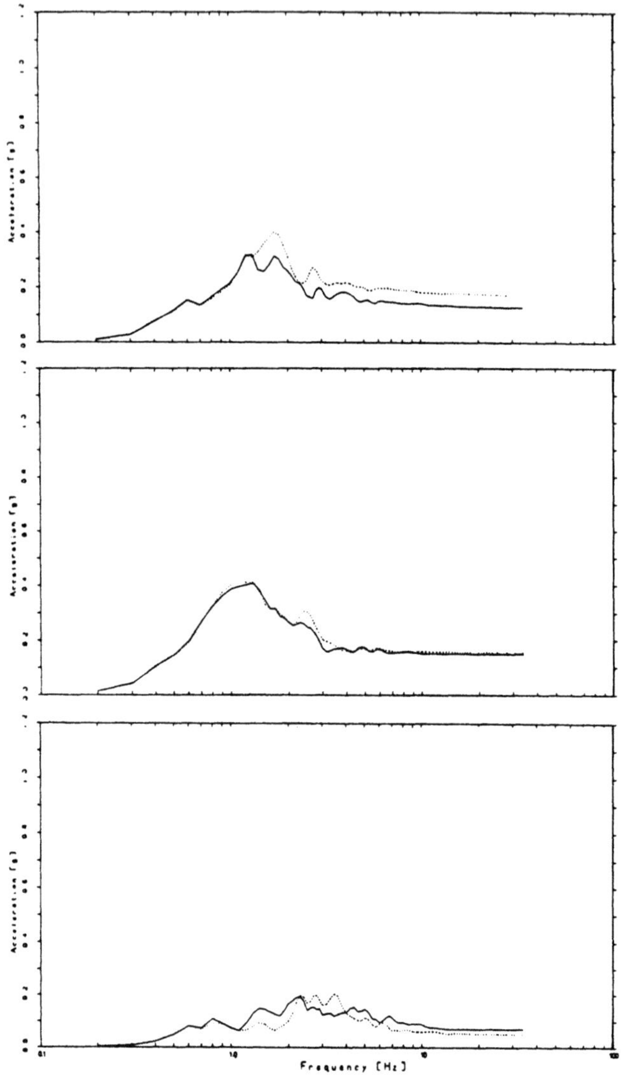

Figure 23. Measured (----) and calculated (- - - -) acceleration spectra at point 113

5.5. CONCLUSIONS

From a visual comparison of the goodness of the match between calculated and measured responses during both the forced vibration tests and the actual earthquake response it is possible to state that the technology was perfectly adequate to predict the response with acceptable margins of error. Any differences are of course attributable to the uncertainties associated with the modeling aspects of the soil medium. At its best, the reduction of complex three-dimensional dynamic response to a equivalent linear medium by way of highly idealized curves such as displayed in Figure 20 must be expected to cause errors. Certainly, there must have been some inherent errors in the basic data supplied to the investigating parties, particularly in relation to the laboratory tests conducted on the geomaterials at the Lotung site.

The value of prior forced vibration testing of structural systems for which a sound basis of modeling is sought was also confirmed through this exercise. Could a more sophisticated way of modeling the soils medium have resulted in better agreement? An answer to that question will be sought in the continuing phase of this investigation.

ACKNOWLEDGMENTS

I thank Basler & Hofmann, Consulting Engineers and Planners of Zürich, Switzerland, for permitting reuse of some of their proprietary data for purposes of this paper. Graduate students A. Yakut and U. Akyüz have helped me in conducting computer runs or in unraveling the finer aspects of word processing software.

REFERENCES

Anderheggen, E., Bartelt, P., Bazzi, G., Elmer, H., Friedrich, T., Klemenz, U., Maag, H., and Theiler, J. (1985) "FLOWERS User's Manual," Swiss Federal Institute of Technology, Zürich, Institute of Informatics, Third Edition, January.

Basler & Hofmann (1986) "BHLUSH/BHBOUND: Computer program for the dynamic analysis of axisymmetric soil-structure interaction problems," User's Manual, Rev. O, Basler & Hofmann, Consulting Engineers, Zürich, December.

Basler & Hofmann (1987a) "Predictive response computations for forced vibration experiments on 1/4-scale containment model in Lotung, Taiwan, Progress Report for Phase 1," No. TB 1471-01, Zürich, July.

Basler & Hofmann (1987b) "Comparison of measured and predicted forced vibration responses and prediction of earthquake response for 1/4-scale containment model in Lotung, Taiwan, Progress Report for Phases 2 & 3," No. TB 1471-02, October.

Basler & Hofmann (1987c) "Comparison of measured and predicted forced vibration responses and earthquake response for 1/4-scale containment model in Lotung, Taiwan, Final Report," No. TB 1471-03, November.

Bathe, K.J., and Wilson, E.L. (1976) Numerical Methods in Finite Element Analysis, Prentice Hall, Englewood Cliffs, New Jersey.

Bathe, K.J. (1982) Finite Element Procedures in Engineering Analysis, Prentice Hall, Englewood Cliffs, New Jersey.

Bolt, B., Loh, C.H., Penzien, J., and Tsai, Y.B. (1982) "Preliminary Report on the SMART-1 Strong Motion Array in Taiwan," Earthquake Engineering Research Center Report No. UCB/EERC-82/13, University of California, Berkeley, August.

Chen, W.F., and Baladi, G. (1985) Soil Plasticity: Theory and Implementation, Elsevier, Amsterdam.

Chen, C.H., Lee, Y.J., Jean, W.Y., Katayama, I., and Penzien, J. (1990) "Correlation of predicted seismic response using hybrid modelling with EPRI/TPC Lotung experimental data," Earthquake Engineering and Structural Dynamics, Vol. 19, pp. 993-1024.

Clough, R.W., and Penzien, J. (1975) Structural Dynamics, McGraw Hill, New York.

Desai, C.S. (1984) Constitutive Laws for Engineering Materials, Prentice Hall, Englewood Cliffs, New Jersey.

Desai, C.S., and Gallagher, R.H. (Ed.) (1984) Mechanics of Engineering Materials, Wiley, New York.

Dungar, R., and Studer, J. (Ed.) (1986) Geomechanical Modelling in Engineering Practice, A.A. Balkema, Rotterdam/Boston.

Dungar, R. (1986) "Linear and non-linear modelling of geomechanical media," in Dungar and Studer (1986).

Graves, H.L., and Philippacopoulos, A.J. (Ed.) (1986) Proceedings of the Workshop on Soil-Structure Interaction, Bethesda, Maryland, June 16-18, 1986, Structural Analysis Division, Department of Nuclear Energy, Brookhaven National Laboratory.

Gudehus, G. (Ed.) (1977) Finite Elements in Geomechanics, Wiley, New York.

Lysmer, J., Udaka, T., Tsai, C.F., and Seed, H.B. (1975) "FLUSH- A computer program for approximate 3-D analysis of soil-structure interaction problems," Earthquake Engineering Research Center Report No. EERC 75-30, University of California, Berkeley, November.

Prater, E.G., and Studer, J. (1986) "Material testing procedures and equipment," in Dungar and Studer (1986).

Seed, H.B., and Idriss, I.M. (1970) "Soil moduli and damping factors for dynamic response analysis," Earthquake Engineering Research Center Report No. EERC 70-10, University of California, Berkeley, December.

Seed, H.B., Wong, R.T., Idriss, I.M., and Tokimatsu, K. (1986) "Moduli and damping factors for dynamic analysis of cohesionless soils," Journal of Geotechnical Engineering, ASCE, Vol. 112, November, pp. 1016-1032.

Tang, H.T. (1986) "EPRI research on soil-structure interaction," in Graves and Philippacopoulos (1986).

Weaver, W., and Johnston, P.R. (1984) Finite Elements for Structural Analysis, Prentice Hall, Englewood Cliffs, New Jersey.

Wittke, W. (Ed.) (1979) Third International Conference on Numerical Methods in Geomechanics, Aachen.

SOIL-STRUCTURE INTERACTION (SSI): OBSERVATIONS, DATA, AND CORRELATIVE ANALYSIS

JAMES J. JOHNSON
ALEJANDRO P. ASFURA
EQE International, Inc.
44 Montgomery Street, Suite 3200
San Francisco, CA 94104

Abstract

Soil-structure interaction (SSI) has been reviewed in light of ground motion and structural response data accumulated to date. Correlations of observations with site response and SSI analyses are presented. The SSI analysis process is discussed according to the component elements: specification of the free-field ground motion, models of soil and structures, SSI analysis, and interpretation of responses. Observed data and correlative analyses help clarify the important issues involved in a SSI analysis, the quality of the assumptions made in modern practice, and their impact on structure response.

1. Introduction

1.1. SSI: A STATEMENT OF THE PROBLEM

The response of a structure during an earthquake depends on the characteristics of the ground motion, the surrounding soil, and the structure itself. For structures founded on rock or very stiff soils, the foundation motion is essentially that which would exist in the soil at the level of the foundation in the absence of the structure and any excavation; this motion is denoted the free-field ground motion. For soft soils, the foundation motion differs from that in the free field due to the coupling of the soil and structure during the earthquake. This interaction results from the scattering of waves from the foundation and the radiation of energy from the structure due to structural vibrations. Because of these effects, the state of deformation (particle displacements, velocities, and accelerations) in the supporting soil is different from that in the free field. In turn, the dynamic response of a structure supported on soft soil may differ substantially in amplitude and frequency content from the response of an identical structure supported on a very stiff soil or rock. The coupled soil-structure system exhibits a peak structural response at a lower frequency than would an identical rigidly supported structure. Also, the amplitude of structural response is affected by the additional energy dissipation introduced into the system through radiation damping and material damping in the soil.

P. Gülkan and R. W. Clough (eds.), Developments in Dynamic Soil-Structure Interaction, 219–258.

The analysis of SSI depends on the specification of the free-field ground motion and the idealization of the soil and structure. Modeling the soil entails its configuration and material properties. Modeling the structure includes the structure itself and its foundation. The calculated responses must be interpreted in light of differences between the idealized system and the real physical situation and the uncertainties known to exist in the phenomena. Figure 1 shows the elements of seismic analysis necessary to calculate seismic response including SSI effects. Table 1 lists the various aspects, including interpretation and recognition of uncertainties in the process.
The state of knowledge of SSI was well documented in 1980 by Johnson [1]. Reference 1 contains solicited contributions from key researchers (Luco [2], Roesset [3], Seed and Lysmer [4]) and drew upon other researchers and practitioners as well (Veletsos [5], Chopra [6]). Reference 1 provided a paradigm for discussions over the decade of the 1980s.

The ensuing decade was characterized most by the accumulation of substantial data supporting and clarifying the roles of the various elements of the SSI phenomena. It is this data and its evaluation that is the primary focus of this paper. In addition to data acquisition and evaluation, significant progress has been made in SSI analysis techniques which include the implementation of boundary element methods. Also, the 1980s has seen a revision of important regulatory practice for nuclear power plants in the U.S. to conform to the current state of knowledge.

The remainder of Sec. 1 describes the specification of the free-field ground motion and the idealization of the soil and structure, in general terms, providing the basis for subsequent discussions of observational data and its evaluation. Finally, a preview of the remaining sections and the introduction of the most comprehensive SSI experiment performed to date is presented.

1.2. SPECIFICATION OF THE FREE-FIELD GROUND MOTION

Describing the free-field ground motion at a site for SSI analysis purposes entails specifying the point at which the motion is applied (the control point), the amplitude and frequency characteristics of the motion (referred to as the control motion and typically defined as ground response spectra, power spectral density functions, and/or time histories), the spatial variation of the motion, and, in some cases, duration, magnitude, and other earthquake characteristics. In terms of SSI, the variation of motion over the depth and width of the foundation is relevant. For surface foundations, the variation of motion on the surface of the soil is important; for embedded foundations, the variation of motion over both, the embedment depth and the foundation width, should be known.

1.3. MODELING THE SOIL-STRUCTURE SYSTEM

Modeling the soil-structure system entails modeling:
Soil Profile. Soil/rock layering in the site vicinity and the dynamic soil/rock properties.
Soil Nonlinear Behavior. The variation of dynamic soil/rock properties as a function of the strain induced by the seismic waves. Soil shear modulus and material damping vs. shear strain relationships define the variation at a point.
Foundation. Three aspects are important -- embedment, stiffness, and geometry.
Structure. Dynamic behavior, including linear and nonlinear aspects.
SSI Parameters and Analysis. Several ways of categorizing SSI analysis methods have been used. Two ways are (1) direct methods which analyze the soil-structure system in a single step and (2) the substructure approach which treats the problem in a series of steps, e.g., determination of the foundation input motion and the foundation impedances, modeling of the structure, and the

analysis of the coupled system. In the context of this paper, it is informative to view the SSI phenomenon in the steps of the substructure approach. Figure 1 showed schematically the substructure approach. The key elements not previously discussed are:
Foundation input motion. Foundation input motion differs from the free-field ground motion in all cases, except for surface foundations subjected to vertically incident waves, provided the spatial variation of the free-field ground motion is taken into account. This is due, first, to the variation of free-field motion with depth in the soil and second, due to the scattering of waves from the soil-foundation interface because points on the foundation are constrained to move according to its geometry and stiffness.
Foundation impedances. Foundation impedances describe the force-displacement characteristics of the soil. They depend on the soil configuration and material behavior, the frequency of the excitation, and the geometry of the foundation. In general, for a linear elastic or viscoelastic material and a uniform or horizontally layered soil deposit, each element of the impedance matrix is complex-valued and frequency dependent. For a rigid foundation, the impedance matrix is 6x6, which relates a resultant set of forces and moments to the six rigid-body degrees-of-freedom.
Soil-structure interaction analysis. The final step in the substructure approach is the actual analysis. The result of the previous steps: foundation input motion; foundation impedances; and structure models are combined to solve the equations of motion for the coupled soil-structure system. The entire process is sometimes referred to as a complete interaction analysis which is separated into two parts: kinematic interaction described by the scattering matrix; and inertial interaction comprised of the effects of vibration of the soil and structure.
In terms of SSI parameters, the scattering functions model kinematic interaction and the foundation impedances model inertial interaction. It is these two parameters which are highlighted in the ensuing sections.

1.4. ORGANIZATION AND PERSPECTIVE

This paper is organized according to the elements of SSI as depicted in Table 1. Each of these elements has been introduced previously and the ensuing sections focus on observational data and its analytical treatment.

Section 2 contains control motion, earthquake characteristics, control point location, and the combined topics of spatial variation of ground motion and kinematic interaction effects; important aspects of the specification of the free-field ground motion. Section 3 discusses soil modeling for site response and SSI analyses. Section 4 discusses the remaining aspects of SSI including foundation impedances, the response of soil-structure systems, and special issues such as structure-structure interaction. Section 5 presents conclusions and perspectives. This approach is similar to that of Johnson and Chang [7] and includes selected data from their work.

Before proceeding, an important SSI experiment is discussed. Very few opportunities exist to record free-field motion and structure response for an earthquake. In the mid 1980s, the Electric Power Research Institute (EPRI), in cooperation with Taiwan Power Co. (TPC), constructed two-scale model reinforced concrete containment buildings (1/4 and 1/12 scale). The experiment was extensively instrumented in the free-field and in the structure. It was located within a strong-motion array (SMART-1, Strong Motion Array Taiwan, Number 1) sponsored by the US National Science Foundation and maintained by the Institute of Earth Sciences of Academia Sinica of Taiwan. The goal of the experiment was to measure the responses at instrumented locations due to vibration tests and actual earthquakes.

Using this data base, a cooperative program to validate SSI analysis methodologies was sponsored by EPRI, TPC, and the US Nuclear Regulatory Commission (NRC). Numerous publications document the results of the SSI analysis studies. A two-volume EPRI report [8] contains the proceedings of a workshop held in Palo Alto, CA, USA on December 11-13, 1987 to discuss the experiment, data collected, and analyses performed to investigate SSI analysis methodologies and their application. Johnson et al [9] performed one set of analyses from which results are presented here to demonstrate various aspects of the SSI phenomenon. Recently, a summary of lessons learned was published [10].

When appropriate throughout this paper, data from the Lotung experiment is presented and discussed.

2. Specification of the Free-field Ground Motion

Specification of the free-field ground motion for SSI analysis of structures entails specifying the control motion, the control point, the spatial variation of the motion, and descriptors of the earthquake such as magnitude, duration, location of the earthquake relative to the site, etc. The control motion and the descriptors of the earthquake are discussed next. Control point location, spatial variation of the free-field ground motion, and selected aspects of kinematic interaction are discussed in a later subsection.

2.1. Control Motion and Earthquake Characteristics

The control motion is defined by specifying the amplitude and frequency content of the earthquake to be considered. Two purposes exist for SSI analysis of a structure: design or evaluation of a facility for a specified earthquake level; or evaluation of a structure for a specific event. In the former case, statistical combinations of recorded or predicted earthquake motions are typically the bases. In the latter case, recorded earthquake acceleration time histories typically comprise the free-field ground motions.

The frequency content of the motion is one of the most important aspects of the free-field motion as it affects structure response. For linear structural behavior and equivalent linear SSI, the frequency content of the free-field motion compared to important frequencies of the soil-structure system determines response. For structures expected to behave inelastically during the earthquake (and, in particular, structures for which SSI is not important), structure response is determined by the frequency content of the free-field motion, i.e., in the frequency range from the elastic frequency to a lower-frequency corresponding to a certain amount of inelastic deformation.

2.1.1. *Aggregated Ground Motions.* A wide variety of ground response spectra have been specified for design of major facilities, such as nuclear power plants, depending on the vintage of the plant and the site soil conditions. The majority of these have been relatively broad-banded spectra representing a combination of earthquakes of different magnitudes and distances from the site. Construction of such design spectra are usually based on a statistical analysis of recorded motions and frequently targeted to a 50% or 84% nonexceedance probability (NEP). Three points are important relative to these broad-banded spectra. First, earthquakes of different magnitudes and distances control different frequency ranges of the spectra. Small magnitude earthquakes contribute more to the high frequency range than to the low frequency range and so forth. Second, it is unlikely that a single earthquake will have frequency content matching the design ground response spectra. Hence, a degree of conservatism is added when a broad-banded response spectra defines the control motion. Third, a single earthquake can have frequency content that exceeds the design ground response spectra in selected frequency ranges. The likelihood of the exceedance depends on the NEP of the design spectra. Figure 2 compares several ground response spectra used in the design or evaluation process. U.S. NRC Regulatory Guide 1.60 [11] response spectra defined design criteria for nuclear power plants designed after about 1973. These spectra were targeted to an 84% NEP of the data considered but exceed this target in selected frequency ranges. Figure 2 shows an additional set of broad-banded spectra for rock and alluvial sites and for 50% and 84% NEP. U.S. NRC NUREG/CR-0098 [12] is the source. These spectra are currently being used extensively for defining a seismic margin earthquake for which seismic margin assessments are being performed for U.S. nuclear power plants. Also, the broad-banded spectral shape is being used to define design criteria for new design.

In addition to the site independent ground response spectra discussed above, two additional forms of ground response spectra are being generated and used for site specific design or evaluations. First, site specific spectra are generated by accumulating recorded data that meets the design earthquake characteristics and local site conditions, analyzing the data statistically, calculating ground response spectra of various statistical attributes, and selecting response spectra for design or reevaluation. Figure 3 shows an example for a specific site; the 84% NEP was selected. Second, seismic hazard studies are performed to generate families of seismic hazard curves which yield estimates of the probability of exceedance of earthquakes with specified peak ground acceleration (PGA) values or greater. Confidence limits for these seismic hazard estimates are derived from the family of curves. Companion to the seismic hazard curves are uniform hazard spectra (UHS) which are ground response spectra generated for a specified return period or probability of exceedance level and with various confidence limits. Figure 4 shows example UHS for a specific site, 10,000 year return period, and 15%, 50%, and 85% confidence limits. Such spectra are generated by the same probabilistic seismic hazard methodology as is used in generating the seismic hazard curves for PGAs.

In almost all cases, design ground response spectra are accompanied by ground motion acceleration time histories -- artificial acceleration time histories generated with response spectra that match or exceed the design ground response spectra. Due to the enveloping process, additional conservatism is introduced. Ground motion time histories are used to generate in-structure response -- loads, response spectra, displacements, etc.

2.1.2. *Individual Recorded Events.* The previous sections discussed aggregated motions as derived from recorded data or empirical models based on recorded data. These aggregated motions do not represent a single event. They have been developed for various design and evaluation purposes. It is informative to present recorded motions from actual earthquakes and visualize differences between a single event and the aggregated motions. Two earthquakes of note from an SSI standpoint were the May 20, 1986 and November 14, 1986 events which affected the Lotung scale model structures. Figure 5 contains response spectra generated from the acceleration time histories recorded on the soil-free surface for the May 20, 1986 event. In the literature, these two events are denoted numbers 7 and 16, respectively. Each earthquake produced PGAs greater than about 0.2g in the horizontal direction with principally low frequency motion, i.e., less than about 5 Hz. Subsequent sections will discuss these and other motions recorded in this area.

2.1.3. *Magnitude Effects.* Ground motion frequency content is strongly dependent on specific factors of the earthquake and site. Two particularly important characteristics of the earthquake are its magnitude and epicentral distance from the site. Small magnitude events are characterized by narrow-banded response spectra and high frequency in comparison to moderate magnitude events. Figure 6 [13] illustrates the effect of magnitude on response spectral shape. In the figure, the response spectral shape obtained from a series of small magnitude ($M_L < 4$) earthquakes is compared with the response spectral shape from a moderate magnitude ($M_L \approx 6\ 1/2$) event. As shown, the small magnitude earthquakes are characterized by a narrow-banded response spectral shape and greater high frequency content than the moderate magnitude event.

2.1.4. *Soil vs. Rock Sites.* One of the most important parameters governing amplitude and frequency content of free-field ground motions is the local site conditions, i.e., soil vs. rock and shallow, soft soil overlying a stiff soil or rock. Two sets of data demonstrate dramatically the difference in motions recorded on rock vs. soil for the same earthquake and sites in close proximity.

The Ashigara Valley in Japan is located about 80 km southwest of Tokyo. A digital strong-motion accelerograph array network [14] was installed in this seismically very active area. The geological profile of the Ashigara Valley is shown in Fig. 7 [24]. The valley is an alluvial basin with rock outcrops at the mountain side and soft sedimentary soil layers at the basin. Accelerographs were installed in the rock outcrop (KR1), at the surface of the soft layers (KS1 and K2S) and inside the soil (KD1 and KD2). Figures 8 and 9 [15] show response spectra of motions recorded at the rock outcrop (KR1) and at the surface of soft layers (K2S) for an earthquake occurring on August 5, 1990. These figures clearly show differences in motions due to different site conditions. High frequencies of about 10 Hz and greater are dominant at the rock site and frequencies below about 5 Hz are dominant at the soil surface. Figures 8 and 9 clearly show the large amplification of the low frequency waves due to the soil response and the deamplification of the high frequency waves due to the filtering effect of the soft soil.

The Loma Prieta earthquake, October 17, 1989, with epicenter near San Francisco, CA, provided the opportunity to collect and evaluate recorded motions on rock and soft soil sites, in some cases immediately adjacent to each other. Figures 10 and 11 [15] compare response spectra for recorded motions on rock (Yerba Buena Island) and on soil (downtown Oakland). Significant differences in amplification are obvious with the rock motions being less. Figure 12 [16] presents amplification factors for soil (Treasure Island) vs. rock (Yerba Buena Island). For horizontal motions, significant amplification of soil over rock is apparent. For vertical motions, rock values are higher than soil which requires additional evaluation. Note, Treasure Island and Yerba Buena Island are the north and south ends of a single land mass. Figures 13 and 14 [16] compare amplification factors for a number of soft soil sites with expected rock or stiff soil motions corrected for distance according to appropriate attenuation laws. Significant amplification for soft soil vs. stiff soil or rock is apparent for horizontal and vertical motions except for the Treasure Island vertical component, discussed above.

The evidence supporting the differences between rock and soft soil motions continues to mount emphasizing the effect of local site conditions on the amplitude and frequency content of the motion.

2.2. Control Point

The term control point designates the location at which the control motion is defined. The control point should always be defined on the free surface of soil or rock at the site of interest. Specifying the control point at locations other than a free surface ignores the physics of the problem and the source of data used in defining the control motion. Past regulatory practice has specified the control point to be at foundation level in the free-field, which is technically untenable. It not only ignores the physics of the problem and the source of recorded data, it also results in motions on the free surface whose response spectra display peaks and valleys associated with frequencies of the embedment layer which are totally fictitious and unrealistic. The frequencies of these peaks correspond to the frequencies of the soil layer between the foundation and the free surface. These free surface motions are dependent on the foundation depth rather than free-field site characteristics. In addition, the peak acceleration of the resulting free surface motion is typically calculated to be significantly greater than the control motion. Hence, by all seismological definitions, a design or evaluation earthquake is increased. A 0.25g earthquake may become a 0.35g earthquake or greater depending on the embedment depth, soil properties, and control motion characteristics. Finally, specification of the control point at foundation level rather than the soil free surface effectively penalizes partially embedded structures compared to surface-founded structures which contradicts common sense and observations.

Simplistic SSI analyses frequently ignore kinematic interaction for embedded foundations. In so doing, the foundation input motion is assumed to be identical to the control motion. Implicit in this assumption is the definition of the control point as any point on the foundation and no spatial variation of ground motion. This assumption is almost always conservative and frequently extremely conservative. Recognition of this conservatism is important in interpreting the results of the SSI analysis.

2.3. Spatial Variation of Motion and Kinematic Interaction

Spatial variations of ground motion refer to differences in amplitude and/or phase of ground motions with horizontal distance or depth in the free-field. Spatial variations of ground motion are associated with different types of seismic waves and various wave propagation phenomena including reflection at the free surface, reflection and refraction at interfaces and boundaries between geological strata having different properties, and diffraction and scattering induced by nonuniform subsurface geological strata and topographic effects along the propagation path of the seismic waves. A vertically incident body-wave propagating in such a medium will include ground motions having identical amplitudes and phase at different points on a horizontal plane (neglecting source to site attenuation effects over short horizontal distances). A plane wave propagating horizontally at some apparent phase velocity will induce ground motion having identical amplitudes but with a shift in phase in the horizontal direction associated with the apparent horizontal propagation velocity of the wave. In either of these ideal cases, the ground motions are considered to be coherent in that amplitudes (acceleration time histories and their response spectra) do not vary with location in a horizontal plane. Incoherence of ground motion, on the other hand, may result from wave scattering due to inhomogeneities of soil/rock media and topographic effects along the propagation path of the seismic waves.

In terms of the SSI phenomenon, variations of the ground motion over the depth and width of the foundation are the important aspects [1]. For surface foundations, the variation of motion on the surface of the soil is important; for embedded foundations, the variation of motion on both the embedded depth and foundation width is important. Spatial variations of ground motions are discussed in terms of variations with depth and horizontal distances.

2.3.1. *Variation of Ground Motion with Depth of Soil.* For either vertically or non-vertically incident waves, ground motions vary with depth. These variations can generally be expressed in terms of peak amplitudes, frequency content, and phase. Variations of ground motion with depth due to vertically and non-vertically incident body waves and surface waves have been extensively studied analytically by many investigators, Chang et al [13]. These studies, based on the physics of plane wave propagation through layered media, all indicate that, due primarily to the free-surface effect, ground motions generally decrease with depth. The nature of the variation is a function of frequency content and wave type of the incident waves, soil layering, and dynamic soil properties of each soil layer, including shear and compressional wave velocities, damping ratio, and mass density.

Free-field Motion. A review and summary of observational data on the variations of earthquake ground motion with depth was conducted by Seed and Lysmer [4] and Chang et al [13] reflecting the state of knowledge as of 1980 and 1985, respectively. Recordings from two downhole arrays in Japan were analyzed [13] and a review of published data from a number of downhole arrays in Japan and the U.S. was conducted. Based on the review of these data on the variations of ground motion with depth, it was concluded: there is a good body of a data to show that, in general, both peak acceleration and response spectra decrease significantly with depth in the range of typical embedment depths of structures, i.e., less than approximately 25m; and it appears that deconvolution procedures assuming vertically propagating shear waves provide reasonable estimates of the variations of ground motion with depth.

Table 2 summarizes the sources of data evaluated by Seed and Lysmer [4] and Chang et al [13]. Since 1985, substantial additional data has been recorded and evaluated [20-26]. These sources are included in Table 2 and four of the most relevant are discussed next.

Lotung, Taiwan. As part of the Lotung experiment, downhole free-field data was recorded at depths of 6m, 11m, 17m, and 47m. Figure 15 [8] shows the configuration of the arrays. Extensive studies investigating analytical modeling of the phenomenon have been performed [9, 20, 21, 26]. Figure 16 compares recorded and analytically predicted response for one of the models. For this case and many others, deconvolution was applied with the control motion being the free-surface recorded motion. Equivalent linear soil properties were developed through an iteration process and used. Other studies have investigated nonlinear soil behavior [20] to gain further insight into its importance. This set of recorded data and analytical predictions clearly support the variation of motion with depth in the soil and, in fact, a reduction of motion with depth as expected. In addition, the assumption of vertically propagating waves and an equivalent linear representation of soil material behavior well models the wave propagation phenomenon especially at depths in the soil important to SSI. Nonlinear material behavior was shown to have an effect when starting from recorded motion at depth (17m) and convolving the motion to the surface. Figure 17 [20] demonstrates this effect.

Turkey Flat, Ca. Cramer [22] and the compilation of Ref. 23 report on data recorded at Turkey Flat, Ca. Although the sources of motion were low magnitude, a comparison of the surface and downhole data show clear reductions of peak values with depth in the soil. Turkey Flat includes instruments in rock and soil with consistent results.

Ashigara Valley. An excellent collection of papers is presented in Ref. 23 regarding recorded and predicted responses in the Ashigara Valley network. Figure 7 shows a profile of the site and the location of the instruments. Figure 18 shows peak ground acceleration data recorded and predicted at two surface and one downhole location. Again, the reduction of motion with depth can be clearly seen. Figure 19 compares response spectra recorded and predicted at the surface and at depth in the soil. Midorikawa [24] gives a summary of the studies performed for the Ashigara Valley.

Garnes Valley, Ca. Seale and Archuleta [25] present data recorded through the end of April 1990 in the Garnes Valley surface and downhole array located at the San Jacinto fault zone. They present individual data from several events that demonstrate the variation of motion with depth of soil. Reductions in amplitude are apparent.

Foundation Response. A comparison of motions recorded in the free-field and on the base of partially embedded structures provides excellent data validating the effects of kinematic interaction and the spatial variation of ground motion. All recorded data on structures includes the effects of SSI to some extent. Section 1.3 introduced the concept of kinematic and inertial interaction. For purposes of validating the spatial variation of motion with depth in the soil, observations of kinematic interaction are sought. The ideal situation is one where free-field surface motions and foundation motions are recorded for a structure whose embedded portion is stiff, approaching rigid behavior, and the dynamic characteristics of the structure are such that inertial interaction is a minimal effect. Two types of data are available.

Free-field surface motions vs. foundation response. Typically, differences in peak values, time histories, and response spectra were observed and transfer functions relating foundation response to free-field surface motion were generated. These frequency-dependent transfer functions are, in essence, one element of the scattering matrix when inertial interaction effects are minimal, i.e., the component relating foundation input motion to horizontal free-field surface motion. In no cases, was enough information available to generate the rocking component of the scattering matrix from recorded motions. To do so requires recorded rotational acceleration time histories or the ability to generate them from other measurements.

LNG tanks, Japan. Eighteen LNG tanks of varying dimensions and deeply embedded were instrumented and motions recorded on their foundations and in the free-field for a large number of earthquakes. Many of the earthquakes were microtremors but detailed response for at least one larger event was presented. Transfer functions between free-field surface motions and foundation response were generated and compared with a calculated scattering element. Results compared well. A significant reduction in foundation response from the free surface values was observed. The mass and stiffness of the tanks was such that kinematic interaction dominated the soil-structure interaction effects. Figure 20 presents the data [27].

Humboldt Bay Nuclear Power Plant. The Humboldt Bay Nuclear Power Plant has experienced numerous earthquakes over the years. Four of note are the Ferndale earthquake of June 7, 1975 and the recent Lost Coast earthquake sequence of April 25, 1992 [28]. Accelerometers in place in the free-field and on the base of a deeply embedded caisson structure (~80 ft.) recorded acceleration time histories. Figure 21 [4] compares free-field surface response spectra with those recorded on the caisson's base for the 1975 event. A comparison of peak accelerations shows:

Earthquake	Peak Accelerations Surface/Caisson E-W	N-S	Vertical
6/7/75	0.35/0.16g	0.26/0.12g	0.06/0.10g
4/25/92 (11:06 PDT)	0.22/0.14g	0.22/0.11g	0.05/0.08g
4/26/92 (00:14 PDT)	0.25/0.12g	0.23/0.12g	0.05/0.12g
4/26/92 (04:18 PDT)	0.13/0.07g	0.098/0.057g	0.031/0.037g

For horizontal motions, significant reductions are apparent, i.e., reductions up to 55%. For vertical motions, a slight increase is observed which is inconsistent with other data and requires further investigation. The dominant SSI phenomenon here, as for the LNG tanks, is likely to be kinematic interaction due to the deep embedment of the caisson.

Hollywood Storage Building, California. The Hollywood storage building, a 14-story reinforced concrete structure with basement embedded 9 ft. and supported on piles, was subjected to the Kern County earthquake of 1952 and the San Fernando earthquake of 1971. Recorded motions in the basement and an adjoining parking lot permit a comparison of free-field surface motion and foundation motion. Reduced motion on the foundation is observed (Fig. 22).

Fukushima Nuclear Power Station. The Fukushima Nuclear Power Station was subjected to the Miyagi-Ken-Oki, Japan earthquake of June 12, 1978. Motions were recorded in the free-field and in Units 1 and 6 reactor buildings. A comparison of peak surface free-field horizontal accelerations with those on the foundation showed 0.10g vs. 0.08g in the north-south direction and 0.13g vs. 0.06g in the east-west direction. Unfortunately, time histories or response spectra are not generally available in the open literature to see the variations with frequency. One such comparison is available which shows a reduction in spectral acceleration over the entire frequency range, i.e., foundation response is less than free-field values. Both kinematic and inertial interaction effects are likely to be important for this case.

Pleasant Valley Pumping Station. The Pleasant Valley Pumping Station on the California aqueduct was subjected to the Coalinga, CA earthquake sequence of May 1983. The structure is embedded 22 ft. below plant grade and is instrumented at three locations. A comparison of foundation response with a fourth instrument considered to be free-field shows a significant reduction in motion. A complication in this case, however, is that this free-field instrument is located in an instrument shelter on top of a slope 70 ft. above plant grade. Consequently, topographical effects are likely to have influenced this free-field record. A direct comparison of it with the foundation motion demonstrates the combined effects of spatial variation of motion and scattering. Figure 23 shows a cross-section through the aqueduct and pumping station and instrument locations. Figure 24 compares surface free-field motions with foundations response for one horizontal direction.

Buildings With and Without Basements. A series of buildings in close proximity to each other, with and without basements, subjected to the San Fernando earthquake of 1971 were investigated.

Comparing recorded basement motions for buildings with and without embedment documents the effects of kinematic interaction. The results show a definite reduction in motion of embedded foundations compared to surface foundations or free-field surface values. Numerous comparisons are presented by Chang et al. [13]; one is shown in Figure 25 for illustration.

Model Structures. Two model structures in Japan and the Lotung scale model structures in Taiwan have been instrumented and have recorded ground motions from a number of earthquakes. One model structure is located at the Fukushima Nuclear Power Plant site. Tanaka et al. [19] report on data recorded in the free-field and structure. A reduction in peak acceleration of about 20% is observed from free-field surface motion to foundation [19]. The second model structure resides at Chiba Field Station, Institute of Industrial Science, University of Tokyo. The purpose of the model was to instrument it for earthquakes, record earthquake motions, and investigate variability in ground motion and response. Figure 26 shows a cross-section [29,30] through the scale model structure including components (piping, hanged tank, and vessel) and instrument locations. As of 1978, Shibata [29] shows graphically that the mean response factor for peak acceleration on the foundation compared to the free-field is about 0.68 with a coefficient of variation of 0.125 in the horizontal direction and 0.83 with a coefficient of variation 0.136 in the vertical direction. A clear reduction in response of the foundation is observed. The third model structure is the Lotung, Taiwan case. Comparing the May 20, 1986 measured free-field surface response with foundation response demonstrates the reduction in free-field motion with depth. Figure 27 shows a schematic of the Lotung one-quarter scale model with instrument locations identified. Figure 28 [9] shows response spectra of motions recorded on the foundation (F4LS) to be compared with Figure 5.

2.3.2 *Variation of Ground Motion in a Horizontal Plane*. Variations observed in the motions of two points located on a horizontal plane are mainly due to the difference in arrival times of the seismic waves at the two points and to the amplitude and frequency modification that those seismic waves undergo due to the geotechnical characteristics of the media between the two points. With the installation of several dense accelerograph arrays (e.g., SMART 1 and LSST arrays in Taiwan, El Centro and Parkfield in California, and Chusal differential array in the former USSR) and with the recorded data collected from them, the spatial variation of ground motions have been quantified by defining a complex coherency function [31-38]. The coherency function of two motions is normally defined as their cross-power spectrum divided by the square root of the product of the power spectra of the individual motions. As shown in Figs. 29 [31] and 30 [34], this function strongly depends on the distance between recording locations and on the frequency of the seismic waves. This spatial variation of ground motions is a critical element in the seismic analysis of structures on large foundations and in the analysis of long structures on multiple foundations. For these cases, coherency functions can be used to develop average motions for large foundations or to develop the proper motions (and their cross-correlation) at each support for long multiply-supported structures.

These data support a "base averaging" effect on free-field ground motions for large stiff foundations. In the absence of performing detailed SSI analyses incorporating the coherency functions, a simpler approach may be taken, i.e., a filtering of motions. For a 50m plan dimension foundation, reductions in spectral accelerations of 20% in the greater than 20 Hz frequency range, of 10% to 15% in the 10 to 20 Hz range, and of 5% in the 5 to 10 Hz range are supported by the data.

3. Soil Modeling

For soil sites describing the soil configuration (layering or stratigraphy) and the dynamic material properties of soil are necessary to perform SSI analysis and predict soil-structure response. Determining soil properties to be used in the SSI analysis is the second most uncertain element of the process; the first being specifying the ground motion. Modeling the soil can be visualized in two stages: determining the low strain in-situ soil profile, and associated material properties; and defining the dynamic material behavior of the soil as a function of the induced strains from the earthquake and soil-structure response. In general, dynamic stress-strain behavior of soils is nonlinear, anistropic, elasto-plastic, and loading path dependent. It is, also, dependent on previous loading states and the degree of disturbance to be expected during construction. Practically speaking, these effects are not quantifiable in the current state-of-the-art and, hence, contribute to uncertainty in describing soil stress-strain behavior. Soil is modeled as a linear or equivalent linear viscoelastic medium in SSI analyses for earthquake motion, except in a few instances where nonlinear analysis has been performed for research objectives. A linear viscoelastic material model is defined by three parameters -- two elastic constants (frequently shear modulus and Poisson's ratio although shear modulus and bulk or constrained modulus may be more appropriate) and material damping. The equivalent linear method approximates the nonlinear stress-strain relation with a secant modulus and material damping values selected to be compatible with the average shear strain induced during the motion using an iterative procedure. Requirements for this model are low strain shear modulus and Poisson's ratio (or bulk or constrained modulus), material damping values, and their variations with strain. The following discussion assumes an equivalent linear viscoelastic material model for soil.

Three aspects of developing soil models are: field exploration, laboratory tests, and correlation of laboratory and field data.

Field Exploration. Field exploration, typically, relies heavily on boring programs which provide information on the spatial distribution of soil (horizontally and with depth) and produce samples for laboratory analysis. In addition, some dynamic properties are measured in situ, for example, shear wave velocity which leads to a value of shear modulus at low strains.

Reference 39 provides a summary of field exploration, in general, and of boring, sampling, and in-situ testing, in particular. Low strain shear and compressional wave velocities are typically measured in the field. Various field techniques for measuring in situ shear and compressional wave velocities exist including the seismic refraction survey, seismic cross-hole survey, seismic down-hole or up-hole survey, and surface wave techniques. The advantages and disadvantages of these techniques are discussed in Ref. 39. For sites having a relatively uniform soil profile, all of these techniques are appropriate. However, for sites having layers with large velocity contrasts, the most appropriate technique is the cross-hole technique. The cross-hole data are expected to be the most reliable because of better control over and knowledge of the wave path. The downhole technique does not permit as precise resolution because travel times in any layer are averaged over the layer. Most field seismic survey techniques are capable of producing ground motions in the small shearing strain amplitude range (less than 10^{-3} percent). Hence, field exploration methods yield the soil profile and estimates of important low strain material properties (shear modulus, Poisson's ratio, water table location).

Laboratory Tests. Laboratory tests are principally used to measure dynamic soil properties and their variation with strain: soil shear modulus and material damping. Currently available laboratory testing techniques have been discussed and summarized in Ref. 39. These techniques include resonant column tests, cyclic triaxial tests, cyclic simple shear tests, and cyclic torsional shear tests.

Each test is applicable to different strain ranges. The resonant and torsional shear column tests are capable of measuring dynamic soil properties over a wide range of shear strain (from 10^{-4} to 10^{-2} percent or higher). The cyclic triaxial test allows measurement of Young's modulus and damping at large strain (larger than 10^{-2} percent).

Typical variations of shear modulus with shear strain for clays, compiled from laboratory test data are depicted in Fig. 31 [40]. Generally, the modulus reduction curves for gravelly soils and sands are similar. Figure 32 shows material damping as a function of shear strain, also for clays. Shear modulus decreases and material damping increases with increasing shear strain levels.

Correlation of Laboratory and Field Data. Once shear modulus degradation curves and material damping versus strain curves have been obtained, it is necessary to correlate laboratory determined low strain shear modulus values with those in situ. Laboratory-measured values of shear moduli at low levels of strain are typically smaller than those measured in the field. Several factors have been found to contribute to lower moduli measured in the laboratory. These factors include effects of sampling disturbance, stress history, and time (aging or period under sustained load). Thus, when the laboratory data are used to estimate in situ shear moduli in the field, considerations should be given to these effects. When laboratory data do not extrapolate back to the field data at small strains, one of the approaches used in practice is to scale the laboratory data up to the field data at small strains. This can be done by proportionally increasing all values of laboratory moduli to match the field data at low strain values or by other scaling procedures.

Equipment Linear Soil Properties. Given the low strain soil profile determined from the combination of field and laboratory tests, and the variation of material parameters with strain level, a site response analysis is performed to estimate equivalent linear soil properties for the SSI analysis.

Uncertainty in Modeling Soil/Rock at the Site. There is uncertainty in each aspect of defining and modeling the site soil conditions for SSI analysis purposes.

The soil configuration (layer or stratigraphy) is established from the boring program. Even after such a program, some uncertainty exists in the definition of the soil profile. Soils are seldom homogeneous, and they seldom lie in clearly defined horizontal layers -- the common assumptions in SSI analysis. In general, complicated soil systems introduce a strong frequency dependence in the site behavior.

Modeling the dynamic stress-strain behavior of the soil is uncertain in two respects. First, modeling soil as a viscoelastic material with parameters selected as a function of average strains is an approximation to its complex behavior. Second, there is uncertainty in defining low strain shear moduli and in defining the variations in shear modulus and observed damping with strain levels. Variability can be observed in reporting test values for a given site and for reported generic test results [40,41,42]. The range of coefficients of variation on "stress-strain behavior" estimated in Ref. 43 is 0.5 to 1.0. Considering only uncertainty in soil shear modulus itself and the equivalent linear estimates to be used in the SSI analysis, a coefficient of variation of 0.35 to 0.5 on soil shear modulus is estimated. Uncertainty in low strain values is less than uncertainty in values at higher strains.

The Lotung SSI experiment provides a unique opportunity to quantify the uncertainty in estimates of equivalent linear soil properties. The bases for determining the equivalent linear soil properties included field and laboratory tests and the results of forced vibration tests on the structure. The latter responses permitted system identification techniques to be employed to refine estimates of the low strain profile.

Soil properties data was extracted from Ref. 8 and plotted in Figs. 33 and 34. Ten independent sets of data were available and their variability is apparent from the figures. To quantify this variability, the data was evaluated statistically and a weighted COV calculated for soil shear modulus and damping; the weighting factors were soil thickness to a depth of about 100 feet. The resulting COVs for shear modulus and damping were 0.48 and 0.39, respectively.

Hence, even for a most ideal situation, significant variability in soil properties as used in the SSI analysis should be expected reflecting variability innate in the soil, its behavior, and the modeling thereof. These estimates are for a single earthquake, the May 20, 1986 event.

4. Soil-structure Interaction Analysis

4.1 Foundation Modeling

Three aspects of modeling structure foundations are important: stiffness, embedment, and geometry.

Stiffness. The stiffness of a structural foundation is of importance because in almost all SSI analyses, by either direct or substructure methods, foundation stiffness is approximated. Most substructure analysis approaches assume the effective foundation stiffness to be rigid. For direct methods, various representations of foundation stiffness have been used depending on the geometry and other aspects of the structure-foundation system.

Most foundations of the type common to major building structures cannot be considered rigid by themselves. However, structural load resisting systems, such as shear wall systems, significantly stiffen their foundations. Hence, in many instances, the effective stiffness of the foundation is very high and it may be assumed rigid. One study which has investigated the effects of foundation flexibility on structure response for a complicated nuclear power plant structure of large plan dimension was performed [44]. In this study, the stiffening effect of the structure on the foundation was treated exactly. Even though the structure had nominal plan dimensions of 350 feet by 450 feet, the effect of foundation flexibility on structure response was minimal. The largest effect was on rotational accelerations of the foundation segment as one would expect. Translational accelerations and response spectra, quantities of more interest, on the foundation and at points in the structure were minimally affected (5%). This study emphasized the importance of considering the stiffening effect of walls and floors on the foundation when evaluating its effective stiffness.

Hence, practically speaking, for design and evaluation of major structures with stiff load resisting systems, the assumption of rigid foundation behavior is justified. For predicting the response of a structure to a single recorded event, more refined modeling of the foundation may be required.

Embedment. Foundations are typically fully embedded, and structures are partially embedded. Foundation embedment has a significant effect on SSI. Both the foundation input motion and the foundation impedances differ for an embedded foundation compared to a surface foundation. Foundation input motion for embedded foundations was discussed extensively in Sec. 2. Foundation impedances comprise a second aspect of foundation modeling. As discussed above, a common and appropriate assumption for many cases is rigid foundation behavior. For a rigid foundation, the impedance matrix describing force-displacement characteristics is, at most, a 6X6, complex valued, frequency dependent matrix. The literature [1] contains numerous exact and approximate analytical representations of foundation impedances. Also, many analytical/experimental correlations exist with relatively good results reported.

Figure 35 [45] shows one such example. Figure 36 [46] demonstrates the effect of embedment on foundation response for embedded and nonembedded configurations of an assumed rigid foundation. Forced vibration tests were performed. The results clearly demonstrate the effect on inertial interaction of embedment.

Geometry. The geometry of major structure foundations can be extremely complicated. Fortunately, many aspects of modeling the foundation geometry are believed to have second order effects on structure response; in particular, modeling the precise shape of the foundation in detail. However, other overall aspects such as nonsymmetry which lead to coupling of horizontal translation and torsion, and vertical translation and rocking, can be very important and need to be considered.

Modeling the foundation is important with respect to structure response. This is one area where knowledge and experience of the practitioner is invaluable. Complicated foundations must be modeled properly to calculate best estimates of response, i.e., including the important aspects as necessary. Foundation modeling plays a key role in assessing whether two-dimensional models are appropriate or three-dimension models are required.

4.2 STRUCTURE MODELING

4.2.1. *Linear Dynamic Behavior.* Structures for which SSI analysis is to be performed require mathematical models to represent their dynamic characteristics. The required detail of the model is dependent on the complexity of the structure or component being modeled and the end result of the analysis. Structure models are typically of two types: lumped-mass, stick models; and finite element models. Lumped-mass, stick models are characterized by: lumped masses defining dynamic degrees-of-freedom, usually at floor slab elevations; simplified assumptions as to diaphragm or floor behavior are made, in particular floors are frequently assumed to behave rigidly, although diaphragm flexibility can be modeled if necessary; connections between lumped masses are usually stiffness elements whose stiffness values represent groups of walls running between the floor slabs; and, in some instances, an offset is modeled between the center of mass and center of rigidity at each floor elevation to account for coupling between horizontal translations and torsion. Finite element models are used to represent complex structures. They permit a more accurate representation of complicated situations without requiring significant simplifying assumptions. Either model type may be adequate depending on the structural configuration, the detail included in the model, and the simplifying assumptions. Lumped-mass, stick models are simpler mathematically than finite element models and usually have a smaller number of degrees-of-freedom. Lumped-mass, stick models take advantage of the judgement of the analyst as to the expected behavior of the structure.

Modeling methods and techniques will not be discussed in detail here. The ASCE Seismic Analysis Standard and Commentary [47] present many aspects of modeling and a bibliography from which additional information may be obtained.

One point of note for shear wall structures is that over the last half of the 1980s, testing of shear walls as individual elements and as a portion of a structure assemblage has been performed. One result of these tests is the apparent reduction in stiffness from the linearly calculated values due to small cracks and other phenomenon. Reference 48 evaluated the relevant data and recommended approaches to account for increases and decreases in stiffness previously not treated explicitly. Increases result from items such as increased concrete strength due to aging and achieving minimum specified strengths in a conservative manner.

4.2.2. *Nonlinear Structure Models.* The nonlinear behavior of structures is important in two regards -- evaluating the capacity of structural members and the structure itself; and estimating the environment (in-structure response spectra and structural displacements) to which equipment and commodities are subjected. Nonlinear structural behavior is characterized by a shift in natural frequencies to lower values, increased energy dissipation, and increased relative displacement between points in the structure. Four factors determine the significance of nonlinear structural behavior to dynamic response.

Frequency content of the control motion versus the frequencies of the structure. Consider a rock-founded structure. Depending on the elastic structure frequencies and characteristics of the control motion, the shift in structure frequency due to nonlinear structural response may result in a relatively large reduction in structure response with a substantial reduction in input to equipment when compared to elastic analysis results. If the structure elastic frequency is located on the peak or close to the peak of the control motion's response spectra, the frequency shift will result in a much large decrease in response than if the elastic frequency is higher that the peak of the control motion's response spectra such that the shift in frequency tends to result in increased response. The same type of reduction occurs for structures excited by narrow band spectra earthquakes where the structure elastic frequency tends to coincide closely with the peak of the ground response spectra.

Soil-structure interaction effects. The effect of nonlinear structure behavior on in-structure response (forces, accelerations, and response spectra) appears to be significantly less when SSI effects are important at the site. This is principally due to the potential dominating effect of SSI on the response of the soil-structure system. The soil can have a controlling effect on the frequencies of the soil-structure system. Also, if SSI is treated properly, the input motion to the system is filtered such that higher frequency motion is removed., i.e., frequency content which may not be suppressed by nonlinear structural behavior if the structure were founded on rock. SSI can have a significant effect on the energy dissipation characteristics of the system due to radiation damping and material damping in the soil. Accounting for the effect of the inelastic structural behavior on structure response must be done carefully for soil-founded structures to avoid double-counting of the energy dissipation effects.

Degree of structural nonlinearity. The degree of structural nonlinearity to be expected and permitted determines the adequacy assessment for the structure and can have a significant impact on in-structure responses. Past reviews of testing conducted on shear walls has indicated that element ductilities of up to about four or five can be accommodated before significant strength degradation begins to occur. However, the allowable achieved ductility for many evaluations will be substantially less. The effect of nonlinear structure behavior on in-structure response spectra has been considered to only a limited extent. In general, increased levels of nonlinearity lead to increasing reduced in-structure response spectra for a normalized input motion. However, in some instances, higher frequency, i.e., higher than the fundamental frequency, peaks can be amplified. This is an area for which research is currently being performed and will provide guidance in the future.

Magnitude effects. Earthquake magnitude as it affects the control motion has been discussed earlier. Recall, however, smaller magnitude, close-in earthquakes may be narrow-banded which are significantly affected by nonlinear behavior as described above.

4.3 SSI RESPONSE

Very few cases exist where the data necessary to effectively analyze a soil-structure system has been developed or measured. In addition, for those with data, typically, not all aspects of the SSI phenomenon are important. One case in point is the Lotung one-quarter scale model. All appropriate data has been developed and measured but the high stiffness of the structure and the very soft soil conditions eliminate structure vibration as a significant phenomenon. Figure 28 [9] shows a comparison of measured and calculated response which is an excellent match.

In addition to the Lotung experiment, there is a recognition in the technical community that additional data appropriate for SSI analysis methods benchmarking and development are necessary. An additional experiment called the Hualien Large-Scale Seismic Test for Soil-Structure Interaction Research is being constructed in Taiwan [49]. This is a stiffer site than Lotung and eliminates some of the deficiencies of the Lotung experiment. Lastly, the U.S.G.S. sponsored a workshop on February 1992 which itemized the need for additional measurements on soil and structures for SSI analysis methods development and benchmarking.

5. Conclusions

Many aspects of SSI are understood and any valid method of analysis is able to reproduce them. Sections 2, 3, and 4 itemized the various aspects of the problem and one's ability to model them. Clearly, uncertainties exist in the process; randomness associated with the earthquake ground motion itself and the dynamic behavior induced in soil and structures. Even assuming perfect modeling, randomness in the response of structures and components is unavoidable. Perhaps the best evidence of such randomness is the Chiba Field Station. Shibata [30] reports the results of 20 years of recorded motion on the model structure; 271 events when taken in total. Analyzing the measured responses of the hanged tank (Fig. 26) yields a coefficient of variation of response of about 0.45 conditional on the horizontal peak ground accelerations of the earthquake. Of course, arguments can be made that many of the events were low amplitude, that group events by epicentral area reduced variability, etc. However, the undisputable fact is that significant variability in response of structures and components due to earthquakes is to be expected.

No deterministically exact solution of the SSI problem can be obtained by existing techniques. However, given the free-field ground motion, data concerning the dynamic behavior of soil and structure, reasonable estimates can be made. In addition, uncertainties in each of the elements do not necessarily combine in such a fashion as to always increase the uncertainty in the end item of interest (structure response). Reference 50 reports on a probabilistic response study for the Diablo Canyon Nuclear Power Plant which demonstrates that combining uncertainties due to the square root of the sum of squares rule is conservative and can be extremely conservative. Hence, although uncertainties exist in one's ability to represent and model the SSI phenomenon, they do not always increase the uncertainty in the end item of interest.

6. References

1. Johnson, J.J. (1981) "Soil-Structure Interaction: The Status of Current Analysis Methods and Research, Lawrence Livermore National Laboratory (LLNL)," UCRL-53011, NUREG/CR-1780, prepared for the U.S. NRC.
2. Luco, J.E. (1980) "Linear Soil-structure Interaction," LLNL, Livermore, CA, UCRL-15272, included in Ref. 1.
3. Roesset, J.M. (1980) "A Review of Soil-Structure Interaction," LLNL, Livermore, CA, UCRL-15262, included in Ref. 1.
4. Seed, H.B. and Lysmer, J. (1980) "The Seismic Soil-Structure Interaction Problem for Nuclear Facilities," LLNL, Livermore, CA, UCRL-15254, included in Ref. 1.
5. Veletsos, A.S. (1978) "Soil-Structure Interaction for Buildings During Earthquakes," in Proc. Second International Conference on Microzonation for Safer Construction -- Research and Application, San Francisco, CA, Vol. 1, pp. 111-133.
6. Chopra, A.K. (1980) Personal Communication.
7. Johnson, J.J. and Chang, C.-Y. (1991) "State of the Art Review of Seismic Input and Soil-Structure Interaction," Appendix E in "A Methodology for Assessment of Nuclear Power Plant Seismic Margin (Rev. 1)," EPRI, Palo Alto, CA, EPRI NP-6041-SL.
8. EPRI (1989) Proceedings: EPRI/NRC/TPC Workshop on Seismic Soil-Structure Interaction Analysis Techniques Using Data From Lotung, Taiwan, Electric Power Research Institute, Palo Alto, CA EPRI NP-6154, Vols. 1 and 2.
9. Johnson, J.J., Maslenikov, O.R., Mraz, M.J., and Udaka, T. (1989) "Analysis of Large-Scale Containment Model in Lotung, Taiwan: Forced Vibration and Earthquake Response Analysis and Comparison," included in Reference 8, pp. 13-1 -- 13-44.
10. Tseng, W.S., and Hadjian, A.H. (1991) "Guidelines for Soil-Structure Interaction Analysis," EPRI, Palo Alto, CA EPRI NP-7395.
11. U.S. Atomic Energy Commission (1973) "Regulatory Guide 1.60, Design Response Spectra for Seismic Design of Nuclear Power Plants", Rev. 1.
12. Newmark, N.M. and Hall, W.J. (1978) "Development of Criteria for Seismic Review of Selected Nuclear Power Plants," NUREG/CR-0098, prepared for the U.S. Nuclear Regulatory Commission.
13. Chang, C.-Y., Power, M.S., Idriss, I.M., Somerville, P.G., Silva, W., and Chen, P.C. (1985) "Engineering Characterization of Ground Motion, Task II: Observational Data on Spatial Variations of Earthquake Ground Motion," NUREG/CR-3805, Vol. 3, prepared for the U.S. Nuclear Regulatory Commission.
14. Kudo, K., Shima, E., and Sakane, M. (1988) "Digital Strong Motion Accelerograph Array in Ashigara Valley - Seismological and Engineering Prospects of Strong Motion Observations," Proc. 9th World Conference on Earthquake Engineering, Vol. VIII, pp. 119-124.
15. Chang, C.Y. (1992) Personal Communication.
16. Bozorgnia, Y.B. (1992) Personal Communication.
17. Gazetas, G., and Bianchini, G. (1979) "Field Evaluation of Body and Surface-wave Soil-amplification Theories," Proceedings of the Second U.S. National Conference on Earthquake Engineering," EPRI, Stanford University.
18. Yanev, P.I., Moore, T.A., and Blume, J.A. (1979) "Fukushima Nuclear Power Station, Effect and Implications of the June 12, 1978, Miyagi-ken-Oki, Japan, Earthquake," prepared by URS/John A. Blume & Associates, Engineers, San Francisco, California.

19. Tanaka, T. et al (1973) "Observations and Analysis of Underground Earthquake Motions," Proceedings of the 5th World Conference on Earthquake Engineering, Rome, Vol. 1, pp. 658-667.
20. Chang, C-Y. et al (1990) "Equivalent Linear Versus Nonlinear Ground Response Analyses at Lotung Seismic Experiment Site," Proceedings Fourth U.S. National Conference on Earthquake Engineering, Palm Springs, CA.
21. Chang, C-Y. et al (1991) "Development of Shear Modulus Reduction Curves Based on Lotung Downhole Ground Motion Data," Proceedings Second International Conference on Recent Advances in Geotechnical Earthquake Engineering and Soil Dynamics, St. Louis.
22. Cramer, C.H. (1991) "Turkey Flat, USA Site Effects Test Area, Report 6 Weak-Motion Test: Observations and Modeling," Technical Report No. 91-1, California Department of Conservation, Division of Mines and Geology, Earthquake Shaking Assessment Project.
23. Proceedings of the International Symposium on the Effects of Surface Geology on Seismic Motions," (1992) Odawara, Japan.
24. Midorikawa, S., (1992) "A Statistical Analysis of Submitted Predictions for the Ashigara Valley Blind Prediction Test," (Subcommittee for the Prediction Criteria of the Ashigara Valley Blind Prediction Test). Proceedings of the International Symposium on the Effects of Surface Geology on Seismic Motion, Odawara, Japan.
25. Seale, S.H., and Archuleta, R.J. (1991) "Analysis of Site Effects at the Garnes Valley Downhole Array Near the San Jacinto Fault," Proceedings of the Second International conference on Recent Advances in Geotechnical Earthquake Engineering and Soil Dynamics, St. Louis, Missouri, Paper No. 8.13.
26. Hadjian, A.H., et al "A Synthesis of Predictions and Correlation Studies of the Lotung Soil-Structure Interaction Experiment" (1991) EPRI Report No. NP-7307-M.
27. Ishii, K., Itoh, T., and Suhara, J. (1984) "Kinematic Interaction of Soil-structure System Based on Observed Data," Proceedings of the 8th World Conference on Earthquake Engineering, San Francisco, Vol. III, pp. 1017-1024.
28. Brady, F.W. (1992) Personal Communications.
29. Shibata, H. "On the Reliability Analysis for Structural Design Including Pipings and Equipment," Presented at the Seminar on Probabilistic Seismic Analysis of Nuclear Power Plants, January 16-19, 1978, Berlin.
30. Shibata, H. (1991) "Uncertainty in Earthquake Engineering in Relation to Critical Facilities," Bulletin of Earthquake Resistant Structure Research Center, Institute of Industrial Science, University of Tokyo, No. 24, pp. 93-104.
31. Abrahamson, N.A., Schneider, J.F., and Stepp, J.C. (1990) "Spatial Variation of Strong Ground Motion for Use in Soil-Structure Interaction Analyses," Proceedings, 4th U.S. National Conference on Earthquake Engineering, Palm Springs, California.
32. Abrahamson, N.A., Schneider, J.F., and Stepp, J.C. (1991) "Empirical Spatial Coherency Functions for Application to Soil-Structure Interaction Analysis," Earthquake Spectra, Vol. 7, No. 1.
33. Schneider, J.F., Abrahamson, N.A., and Stepp, J.C. (1992) "The Spatial Variation of Earthquakes Ground Motion and Effects of Local Site Conditions," Proceedings, 10th World Conference on Earthquake Engineering, Madrid, Spain.
34. Schneider, J.F., et. al (1990) "Spatial Variation of Ground Motion from EPRI's Dense Accelerograph Array at Parkfield, California," Proceeding, 4th U.S. National Conference Earthquake Engineering, Palm Springs, California.

35. Hao, H., Oliveira, C.S., and Penzien, J. (1989) "Multiple-station Ground Motion Processing and Simulation Based on SMART-1 Array Data," Nuclear Engineering and Design 111, pp. 293-310.
36. Harichandran, R.S. (1988) "Local Spatial Variation of Earthquake Ground Motion," Proceedings, Earthquake Engineering and Soil Dynamics ASCE Specialty Conference, Park City, Utah.
37. Somerville, P.G., et. al (1988) "Site Specific Estimation of Spatial Incoherence of Strong Ground Motion," Proceedings, Earthquake Engineering and Soil Dynamics ASCE Specialty Conference, Park City, Utah.
38. Loh, C.-H., and Chen, S.-J. (1990) "Directionality in Spatial Variation of Earthquake Ground Motion," Proceedings, 4th U.S. National Conference on Earthquake Engineering, Palm Springs, California.
39. Woods, R.D. (1978) "Measurement of Dynamic Soil Properties," ASCE Specialty Conference on Earthquake Engineering and Soil Dynamics, Pasadena, California, Vol. 1, pp. 91-178.
40. Sun, Joseph I., Golesorkhi, R., Seed, H. Bolton (1988) "Dynamic Moduli and Damping Ratios for Cohesive Soils," Report No. UCB/EERC-88-15, Earthquake Engineering Research Center, University of California, Berkeley, California.
41. Seed, H.B., Wong, R.T., Idriss, I.M., and Tokimatsu, K. (1984) "Moduli and Damping Factors for Dynamic Analyses of Cohesionless Soils," Report No. UCB/EERC-84/14, Earthquake Engineering Research Center, University of California, Berkeley, California.
42. Seed, H.B., and Idriss, I.M. (1970) "Soil Moduli and Damping Factors for Dynamic Response Analyses," Report No. EERC 70-10, Earthquake Engineering Research Center, University of California, Berkeley, California.
43. ASCE Committee on Reliability of Offshore Structures, Subcommittee on Foundation Materials (1979) Probability Theory and Reliability Analysis Applied to Geotechnical Engineering of Offshore Structure Foundations.
44. Johnson, J.J., Maslenikov, O.R., and Benda, B.J. (1984) "SSI Sensitivity Studies and Model Improvements for the U.S. NRC Seismic Safety Margins Research Program," NUREG/CR-4018, Prepared for the U.S. Nuclear Regulatory Commission.
45. Hijikata, K. et al (1987) "Dynamic Soil Stiffness of Embedded Reactor Buildings," SMiRT 9 Transactions, Vol. K1, August 1987, Lausanne, Switzerland.
46. Kobayashi, T. et al (1991) "Forced Vibration Test on Large Scale Model on Soft Rock Site (Embedment Effect Test on Soil-structure Interaction)," SMiRT 11 Transactions, Vol. 7, August 1991, Tokyo, Japan.
47. ASCE Seismic Analysis of Safety-Related Nuclear Structures Standards Working Group (1985) "Standard for the Seismic Analysis of Safety-Related Structures," Standard and Commentary, Draft.
48. ASCE (1992) Dynamic Analysis Committee Working Group Report on Stiffness of Concrete.
49. Tang, Y.K. et al (1991) "The Hualien Large-Scale Seismic Test for Soil-Structure Interaction Research," Transactions, SMiRT 11, Vol. K, pp. 69-74.
50. Bozorki, G. et al (1990) "Review of the Diablo Canyon Probabilistic Risk Assessment," Prepared for U.S. Nuclear Regulatory Commission, NUREG/CR-5726.

Table 1
ELEMENTS OF A SEISMIC RESPONSE ANALYSIS
INCLUDING SOIL-STRUCTURE INTERACTION

Specification of the Free Field Ground Motion
- Control point
- Control motion (peak ground acceleration and response spectra)
- Spatial variation of motion (wave propagation mechanism)
- Magnitude, duration

Models of Soil and Structures
- Soil properties
  - Nominal properties (low and high strain)
  - Variability
- SSI parameters
  - Kinematic interaction
  - Foundation impedances
- Structure models
  - Important features (torsion, floor flexibility, frequency reduction with strain, etc.)
  - Variability (frequency and mode shapes, damping)
  - Nonlinear behavior

SSI Analysis

Interpretation of Responses

Table 2
FREE-FIELD DOWNHOLE ARRAYS

Location	Instrument Depth (m)	Reference
Narimasu, Tokyo	-1, -5, -8, -22, -55	13
Waseda, Tokyo	-1, -17, -67, -123	13
Menlo Park, CA	GL, -12, -40, -186	13
Richmond Field Station, CA	GL, -14.5, -40	13
Ukishima Park, Tokyo	GL, -27, -67, -127	13
Futtsu Cape, Chiba	GL, -70, -110	13
Kannonzaki, Yokosuka City	GL, -80, -120	13
Tokyo International Airport	GL, -50; GL, -65	13
Ohgishima Station, Kawasaki City	GL, -15, -38, -150	17
Earthquake Research Institute,	GL, -82	13
Miyako, Tokyo	-.5, -18, -26.5	13
Tomakomai, Hokkaido	GL, -30, -50	13
Tateyama, Tokyo	-26; -38, -100	13
Higashi-Matsuyama City, Saitama	-1, -58, -121;	13
Shuzenji-cho, Shizuoka	-36, -100; -49; -74	13
Choshi City, Chiba	GL, -18	13
Beatty, Nevada	GL, -41	13
Fukushima Nuclear Power Plant	GL, -50; -.5, -3, -8.5, -23.5	18, 19
Lotung, Taiwan, ROC	GL, -6, -11, -17, -47	20, 21, 26
Turkey Flat, CA	GL, -10, -20	22, 23
Ashigara Valley, Japan	GL, -30, -95	23, 24
Garnes Valley, CA	GL, -6, -15, -22, -220	25

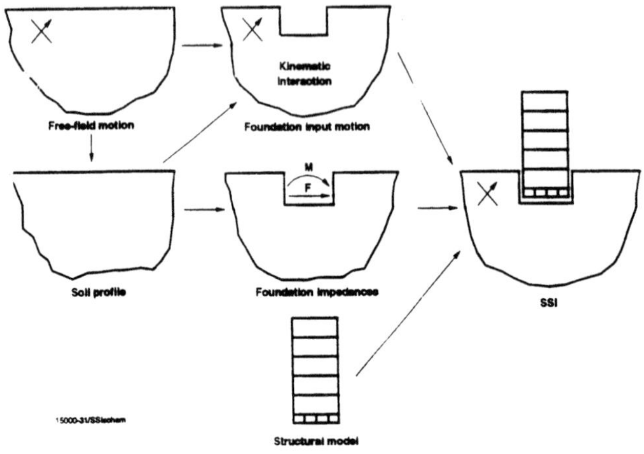

Figure 1: Schematic Representation of the Elements of Soil-structure Interaction

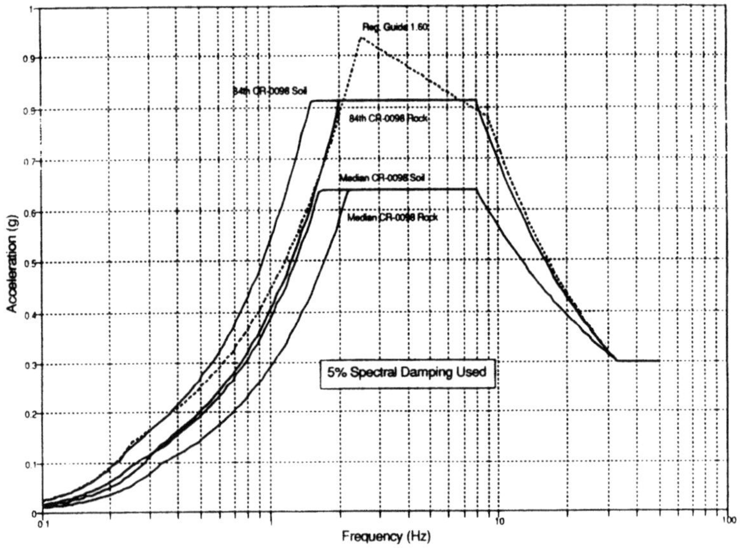

Figure 2: Examples of Aggregated Ground-motion Response Spectra

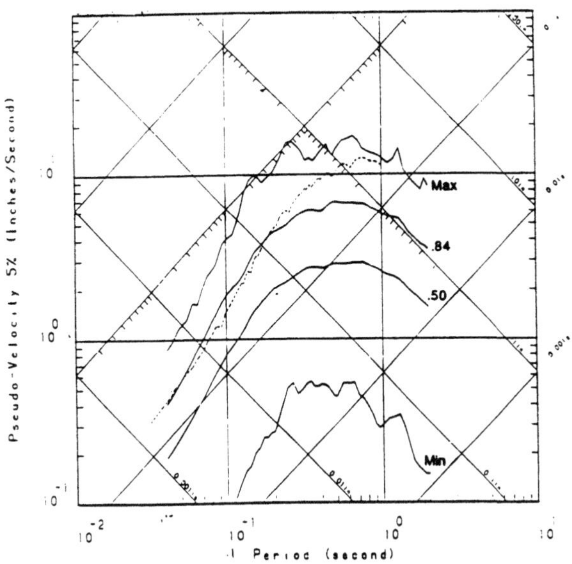

Figure 3: Site-specific Response Spectrum

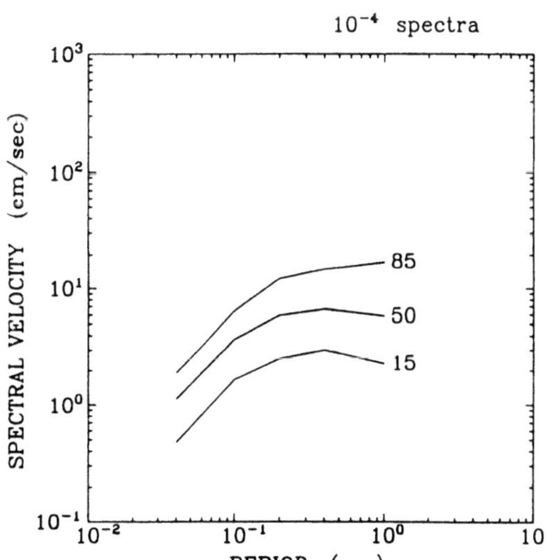

Figure 4: Uniform Hazard Spectra

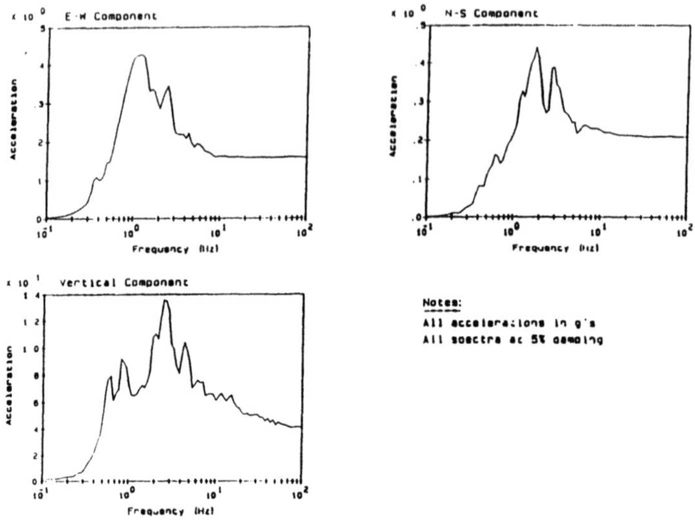

Figure 5: Response Spectra of Free-field Ground Surface Motions of the May 20, 1986 Earthquake

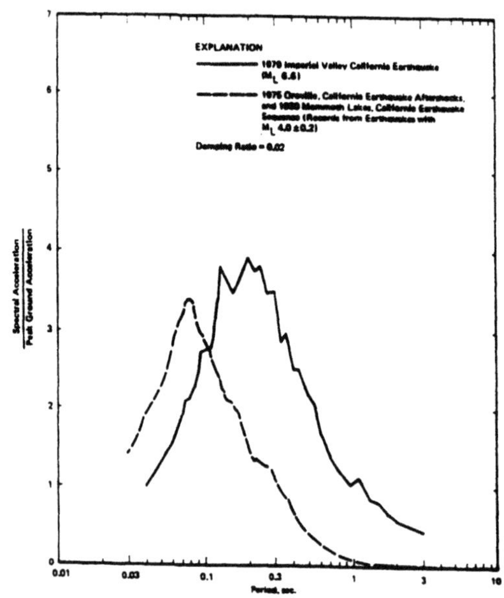

Figure 6: Illustration of Effect of Earthquake Magnitude on Response Spectral Shape Obtained from Statistical Analysis

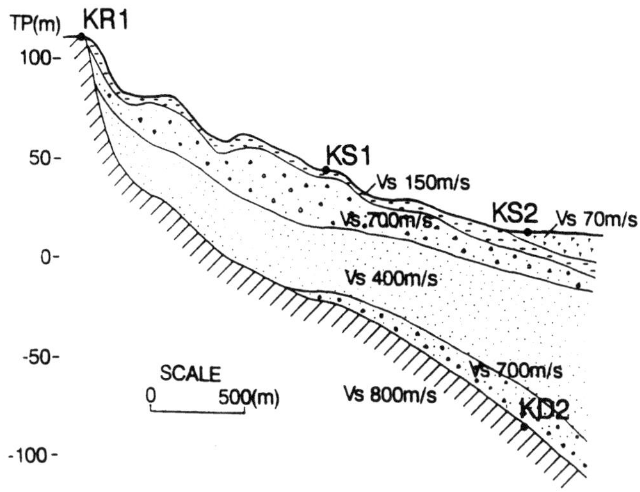

Figure 7: Ashigara Valley, Japan

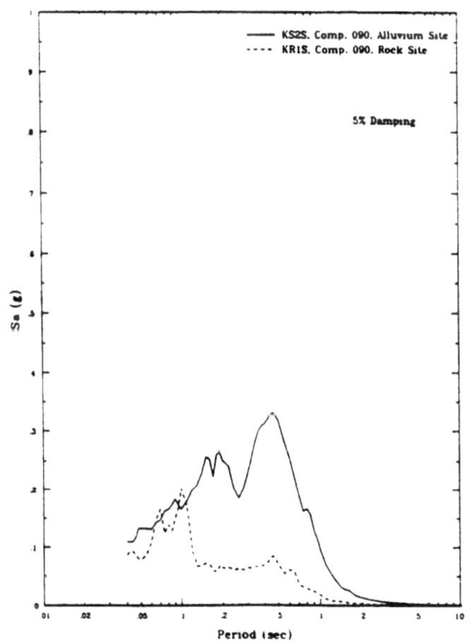

Figure 8: Response Spectra at Rock and Alluvium Sites - Ashigara Valley, Japan

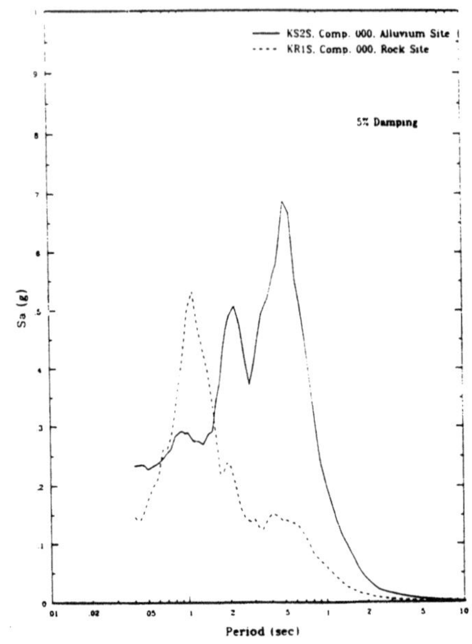

Figure 9: Response Spectra at Rock and Alluvium Sites - Ashigara Valley, Japan

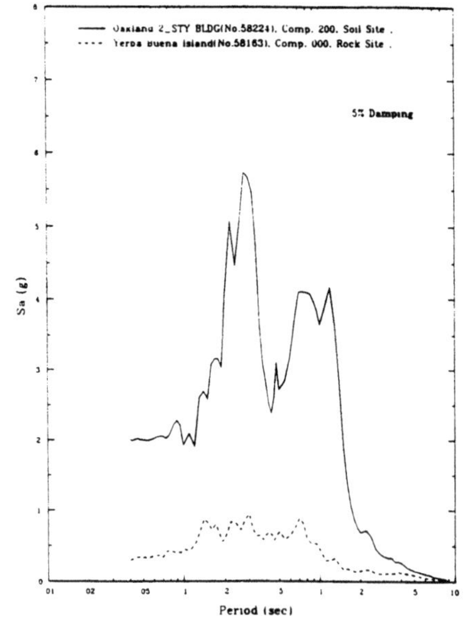

Figure 10: Response Spectra at Rock and Soil Sites - Loma Prieta Earthquake

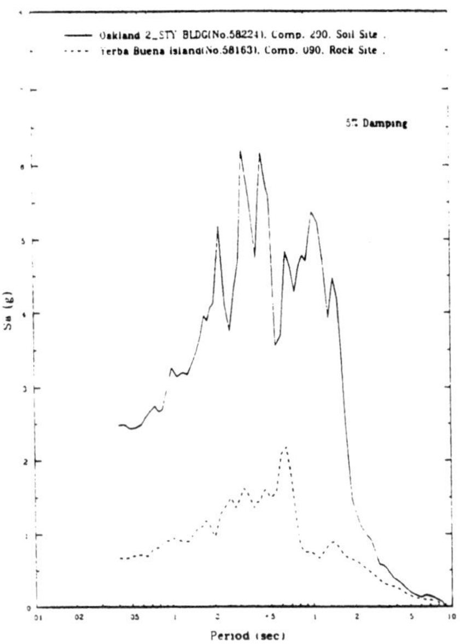

Figure 11: Response Spectra at Rock and Soil Sites - Loma Prieta Earthquake

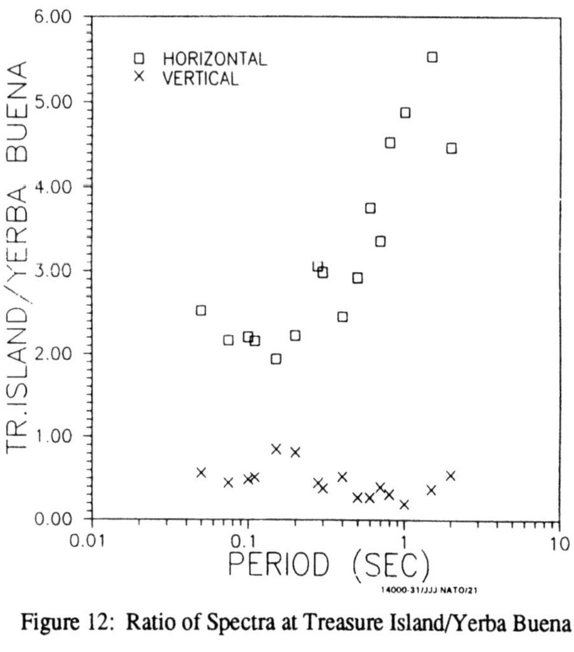

Figure 12: Ratio of Spectra at Treasure Island/Yerba Buena Island

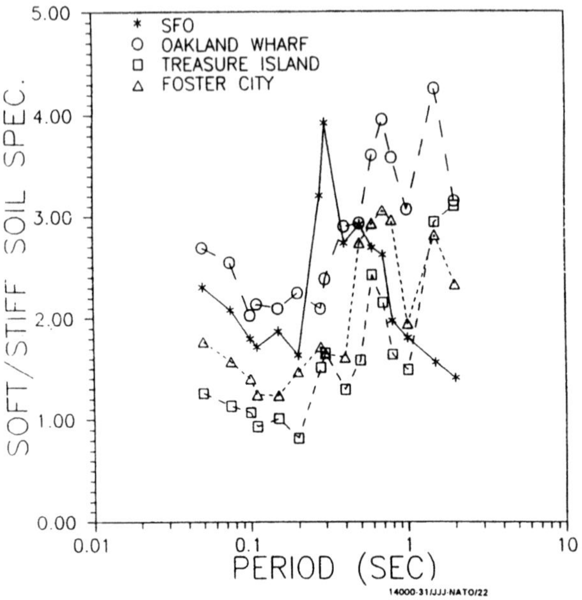

Figure 13: Horizontal Spectra Amplification at Soft Soil/Stiff Soil

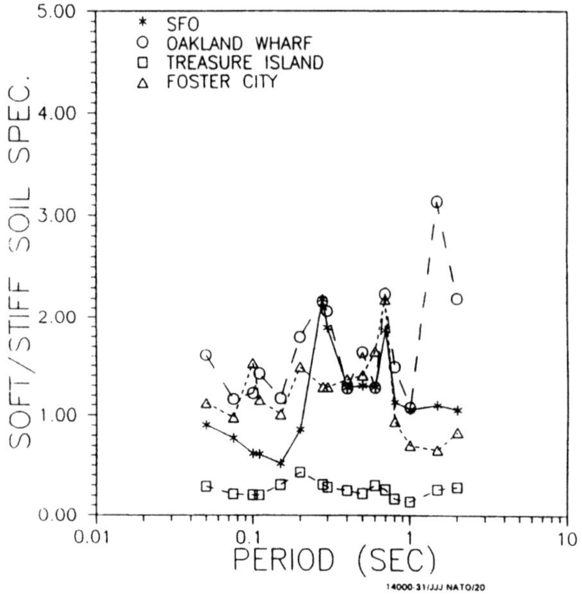

Figure 14: Vertical Spectra Amplification at Soft Soil/Stiff Soil

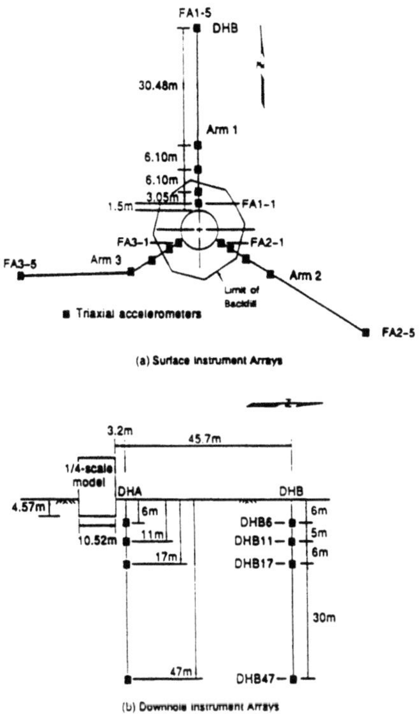

Figure 15: Location of (a) Surface and (b) Downhole Accelerographs [8]

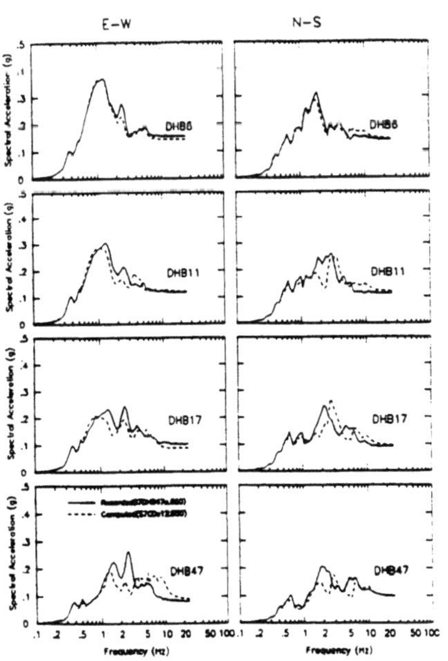

Figure 16: Comparison of Recorded and Computed Response Spectra (5% Damping), Deconvolved with Iterated Strain-Compatible Properties, Event LSST07 [8]

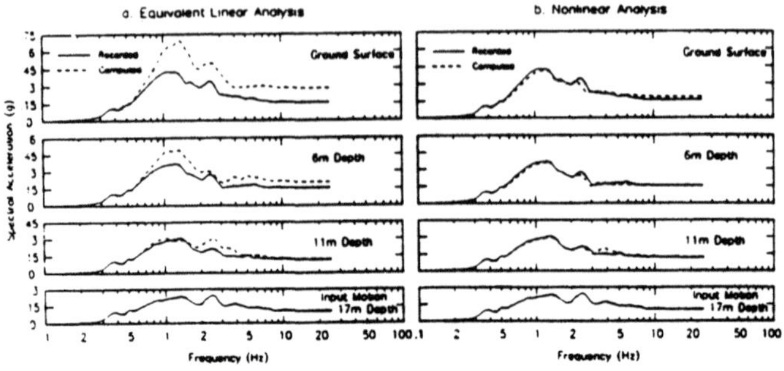

Figure 17: Comparisons of Response Spectra (5% Damping) of Recorded and Computed Ground Surface Motions Using Recorded Motion at 17m Depth as Input Motion, Event LSST07 [20]

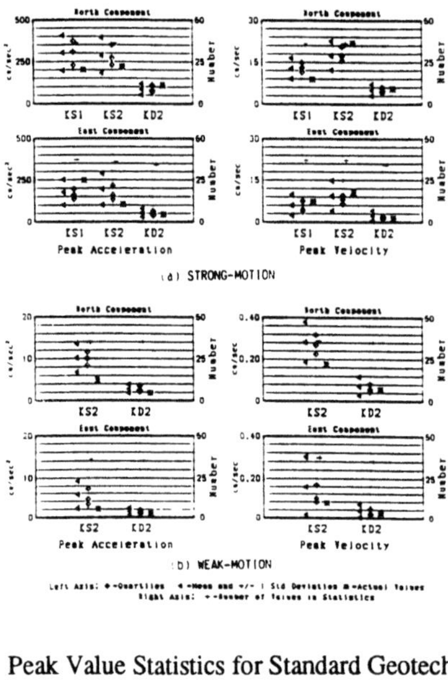

Figure 18: Peak Value Statistics for Standard Geotechnical Model Predictions

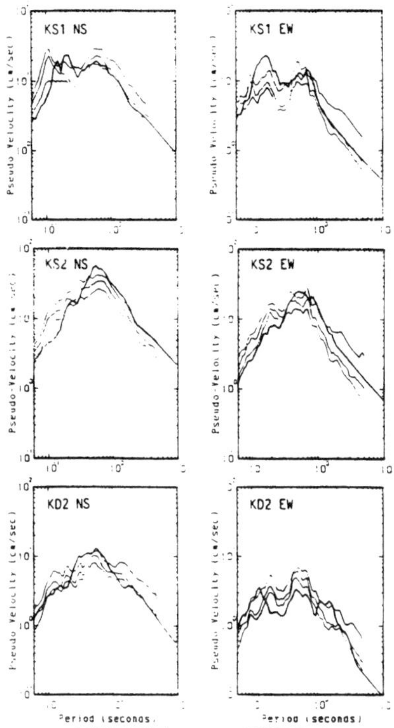

Figure 19: Response-Spectra Quartiles for Strong-Motion Standard Geotechnical Model Predictions

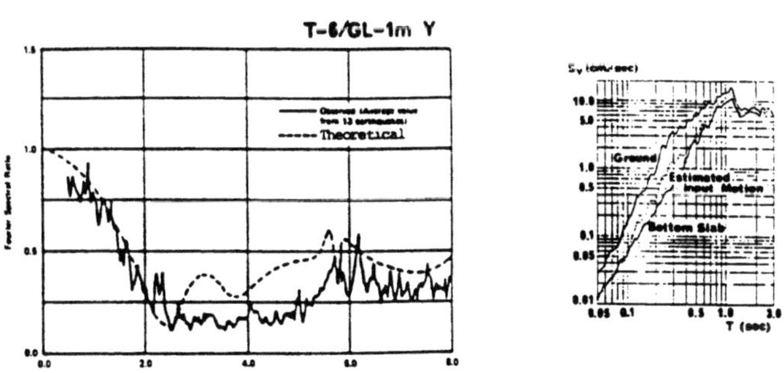

Figure 20: Observed Transfer Function Between Foundation Level Motion (-26.2m) of a Large-Scale Below-ground LNG Tank and Free-field (-1m) Horizontal Motion [27]

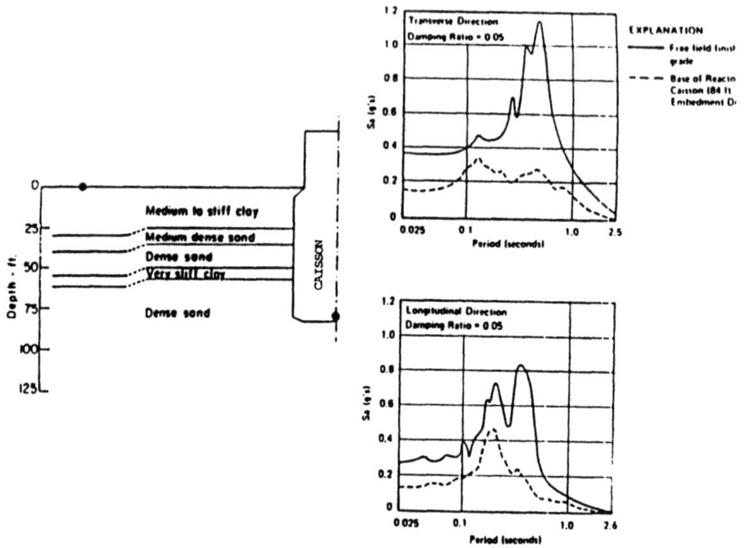

Figure 21: Comparison of Response Spectra of Accelerograms Recorded at Finished Grade in the Free-field and at the Base of the Reactor Caisson at the Humboldt Bay Plant During the June 6, 1975, Ferndale, California Earthquake [4]

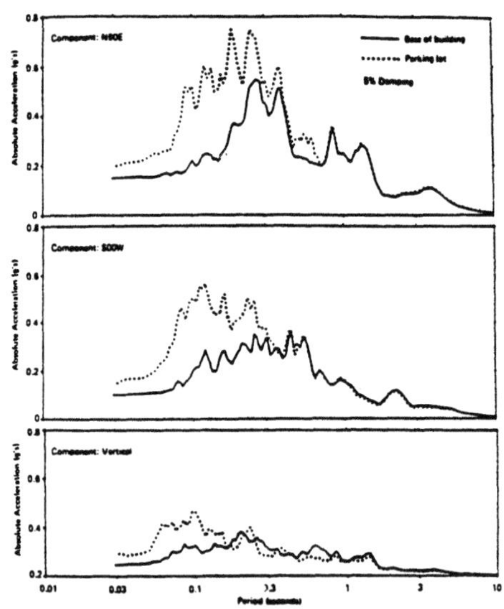

Figure 22: Comparison of Response Spectra of Motions Recorded at the Base of the Hollywood Storage Building and in the Adjacent Parking Lot During the 1971 San Fernando Earthquake [13]

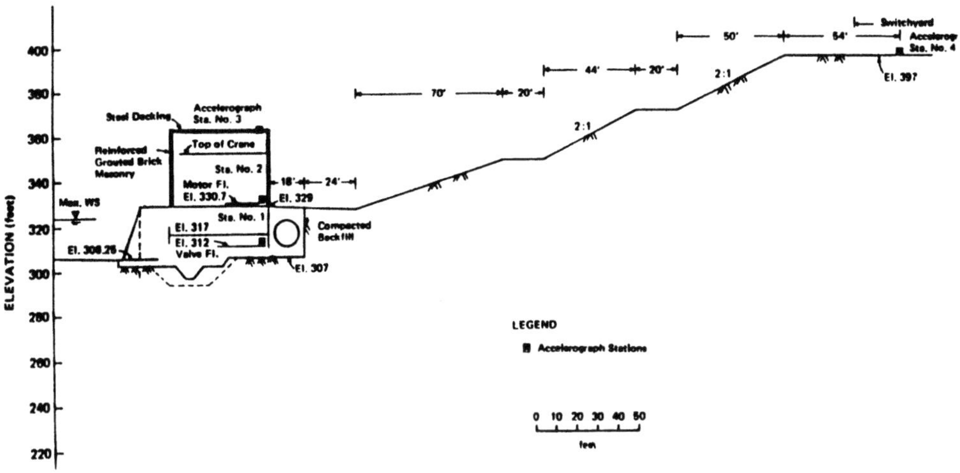

Figure 23: Cross Section Through Pleasant Valley Pumping Station [13]

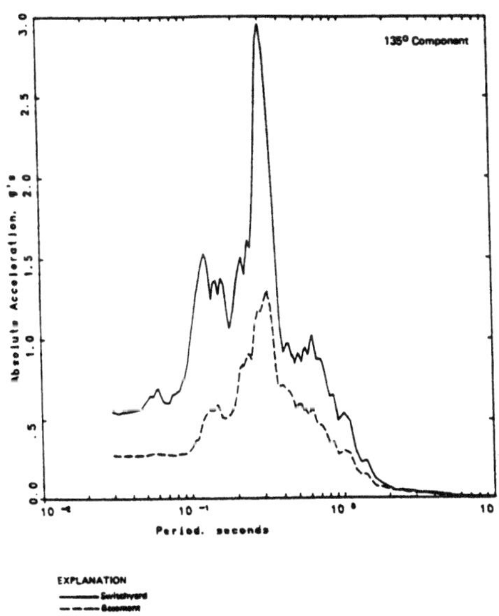

Figure 24: Comparison of Horizontal Response Spectra of Motions Recorded at the Switchyard and in the Basement of the Pleasant Valley Pumping Plant During the May 2, 1983 Coalinga Earthquake Mainshock [13]

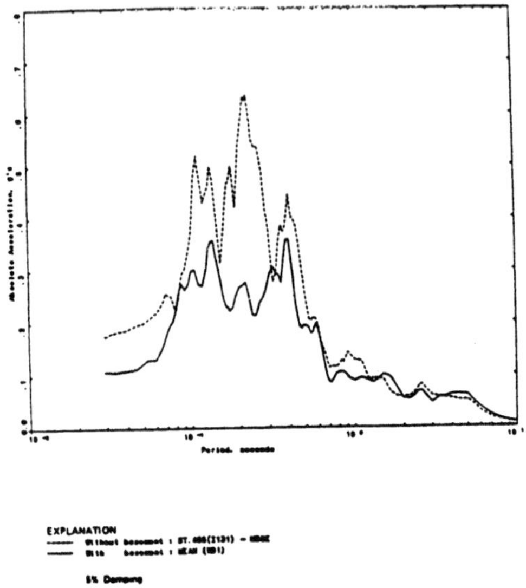

Figure 25: Comparison of Response Spectra Adjacent Buildings With and Without Basement, San Fernando Earthquake [13]

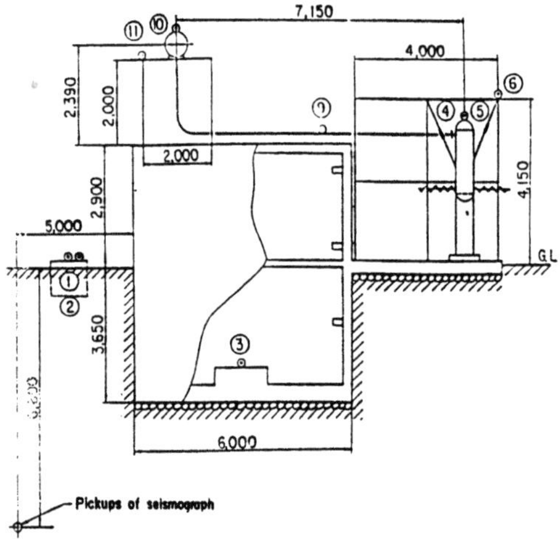

Figure 26: Cross-section Through Chemical Engineering Plant Model, Chiba Field Station [29]

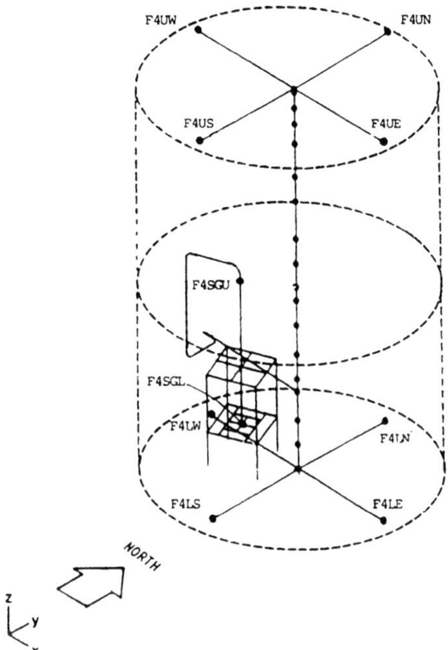

Figure 27: Isometric View of One-quarter Scale Model Showing Response Location and Station Codes

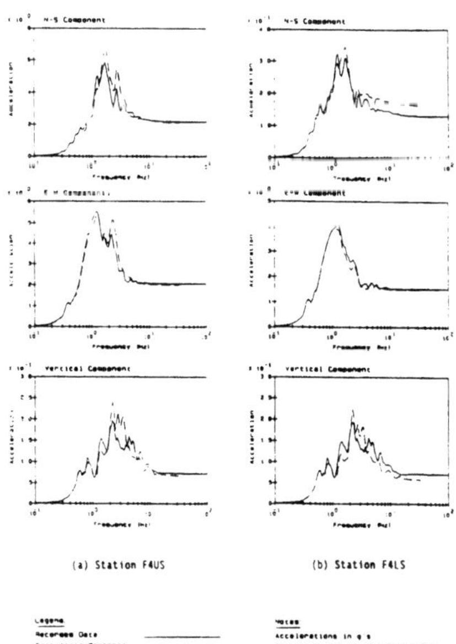

Figure 28: Recorded and Calculated In-structure Response Spectra, Base (F4LS) and Top (F4US), May 20, 1986 Earthquake [9]

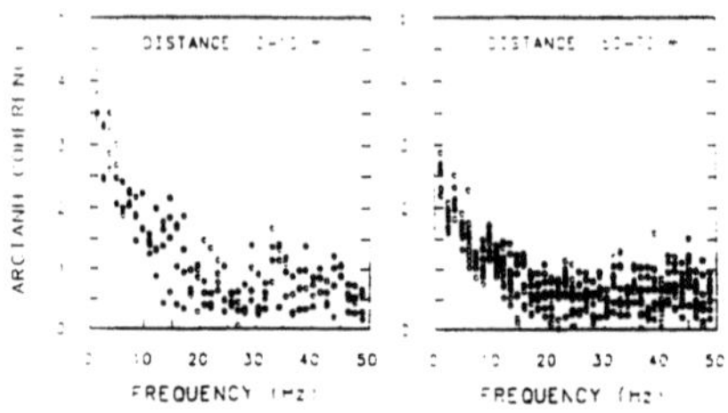

Coherencies from LSST Event 12. (A) Separation Distances of 6-10 Meters. (B) Separation Distances of 50-70 Meters.

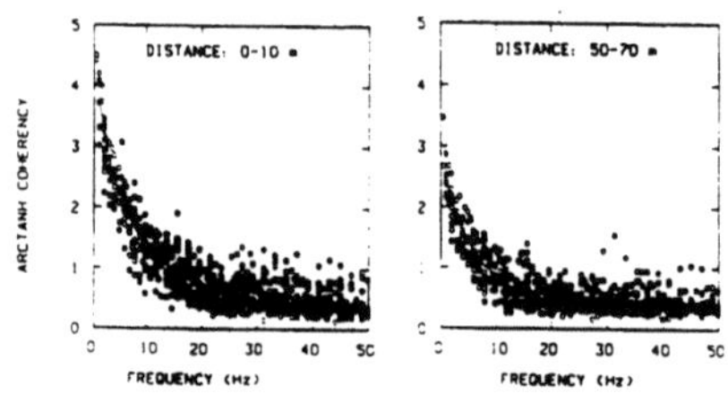

Mean Coherencies from each of the 15 LSST Events. (A) Separation Distances of 6-10 Meters. (B) Separation Distances of 50-70 Meters.

Figure 29: Variation Coherency with Separation Distance [31]

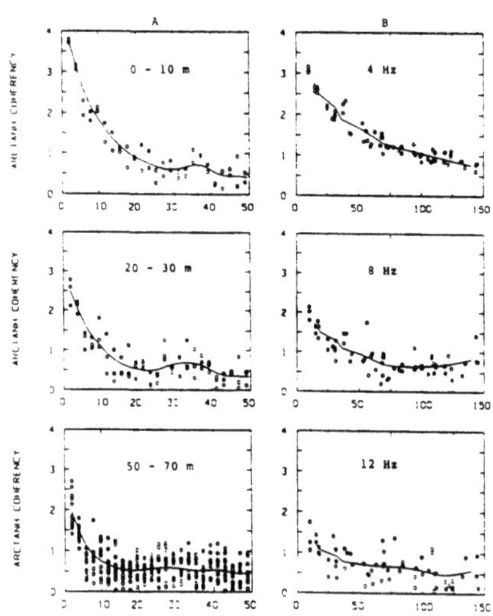

Figure 30: Coherency Estimates and Smoothed Nonparametric Curves for 4 Hz Frequency Bands (A) [34]

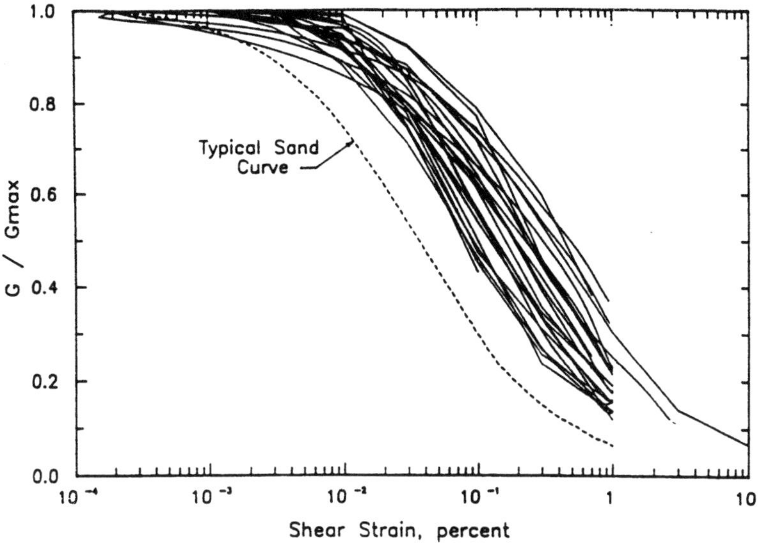

Figure 31: Normalized Modulus Reduction Relationship for Clays with Plasticity Index Between 40 to 80 [40]

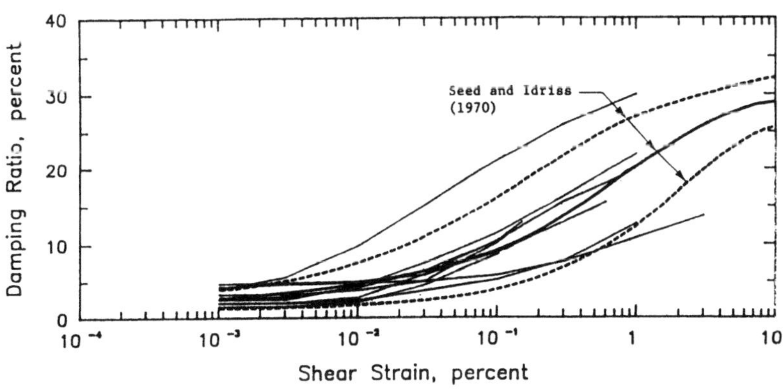

Figure 32: Strain Dependent Damping Ratios for Clays [40]

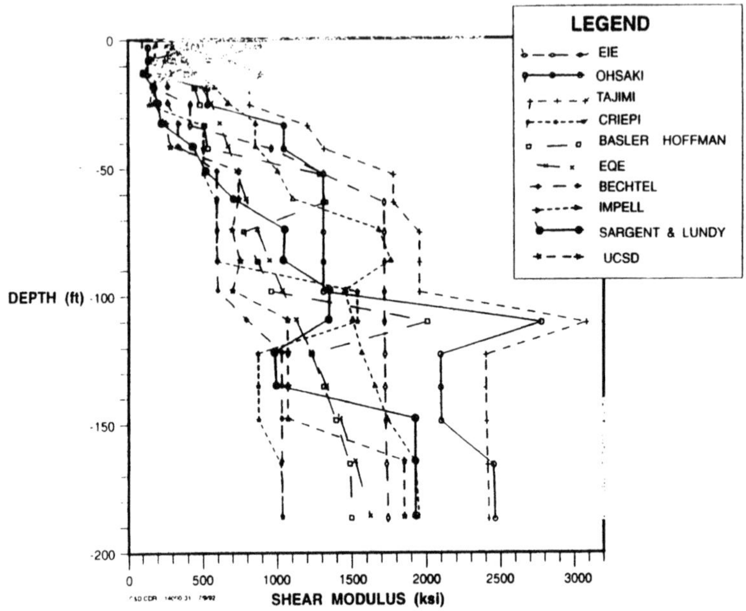

Figure 33: Variability of Equivalent Linear Shear Modulus Due to Different SSI Analyses, May 20, 1986, Lotung, Taiwan

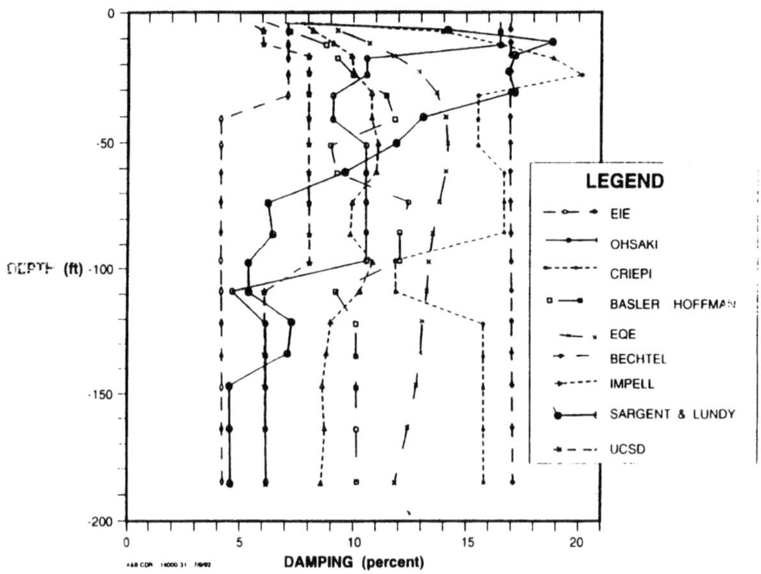

Figure 34: Variability of Equivalent Linear Damping Due to Different SSI Analyses, May 20, 1986, Lotung, Taiwan

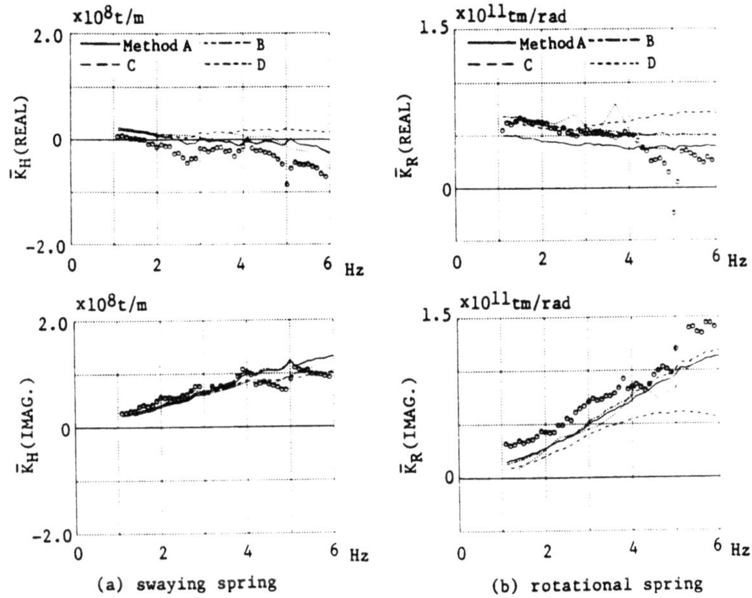

Figure 35: Analytical/Experimental Foundation Impedances [45]

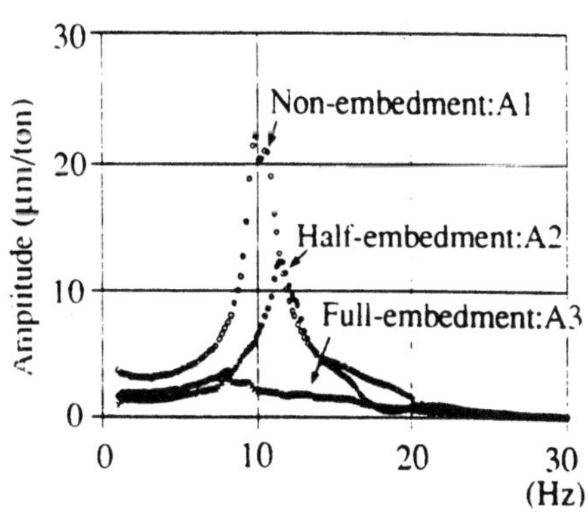

Figure 36: Comparison of Horizontal Displacement Resonance Curves at Foundation Bottom, Forced Vibration Test [46]

Acknowledgement

The authors wish to acknowledge the contributions of Dr. C-Y Chang and Mr. Maurice S. Power of Geomatrix Consultants, San Francisco, California, USA; Dr. Norman A. Abrahamson, San Gabriel, California, USA; and Dr. Yousef Bozorgnia of EQE International, San Francisco, California, USA.

DYNAMIC TESTS ON LARGE STRUCTURES FOR ANALYSIS OF SOIL-STRUCTURE INTERACTION: EXPERIMENTAL METHODS AND EXPERIMENTAL TECHNIQUES.

A.CASTOLDI
ISMES
viale G. Cesare,29
24100 Bergamo
Italy

1. INTRODUCTION

The soil-structure interaction phenomenon has been the object, in the recent years, of an increasing interest, mainly due to the need of a more refined analysis of the seismic behaviour of structures of great social and economical importance as, for example, the nuclear power plants.

Therefore, the various aspects of the problem are, today, well understood, a comprehensive theory of the phenomenon is available and numerical codes have been developed, adequate to face this analysis with the required accuracy.

This constant progress of the numerical methods, however, has not reduced the importance of the experimental techniques, that still continue to keep a fundamental role both as a mean to validate codes and as a tool for a deeper investigation of specific cases.

In fact, the experimental techniques too, have rapidly evolved and are now an extremely efficient research mean, even in a field complex like that of soil-structure interaction analysis; beside the improvement of the loading equipment and of the measuring instrumentation, the main aspect of this development has certainly been the introduction of more efficient data processing methods, made possible by a tight interaction between numerical and experimental techniques.

For a general discussion of the role that the experimental methods can have in the study of the soil-structure interaction it is convenient to recall the fundamental aspects of the problem.

The motion equation of a structure excited by an earthquake and loaded by an external system of forces can be written as:

$$\begin{bmatrix} [Z_{ss}] & [Z_{sb}] \\ [Z_{bs}] & [Z^{s}_{bb}] + [Z^{g}_{bb}] \end{bmatrix} \cdot \begin{Bmatrix} \{Q^{t}_{s}\} \\ \{Q^{t}_{b}\} \end{Bmatrix} = \begin{Bmatrix} \{P_{S}\} \\ [Z^{g}_{bb}] \cdot \{Q^{g}_{b}\} \end{Bmatrix} \qquad 1)$$

P. Gülkan and R. W. Clough (eds.), Developments in Dynamic Soil-Structure Interaction, 259–283.

The previous equations refer to a discretization of the structure in which structural nodes are denoted by "s" and those belonging simultaneously to the structure and to the foundation by "b" and are written in the frequency domain under the assumption of a linear behaviour of both the structure and the soil.

The meaning of the different terms is as follows:

- $[Z_{ss}]$, $[Z_{sb}]$, $[Z_{bs}]$ $[Z^s_{bb}]$ are the dynamic stiffness matrices of the structure, where $[Z] = [K].(1+2\varsigma i)-\omega^2[M]$,
- $[Z^g_{bb}]$ is the dynamic stiffness of the soil,
- $\{P_s\}$ is the vector of forces acting on the structure,
- $\{Q^g_b\}$ is the vector of displacement amplitudes of soil with escavation for an earthquake excitation ("scattered motion"),
- $\{Q^t_s\}$ and $\{Q^t_b\}$ are the vectors of the total displacements of the structure and of the soil.

A critical analysis of the previous equations allow us to easily find out the most important contributions of the experimental method, that are:

a) Determination of material properties

The knowledge of the physical and mechanical properties of the materials of the structure and soil is, obviously, fundamental for the determination of the dynamic-stiffness matrices; many geotechnical and geophysical methods are today available to investigate the soil and allow reliable results to be obtained; for the structure, usual laboratory tests are used.

b) Characterization of the dynamic behaviour of the system

It is done by determining the natural modes of vibration (frequencies, modal shapes and damping coefficients); the study is generally carried out by applying sinusoidal controlled forces to the structure by means of mechanical vibrators.

The results of the experimental analysis are used to calibrate the mathematical model.

c) Determination of the soil dynamic stiffness matrix $[Z^g_{bb}]$

While a large experience exists in modelling the structure, the uncertainty associated with the construction of the matrix $[Z^g_{bb}]$ is quite high, due not only to an often insufficient knowledge of the soil characteristics but also to an inadequate reproduction of the wave pattern, on which the numerical expression of $[Z^g_{bb}]$ depends.

$[Z^g_{bb}]$ can be determined in an indirect way through a dynamic test in which forces applied to the foundation (inertial forces of the structure and external forces) are computed and the components of the foundation motion are measured.

d) Determination of the "scattered" motion $\{Q^g_b\}$
The knowledge of the scattered motion is required to compute the seismic response according to the equations 1).
As well known, the earthquake is generally described through the motion recorded in a "control point" (on the ground surface or at an assumed rock outcrop); however this control motion does not contain all the necessary informations to determine in a univocal manner the free-field motion $\{Q^f_b\}$ and the scattered motion $\{Q^g_b\}$.
In fact, $\{Q^f_b\}$ depends on the wave pattern that is supposed to have generated the control motion and therefore on the direction of propagation, on the orientation of planes of incidence and, in conclusion, on the location of the site relative to the source of the earthquake.
For this reason experimental techniques alone are not able to determine $\{Q^g_b\}$ starting from the control motion; however, they can advantageously be used to study the site amplification and the propagation law.

e) Analysis of non-linear behaviour
Nonlinearity effects due to the soil behaviour can be taken in to account by means of experimental methods only in some particular cases.
The now available loading equipments can excite large structures with significant motion amplitude, often of the same level of that produced by the expected site earthquake; in this case the inertia forces generated and, therefore, the interaction with soil are correctly simulated; however it is never possible to reproduce the stress state of soil due to the passage of seismic wave.

In what follows the different experimental techniques, the criteria adopted for the data processing and the results that can be obtained are analyzed in detail and illustrated with some examples taken from the activity of ISMES.

2. THE BASIC CRITERIA FOR THE EXPERIMENTAL ANALYSIS

2.1. Determination of modal characteristics of the system

The determination of modal parameters is a particular case of the more general problem of the structural identification.

A large literature exists and the different techniques now in use are well documented; without entering in details, the experimental process can be summarized as follows:

In case of linear system, the parameters that can be used to describe the dynamic behaviour are:

- the coefficients of the matrices [M], [K], [C] of the motion equations,
- the modal parameters: natural frequencies, modal shapes and dampings,
- the transfer functions that relate, in the frequency domain, any output quantity (displacement or stress in a point) and any other input quantity (force or motion at base),
- impulse responses, defined as Fourier transforms of the relevant transfer functions.

Although from a theoretical point of view the four methods are completly equivalent, in practice, taking into account the operational problems posed by actual equipments and the requirements for accurate measurements, tests are usually performed with the aim of determinining, as a first step, the transfer functions.

As it is well known, given a system characterized by the coordinates q(t), the transfer function between the response measured in the point s-th: $q_s(t)$ and the force applied in the point r-th: $f_r(t)$ is defined as the ratio of the Fourier transforms $Q_r(\omega)$ and $F_s(\omega)$ of $q_s(t)$ and $f_r(t)$, that is:

$$h_{r,s}(\omega) = Q_r(\omega)/F_s(\omega) \qquad 2)$$

Tests are usually carried out applying sinusoidal forces by means of mechanical vibrators and measuring the modulus and the phase difference of the response; the exciting frequency is changed continously or stepwise.

In reality, the excitation techniques and the commercial control, measuring and data processing systems are much more complex and allow the use, beside sinusoidal forces, of random stationary excitation, impulsive excitation and multiple excitation: it is also possible to utilize the response of the structure to a natural, not directly measurable, excitation (for example: wind, microtremors etc.) by adopting suitable hypotheses on the characteristics of the excitation (stationarity, power spectral density shape etc.).

Starting from the knowledge of the transfer functions is then possible to identify the modal parameters by separating the contribution of each mode to the global response; this identification is based on the relationship that relates transfer functions to modal parameters:

$$h_{r,s}(\omega) = \Sigma_{k=1,N}\, \Phi_r^{(k)} . \frac{\Phi_s^{(k)}}{\Sigma_{j=1,N}\, m_j . (\Phi_j^{(k)})^2} . \frac{1}{\omega_k^2 . [(1-\omega^2/\omega_k^2) + i2\zeta_k\omega/\omega_k]}$$

Since the previous expression is not linear in the unknown parameters, special numerical algorithms are used: today numerous commercial systems, characterized by an high interactivity with the operator are available for this identification.

2.2 Determination of the soil dynamic stiffness

According to the basic definition of the matrix $[Z^g_{bb}]$ each column r-th could be experimentally determined performing a test in which the soil is simultaneously loaded in each node "b" by a force vector $\{N(\omega)\}$ so that the resulting displacements are:

$$Q_r = 1 \quad \text{and } Q_j = 0 \text{ for } j = r$$

This loading condition, however, is very complex and difficult to be accomplished: usually it is more advantageous to determine the inverse matrix called dynamic flexibility:

$$[G^g_{bb}] = [Z^g_{bb}]^{-1}$$

In this case the tests are performed by loading one node at a time and measuring the displacement in all the nodes.

If the foundation is flexible this method results so cumbersome to be not feasible; from a practical point of view it is more interesting the case, quite frequent however, of a rigid foundation that gives rise to a 6x6 soil stiffness matrix, simply indicated in the following by $[Z(\omega)]$.

In this case the soil stiffness matrix $[Z(\omega)]$ can be easily evaluated from the results of the tests carried out to determine the modal parameters according to the following procedure:

- let $\{Q_O(\omega)\}$ be the vector that represents the six independent movements of the point "O" at the bottom of the building foundation considered as a rigid body: $\{Q_O(\omega)\}$ represents also the soil movement under the hypothesis that no relative movement can exist between soil and foundation,
- let $\{N(\omega)\}$ be the vector representing the forces acting on the soil when external forces $\{F(\omega)\}$ are applied to the structure. This vector has the following expression:

$$\{N(\omega)\} = [T].(\{F(\omega)\} - [M_s].\{\ddot{Q}_s(\omega)\}) \qquad 3)$$

where $\{F(\omega)\}$ are the external forces and $[M_s].\{\ddot{Q}_s(\omega)\}$ are the inertia forces, while [T] is a topological matrix. More precisely $[M_s]$ is the mass matrix of the structure and $\{\ddot{Q}_s(\omega)\}$ represents the absolute accelerations of the lumped masses by which the structure is described,

- in this way the soil dynamic stiffness matrix is related to vectors $\{N(\omega)\}$ and $\{Q_0(\omega)\}$ by the following relation:

$$\{N(\omega)\} = [Z(\omega)].\{Q_0(\omega)\}$$

and therefore, for each excitation test carried out, the following equation can be written:

$$[Z(\omega)].\{Q_0(\omega)\} = [T]\ (\{F(\omega)\} - [M_s].\{\ddot{Q}_s(\omega)\}) \qquad 4)$$

Obviously since in general the unknown functions are 36 (21 if the symmetry of the matrix $[Z(\omega)]$ is considered), at least six independent excitation force vectors are needed.

- each functions $z_{i,j}(\omega)$ can be written in the form:

$$z_{i,j}(\omega) = k_{i,j}(\omega) + i\omega b_{i,j}(\omega)$$

where :

$$k_{i,j}(\omega) = PR[z_{i,j}(\omega)] \quad \text{= equivalent stiffness}$$

$$b_{i,j}(\omega) = PI[z_{i,j}(\omega)]/x \quad \text{= equivalent viscous damping} \qquad 5)$$

and PR[] and PI[] are the real and imaginary parts of the stiffness functions.

It is worthwhile to underline that with the previously described procedure it is possible to take advantage of the dynamic amplification in resonance conditions to considerably increase the forces acting on the foundation and therefore to study the non linear behaviour of soil.

3. SOME EXAMPLES

To better understand the criteria previously described and the actual capabilities of the experimental techniques in this field, three significant examples will be illustrated.

Each example will cover a particular case in which the soil-structure interaction plays an important role: the response of the reactor building to an earthquake, the dynamic behaviour of a pile foundation of a turboalternator and the vibrations of buildings produced by railroad traffic.

3.1 On site experimental dynamic analysis of the soil-structure interaction and of the seismic behaviour of the Italian PEC Fast Reactor building.

The PEC reactor building has a height of about 28 m and a diameter of about 23 m, and is housed in a steel cylindrical container resting on a reinforced concrete foundation raft. Heavy concrete solid blocks were poured inside the container up to about half container height (fig.1).

Thus the building structure supporting the components is rather massive. It should be noted that the building structure is far from being axisymmetric and that several other structures of huge mass have been built around the reactor building (a schematic plan of the power plant buildings is shown in fig.2).

To excite the reactor building use has been made of three different excitation methods:

- a two eccentric counter-rotating mechanical vibrator, capable of delivering sinusoidal forces (up to a maximum value of 100 kN) of slowly adjustable frequency within the seismic range.
  The equipment was rigidly connected with the structure at the concerned floors (fig.1),
- hydraulic actuators, installed between the foundations of the reactor building (ER) and those of the adjacent fuel element handling building (EMC), used to apply sinusoidal forces up to a maximum value of 2500 KN,
- blasting in bore-hole by explosions of two charges of 300 kg TNT at about 2 km from the reactor.

In situ tests have been performed from 1983 to 1985; within this time span the construction of the reactor building was completed as far as the civil works are concerned and some important mechanical components were installed. Therefore structural responses resulted to be considerably different (both in terms of natural frequencies and amplification) in the three successive periods in which the structure was tested.

Determination of vibration mode (modal analysis)

A detailed analysis of the transfer functions (fig.3) obtained during the final tests allowed the natural frequencies and modal shapes of the first five modes of the structure to be determined. Comparing the response at various monitored elevations, it was concluded that these amplifications do not correspond to rigid body motion, but elastic deformation of the building are also present.

The first five experimental modal shapes, which have been described by the motions of the nine lumped masses constituting the model by which the structure has been represented are plotted in fig.4, where are compared with those obtained from the numerical model.

<u>Building to building interaction</u>.

In order to study a possible interaction among the reactor building and the neighbouring buildings, each of these was instrumented with a three components seismometer placed at the top.

A careful analysis of the transfer functions so obtained, has confirmed that such interaction takes place, as demonstrated by the presence, beside the principal peaks corresponding to the natural modes of the building, of some less marked amplifications at the natural frequencies of the other buildings.

In particular fig.3 illustrates how the reactor building response feels the effect of the dynamic motions of the adjacent buildings, by showing the second mode of the General Services and Control building (15.3 Hz) and the third mode of the Fuel Element Handling building (16.1 Hz).

In the same manner the transfer functions relevant to the seismometer network installed at the three adjacent building (fig.5) contains peaks at the frequency of the first mode of the Reactor building.

<u>Determination of the "experimental dynamic soil stiffness matrix"</u>

By exploiting the equation 4) the functions $z_{i,j}(\omega)$ (where $i,j = x,y,z,\phi_x,\phi_y,\phi_z$) of the soil stiffness matrix have been evaluated; in fig.6 the real and the imaginary parts of the six elements of the main diagonal of the matrix are shown.

The stiffness functions are in general characterized by a sufficiently regular behaviour; the real part is quasi-constant with increasing frequency, while the imaginary part varies linearly.

This pattern can be described by the following analytical expression:

$$z_{i,j}(\omega) = k_{i,j} + i\omega b_{i,j}$$

that corresponds to a simple spring-damper model of the soil, characterized by a constant equivalent stiffness $k_{i,j}$ and a constant equivalent viscous damping $b_{i,j}$.

Therefore, according to equations 5) the values of $k_{i,j}$ and $b_{i,j}$ were obtained, through the least square method from the relations:

$$k_{i,j} = PR[z_{i,j}(\omega)] \qquad b_{i,j} = PI[z_{i,j}(\omega)]/\omega$$

These values have been used to derive the "analytical" stiffness functions plotted in fig.6 together with the experimental ones.

The analysis of the non linear behaviour of soil was made plotting (fig.7) the experimental correlation between applied moments $M_x(\omega)$ and $M_y(\omega)$ and foundation rotations $\phi_x(\omega)$ and $\phi_y(\omega)$ according to the following equations:

$$k_{\phi x,\phi x}(\omega).\phi_x(\omega) = PR[M_x(\omega)] \qquad b_{\phi x,\phi x}(\omega).\phi_x(\omega) = PI[M_x(\omega)]/\omega$$
$$k_{\phi y,\phi y}(\omega).\phi_y(\omega) = PR[M_y(\omega)] \qquad b_{\phi y,\phi y}(\omega).\phi_y(\omega) = PI[M_y(\omega)]/\omega$$

The values of the coefficients "$k_{i,j}$" and "$b_{i,j}$" obtained from the experimental data, which have been approximated by means of straight lines using the least square method, confirm that soil behaviour, at least within the limits, rather broad however, reached by the loads during the tests, is practically linear.

Seismic analysis of the reactor building

The seismic response of the building and, in particular the floor response spectra have been determined both using a numerical model and through a suitable processing of the experimental data.

For what concerns the numerical method, the usual procedure has been adopted, based on:

- computation of the natural modes of the fixed-base structure,
- use of the experimentally determined soil stiffness matrix to determine the natural modes of the global system (soil and structure),
- calibration of the model (fig.4) and computation of the floor response spectra (fig.8).

The direct evaluation of the seismic response is based on the determination of the transfer matrix $[R(\omega)]$ which relates the absolute acceleration response $\{\ddot{Q}^t{}_s\}$ of the structure to the "scattered motion" acceleration $\{\ddot{Q}^g{}_0\}$ by the equation:

$$\{\ddot{Q}^t{}_s(\omega)\} = [R(\omega)].\{\ddot{Q}^g{}_0(\omega)\}$$

$[R(\omega)]$ can be evaluated from the experimental results according to the following procedure:

- let $[H_r(\omega)]$ be the matrix which relates the relative accelerations $\{\ddot{Q}^r{}_s(\omega)\}$ of the reactor building to the column matrix $\{\ddot{Q}^t{}_0(\omega)\}$. In this way the absolute accelerations $\{\ddot{Q}^t{}_s\}$ can be calculated by the use of the equation:

$$\{\ddot{Q}^t{}_s(\omega)\} = [H_r(\omega)].\{\ddot{Q}^t{}_0(\omega)\} +[T].\{\ddot{Q}^t{}_0(\omega)\}$$

- the six independent seismic movements of the reactor building foundations are given by:

$$\{\ddot{Q}^t{}_0(\omega)\} = \{\ddot{Q}^g{}_0(\omega)\} - \omega^2[Z(\omega)]^{-1}.(-[T]^T[M_s].\{\ddot{Q}^t{}_s(\omega)\})$$

- on the basis of the previous two equations, the transfer matrix $[R(\omega)]$ can be expressed by:

$$[R(\omega)] = ([I]+([H_r(\omega)]+[T]).(-\omega^2).[Z(\omega)]^{-1}[T]^T[M_s])^{-1}.([H_r(\omega)]+[T])$$

To make use of this equation the knowledge of the transfer matrix $[H_r(\omega)]$ is required. It can be calculated by the equation:

$$[H_r(\omega)] = \Sigma_{k=1,N}\ \{\phi_k\} \cdot \frac{\{\phi_k\}^T.(-[M_s][T])}{\{\phi_k\}^T[M_s]\{\phi_k\}} \cdot \frac{-\omega^2}{[(1-(\omega/\omega_k)^2) + 2i\zeta_k(\omega/\omega_k)]}$$

where ω_k, f_k, $\{\phi_k\}$ are the modal parameters of fixed base building; an experimental-numerical method allowed these modal parameters to be determined on the basis of the experimental data.

The floor response spectra obtained via the experimental results are plotted in fig.8, where are compared with those determined through the numerical model.

3.2 Dynamic characterization of foundation piles

Two of the major problems in the design of the foundations of rotating machines or of equipments (pumps, fans etc.) capable of generating dynamic forces, are the reduction of the amplitude of vibrations that could endanger the operational life of the machine and the limitation of the level of disturbances transmitted to the neighbouring buildings.

To succefully face this problem it is essential to know the dynamic stiffness of soil and piles.

In this respect, particularly significant are the field full-scale tests, where real piles are installed in their natural soil deposit and elastic waves propagation is not obstructed.

As an example the tests carried out to study the turboalternator foundation of a thermal power plant are described.

The tests aimed at determining the dynamic behaviour (in particular the dynamic stiffness matrix) of each single pile of a group of four piles and at validating a new data processing method capable of predicting the dynamic stiffness of the group.

The tests were performed on four piles (fig.9) with a diameter of 800 mm and a length of 45 m by applying sinusoidal forces by means of a mechanical vibrator rigidly connected to the top of the piles; all the piles were excited , not at the same time, by dynamic horizontal loads and axial and eccentric vertical forces, of slowly adjustable frequency within a wide range (1-25 Hz) up to a maximum value of 100 kN.

The response measurement was carried out by seismometers and accelerometers installed on the head of the piles. The seismometers more sensitive than the accelerometers were positioned on the head of the pile, at the moment, not directly excited.

The processing of the data has been performed according to the following procedure (fig.9):

- supposing that the load $\{F\}$ is applied to the pile 1, the direct flexibility matrix of the loaded pile $[G^{1,1}(\omega)]$ and the indirect flexibility matrices $[G^{k,1}(\omega)]$ of the other piles of the group were first evaluated according to the relations:

  $$\{F^1(\omega)\} = [G^{1,1}(\omega)].\{Q^1(\omega)\}$$

  $$\{F^1(\omega)\} = [G^{k,1}(\omega)].\{Q^k(\omega)\} \quad \text{for} \quad k=2,3,4$$

  Since five degrees of freedom have been considered (the two horizontal and the vertical displacements Q_x, Q_y, Q_z and the two rotations $Q_{\phi x}$, $Q_{\phi y}$), $[G(\omega)]$ is a 5x5 matrix and at least three independent tests have to be performed.
  The translational terms of the matrices, obtained averaging all available data under the hypotesis of a complete symmetry with respect to x and y axes, are plotted in fig. 10.

- the dynamic stiffness of the loaded pile $[Z^{1,1}(\omega)]$ was then obtained by the simple inversion of the flexibility matrix:

  $$[Z^{1,1}(\omega)] = [G^{1,1}(\omega)]^{-1}$$

  Assuming for the soil-pile system a visco-elastic linear model, each stiffness functions can be described by the following analytical expression:

  $$z_{i,j}(\omega) = (k_{i,j} - \omega^2 m_{i,j}) + i\omega b_{i,j}$$

  where

  $$PR[z_{i,j}(\omega)] = k_{i,j} - \omega^2.m_{i,j} \qquad PI[z_{i,j}(\omega)] = \omega b_{i,j}$$

  The coefficients: $m_{i,j}$, $k_{i,j}$, $b_{i,j}$ were then obtained by fitting the experimental data with the least square method; examples of the experimental and "fitted" stiffness functions are plotted in fig. 11.

- the equivalent dynamic stiffness matrix of the group of four piles connected at top by a rigid cap , was determined by:

- assembling first, the flexibility matrices of all the piles of the group in the form:

$$[G^P] = \begin{bmatrix} [G^{11}] & [G^{12}] & \dots \\ [G^{21}] & [G^{22}] & \dots \\ \dots & \dots & \dots \end{bmatrix} \qquad 6)$$

- imposing congruence conditions among the motion components of the head of each pile $\{Q^k\}$ and the motion components $\{Q_0\}$of the centre 0 of the rigid cap base; this leads to the relation:

$$\begin{Bmatrix} \{Q^1\} \\ \vdots \\ \{Q^4\} \end{Bmatrix} = [R].\{Q_0\} \qquad 7)$$

- imposing the equilibrium of forces $\{F^k\}$ acting on the piles heads and the external forces applied to the cap and referred to the point 0, from which:

$$\{P_{est}\} = [R'].\begin{Bmatrix} \{F^1\} \\ \vdots \\ \{F^4\} \end{Bmatrix} \qquad 8)$$

- finally, substituting the previous equations 6),7),8) into the dynamic stiffness expression:

$$\{P_{est}\} = [Z(\omega)].\{Q_0\}$$

Fig.12 shows the horizontal, vertical and rotational stiffness functions of the pile group, obtained trough the described processing scheme.

To validate the previous procedure the pile heads were then connetted with a rigid cap and a test was carried out to directly measure the global dynamic stiffness of the group; the results so obtained, are compared with the computed ones in fig.12 and show a good agreement.

Recently a new testing equipment which represents an important improvement in this field has been developed; the equipment (see the general scheme in fig.13) allows to perform the usual field dynamic tests loading, at the same time, the pile with a static force and is thus capable of reproducing the real load condition of a pile.

This new equipment has been used to test the previous pile and to study the influence of the static load on the dynamic stiffness; as an example the pile vertical stiffness is plotted in fig.14 for a static load of 2100 KN and for two different intensities of dynamic force.

Fig.15 show a synthesis of all the results obtained from which it can be concluded that, at least in this case, the static load has a little influence.

3.3 PREDICTION OF VIBRATIONS IN BUILDINGS DUE TO RAILROAD TRAFFIC

The design and construction of a urban railway network is carried out in such a way to minimize the impact on local business, on traffic, on the city "life", both during the construction works and the operation of the lines. In connections with these problems, mainly related to the operation of the system, special attention is paid to possible vibratory disturbances caused by the crossing of the trains.

A numerical-experimental technique has been set up to allow the prediction to be made of the vibration level which will be produced on the neighbouring buildings by the passing of trains.

This procedure can be briefly described as follows:

a) Experimental determination of vibrations (or forces) generated by the carriages on the railroad superstructure or on its foundation slab, carried out through measurements on the superstructure of operating lines having characteristics similar to those of the new one.

b) Experimental evaluation, by means of forced vibration tests, of the transfer functions between the vibrations induced on a short stretch of the new superstructure (or the forces directly applied to its foundations) and the vibrations measured on the nearby buildings.

c) Prediction of the characteristics (r.m.s. values, frequency spectrum etc.) of the vibrations which will be induced by trains, starting from the results of the experimental investigations as per a and b.

<u>Basic assumptions and scheme for data processing</u>

With reference, as a first approximation approach, to a train travelling at constant speed v_t and assuming that the tunnel and the superstructure are uniform throughout, then the dynamic load may be represented as a system of concentrated forces acting in correspondence to the track support involved in that moment by the passage of the train (fig.17). In general terms, the analytical expression of each force acting in corrispondence of the j-th track support is

$$r(x_j,t) = r(x_j - v_t.t) = r(t - x_j/v_t)$$

while its Fourier transform is given by:

$$R(x_j,\omega) = R(x).e^{-i\omega(xk-xj)/vt}$$

Each of the forces $r(x_k,t)$ generated by the train sets up vibrations that are spread out in the ground until they reach , with a delay depending on the wave propagation velocity, the neighbouring buildings, setting them in oscillation.

If $H_{i,k}(\omega)$ is the transfer function between the building vibration $Q_i(\omega)$ in position "i" and the force in position "k" along the railway, then the expression, in the frequency domain, of the oscillation in a point i of the building is given by the sum:

$$Q_i(\omega) = \Sigma_{k=1,N} \quad H_{i,k}(\omega).R(x_k,\omega) \qquad 9)$$

The summation is obviously extended to N positions within the tunnel, whose distance from the building is such as to generate on it significant disturbances.

Actually, however, the dynamic component of the loads transmitted by the trains can be described as a transient whose central part is a stationary random process (fig.16); then, from an engineering point of view, is more useful to refer to the statistical properties of the phenomenon (r.m.s values, spectral densities, etc.) than to single time histories.

Besides this, there is a time interval (correlated to the "significant" stretch of the tunnel, and to the speed and the length of the train) during which the vibrations reaching the building from different points of the tunnel, can be practically considered as a steady-state phenomenon (see fig.17).

In such occurrence, relationship 9) can be expressed more conveniently as follows:

$$S_{Qi}(\omega) = \Sigma_{k=1,N} \left|H_{i,k}(\omega)\right|^2.S_{Rk}(\omega)$$

where:

- $S_{Qi}(\omega)$ and $S_{Rk}(\omega)$ are the power spectral density of the response $q_i(t)$ and of the excitation $r(x_k,t)$ respectively
- $|H_{i,k}(\omega)|$ is the modulus of the transfer function $H_{i,k}(\omega)$

In this sum, which obviously refers only to the points near the buildings, the power spectral density of the forces does not depend on the point "k" on the line.

<u>Vibration prediction</u>

The input functions $S_{R,k}(\omega)$ can be indifferently defined in terms of force or vibration.

The determination of the forces generated by trains on the superstructure foundation is obtained in an indirect way through laboratory tests carried out on full size mock-ups of the superstructure by a suitable dynamic compression testing machine that is controlled so as to reproduce actual recordings of the settlements produced by the train on an operating superstructure similar to that under testing; the required force is measured during the test.

This technique, applied to different types of superstructures, has led to the diagrams of fig.18.

Vibrations induced on the superstructure by the trains are measured on railway stretches in which the superstructure and the type of trains are similar to those intended for the new line, for which the prediction is requested (fig.19).

The determination of the transfer functions was made as usual, by applying sinusoidal forces in several excitation stations and by measuring the response of the building by means of velocity transducers.

The loading device is a 15 KN hydraulic actuator installed on a carriage, on which steel masses are in turn mounted through a low frequency spring suspension. The total weight of the carriage is equal to that of an axle of the train cars to account for nonlinear behaviour of the superstructure.

This technique has been applied to the study of the line 2 of Milan underground in two sections where the tunnel passes very close or under existing buildings (fig.20).

Of course, according to the type of input chosen (force or vibration), two different series of transfer functions are obtained: building vibration over force on rail (fig.21) or building vibration over vibration of the track (fig.22).

The assumption of linearity of the system which is at the base of the convolution method, was checked performing tests at different intensity of the applied force (fig 21).

For an overall examination of the various transfer functions and in order to define the stretch of railway line in which the action of the train is significant, the r.m.s. value $V_{i,k}$ of the response in the point "i" for a force having a "unit" power spectral density $S_{Rk}(\omega)$ in the considered frequency range is computed (fig.23).

The spectral density $S_{Rk}(\omega)$ of the forces applied to the slab being known, as also the transfer functions $H_{i,k}(\omega)$, the spectral densities of the vibrations of the buildings during the passage of the trains are predicted through the procedure described.

The processing is carried out in terms of velocity. Starting from the relevant spectral densities the r.m.s values of the analysis at 1/3 octave were then computed.

By way of an example of the results obtained, fig.24 shows the r.m.s. values (1/3 octave) predicted in a measuring position in the foundation of a building.

When the railroad became operative, the actual vibrations were measured in the same position of the building; measurements were taken during the night, that is without surface traffic.

The results of these measurements are plotted in the same fig.24 and compared with previous ones; as may be seen, they are in very satisfactory agreement with those measured during the passage of trains.

4. CONCLUSIONS

The engineering problems requiring an analysis of soil-structure interaction are quite numerous and different in nature; as demonstrated by the examples so far described, they often need a recourse to the experimental techniques for a their reliable and accurate solution, despite of the impressive progress of numerical methods.

For the future the contribution of the experimental techniques will certainly be even more important, keeping in mind, on the one hand, the more stringent requirements for safety of the new structures and, on the other, the continous developments of these techniques.

For what concerns the latter aspect the availability of more and more sophisticated loadings and measuring equipments like hydraulic servocontrolled exciters and optoelectronic sensors, will made easier and less expensive the carrying out of tests, while the development of new processing techniques and a more tight interaction between tests and numerical modelling will give the single experimental data a wider meaning and made possible a deeper insight into the problem being studied.

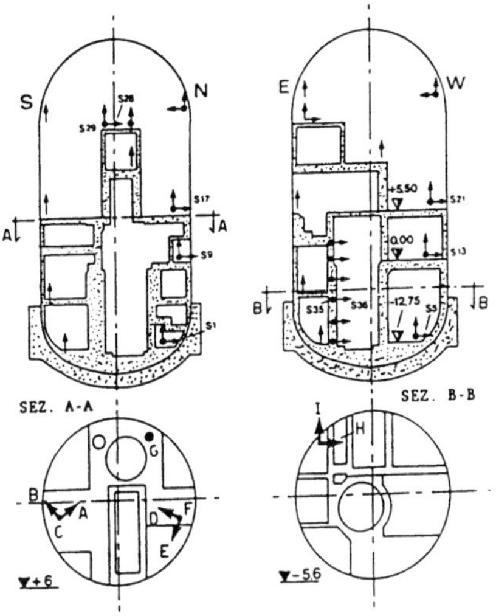

Fig. 1: Sketch of the PEC Reactor Building. Positions and directions of mechanical shaker and of the measuring instrumentation.

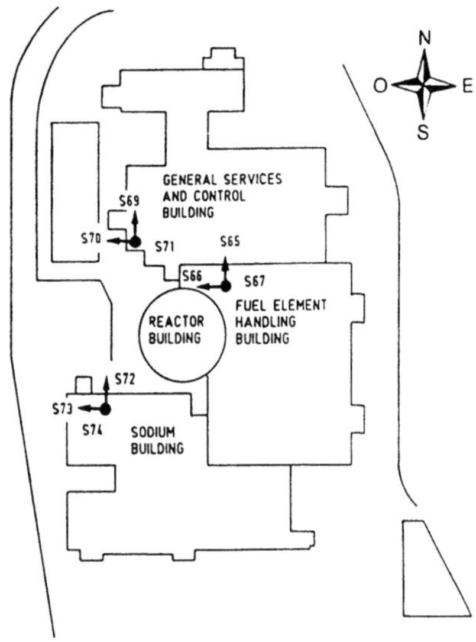

Fig. 2: Schematic plan of the power plant

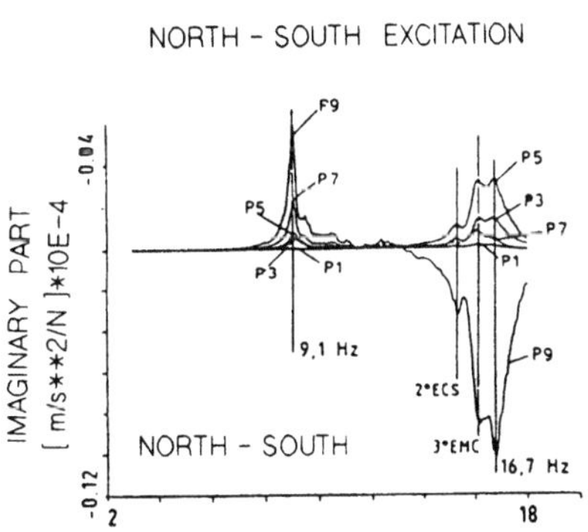

Fig. 3: Examples of experimental transfer functions.

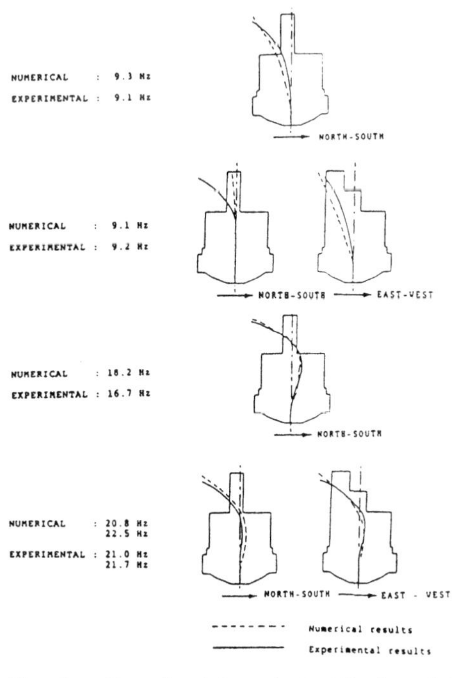

Fig. 4: Experimental and numerical modal shapes of the first five modes.

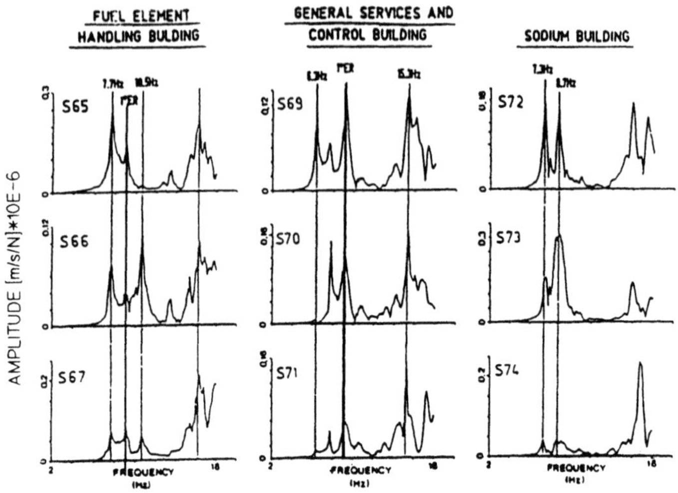

Fig. 5: Transfer functions relevant to the 3-components seismometers located at the top of the three adjacent buildings.

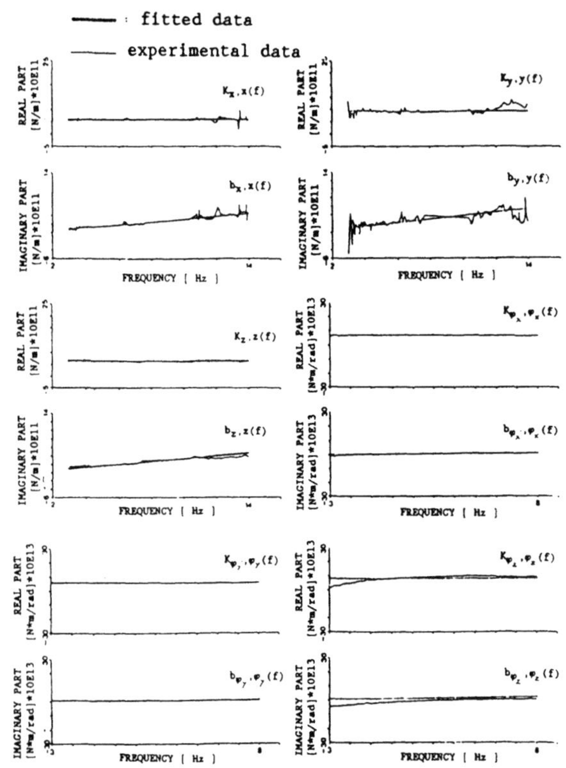

Fig. 6: Real and imaginary parts of the six elements of the main diagonal of the stiffness matrix.

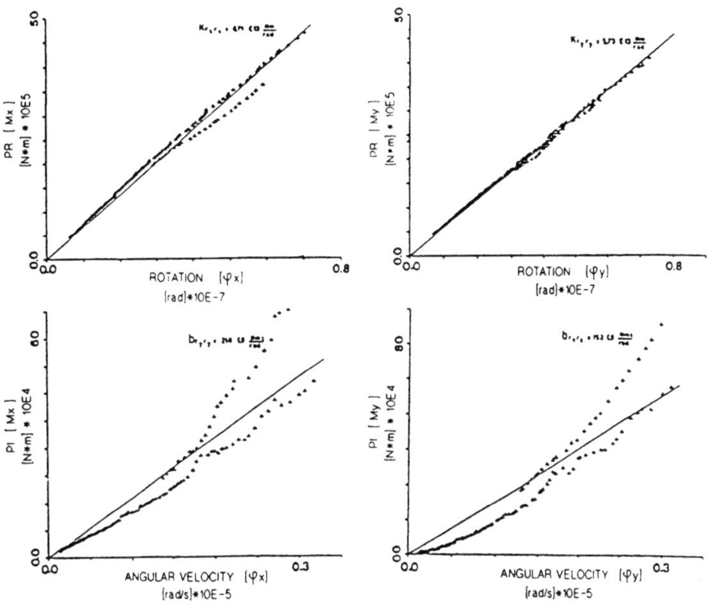

Fig. 7: Correlation between applied loads and foundation motions.

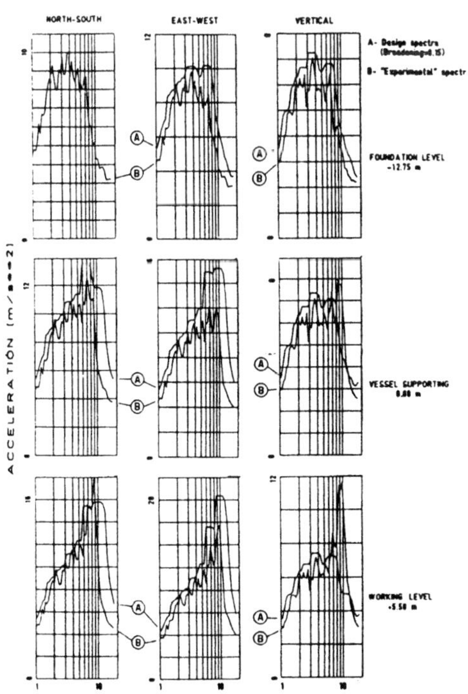

Fig. 8: "Experimental" and numerical floor response spectra.

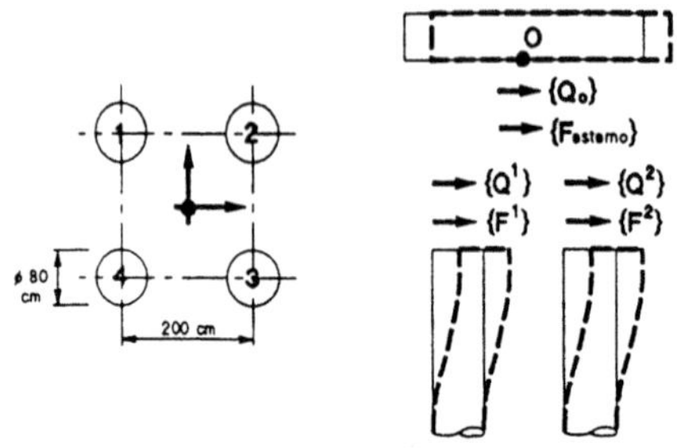

Fig. 9: Sketch of the experimental data. Data processing reference scheme.

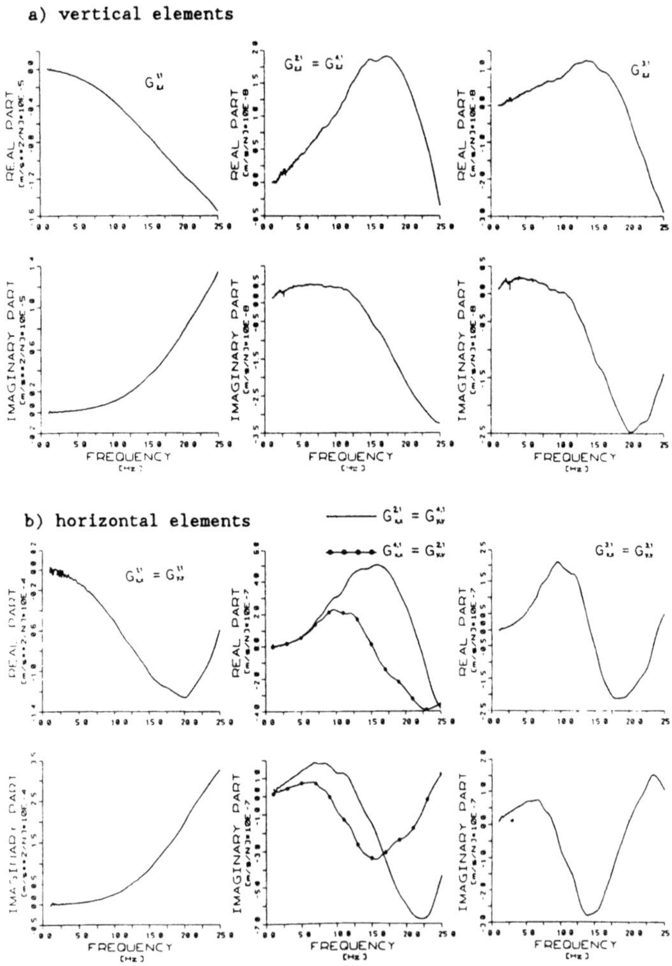

Fig. 10: Translational elements of the flexibility matrices of the four piles.

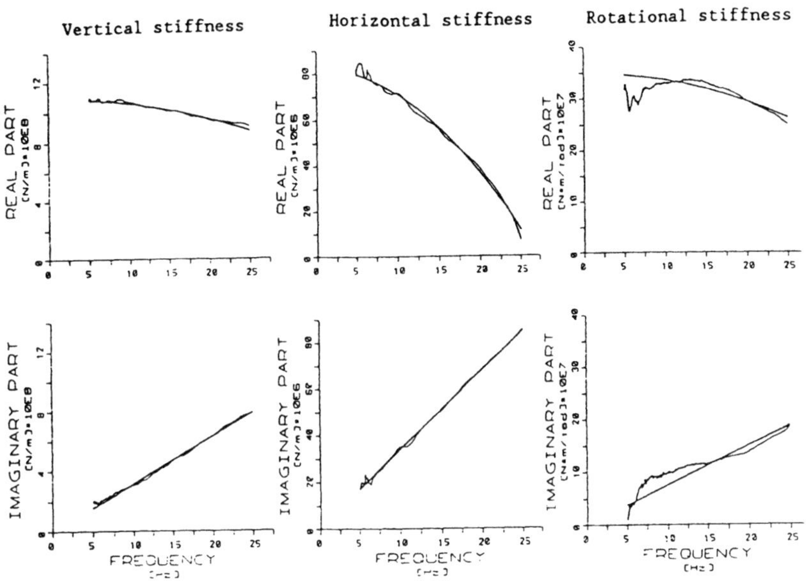

Fig. 11: Dynamic stiffness functions of the loaded pile.

experimental stiffness functions

analitical fitting

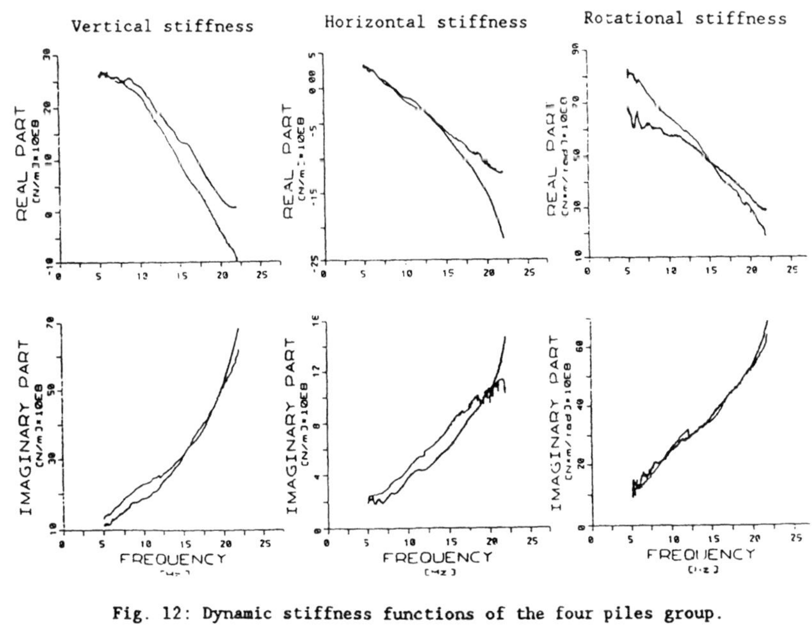

Fig. 12: Dynamic stiffness functions of the four piles group.

experimental stiffness functions

stiffness functions derived from data obtained on single piles

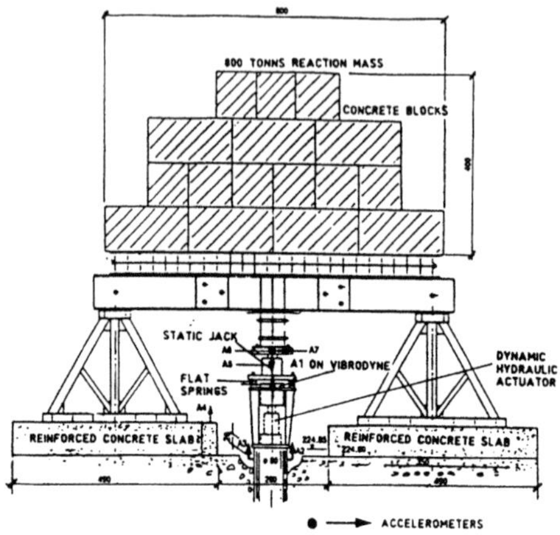

Fig. 13: Test equipment for a simultaneous static and dynamic loading of a pile.

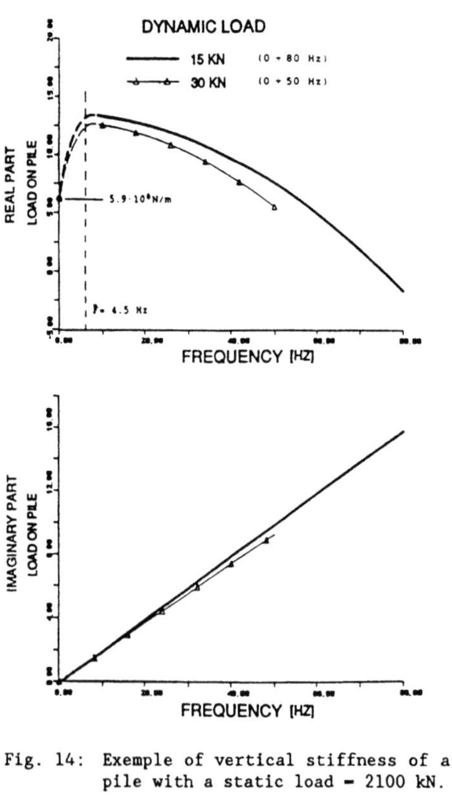

Fig. 14: Exemple of vertical stiffness of a pile with a static load = 2100 kN.

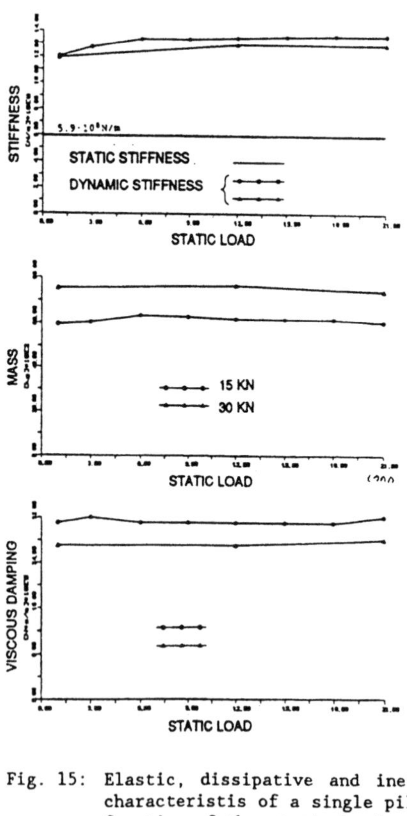

Fig. 15: Elastic, dissipative and inertial characteristis of a single pile as function of the static load.

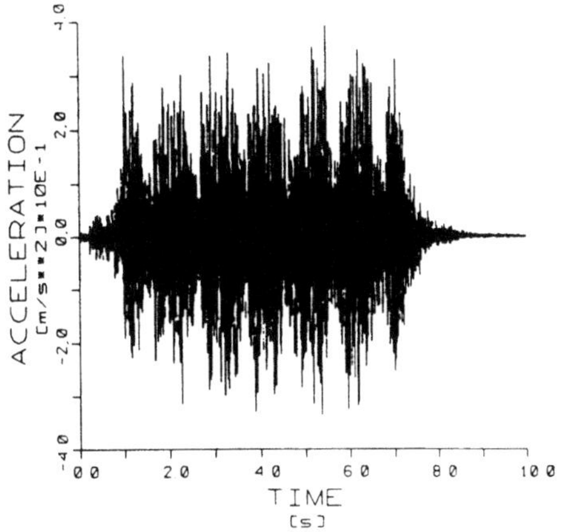

Fig. 16: High frequency component of the vibration generated by train on the superstructure.

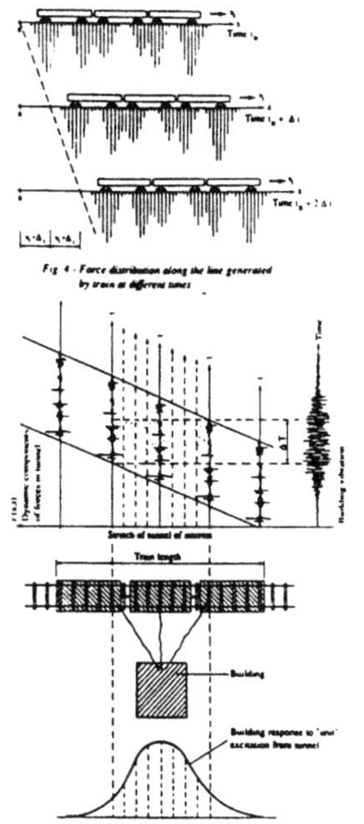

Fig. 17: Schematic representation of forces and vibrations generated by train.

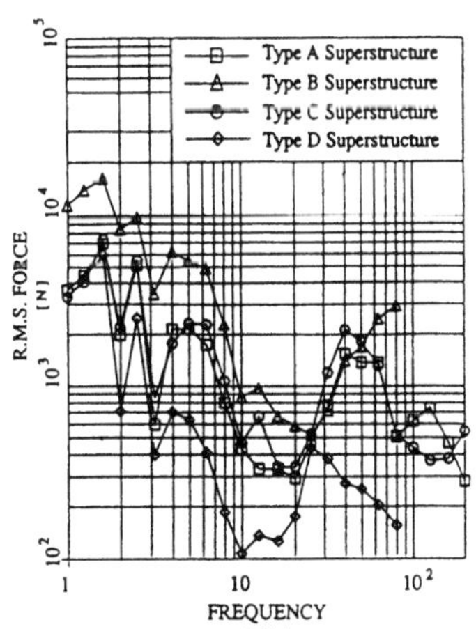

Fig. 18: 1/3 octave analysis of forces generated by trains (speed 50 Km/h approx) on the foundation of superstructures.

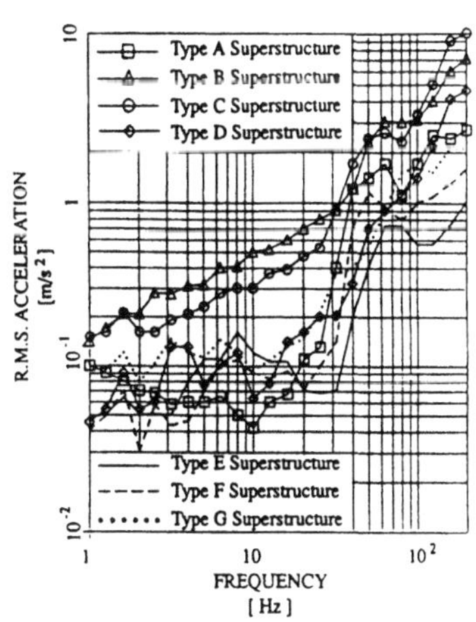

Fig. 19: 1/3 octave analysis of vibrations recorded at the foundation of a building in the vicinity of the Milan underground.

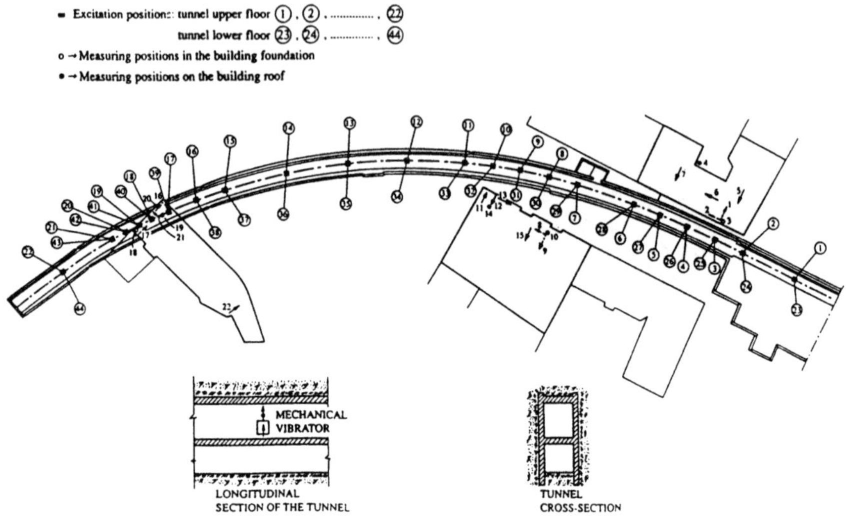

Fig. 20: Exciting and measuring positions for the determination of the transfer functions between superstructure foundation and three nearby building (Line 2 - Milan underground).

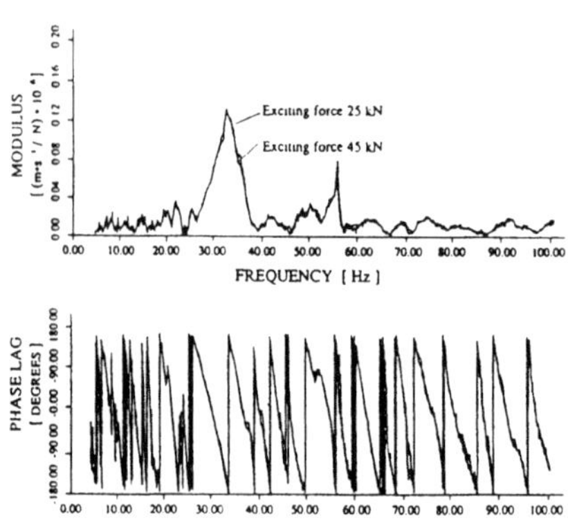

Fig. 21: Example of transfer function between a force applied to the superstructure foundation and a nearby building.

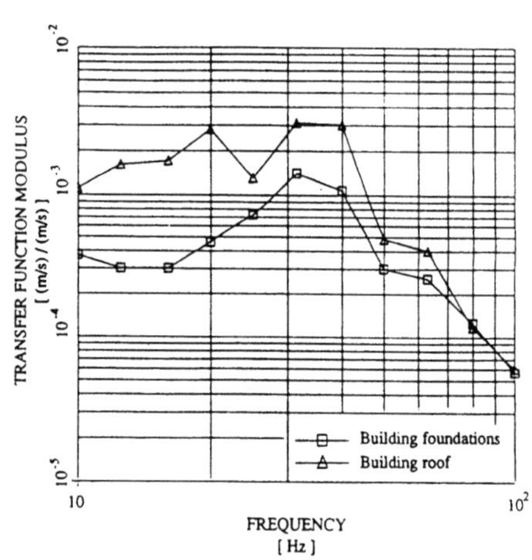

Fig. 22: Example of a 1/3 octave analysis of transfer functions between the building motions and the track motions.

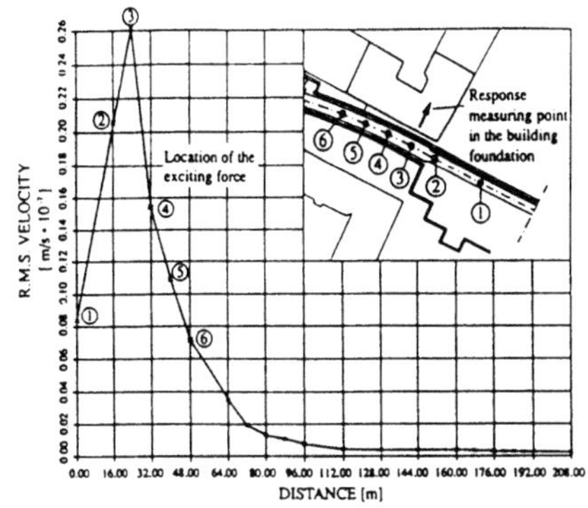

Fig. 23: Response of a building to a "unit" exciting force acting on the superstructure foundation. (Line 2 - Milan underground network).

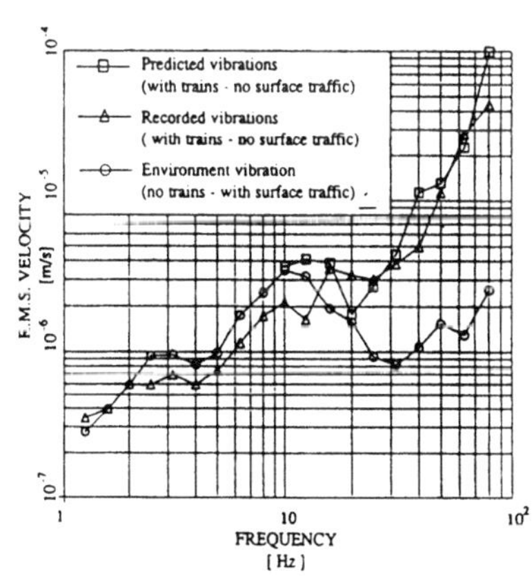

Fig. 24: 1/3 octave analysis of vibrations recorded at the foundation of a building in the vicinity of the Milan underground.

SIMPLIFIED SOIL DYNAMICS USING CONE MODELS

JETHRO W. MEEK
Fr. Holst Construction Co.
Ellerholzweg 14
2102 Hamburg 93
Germany

ABSTRACT. In lieu of rigorous elastodynamical analysis, a wide variety of soil-structure interaction problems may be formulated and solved approximately using simple cone models. The different types of problems amenable to the cone approach are identified step by step, each time evaluating the accuracy by comparison to exact solutions. A practical example illustrates the steps of calculation for cone models in both the frequency and time domains.

1. Introduction

In structural mechanics engineers are accustomed to using the term "theory of elasticity" to denote rigorous solutions and the term "strength of materials" to characterize approximate solutions based on assumed variations of displacements. In the related context of soil dynamics the boundary-integral and boundary-element methods belong to the category "theory of elasticity". Unfortunately, the mathematical complexity of such rigorous solutions tends to obscure physical insight and intimidate practitioners who would prefer a simpler approach which works in the time domain of everyday experience. (Rigorous methods almost always require an excursion into the more abstract frequency domain.) In addition, the question arises whether "exact" solutions for an elastic medium are "accurate" for real soil, which is certainly not a homogeneous, perfectly elastic material.

These considerations have motivated the author and his colleague J. P. Wolf to develop in a three-year joint effort a unified "strength of materials" approach to soil dynamics based on the cone models described herein. The emphasis is placed on comparing results with exact solutions; for mathematical details the reader is referred to other publications.

The cone models should not supplant but rather supplement rigorous solutions. They may be used to check and verify an "exact" analysis. For example, a parameter study involving upper and lower bounds of soil properties may be performed with a small fraction of the computational effort required for a single rigorous analysis, leading to better engineering insight. As a guideline, results obtained via cone models usually deviate from rigorous solutions by less than 10%.

2. Cone Models for Surface Foundations on a Homogeneous Halfspace

Figure 1 shows the basic cone models for translational and rotational motion. The apex of the cone is located at height z_0 above the rigid, massless basemat with area A_0 and area moment I_0. For conceptual

P. Gülkan and R. W. Clough (eds.), Developments in Dynamic Soil-Structure Interaction, 285–306.

purposes it is convenient to view the basemat as a circular disk with radius r_0; if the basemat is non-circular in plan, the equivalent radius r_0 is back-calculated from A_0 (translational motion) or I_0 (rotational motion). The aspect ratio of the cone, specified by the ratio z_0/r_0, is chosen to match the static stiffness of the cone to the corresponding rigorous solution. Although Fig. 1(a) is drawn for the case of vertical displacement and Fig. 1(b) illustrates the rocking case, analogous equations of motion are obtained for horizontal displacement and torsion.

The translational cone was first proposed half a century ago by Ehlers [1], the rotational cone almost twenty years ago by Meek and Veletsos [3]. Shortly thereafter Veletsos and Nair [9] showed that the original rocking cone is also applicable to the torsional case. Gazetas and Dobry [2] employed cone models to obtain insight into the phenomenon of radiation damping. Although the cones are by no means new, their direct use in engineering practice has been overshadowed by a preoccupation with exact solutions. Now that an extensive compendium of rigorous solutions has accumulated, it seems appropriate to step back and evaluate the accuracy obtainable with the cone approach, which a decade ago may well have appeared to be an oversimplification of reality.

All cone models may be described mathematically by the same canonical equation of motion

$$\frac{d^2 w}{dz^2} + \frac{n}{z}\frac{dw}{dz} = \frac{1}{c^2}\frac{d^2 w}{dt^2} \tag{1}$$

The particular member of the cone family is specified by the integer n in Eq.(1), with n = 1 for a 2-dimensional strip in translation, n = 2 for a 3-dimensional disk in translation, n = 3 for a 2-dimensional strip in rocking, and n = 4 for a 3-dimensional disk in rocking or torsion. Observe that the trivial case n = 0 corresponds to the wave equation of a prismatic bar in translation. In the following, attention is restricted to the cases n = 2 and 4, which are most important in engineering practice. (As pointed out in [6], great caution should be exercised in using infinitely long strips, n = 1 and 3, even to represent very slender foundations. Strips tend to overestimate the radiation damping, often by a factor of two or more; this potentially dangerous effect is addressed later in the numerical example.) In Eq. (1), w represents the displacement (or rotation) measured at the vertical distance z away from the apex of the cone. The quantity c is the appropriate velocity of wave propagation, with c = the shear-wave velocity c_s for horizontal translation and torsion and c = the dilatational-wave velocity c_p for vertical translation and rocking. Recent study of the nearly incompressible case (Poisson's ratio between 1/3 and 1/2) has revealed that the effective upper bound of c_p should be limited to 2 c_s.

It is important to note that the mass of the basemat is not incorporated into the cone model; it must be included in the dynamic model of the superstructure. In the nearly incompressible case just mentioned, a portion of soil "trapped" under the footing moves in phase with it. It is advantageous to assign this added mass to the basemat; this "trick" enables cone models to be realized in the same way for all values of Poisson's ratio.

To effect the realization, Eq.(1) is solved for a Dirac-delta impulse of force or moment. For n = 2 or 4 the analysis entails making certain substitutions in Eq.(1) which remove the second term on the left side

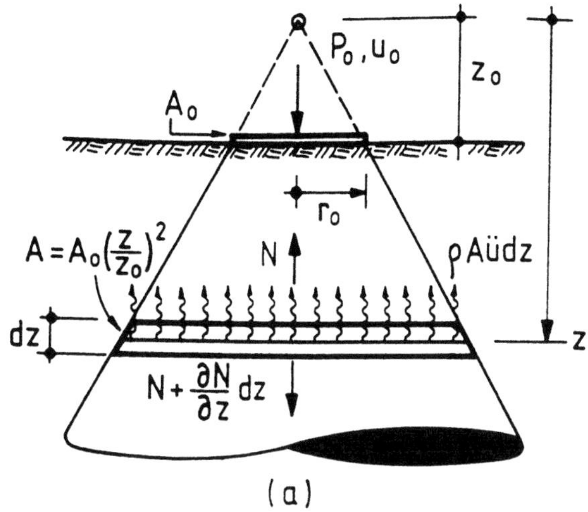

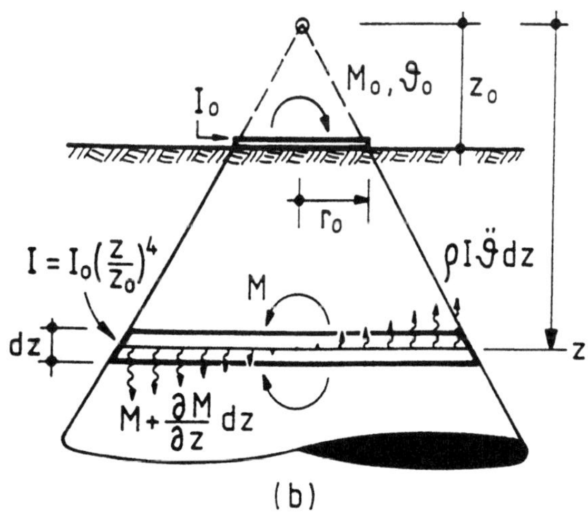

Figure 1. Cone models (a) for translation and (b) for rotation.

and thus yield the wave equation. The results in Fig. 2 are seen to be in excellent agreement with the associated rigorous solutions. Then it is easy to verify that the impulse response functions of the cones are exactly identical to those of the discrete models presented in Fig. 3 [4]. The values of the discrete parameters, which depend on Poisson's ratio, may be computed from Table 1. When using the table, the first step is to compute the cone's apex height z_0 according to column 4. In the cases requiring $c = c_p$, recall that $c_p = 1.414\ c_s$ for $\nu = 0$, $c_p = 1.732\ c_s$ for $\nu = 1/4$ and $c_p = 2.000\ c_s$ for $\nu = 1/3$. The springs K and K_ϑ are obviously just the static stiffnesses. The rotational model includes an inertial "monkey tail" which turns out to be $I_\vartheta = \rho z_0 I_0$, with ρ = the mass density of the soil. Of particular interest are the dashpots C and C_ϑ, which capture the energy radiated away from the foundation. In both instances the dashpots are calculated by multiplying A_0 (or I_0) by the so-called specific impedance ρc. This amount of radiation damping per unit contact area is asymptotically exact for high-frequency excitation (taking into account the restriction $c_p = 2\ c_s$ for incompressible soil in vertical translation or rocking). Thus the cone models are doubly-asymptotic approximations, correct both for zero frequency (static case) and for frequencies tending to infinity. This explains the high accuracy of cone models, even though they do not incorporate Rayleigh surface waves, which transmit 2/3 of the energy of low-frequency vibrations. The importance of Rayleigh waves diminishes to zero with increasing frequency, a fact often overlooked.

One way to implement cone models is simply to attach the discrete-parameter systems in Fig. 3 to the undersurface of the model of the superstructure, then numerically integrate in the time domain using a general-purpose computer program. As an alternative, it might appear feasible to use the impulse-response functions and evaluate the Duhamel (convolution) integral; however, this classical approach is computationally inefficient. Instead, the preferable way to proceed is to use recursive relationships of the form

$$y_n = a_1 y_{n-1} + a_2 y_{n-2} + b_0 x_n + b_1 x_{n-1} + b_2 x_{n-2} \tag{2}$$

in which $y_n = y(n\Delta t)$ = the response at time $t = n\Delta t$ and $x_n = x(n\Delta t)$ = the excitation at time $t = n\Delta t$. Expressions for the recursion constants a_j and b_j are presented in [4]. It should be emphasized that the recursive relatonship, Eq. (2), is not an approximate numerical integration formula. The results are exact, subject only to the assumption that the excitation function x(t) is a polygon formed by joining the samples x_n by straight lines. The recursive method, particularly well-suited for hand calculations, is applicable both to stiffness problems, in which the input samples x_n are displacements (or rotations) and the output samples y_n are forces (or moments), and to flexibility problems, in which the roles of input and output are reversed. The stiffness formulation for rotational cones is presented in the appendix of this paper and demonstrated in the numerical example.

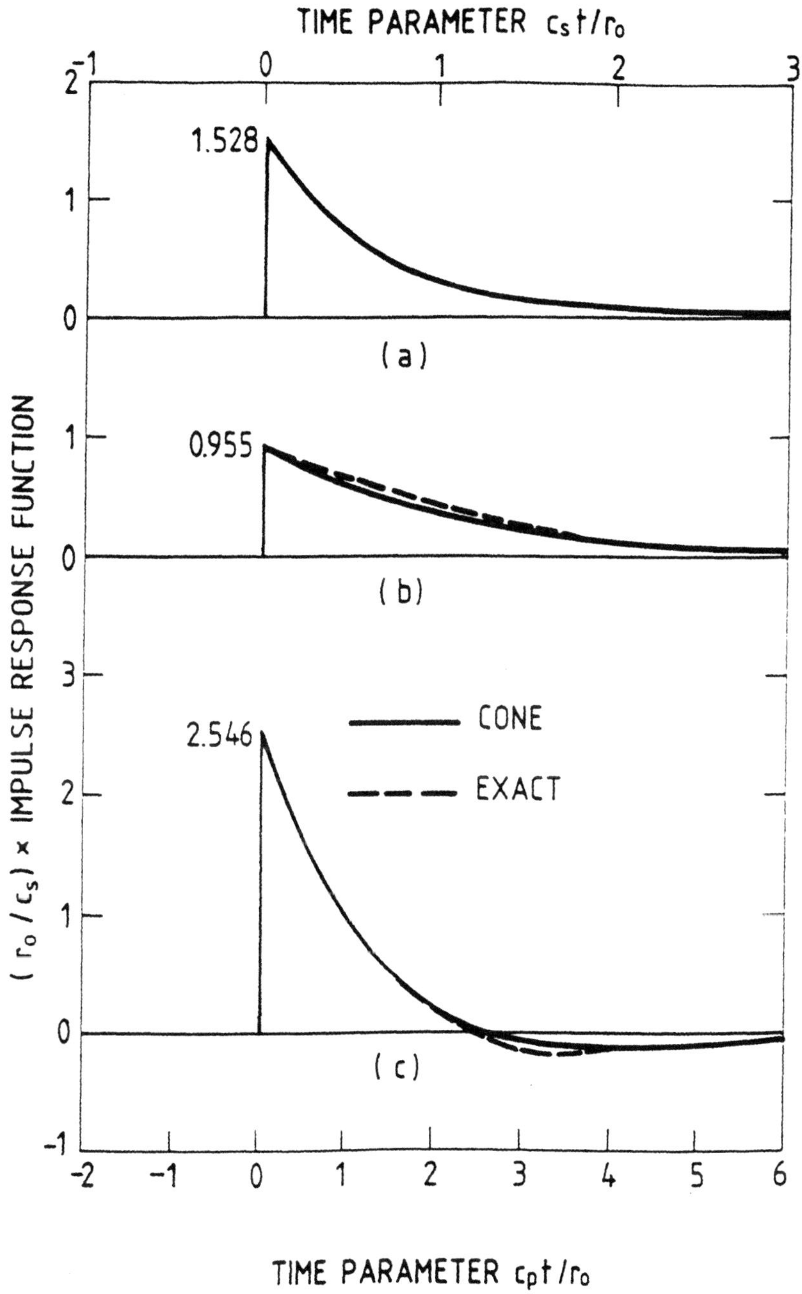

Figure 2. Impulse-response functions (a) for horizontal translation, (b) for vertical translation, and (c) for rocking.

TABLE 1. Properties of cone models

Cone Type	Motion	r_0	z_0/r_0	ν	c	Trapped Mass
(1)	(2)	(3)	(4)	(5)	(6)	(7)
Translational $K=\rho c^2 A_0/z_0$ $C=\rho c A_0$	Vertical	$\sqrt{\frac{A_0}{\pi}}$	$\frac{\pi}{4}(1-\nu)\left(\frac{c}{c_s}\right)^2$	$\leq 1/3$	c_p	0
				$> 1/3$	$2c_s$	$2.4(\nu-1/3)\rho A_0 r_0$
	Horizontal	$\sqrt{\frac{A_0}{\pi}}$	$\frac{\pi}{8}(2-\nu)$	all ν	c_s	0
Rotational $K_\vartheta=3\rho c^2 I_0/z_0$ $C_\vartheta=\rho c I_0$ $I_\vartheta=\rho I_0 z_0$	Rocking	$\sqrt[4]{\frac{4I_0}{\pi}}$	$\frac{9\pi}{32}(1-\nu)\left(\frac{c}{c_s}\right)^2$	$\leq 1/3$	c_p	0
				$> 1/3$	$2c_s$	$1.2(\nu-1/3)\rho I_0 r_0$
	Torsion	$\sqrt[4]{\frac{2I_0}{\pi}}$	$\frac{9\pi}{32}$	all ν	c_s	0

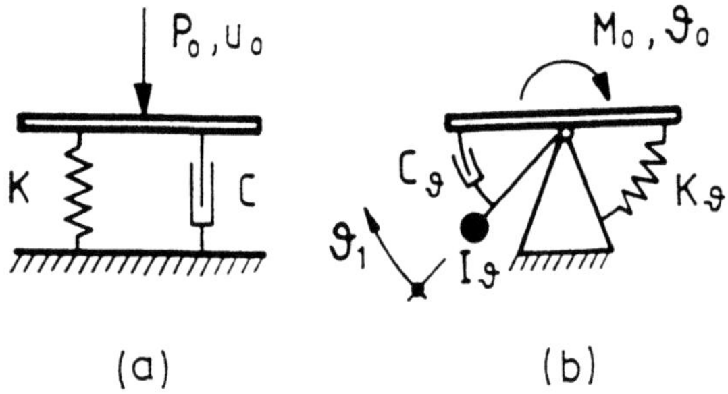

Figure 3. Discrete models (a) for translation and (b) for rotation.

3. Cone Models for a Soil Layer on Rigid Rock

To model layered soil or embedded foundations, a double cone is used to represent a disk in the fullspace. As shown in Fig. 4, the double cone is simply a pair of halfspace cones back-to-back. If the soil in the lower cone experiences compression, the soil in the upper cone experiences tension of the same magnitude; as a result, the static stiffness is doubled. For an infinite stack of double cones, excited as in Fig. 5, symmetry arguments reveal that the boundary conditions for a layer on rigid rock (zero stress on the surface of the layer, zero displacement on the surface of the rock) are satisfied. If then $w_o(t)$ denotes the displacement of a disk on the unlayered halfspace, the desired displacement $u_o(t)$ of the disk on the soil layer may be constructed from $w_o(t)$ as

$$u_o(t) = w_o(t) + e_1 w_o(t-T) + e_2 w_o(t-2T) + \ldots \qquad (3)$$

in which $T = 2d/c$ is the travel time from the surface down to the rock and back. An analogous expression also applies for forces. Physically speaking, Eq. (3) may be viewed as the sum of the immediate response $w_o(t)$, augmented by a sequence of echoes. The echo constants e_j, calculated according to [5], die off rapidly with increasing j, so that it suffices to truncate the sum after about 20 terms.

In the static case, in which u_o and w_o are constants, Eq. (3) simplifies to

$$u_o = (1 + e_1 + e_2 + e_3 + e_4 + e_5 + \ldots +)w_o \qquad (4)$$

in which a typical echo constant for translational motion, e_j, is evaluated by the formula

$$e_j = \frac{2\,(-1)^j}{1 + jK} \qquad (5)$$

in terms of the dimensionless parameter $K = 2d/z_o$. For the particular example of a vertically excited disk on a layer with depth $d = r_o$ and Poisson's ratio 1/3, $z_o = 2.094\, r_o$ from Table 1; and thus the value $K = 2/2.094 = 0.955$ is obtained. Equation (4) takes on the numerical values

$$u_o = (1 - 1.023 + 0.687 - 0.517 + 0.415 - 0.346 + \\ + 0.297 - 0.260 + 0.231 - 0.208 + \ldots -)w_o \qquad (6)$$

which sums to $u_o = 0.377\, w_o$, in excellent agreement with the rigorous result $u_o = 0.391\, w_o$. The echo constants are thus quite useful for static settlement computations.

For dynamic problems Eq. (3) is evaluated directly in the time domain. However, to assess the accuracy of the method, it is more advantageous to look at the dynamic-stiffness relationship in the frequency domain, conventionally expressed as

$$S(a_o) = K[k(a_o) + i a_o c(a_o)] \qquad (7)$$

in which a_o = the customary dimensionless frequency parameter $\omega r_o/c_s$. The dimensionless spring and damping coefficients $k(a_o)$ and $c(a_o)$ for

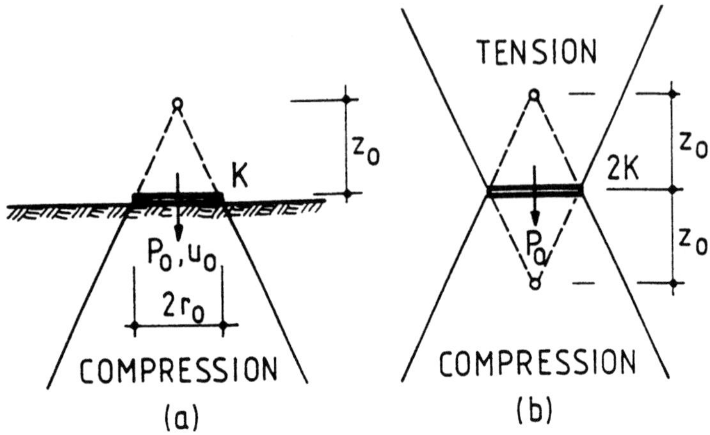

Figure 4. Cone models (a) single cone for halfspace and (b) double cone for fullspace.

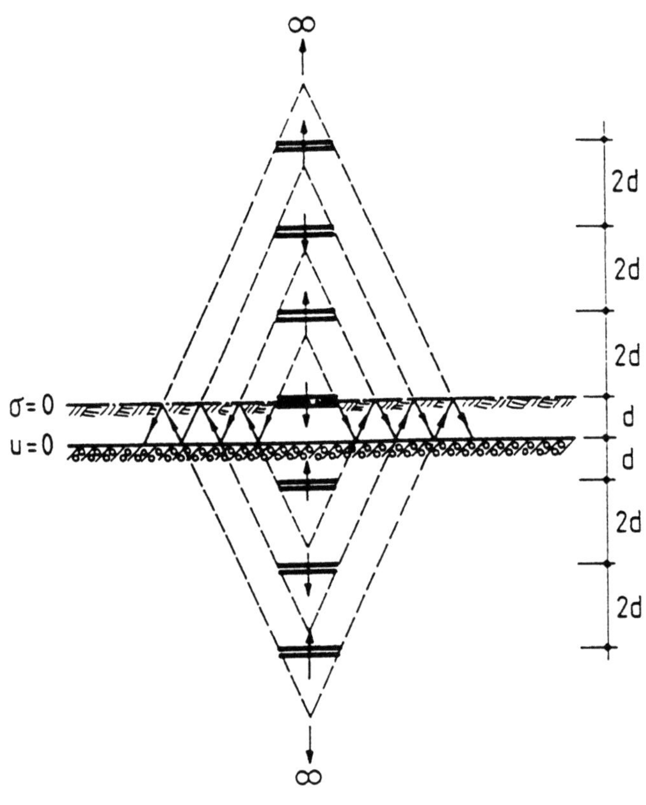

Figure 5. Stack model for layer on rigid rock.

the vertical case are presented in Fig. 6 [5]. The cone is seen to yield smoothed versions of the rigorous jagged curves.

The salient feature of layered soil is the existence of a cutoff frequency, below which the radiation damping vanishes. In the engineering sense the cutoff frequency is equal to the fundamental resonant frequency of the layer in vertical motion, corresponding to $\omega T = \pi$. In the mathematically strict sense the cutoff frequency is actually located at $\omega T = \pi/2$, because shear waves also transmit a small portion of energy in vertical and rocking motion. As shown in the figure, the cone indeed captures the phenomenom of the cutoff frequency. It is quite amazing that the simple, physically appealing concept of echoes can account so well for the complicated behavior of layered soil. Without going into details, it should be mentioned that the underlying substratum need not be perfectly rigid. By a slight modification of the echo constants, the flexibility of the bedrock may be incorporated [10].

4. Cone Models for Embedded Foundations

Symmetry arguments similar to those for a soil layer enable embedded foundations also to be represented by a stack of double cones, Fig. 7. For each disk in the lower halfspace a mirror-image disk is introduced in the upper halfspace. Provided each pair (disk + mirror disk) executes the same motion, the middle surface of the fullspace remains stress-free and thus simulates the upper surface of the soil.

The displacement of a double cone in translation dies off in inverse proportion to the distance from the apex. For the double cone in rotation the decay law is somewhat more complicated. In any case it is easy to formulate a flexibility matrix for the array in Fig. 7, invert it to obtain a stiffness matrix, and then apply the rigidity constraints to compute the dynamic-stiffness relationship in the form of Eq. (7) for an embedded foundation. Fig. 8 shows the results for rocking motion; again, the agreement with the rigorous solution is quite satisfactory [7].

5. Cone Models for Piles

Piles constitute an extreme case of embedded foundations, in which the embedment (pile length) is very large in comparison to the radius. In addition, the pile is no longer perfectly rigid. Fig. 9 shows a pile group consisting of an array of 3 x 3 piles. The dynamic-stiffness relationship of the pile group in vertical motion is presented in Fig. 10. The results calculated using cone models are highly accurate [11].

6. Approximate Green's Functions for Surface Foundations

Fig. 11, taken from Wolf's text [12], shows that surface foundations of arbitrary geometry may be idealized by an assemblage of small subdisks. In nuclear-engineering applications such as this, the displacement field (Green's function) for the incremental subdisk is calculated rigorously. However, for less critical applications it is sufficient to specify the decay law for displacements away from the subdisk *a priori*. It is well known that the surface displacement dies off in inverse proportion to the distance in the immediate vicinity of the

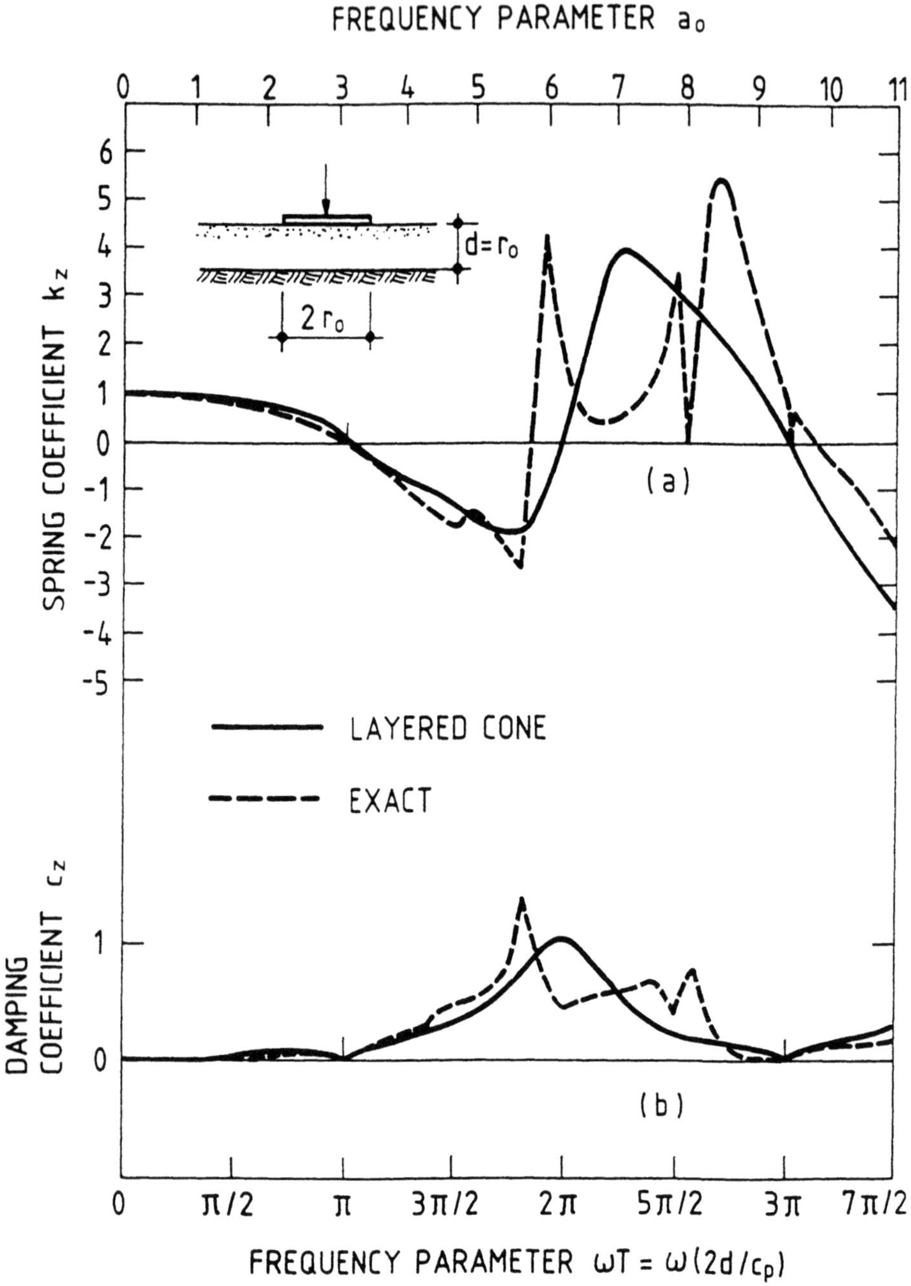

Figure 6. Dynamic-stiffness relationship for layer on rigid rock (a) spring coefficient and (b) damping coefficient.

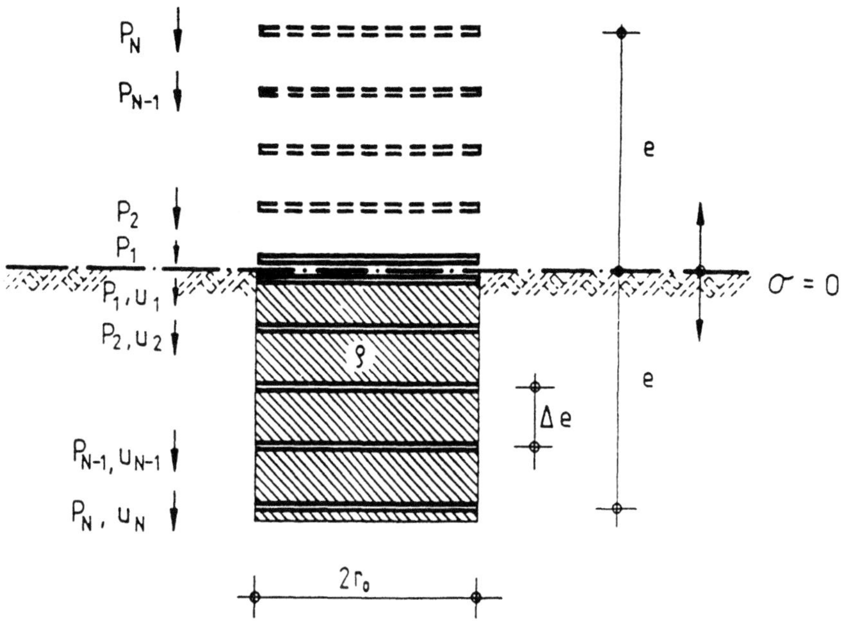

Figure 7. Stack model for embedded foundation.

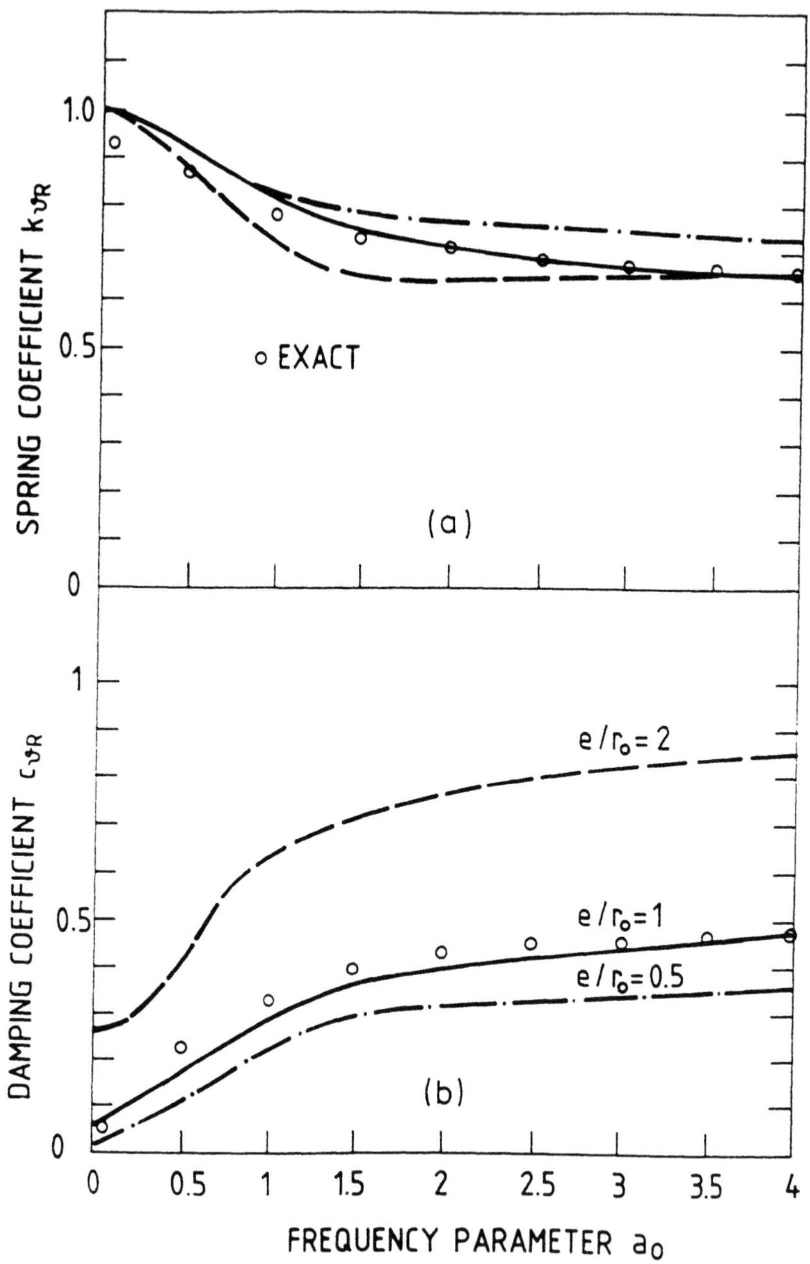

Figure 8. Dynamic-stiffness relationship for embedded foundation in rocking motion (a) spring coefficient and (b) damping coefficient.

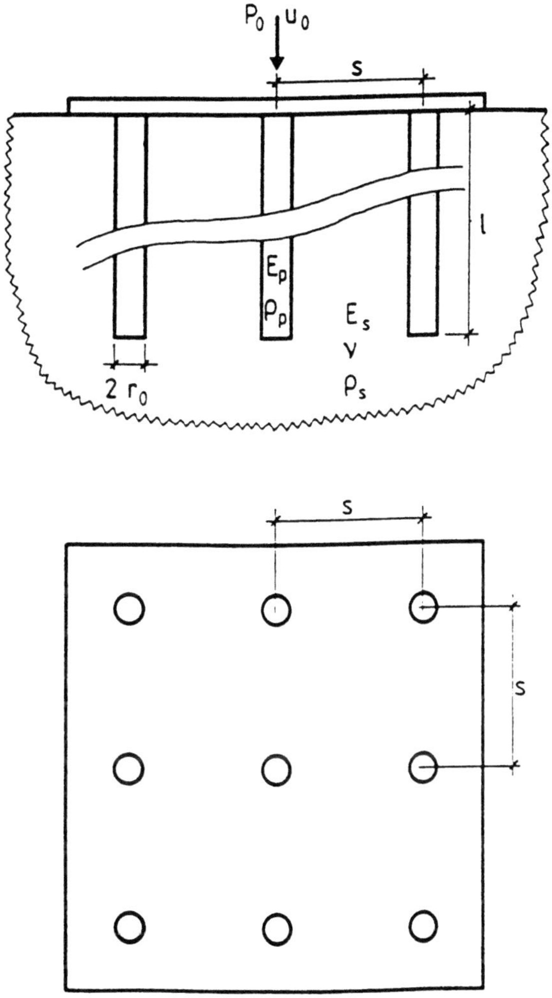

Figure 9. 3 × 3 pile group.

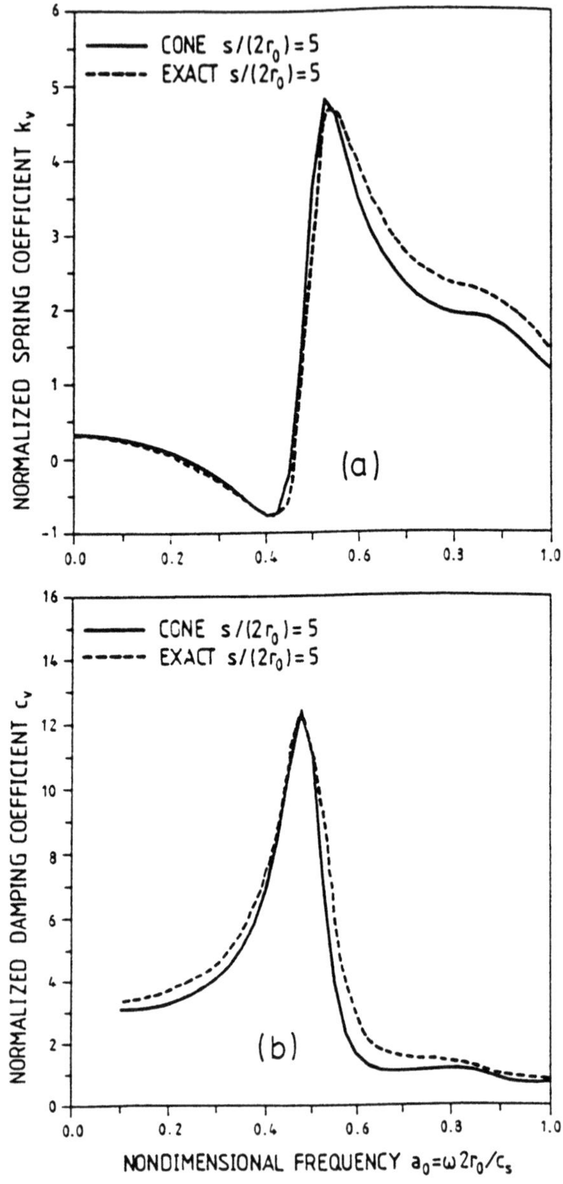

Figure 10. Dynamic-stiffness relationship for pile group in vertical motion (a) spring coefficient and (b) damping coefficient.

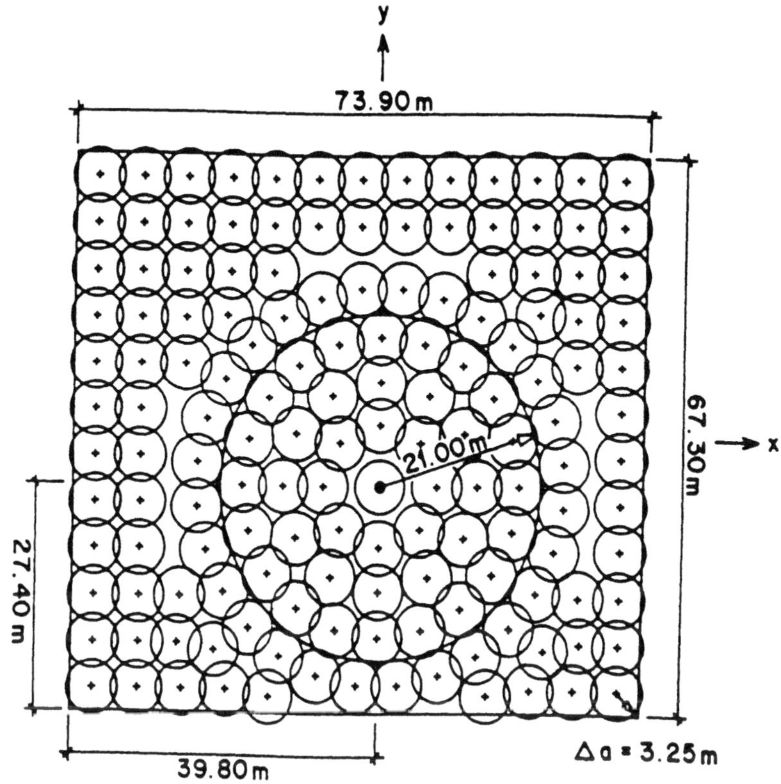

Figure 11. Subdisk discretization for nuclear powerplant.

disk (near field) and in inverse proportion to the square root of the distance farther out (far field). This simplified decay law defines an approximate Green's function, which may be used to assemble the flexibility matrix of the total array. As for the embedded foundation, the flexibility matrix is inverted; and the rigid-body constraints are applied to the resulting stiffness matrix to obtain the dynamic-stiffness relationship in the form of Eq. (7). The dimensionless spring and damping coefficients for a square basemat in vertical and rocking motion are presented in Fig. 12. The accuracy obtained using the approximate Green's function is indeed adequate for practical applications [8].

7. Example

To help the reader become familiar with the methods of computation for cone models, a practical example is investigated. Consider a cylindrical structure with radius r_0 = 10 m and height h = 30 m. The average mass density of the structure is 300 kg/m³, a typical value both for multistory buildings and for reactor containments. Now if the example structure represents a reactor containment, it will be very stiff and may be viewed to a first approximation as a rigid body with mass $M = 2.83 \times 10^6$ kg and mass moment of inertia $J = 9.18 \times 10^8$ kgm² about the base diameter.

The cylinder rests on the surface of homogeneous soil with mass density $\rho = 2000$ kg/m³ and Poisson's ratio 1/3. For relatively stiff sand typical values of the wave-propagation velocities would be c_s = 250 m/s and $c_p = 2\ c_s$ = 500 m/s. Because the structure is taller than it is wide, the soil-structure interaction response will be dominated by rocking; and attention will be restricted to this component of motion, for which the appropriate wave velocity is $c = c_p$ = 500 m/s. The specific damping per unit contact area of the basemat is $\rho c = 1 \times 10^6$ Ns/m³. Stated in more familiar units, the damping force produced by a vertical velocity of 1 m/s is 1 MN/m², a large value indeed. The confined dynamic modulus of elasticity of the soil, $E_s = \rho c^2$ = 500 MN/m², is also quite large, perhaps five times the static modulus used in ordinary settlement computations.

From Table 1 the apex height of the rocking cone is calculated to be z_0 = 23.6 m. The area moment of inertia of the basemat is $I_0 = \pi r_0^4/4$ = 7850 m⁴. In combination with the soil properties these geometric data determine the rotational spring and dashpot of the discrete-parameter model in Fig. 3(b) as $K_\vartheta = 3\ E_s I_0/z_0 = 5 \times 10^5$ MNm/rad and $C_\vartheta = \rho c I_0$ = 7850 MNms/rad. [The third discrete parameter in Fig. 3(b), the "monkey-tail" rotatory inertia, numerically equal to $I_\vartheta = \rho I_0 z_0 = 3.71 \times 10^8$ kgm², is not used explicitly in the following analysis; instead, its influence will be captured implicitly via recursion constants. It should be emphasized, however, that the recursive approach is dynamically equivalent to numerical integration of the mechanical model in Fig. 3(b).]

If the "monkey tail" were not present, the natural frequency of the coupled soil-structure system could be evaluated from the standard formula $\omega = (K_\vartheta /J)^{1/2} = (5 \times 10^{11}/9.18 \times 10^8)^{1/2}$ = 23.3 rad/s, i.e. 3.71 Hz. This relatively low value by comparison to the fixed-base natural frequency of a concrete cylinder verifies that it is indeed correct to idealize the structure as a rigid body.

The inclusion of the monkey tail (consisting of C_ϑ and I_ϑ connected in series) leads to a slight reduction of the effective stiffness.

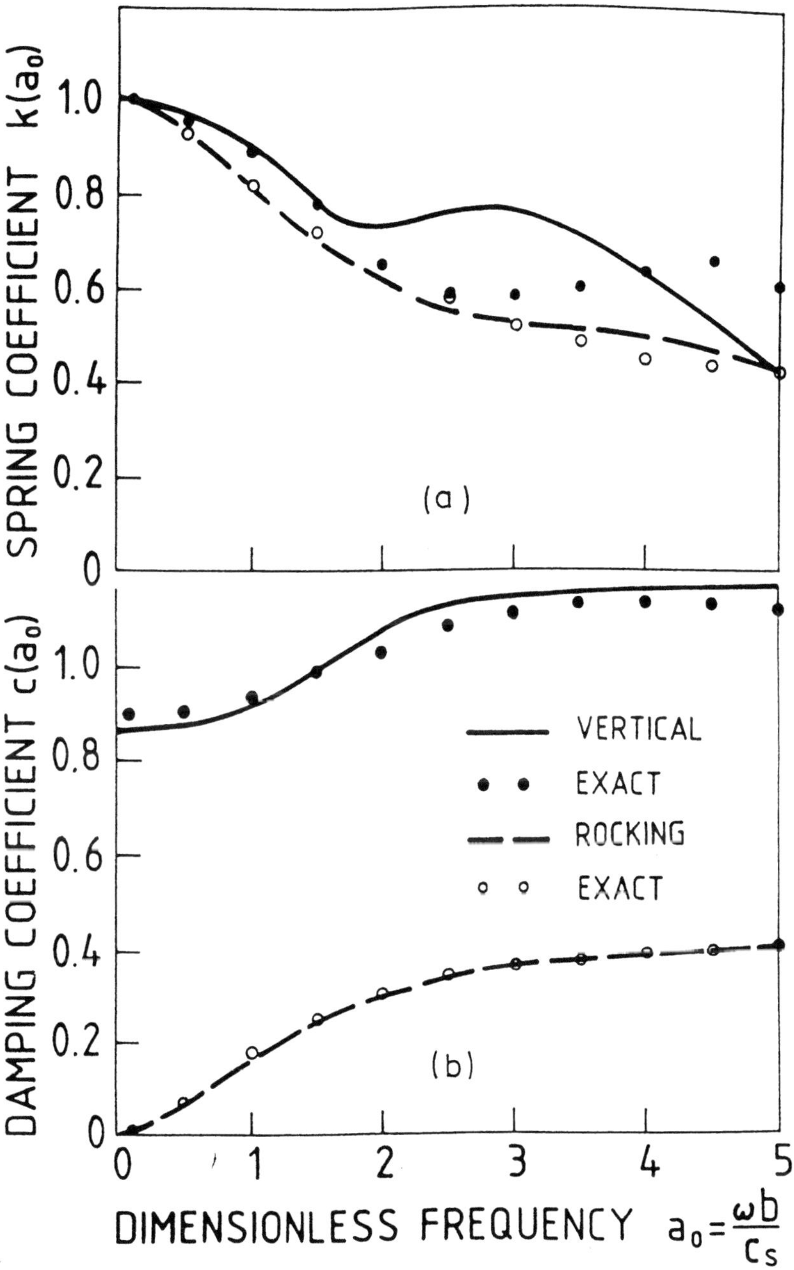

Figure 12. Dynamic-stiffness relationship for square foundation with Poisson's ratio 1/3 (a)spring coefficient and (b) damping coefficient.

This effect may be quantified with the help of Fig.13, which presents the dynamic-stiffness coefficients for all members of the cone family characterized by the governing Eq. (1). For the cases n = 2 (disk in translation) and n = 4 (disk in rotation) the curves can be computed quite simply by setting up and solving the equations of harmonic motion for the discrete-parameter models in Fig. 3. Unfortunately, however, no simple mechanical models exist for the cases n = 1 and 3, corresponding to long strips in translation and rocking. In these instances the curves must be evaluated by direct solution of Eq. (1), which entails Bessel functions [2].

In Fig. 13 it is important to observe that the dimensionless frequency parameter along the abscissa is defined as $a_0 = \omega z_0/c$ [instead of the conventional definition $a_0 = \omega r_0/c_s$ introduced in conjunction with Eq. (6)]. Suppose now that the coupled soil-structure system is subjected to a harmonically varying rocking moment M_0, the frequency of which gradually increases, as would be produced by a machine during the start-up phase. The greatest response obviously occurs when the operating frequency corresponds to the natural frequency of the system. Taking as a first approximation the frequency ω = 23.3 rad/s calculated previously, the value of a_0 at resonance would be 23.3·23.6/500 = 1.1. From Fig. 13(a) at a_0 = 1.1 it follows that the effective dynamic spring has decreased from the static value 3 $E_s I_0/z_0$ to about 2.5 $E_s I_0/z_0$ = 0.833 K_ϑ = 4.17 × 10^5 MN/rad. In view of this reduced stiffness, the new estimate of the resonant frequency is $\omega = (4.17 \times 10^{11}/9.18 \times 10^{8})^{1/2}$ = 21.3 rad/s (3.39 Hz). For the second iteration the slightly lower value of a_0 at resonance is 21.3·23.6/500 = 1.0, essentially the same as before. The dynamic spring need not be modified further, and from Fig 13(b) it follows that the dynamic dashpot is 0.5 C_ϑ = 3925 MNms/rad. For a single-mass oscillator the effective damping ratio is computed via the well-known formula $\zeta = \omega C_\vartheta(a_0)/[2K_\vartheta(a_0)]$ = 21.3·3925/(2·4.17 × 10^5) = 0.100. The result may also be read directly from Fig. 13(c), which additionally enables ascertation of the potential error implicit in modeling a 3-dimensional disk by a 2-dimensional strip. As indicated by the multiplier "×2", the damping ratio of the strip is *double* that of the disk when a_0 = 1! A nonconservative error of this magnitude is obviously unacceptable and even potentially dangerous.

If the harmonic exciting moment were applied *quasistatically*, the induced angle of rotation would be M_0/(4.17 × 10^5 MN/rad). Note that the reduced stiffness corresponding to a_0 = 1 appears in the denominator, not the static rotational stiffness K_ϑ. Due to resonance, the quasistatic rotation is multiplied by the amplication factor familiar from elementary dynamics, AF = $1/(2\zeta)$ = 5. The dynamic rotation is thus 5 M_0/(4.17 × 10^5 MN/rad) = 6 M_0/K_ϑ. This implies that if the rotation is computed using the *static* stiffness K_ϑ (instead of the *quasistatic* stiffness 0.833 K_ϑ), the apparent amplication factor is 6, not 5.

The same result may be calculated directly in the time domain. The dynamic stiffness of the coupled system is the sum of the inertial contribution $J\ddot{\vartheta}_0$ from the structure plus the soil reaction computed via Eq. (15).

$$M_0 = J\ddot{\vartheta}_0 + K_\vartheta \vartheta_0 + C_\vartheta \dot{\vartheta}_0 - C_\vartheta \dot{\vartheta}_1 \tag{8}$$

At the resonant frequency ω = 21.3 rad/s the rotation, angular velocity, and angular acceleration are given by

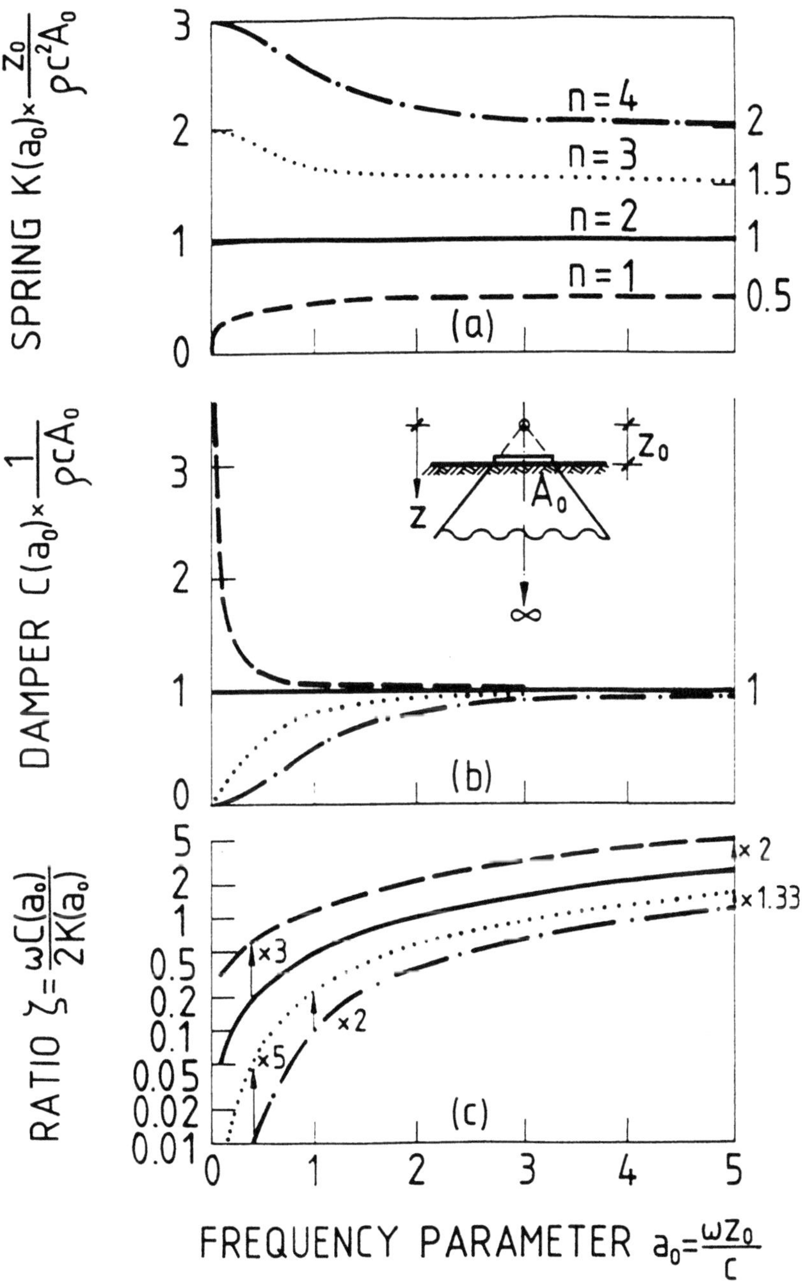

Figure 13. Dynamic-stiffness relationship for cone-wedge family (a) frequency-dependent spring, (b) frequency-dependent damper, and (c) damping ratio.

$$\vartheta_0 = \sin(21.3t) \tag{9}$$

$$\dot{\vartheta}_0 = 21.3 \cos(21.3t) \tag{10}$$

$$\ddot{\vartheta}_0 = -454 \sin(21.3t) \tag{11}$$

For any value of $t = n\ \Delta t$ the first three terms in Eq. (8) may be evaluated directly. To obtain a sufficiently fine discretization of the response, Δt is taken to be 1/20 of the period of the harmonic motion, i.e. $\Delta t = 0.01475$ s, so that $21.3t = 0.31416\ n$. The fourth term in Eq. (8) is computed by recursion. The dimensionless quantity $\beta = c\ \Delta t/z_0$ is numerically equal to $500 \cdot 0.01475/23.6 = 0.3125$; and according to Eqs. (18) through (20), the recursion constants are $a_1 = 0.73162$, $b_0 = 0.14117$, and $b_1 = 0.12721$. Insertion of the numerical values into Eqs. (16) and (8) leads first to the updated value of the angular velocity $\dot{\vartheta}_1$ at time step n

$$\dot{\vartheta}_{1n} = 0.73162\ \dot{\vartheta}_{1n-1} + 3.0069 \cos(0.314n) + 2.7096 \cos[0.314(n-1)] \tag{12}$$

and then to the expression for the current moment

$$\frac{M_{0n}}{K_\vartheta} = 0.167 \sin(0.314n) + 0.333 \cos(0.314n) - 0.0157\ \dot{\vartheta}_{1n} \tag{13}$$

If these formulas are evaluated up to n = 100 (five periods of motion), the steady state is reached; and the expected result, namely $\vartheta_0 = 6\ M_0/K_\vartheta$, is obtained.

This example shows that the time-domain analysis by recursion yields the same response as in the frequency domain (Fig. 13). Of course, the final accuracy of the results depends on whether or not the fundamental assumption "disk on a homogeneous halfspace" is physically justified. If the sandy soil is in fact a layer of finite thickness d over rigid bedrock, a cutoff frequency = $c_p/(4d)$ exists. For c_p = 500 m/s the cutoff frequency is greater than the resonant frequency of the coupled soil-structure system (3.39 Hz) unless the layer depth exceeds 37 m. Because the radiation damping *vanishes* below the cutoff frequency (see Fig. 6), the resonant amplitude of rotation computed for the example structure on a sand layer d < 37 m tends not to $6\ M_0/K_\vartheta$ but rather to *infinity* (limited only by the material damping of the soil, not considered herein)! Obviously, then, it is of utmost importance to beware of cutoff frequencies. When in doubt, it is quite easy to take the effect of layering into account via the echo constants described in Section 3.

8. Appendix: Recursive Realization of Rotational Cone

Whereas the dynamic stiffness of the translational cone [spring-dashpot model in Fig. 3(a)] is simply

$$P_0 = K\ w_0 + C\ \dot{w}_0, \tag{14}$$

the corresponding relationship for the rotational cone is slightly more complicated. As might be expected, the dynamic stiffness involves the static spring K_ϑ (multiplied by the rotation ϑ_0) as well as the high-frequency damper C_ϑ (multiplied by the angular velocity $\dot{\vartheta}_0$); however, an additional third contribution arises due to the "monkey tail". Notice in Fig. 3(b) that the force in the dashpot C_ϑ is produced by the *difference* between the surface angular velocity $\dot{\vartheta}_0$ and the angular velocity $\dot{\vartheta}_1$ experienced by the rotatory inertia I_ϑ, i.e.

$$M_0 = K_\vartheta \vartheta_0 + C_\vartheta \dot{\vartheta}_0 - C_\vartheta \dot{\vartheta}_1 \tag{15}$$

Physical reasoning reveals the value of ϑ_1 in two limiting cases. For very low-frequency motion the inertial force generated by I_ϑ is very small, and therefore ϑ_1 is essentially equal to ϑ_0. On the other hand, for very high-frequency motion I tends to stand still; and ϑ_1 vanishes. For intermediate frequencies the angular velocity $\dot{\vartheta}_1$ is an auxiliary function which may be calculated at the n-th time step by recursion as

$$\dot{\vartheta}_{1n} = a_1 \dot{\vartheta}_{1n-1} + b_0 \dot{\vartheta}_{0n} + b_1 \dot{\vartheta}_{0n-1} \tag{16}$$

The recursion constants depend on the dimensionless time-step parameter

$$\beta = c\,\Delta t/z_0 \tag{17}$$

which is evaluated using the apex height of the cone, z_0, and the appropriate velocity of propagation, c (= c_s for torsion or c_p for rocking). The equations for a_1, b_0 and b_1 are [4]

$$a_1 = e^{-\beta} \tag{18}$$

$$b_0 = 1 - (1 - e^{-\beta})/\beta \tag{19}$$

$$b_1 = 1 - a_1 - b_0 \tag{20}$$

9. References

[1] Ehlers, G. (1942) 'The effect of soil flexibility on vibrating systems' (in German), Beton und Eisen 41, pp. 197-203.

[2] Gazetas, G. and Dobry,R. (1984) 'Simple radiation damping model for piles and footings', Journal of Engineering Mechanics, ASCE, 110, pp.937-956.

[3] Meek, J. W. and Veletsos, A. S. (1974) 'Simple models for foundations in lateral and rocking motion', Proc. 5WCEE, Rome, pp. 2610-2613.

[4] Meek, J. W. and Wolf, J. P. (1992) 'Cone models for homogeneous soil I', Journal of Geotechnical Engineering, ASCE, 118, pp. 667-685.

[5] Meek, J. W. and Wolf, J. P. (1992) 'Cone models for soil layer on rigid rock II', Journal of Geotechnical Engineering, ASCE, 118, pp. 686-703.

[6] Meek, J. W. and Wolf, J. P. (1992) 'Insights on 2D- vs. 3D modelling of surface foundations', Proc. 10WCEE, Madrid, pp. 1633-1638.

[7] Meek, J. W. and Wolf, J. P. (1993) 'Cone models for an embedded foundation', submitted to Journal of Geotechnical Engineering, ASCE.

[8] Meek, J. W. and Wolf, J. P. (1993) 'Approximate Green's functions for a surface foundation', submitted to Journal of Geotechnical Engineering, ASCE.

[9] Veletsos, A. S. and Nair, V. V. D. (1974) 'Torsional vibrations of viscoelastic foundations', Journal of the Soil mechanics and Foundations Division, ASCE, 100, pp. 225-245.

[10] Wolf, J. P. and Meek, J. W. (1993) 'Cone models for a soil layer on flexible rock halfspace', submitted to Journal of Earthquake Engineering and Structural Dynamics.

[11] Wolf, J. P., Meek, J. W. and Song, C. (1992) 'Cone models for a pile foundation', to be published in Proc. ASCE Conference on Pile Dynamics, New York.

[12] Wolf, J. P. (1985) Dynamic soil-structure interaction, Prentice-Hall, Englewood Cliffs, N.J., p. 337.

DESIGN CONCEPTS FOR DYNAMICS OF SOIL-STRUCTURE INTERACTION

A. S. VELETSOS
Department of Civil Engineering
Rice University
Houston, Texas 77251, USA

ABSTRACT. The principal effects of soil-structure interaction on the dynamic response of ground-excited systems are identified, and information and concepts are presented with which these effects may be provided for rationally and cost-effectively in design. Both kinematic and inertial actions are examined. The concepts involved are developed by reference to relatively simple structures, but their application to more involved systems is also highlighted. In addition, a brief account is given of the application of some of these concepts in the formulation of appropriate seismic design provisions for building structures.

1. Introduction

With the methods of analysis and the computer programs now available, it is in principle possible to evaluate the dynamic response of any structure-foundation-soil system to any excitation of the base. Such evaluations, however, can be no better than the assumptions underlying the analyses.

Central to any analysis of the seismic response of a structural system are assumptions and approximations concerning the physical properties of the structure and the supporting soils, the characteristics of the excitation, and the manner in which the ground motion reaches the site and is transmitted to the structure. Major assumptions normally also are involved in the analysis of the idealized system itself. It is important, therefore, that the resulting solutions be interpreted with due appreciation not only for the idealizations embodied in the modeling of the system, but also for the senstivity of the response to possible deviations from the assumed conditions.

Even with the computer facilities now available, mathematically precise analyses of complex structure-foundation-soil systems involve a major computational effort, and are generally too costly for preliminay design purposes. This is particularly true if the responses of the structure and the supporting medium in the inelastic range of deformation are to be evaluated. Compounding the difficulty is the fact that, because of the many factors which influence the dynamic response of such systems, and because of the sensitivity of their response to variations in these factors, it is generally quite difficult, if not impossible, intelligently to interpret the results of a limited number of highly sophisticated, complex analyses. Finally, complex analyses are more prone to misinterpretation and misuse than are simpler methods, and they may discourage the exercise of the educated judgement which is essential to a successful design.

There continues to be a need at this time for simple procedures and concepts with which the effects of soil-structure for ground-excited systems may be provided for readily and reliably in design. It is the objective of this contribution to highlight such concepts and procedures. Both inertial and kinematic actions are examined, and the relative importance of the two actions is eluci-

P. Gülkan and R. W. Clough (eds.), Developments in Dynamic Soil-Structure Interaction, 307–325.

dated. A brief account also is given of the application of some of these concepts in the formulation of appropriate seismic design provisions for building structures.

2. Statement of Problem

It is generally recognized that the motion that is experienced by the foundation of a structure during an earthquake may differ substantially from the free-field ground motion, which is the motion that the ground would experience at its interface with the foundation in the absence of the structure. Two factors are responsible for this difference: (1) The inability of a rigid foundation to conform to the generally non-uniform, spatially varying free-field ground motion; and (2) the interaction or coupling between the vibrating structure, its foundation, and supporting soils.

Several factors contribute to the spatial variability of the ground motion. The seismic waves may emanate from different points of an extended source and may impinge the foundation at different instants or with different angles of incidence, or they may propagate through paths of different physical properties and may be affected differently in both amplitude and phase by the characteristics of the travel paths and by reflections from, and diffractions around, the foundation. Even when the seismic wave front is plane, it may impinge the foundation-soil interface obliquely, leading to ground motions that differ in phase from point to point. The spatial variability of the ground motion due to the propagation of a plane wave is known as the wave passage effect, whereas that due to the other, generally random, factors is known as the ground motion incoherence effect.

The seismic response of a structure is frequently evaluated considering the motion of its base to be equal to the stipulated free-field ground motion at a convenient reference or control point, which is normally taken at the ground surface. No provision is made in this approach for either the spatial variability of the free-field ground motion or for the properties of the supporting medium. The exact analysis requires that the structure be considered to be part of a larger system which includes the foundation and the supporting medium, and that due cognizance be taken of the spatial variability of the ground motion and of the properties of the soils involved.

Such an analysis is implemented in two steps: First, the motion of the foundation is evaluated considering both the foundation and the superimposed structure to be massless. Referred to as the foundation input motion (FIM), the resulting motion generally includes torsional and rocking components in addition to translational components. Next, the response to the FIM of the actual foundation-structure system with mass is evaluated using the actual properties of the supporting medium and providing for the dynamic interaction between its elements. The flexibility of the supporting medium has a two-fold effect: (1) It increases the number of degrees of freedom of the system and lowers its effective stiffness; and (2) it makes it possible for part of the vibrational energy of the structure to be dissipated in the supporting medium by radiation of waves and by hysteretic action in the soil itself. These forms of energy dissipation have no counterpart in a rigidly supported structure.

The difference in the responses of the superstructure computed for the FIM and the free-field control point motion (CPM) represents the kinematic interaction (KI) effect, whereas the difference of the responses computed with and without regard for the flexibility of the supporting medium is known as the inertial interaction (II) effect. The total soil-structure interaction (SSI) is given by the sum of the KI and II effects. More specifically, if R_0(FIM) = the response to the FIM computed considering the supporting medium to be rigid, and R_0(CPM) = the corresponding response to the CPM, then

$$KI = R_o(FIM) - R_o(CPM) \tag{1}$$

Similarly, if R(FIM) = the response to the FIM computed with due regard for the flexibility of the supporting medium, then

$$II = R(FIM) - R_o(FIM) \tag{2}$$

and

$$SSI = KI + II = R(FIM) - R_o(CPM) \tag{3}$$

The starting point for the material presented herein is the assumption that the control point motion of the ground is known. The determination of this motion is probably the most difficult and uncertain step in analyses of the seismic response of structures.

The characteristics of the CPM depend on such factors as the magnitude of the earthquake; the distance from the earthquake source to the site under consideration; the source mechanism, which refers to the details of the fracturing process and to the orientation and direction of propagation of the break; the characteristics of the travel path, which include the size, orientation and physical properties of the surface and subsurface strata through which the waves must travel to reach the site; and the topography, geology, and local soil conditions of the site.

The soil-structure interaction problem is sometimes defined in a generalized sense to include all, or at least some, of the factors involved in the definition of the control point motion. Local site conditions represent the factor most commonly considered in this regard. Such generalized interpretations of soil-structure interaction, however, are believed to be unduly broad; they tend to diffuse the issues involved and to complicate their resolution. A more desirable approach consists in breaking the problem into its component parts and examining each part separately and critically. The presentation here is based on the narrower interpretation and relies heavily on studies conducted at Rice University.

3. System Considered

The concepts involved will be identified by reference to a simple, linear structure of mass m and height h which is supported through a foundation of mass m_o at the surface of a homogeneous, elastic halfspace. The circular natural frequencies of the lateral and torsional modes of vibration of the fixed-base structure are denoted by $p_x = 2\pi f_x$ and $p_\theta = 2\pi f_\theta$, respectively, in which f_x and f_θ are the associated frequencies in cycles per unit of time, and the corresponding percentages of critical damping are denoted by β_x and β_θ. The foundation mat is idealized as a rigid circular plate of negligible thickness and radius R which is bonded to the halfspace so that no uplifting or sliding can occur, and the columns of the structure are presumed to be massless and axially inextensible. Both m and m_o are assumed to be uniformly distributed over identical circular areas. The supporting medium is characterized by its mass density, ρ, shear wave velocity, v_s, and Poisson's ratio, ν. The free-field control point motion is defined at the center of the foundation-soil interface, and it is considered to be a unidirectional, horizontal excitation. This structure may be viewed either as the direct model of a single-story building frame or, more generally, as the model of a multistory, multimode structure that responds as a system with one lateral and one torsional degrees of freedom in its fixed-base condition.

4. Inertial Interaction Effects

First, the effects of a vertically propagating plane seismic wave are examined. The FIM in this case is equal to the CPM, and only inertial interaction effects are present. Under the influence of such an excitation, the foundation of the structure displaces horizontally by an amount x(t) which is generally different from $x_g(t)$, and rocks or rotates by an amount $\psi(t)$ about a horizontal axis normal to the direction of ground shaking. The configuration of the system can then be defined by x(t), $\psi(t)$ and the structural deformation, u(t). The rocking component of foundation motion would be expected to be particularly prominent for tall structures and flexible soils. For a rigidly supported structure, $\psi(t) = 0$ and $x(t) = x_g(t)$.

4.1. POSSIBLE METHODS OF ANALYSIS

The response of this system is governed by a set of three coupled, second-order linear differential equations. The solution of these equations is complicated by the fact that the damping terms which provide for the energy dissipated into the supporting medium are of a form which does not permit the uncoupling of the equations by an expansion of the coordinates in terms of the undamped natural modes of vibration of the system. Also complicating the analysis is the fact that the restraining forces exerted by the supporting medium are, strictly speaking, functions of the frequency of the motion [15, 21, 23]. The governing equations can be solved either in the frequency domain, by use of Fourier transform or Laplace transform techniques, or directly in the time domain. The time-domain analysis may be implemented either in the manner indicated in [8] and [24] by use of the convolution integral and the foundation impulse response functions presented in [16] and [22], or as shown in [25] and [26] by use of foundation models with frequency independent parameters, such as those given in [7] and [16]. These methods may be too elaborate for design applications, however, and simpler approximate procedures have been sought.

Considering that in the conventional analysis of the seismic response of structures the superstructure is presumed to be rigidly supported, it is natural to inquire if a reasonably accurate description of the significant aspects of the response of the flexibly supported structure considered here could not be obtained from a single-degree-of-freedom analysis, that is, by assuming the structure to be rigidly supported. A satisfactory evaluation can indeed be made in this manner, and two different approaches have been used for this purpose [10, 14, 17]. The first involves modifying the stipulated free-field ground motion and evaluating the response of the structure to the modified motion of the foundation, whereas the second involves modifying the dynamic properties of the structure, considering it to be rigidly supported, and evaluating the response of the modified structure to the prescribed free-field ground motion. The second approach, which permits the direct use of response spectra for the specified free-field motion, is the more convenient of the two and is used exclusively here.

4.2. PRINCIPAL EFFECTS

The interaction effects in the latter approach are expressed approximately by an increase in the fixed-base natural period of the structure, and by a change (generally an increase) in the associated damping. The increase in period results from the flexibility of the supporting medium, whereas the increase in damping results from the capacity of the medium to dissipate energy by radiation of waves and by hysteretic action.

If T represents the natural period of the structure in its fixed-base condition, and $\tilde{T}$ represents the period of the modified structure which approximates the flexibly supported system, it can be shown (e.g. , Ref. 10 or 14) that

$$\tilde{T} = T\sqrt{1 + \frac{k}{K_x}\left(1 + \frac{K_x h^2}{K_\psi}\right)} \tag{4}$$

in which K_x = the lateral translational stiffness of the foundation, defined as the horizontal force necessary to displace the foundation by a unit amount; and K_ψ = the rocking stiffness of the foundation, defined as the moment necessary to rotate the foundation by a unit amount about a horizontal centroidal axis. Strictly speaking, these stiffnesses should be evaluated iteratively for a harmonic excitation of period $\tilde{T}$. However, reasonable approximations are obtained by use of their static values, or better still, by use of the values corresponding to the fixed-base natural period of the system, T, or some reasonable estimate of $\tilde{T}$.

For the surface-supported foundation and the uniform soil deposit considered, the **static** values of the translational and rocking are given by

$$K_x = \frac{8}{2-\nu} G R \tag{5}$$

and

$$K_\psi = \frac{8}{3(1-\nu)} G R^3 \tag{6}$$

in which $G = \rho v_s^2$ = the shear modulus of the soil. Eqs. 5 and 6 do not provide for the small coupling between translational and rocking actions.

It can further be shown [8, 17] that if β represents the percentage of critical damping for the fixed-base structure, and $\tilde{\beta}$ represents the corresponding damping of the modified structure that approximates the interacting system, the two quantities are interrelated by

$$\tilde{\beta} = \beta_o + \frac{\beta}{(\tilde{T}/T)^3} \tag{7}$$

in which β_o represents the contribution of the foundation damping, including radiation and soil material damping. Note that β_o and β are not directly additive, and that the effectiveness of the structural damping is reduced by soil-structure interaction, the reduction being substantial when $\tilde{T}/T$ is large. In fact, unless the reduced contribution of structural damping is compensated by the foundation damping, the overall damping of the interacting system will be less than that of the rigidly supported structure.

The three most important parameters that affect the value of β_o are: the period ratio, $\tilde{T}/T$, which is a measure of the relative flexibilities of the foundation medium and structure; the ratio of the height of the structure to the radius of the foundation, h/R; and the hysteretic capacity of the soil itself, defined by the factor

$$\tan\delta = \frac{1}{2\pi}\frac{\Delta W_s}{W_s} \tag{8}$$

in which ΔW_s represents the area of the hysteresis loop in the stress-strain diagram for a soil specimen undergoing harmonic shearing deformation, and W_s represents the strain energy stored in a linear material subjected to the same maximum values of stress and strain. This factor generally increases with increasing intensity of excitation or level of straining.

The variation of β_o with $\tilde{T}/T$ is shown in Fig. 1 for two values of $\tan\delta$. The dashed lines, which refer to systems supported on a purely elastic medium, represent the effect of radiation damping only, whereas the solid lines, which refer to a viscoelastic medium with $\tan\delta = 0.10$, represent the combined effect of radiation and hysteretic soil action. It can he seen that the contribution of the foundation damping may be quite substantial for relatively short, stubby structures, and that the effect of hysteretic soil action may be particularly significant for tall structures for which the radiational effects are generally quite small.

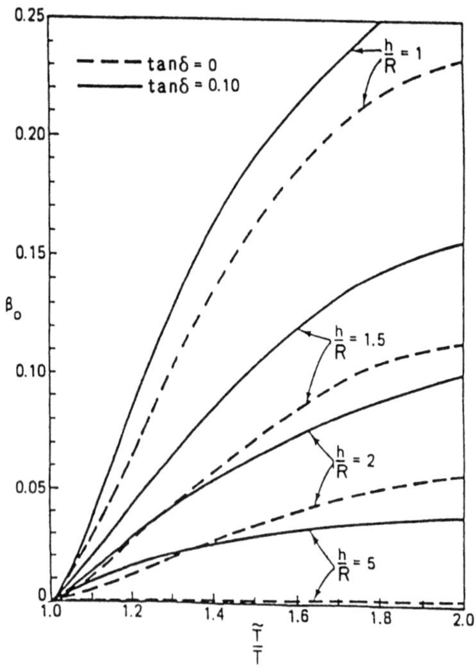

Fig. 1 Foundation damping factor

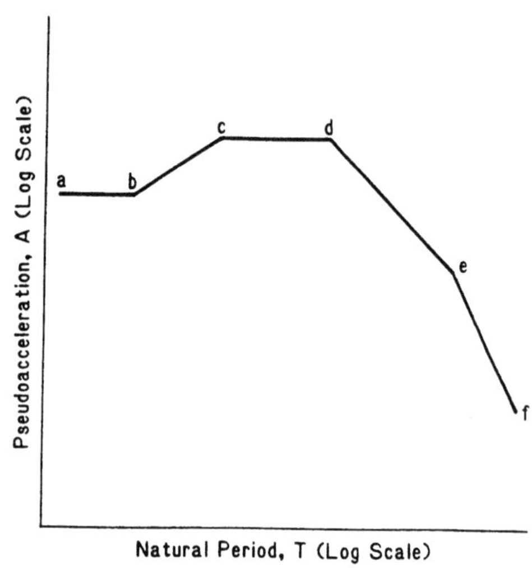

Fig. 2 Design response spectrum

The particular data presented in Fig. 1 are for systems with negligible foundation mass and a structural mass equal to 15 percent of the mass of the structure when filled with soil. The latter value is representative of that for building structures. Additional data are available in [8] and [17].

4.3. SUMMARY OF PROCEDURE

The steps involved in the application of the procedure may be summarized as follows:

1. Compute the fixed-base natural period of the structure, T, and by application of Eq. 4, evaluate the modified period, $\tilde{T}$.
2. Estimate the structural damping factor, β, and the soil material damping factor, $\tan\delta$, which would be appropriate for the anticipated strains in the supporting medium, and by application of Eq. 7 and the data presented in Fig. 1 and/or [8, 10 and 17], determine the effective damping factor, $\tilde{\beta}$. Some of the considerations involved in the choice of the values of β and $\tan\delta$ are discussed later.
3. From the response spectrum for single-degree-of-freedom systems subjected to the stipulated free-field ground motion, evaluate the pseudoacceleration, $\tilde{A}$, corresponding to $\tilde{T}$ and $\tilde{\beta}$.
4. The maximum value of the base shear for the interacting system, $\tilde{Q}$, is then determined from

$$\tilde{Q} = m\tilde{A} \tag{9}$$

the maximum deformation of the structure, u_m, is determined from

$$u_m = \frac{\tilde{Q}}{k} \tag{10}$$

and the maximum displacement of the structure relative to the ground, Δ_m, is determined from

$$\Delta_m = \frac{\tilde{Q}}{k} + \frac{\tilde{Q}h^2}{K_\psi} = u_m\left(1 + \frac{kh^2}{K_\psi}\right) \tag{11}$$

in which the term on the extreme right represents the contribution of the foundation rotation.

For building structures that must be analyzed as multi-degree-of-freedom systems in their fixed-base condition, a reasonable approximation to the maximum response of the structure may be obtained by assuming that soil-structure interaction affects only the response component contributed by the fundamental mode of vibration. This component is computed by the procedure outlined, by interpreting m and h to be the effective mass and effective height of the structure when vibrating in its fundamental mode. The contributions of the higher modes are then evaluated disregarding the interaction effects. The rationale for this approach is explained in [10] and [17].

4.4. CONSEQUENCES OF INTERACTION

A recurring question is whether soil-structure interaction (SSI) increases or decreases the maximum response of a structure. The answer depends on the response quantity under examination, and on the characteristics of the ground motion and the system itself. More specifically, SSI may increase, decrease, or have no effect on the peak response of a system depending on the characteristics of the relevant response spectrum, and the regions of the spectrum to which the fundamental natural periods of the fixed-base and the interacting systems fall.

The various possibilities are illustrated in the following paragraphs by reference to the base shear induced by a free-field ground motion the pseudoacceleration spectrum of which is represented by the piecewise linear diagram shown in Fig. 2.

1. If both T and $\tilde{T}$ fall in the extremely small-period range of the spectrum (to the left of point b), SSI will have no effect on the response, as the pseudoacceleration value in this case is unaffected by changes in either period or damping.
2. If T falls to the right of point c, SSI will reduce the maximum response, the reduction being a

function of the values of T, $\tilde{T}$, β and $\tilde{\beta}$. An increase in damping under these conditions decreases the pseudoacceleration, whereas an increase in period either does not change it or further decreases it.

3. If T falls in the intermediate period range (between points b and c in Fig. 2), or if T falls to the left of point b and $\tilde{T}$ falls to the right of this point, SSI may increase or decrease the response depending on the values of $\tilde{T}/T$ and $\tilde{\beta}/\beta$. An increase in period in this case increases the response, whereas an increase in damping has the opposite effect.

For ground motions characterized by less smooth, jagged response spectra, the consequenses of SSI cannot be defined as readily, but they can be evaluated by the same reasoning.

4.5. DESIGN PROVISIONS

The concepts described in the preceding sections have provided the basis of the SSI design provisions for building structures which have been recommended by the Applied Technology Council [1] and have been adopted recently by the Building Seismic Safety Council [3] in connection with the National Earthquake Hazard Reduction Program (NEHRP).

Two different sets of procedures are recommended in these provisions. The first involves the use of equivalent static lateral forces, whereas the second is a simplified version of the modal superposition method of analysis. Only the first approach is considered herein, with attention being focused exclusively on the evaluation of the total lateral force.

For buildings analyzed without regard for SSI, the total lateral force or base shear, Q, is expressed as

$$Q = C_s W \tag{12}$$

in which W = the total weight of the building, including the dead weight and the effective portion of the design live load, and C_s = the lateral force coefficient which represents the ratio of the effective spectral acceleration for the system, A, to the gravitational acceleration, g. The factor C_s depends on the following: the effective ground acceleration, $(\ddot{x}_g)_e$, which is an index of the severity of the anticipated seismic activity for the region or site under consideration; a soil factor which accounts for the type of local soil conditions; a response modification factor which is a function of the type of structure and materials involved; and, of course, the fundamental natural period of vibration of the structure, T. The damping of the structure does not appear explicitly in these provisions, but a constant value of 5 percent of critical damping was used in the specification of the design response spectrum.

The variation of C_s with T is represented by a function which is initially constant and then decreases with increasing T. As a result, consideration of soil-structure interaction in this case will reduce the design values of the lateral forces, shears and overturning moments below the levels applicable to a rigid-base condition, and it will be conservative to neglect its effects. Because of the influence of foundation rocking, however, the horizontal displacement of the structure relative to the moving base may increase due to interaction, and this increase will, in turn, increase both the required spacing between buildings and the secondary design forces associated with the P-Δ effects. The latter increases, however, are generally small and have only a minor influence on the final design.

The reduction of the base shear in the ATC-NEHRP design provisions, ΔQ, is defined by the equation

$$\Delta Q = \left[C_s - \tilde{C}_s \left(\frac{0.05}{\tilde{\beta}} \right)^{0.4} \right] \overline{W} \tag{13}$$

in which C_s and $\tilde{C}_s$ = the lateral force coefficients corresponding to the fundamental natural periods of the fixed-base and the elastically supported systems, respectively; $\tilde{\beta}$ = the percentage of critical damping for the structure-foundation-soil system; and $\overline{W}$ = the effective weight of the structure. The latter is taken as 70 percent of the total weight, W, except when the weight is effectively concentrated at one level, in which case it is taken as W.

The difference between the values of C_s and $\tilde{C}_s$ in Eq. 13 provides for the effect of the increased period of the elastically supported structure compared to that of the rigidly supported structure; the factor raised to the 0.4 power accounts for the change in the effective damping of the elastically supported structure from the value of β = 0.05 used in the formulation of the basic design spectrum; and the use of the effective weight, instead of the total weight, provides for the fact that soil-structure affects almost exclusively the response component contributed by the fundamental mode of vibration.

Strictly speaking, the relationship between $\overline{W}$ and W depends on the detailed characteristics of the structure. The constant value of $\overline{W}$ = 0.7W was used in the interest of simplicity and because it is a good appproximation for typical buildings. For example, for buildings with a constant weight per floor level and a fundamental mode of vibration that increases linearly with height, the exact value of $\overline{W}$ = 0.75 W.

The maximum permissible reduction in base shear due to the effects of soil-structure interaction was set at 30 percent of the value calculated for a rigid-base condition. It is expected, however, that this limit will control only infrequently, and that the calculated reduction will normally be less.

4.5.1. *Effective Building Period.* The effective period of the building, $\tilde{T}$, is determined from Eq. 4, with k interpreted as the effective stiffness of the structure when vibrating in its fixed-base fundamental mode, and with h interpreted as the height from the base to the centroid of the inertia forces associated with the fundamental mode of vibration of the fixed-base structure.

Denoted by $\bar{k}$ the effective stiffness is defined by

$$\bar{k} = 4\pi^2 \frac{\overline{W}}{gT^2} \tag{14}$$

and the effective height, $\bar{h}$, is taken as 70 percent of the total building height, except when the total weight is effectively concentrated at a single level, in which case it is taken as the height to that level. Strictly speaking, the relationship between $\bar{h}$ and h depends on the characteristics of both of the structure and the supporting soil. The constant value of $\bar{h}$ = 0.7h is used in the interest of simplicity and consistency with the approximation employed in the definition of $\overline{W}$. For structures with the same weight per floor level and a fundamental mode of vibration that increases linearly with height, the exact value of $\bar{h}$ = (2 / 3) h.

For mat foundations of arbitrary shape, the quantity R in Eq. 5 must be interpreted as the radius of a disk which has the area of the actual foundation, and in Eq. 6, it must be interpreted as the radius of a disk the static moment of which about a horizontal centroidal axis is equal to that of the actual foundation in the direction in which the response is being evaluated.

4.5.2. *Effective System Damping.* The effective damping of the structure-foundation-soil system is determined from Eq. 7, with the structural damping factor taken as $\beta = 0.05$ and the minimum value of $\tilde{\beta}$ taken as 0.05.

The foundation damping factor, β_o, is determined from plots similar to those displayed in Fig. 1. Two sets of curves are included in the recommended provisions. The first, recommended for effective ground accelerations $(\ddot{x}_g)_e \leq 0.10$ g, corresponds to a value of $\tan\delta = 0.05$, wheareas the second, which is recommended for values of $(\ddot{x}_g)_e \geq 0.20$ g, corresponds to $\tan\delta = 0.15$. These curves are based on the results of extensive parametric studies and represent average values.

For mat foundations of arbitrary shape, the radius R in Fig.1 should be interpreted as a characteristic length related to the length of the foundation, L_o, in the direction in which the response of the structure is being evaluated. For short, squatty structures for which $h/L_o \leq 0.5$ the overall damping of the structure-foundation system is dominated by the translational action of the foundation, and R must be determined by equating the areas of the actual and circular foundations. On the other hand, for structures with $h/L_o \geq 1$, the interaction effects are dominated by the rocking motion of the foundation, and R must be determined by equating the static moments of the actual and circular foundations. A linear interpolation is proposed for values in between.

Equation 7 may lead to values of $\tilde{\beta}$ which are smaller than the structural damping factor, β. However, since the value of $\beta = 0.05$ used in the development of the design provisions for rigidly supported structures is presumably based on the results of tests on actual buildings, it reflects the damping of the entire structure-soil system, not merely of the component contributed by the superstructure. Therefore, the value of $\tilde{\beta}$ determined from Eq. 7 should not be taken less than β, and a low bound of $\tilde{\beta} = \beta = 0.05$ has been imposed. The use of values of $\tilde{\beta} > \beta$ is justified by the fact that the experimental values correspond to extremely small-amplitude motions and do not reflect the higher soil damping capacities associated with the design ground motions.

4. 5. 3. *Soil Properties.* The soil properties needed in the evaluation of the interaction effects are likely to vary from point to point of a construction site, and it is necessary to use average values for the region of the soil significantly affected fy the forces acting on the foundation. Furthermore, since the stress-strain relations of soils are nonlinear, the values of G and v_s are functions of the severity of ground shaking and should be determined so as to correspond to the dominant strain levels associated with the design ground motion. The recommended values of G and v_s, expressed in terms of their low-amplitude strain values, are listed in the following as a function of the effective ground acceleration, $(\ddot{x}_g)_e$:

Value of $(\ddot{x}_g)_e$ in g's	$\leq$ 0.10	0.15	0.20	$\geq$ 0.30
Value of G / G_o	0.81	0.64	0.49	0.42
Value of v_s / v_{so}	0.9	0.8	0.7	0.65

The normalizing quantity v_{so} in this tabulation represents the average shear-wave velocity for the soils beneath the foundation at strains of the order 10^{-3} percent or less, and G_o represents the associated shear modulus.

The tabulated values of G / G_o and v_s / v_{so} represent first-order approximations. More precise evaluations would require field tests or laboratory tests on undisturbed samples from the site, as well as studies of wave propagation for the site to determine the magnitudes of the maximum soil strains associated with the stipulated design ground motion.

5. Kinematic Interaction Effects

5.1. EFFECTS OF WAVE PASSAGE

The structure in this section is presumed to be excited by a horizontally polarized, plane shear wave propagating in a direction that makes an angle α with the vertical. In addition to a horizontal displacement component, $x_i(t)$, the foundation input motion (FIM) in this case includes a torsional component, $\theta_i(t)$, about a vertical axis. These displacements refer to the foundation motion of the massless foundation-structure system, and should not be confused with those of the actual system with mass, x(t) and $\theta(t)$. For convenience, the torsional component of the FIM will be expressed by the circumferential displacement along the periphery of the foundation, $y_i(t) = R\theta_i(t)$. The absolute maximum values of these displacements will be denoted by x_i, θ_i, and y_i, and of the associated velocities and accelerations will be identified by one and two dot superscripts, respectively.

If use is made of the averaging technique of Iguchi [4] and Scanlan [9], in which the restraining action of the supporting medium is effectively represented by a series of Winkler springs, it can be demonstrated that the interrelationship of the FIM and the CPM is governed by the time τ^* required for the wave front to traverse the foundation radius. Referred to as the effective transit time, this quantity is given by

$$\tau^* = \frac{R}{c} = \frac{R}{v_s}\sin\alpha = \tau \sin\alpha \tag{15}$$

in which v_s = the velocity of shear-wave propagation in the medium, $c = v_s / \sin\alpha$ = the apparent horizontal velocity of wave propagation, and $\tau = R / v_s$. For a harmonic motion of circular frequency ω, the parameters ω and τ^* appear in the solution as a product, and the foundation input motion in this case is defined by the dimensionless frequency parameter

$$a_o^* = \omega\tau^* = a_o \sin\alpha \tag{16}$$

in which $a_o = \omega R / v_s$ = the well known dimensionless frequency parameter used in studies of foundation dynamics. For $\tau^* = a_o^* = 0$, all points of the foundation are excited simultaneously, the FIM = CPM, and there is no kinematic interaction.

5.1.1. *Foundation Input Motion.* The effect of a_o^* on the amplitude of the FIM for harmonically excited systems is displayed in part (a) of Fig. 3. The results are normalized with respect to the amplitude of the free-field CPM. As would be expected, the horizontal component of the FIM generally decreases with increasing a_o^*, the reduction being particularly significant for values of a_o^* of the order of 2 or greater. The torsional component of the FIM, on the other hand, first increases and then decreases.

For the harmonic motion considered, the temporal variations of its displacement, velocity and acceleration traces are identical, and the displacement ratios in Fig. 3(a) may also be interpreted as velocity or acceleration ratios. By contrast, for an actual earthquake ground motion, for which the frequency contents of the acceleration, velocity and displacement traces are different, the peak displacements of the FIM will be related to the peak displacement of the CPM differently from the corresponding velocity and acceleration values. This is demonstrated in part (b) of Fig. 3 which shows the results obtained for the CPM corresponding to the first 6.24 sec. of the N-S component of the 1940 El Centro, California earthquake. The acceleration, velocity and displacement traces of this record are available in [20].

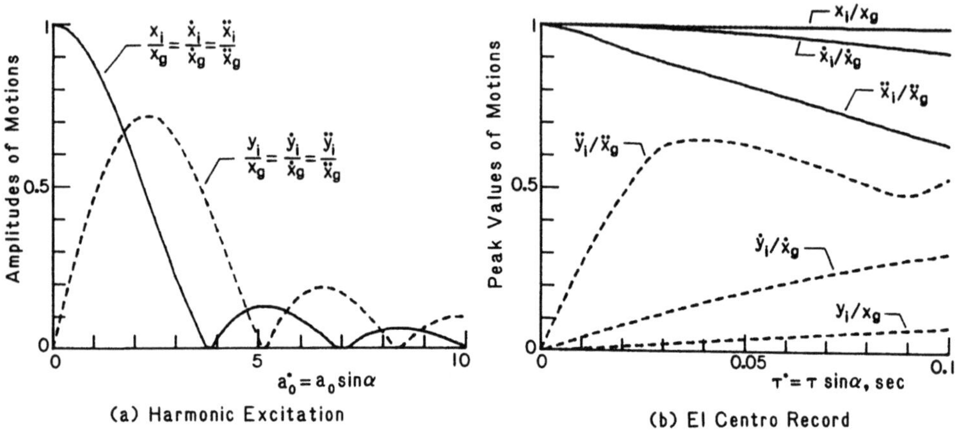

Fig. 3 Peak foundation input motions for obliquely incident plane shear waves

The following observations may be made and inferences drawn from the data presented in Fig. 3:

1. The reduction in the horizontal component of the foundation input motion and the corresponding increase in the torsional component are greatest for acceleration, much smaller for velocity, and almost negligible for displacement. Since the foundation filters the high-frequency wave components more effectively than the low-frequency components, the acceleration traces of the ground motion, which are richer in high-frequency content than the velocity and displacement traces, are influenced more than the latter traces.
2. Considering that high-frequency systems are acceleration-sensitive whereas low-frequency systems are displacement-sensitive, it should be clear that the effects of kinematic interaction on the lateral component of response would be important for high-frequency systems and inconsequential for low-frequency systems. Furthermore, medium-frequency systems, which are velocity-sensitive, would be expected to be affected moderately. That this is indeed the case is confirmed by the data presented in the next section.

5.2. STRUCTURAL RESPONSE

Let U_x be the absolute maximum value of the structural deformation induced by the lateral component of the FIM, and U_y be the corresponding deformation induced at the periphery of the deck by the torsional component. Fig. 4 presents response spectra for these deformations considering damping factors $\beta_x = \beta_\theta = 0.02$. The results are displayed in logarithmic scales, with the abscissa representing the natural frequency of the system for the particular mode of vibration considered, and the ordinates representing the pseudovelocity values, $V_x = p_x U_x$ and $V_y = p_\theta U_y$, normalized with respect to the peak velocity of the CPM, $\dot{x}_g$. Three different values of τ^* are considered, including the limiting value of $\tau^* = 0$, for which there is no kinematic interaction.

As already anticipated, the effect of wave-passage on the lateral component of the response is greatest for high-frequency systems, significantly smaller for medium-frequency systems, and negligible for low-frequency systems. Furthermore, both the magnitudes and the trends of the response spectra for torsional response are consistent with those expected from the torsional components of the FIM presented in Fig. 3(b).

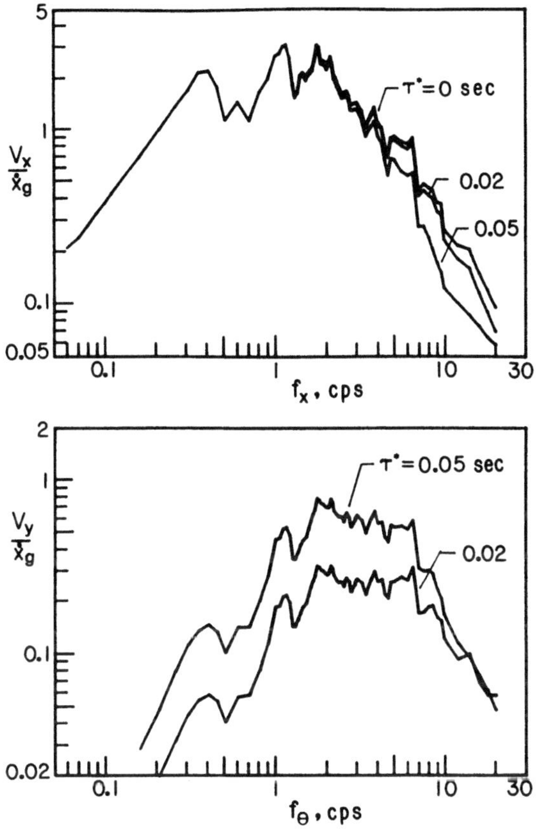

Fig. 4 Effect of wave passage on maximum lateral and torsional deformations

5.3. EFFECTS OF GROUND MOTION INCOHERENCE

The free-field ground motion in this section is considered to be due to horizontally polarized, vertically propagating, incoherent shear-wave trains. Purely horizontal but differing from point to point of the foundation-soil interface, the motion in this case cannot at the present state of knowledge be specified in deterministic terms. It is, instead, specified stochastically in terms of a local power spectral density (psd) function, $S_g(\omega)$, and a cross psd function, $S(\vec{r}_1, \vec{r}_2, \omega)$, in which ω = the circular frequency of the harmonic component of the motion under consideration; and $\vec{r}_1$ and $\vec{r}_2$ are position vectors for two arbitrary points on the foundation-soil interface. The cross psd function may be expressed as

$$S(\vec{r}_1, \vec{r}_2, \omega) = \Gamma(|\vec{r}_1 - \vec{r}_2|, \omega)\, S_g(\omega) \tag{17}$$

in which Γ, referred to as the incoherence function, is a dimensionless, decreasing function of ω and of the distance between points, $|\vec{r}_1 - \vec{r}_2|$. Several different expressions have been proposed for

Γ, and there is no general agreement at this time on the form that may be the most appropriate for realistic earthquakes. In this study, Γ is taken in the form recommended by Luco and Mita [6] as

$$\Gamma(|\vec{r}_1 - \vec{r}_2|, \omega) = \exp\left[-\left(\frac{\gamma\omega|\vec{r}_1 - \vec{r}_2|}{v_s}\right)^2\right] \tag{18}$$

in which γ is a dimensionless factor with a value typically between zero and 0.5.

For a free-field harmonic motion of circular frequency ω and spatially varying amplitudes and phases, the interrelationship between the CPM and FIM depends on the dimensionless frequency parameter

$$\tilde{a}_o = \gamma\, a_o = \gamma\, \frac{\omega R}{v_s} \tag{19}$$

whereas for a transient motion of arbitrary temporal variation, it depends on the dimensional effective transit time

$$\tilde{\tau} = \gamma\, \tau = \gamma\, \frac{R}{v_s} \tag{20}$$

These statements hold true under the same simplifying assumption as that made in the evaluation of the wave passage effects.

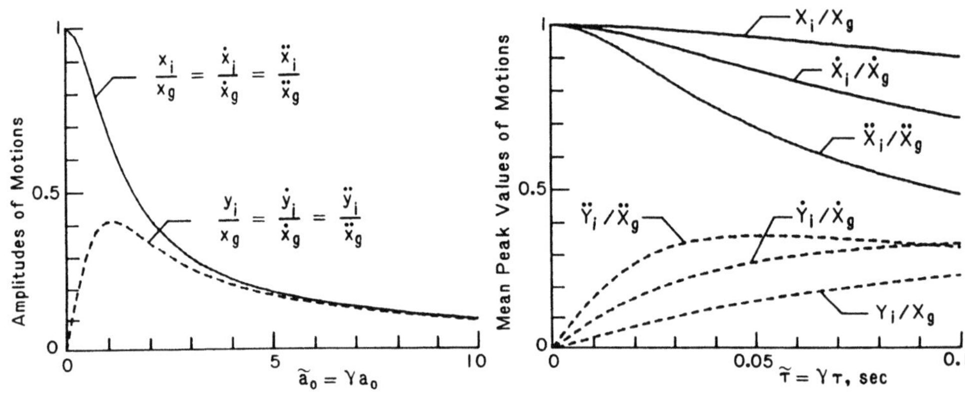

Fig. 5 Peak foundation input motion for vertically incident incoherent waves

In part (a) of Fig. 5 are shown the normalized amplitudes of the horizontal and torsional components of the FIM for harmonic free-field excitations, and in part (b) are given corresponding data for an ensemble of transient control point motions for which the psd function of the acceleration histories is of the form considered in [5] and [18]. The cut-off frequency of the latter function is taken as 8 cps in the present study. The quantities X_i and Y_i in Fig. 5(b) represent the ensemble means of the absolute maximum displacement values of the horizontal and circumferential components of the FIM, and X_g represents the corresponding value of the CPM. One dot and two dot superscripts identify the mean peak values of the corresponding velocity and displacement histories. These results were computed by use of Der Kieureghian's empirical relations [2] considering the duration of the intense portion of the excitation to be 20 sec.

The plots in Fig. 5 are similar to those presented in Fig. 3 for an obliquely incident plane wave, and the conclusions and inferences drawn from the previous plots also apply in this case. This is confirmed by the solid curves in Fig. 6, which represent response spectra for the ensemble means of the absolute maximum values of the lateral deformations, $\hat{U}_x$, and of the circumferential deformations, $\hat{U}_y$, induced along the periphery of the structure by the torsional component of the foundation input motion. The damping values in these solutions are taken as $\beta_x = \beta_\theta = 0.02$. As before, the results are expressed in terms of normalized pseudovelocity values, in which $\hat{V}_x = p_x\hat{U}_x$ and $\hat{V}_y = p_\theta\hat{U}_y$. The reductions in the lateral component of the response due to the incoherence of the ground motion, and the general trends of the spectra for circumferential response are indeed fully compatible with the peak values of the FIM presented in Fig. 5(b).

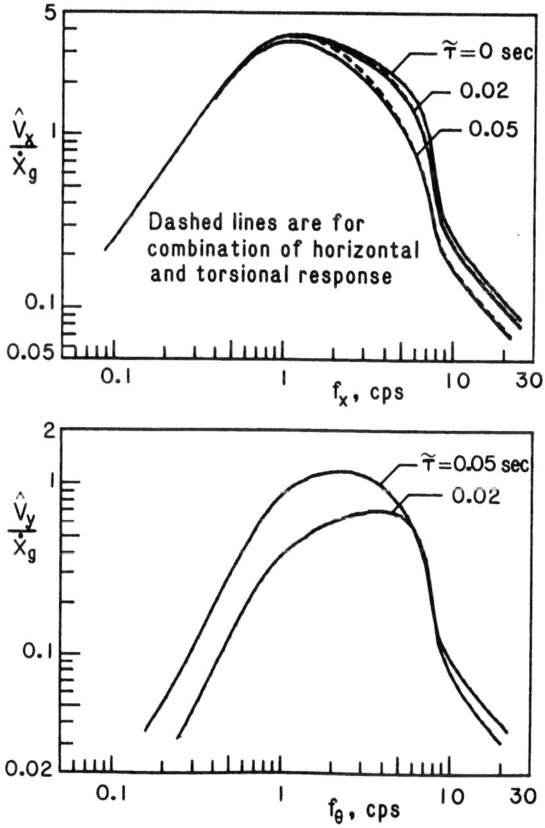

Fig. 6 Effect of ground motion incoherence on mean peak values of lateral and torsional deformations

Also shown in the upper part of Fig. 6 by dashed lines are the mean peak values of the sum of the lateral and torsional components of deformation along the periphery of structures for which f_θ

= 1.5 f_x. The contribution of the torsional component of response is clearly small in this case.

5.4. DESIGN IMPLICATION OF RESULTS

The principal step in the evaluation of the kinematic interaction effects is the computation of the lateral and torsional components of the foundation input motion for the massless structure-foundation system. For both plane and incoherent wave fields, these motions may be computed by Fourier transform techniques, making use of the foundation transfer functions presented in Figs. 3(a) and 5(a). Of particular importance in this regard are the peak (or mean peak) values of the acceleration, velocity and displacement histories of the motions. With these values established, the maximum response of the interacting system may be approximated by application of well established rules relating these input values to the characteristics of response spectra The lateral and torsional effects should be evaluated separately and then combined by the RMS rule, or an appropriate variant of it, giving due regard to the relative values of f_x and f_θ.

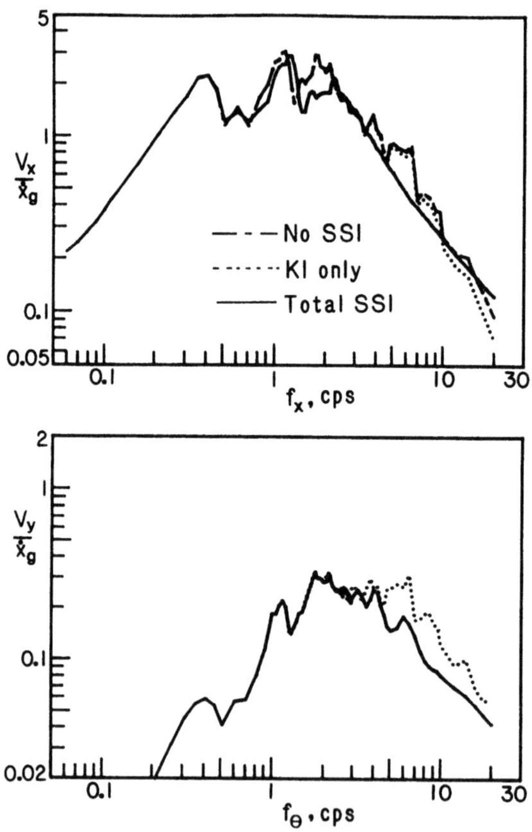

Fig. 7 Response spectra for interacting structures with h/R = 2; obliquely incident El Centro record; $\sin\alpha = 0.4$, $\tau = 0.05$ sec

6. Total Soil-Structure Interaction Effects

As a measure of the relative importance of kinematic and inertial interaction effects, in Figs. 7 and 8 are shown response spectra for the lateral and circumferential responses obtained for systems with a height to base-radius ratio, h/R = 2. The plots in Fig.7 refer to the obliquely incident plane wave considered in Figs. 3 and 4, whereas those in Fig.8 are for the vertically propagating, incoherent wave field considered in Figs. 5 and 6. The principal parameters for these solutions are identified on the figure headings, and the remaining parameters are the same as those for the solutions presented in [18].

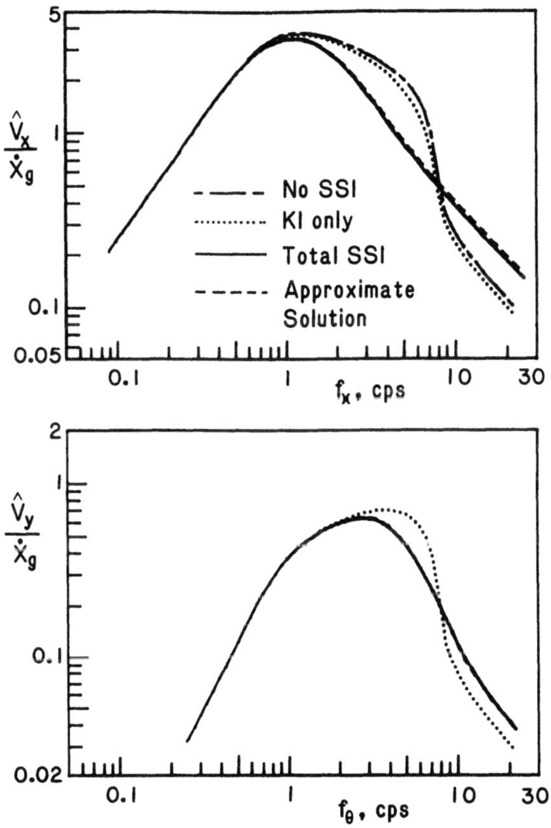

Fig. 8 Response spectra for interacting structures with h/R = 2; vertically incident incoherent waves, $\gamma = 0.4$, $\tau = 0.05$ sec.

Three sets of solutions are displayed in each case: (1) Making no provision for soil-structure interaction, i.e., considering the foundation motion to be the same as the free-field control point motion; (2) providing only for the kinematic interaction effects, i.e., using as base excitation the foundation input motion and analyzing the superstructure without regard for the flexibility of the supporting medium; and (3) providing for both kinematic and inertial interaction effects, i.e., analyzing the structure-foundation-soil system exactly as a coupled system. Also included in Fig. 8

are the results obtained by the simplified analysis presented in the early part of the paper. The base excitation in these solutions was taken equal to the foundation input motion, and the natural frequency and damping of the system for each mode of vibration were modified appropriately.

From these data and additional solutions reported in [18] and [20], the following conclusions are drawn:

1. Like kinematic interaction, inertial interaction influences most the responses of systems in the medium- and high-frequency spectral regions.
2. The II effects are generally more important that the KI effects.
3. Unlike kinematic interaction which generally reduces the lateral response, inertial interaction may reduce or increase the corresponding response.
4. The interaction effects for low-frequency, highly compliant structures are negligible because such systems "see" the supporting halfspace as a very stiff, effectively rigid medium.
5. The concept of modifying the fixed-base natural frequencies and the associated damping values of the system provides a simple and highly reliable means for assessing the inertial interaction effects.

7. Conclusion

The information and concepts presented herein provide simple, practical means for assessing the effects of soil-structure interaction in the seismic design of structures. Although presented for relatively simple structures and somewhat idealized conditions, the concepts involved are applicable to more complex systems as well. As a matter of fact, these concepts have been used effectively in studies of a variety of structures, including buildings, nuclear containment structures, offshore structures, and liquid-containing tanks. However, there continues to be a need for much additional research. Topics requiring further study include the behavior of embedded and underground structures, the behavior of pile-supported structures, and the interaction effects for structures responding in the inelastic range of deformation.

8. Acknowledgment

This contribution is based in part on Ref. [19] and in part on Refs. [12] and [13].

9. References

1. Applied Technology Council (1978), Tentative Provisions for the Development of Seismic Regulations for Buildings, ATC-3-06, Palo Alto, California.
2. Der Kiureghian, A. (1980), 'Structural response to stationary excitation', J. Engrg. Mech., ASCE, 106(6), 1195-1213.
3. Federal Emergency Management Agency (1986), NEHRP Recommended Provisions for the Development of Seismic Regulations for New Buildings, Building Seismic Safety Council, 1015 15th Street, N.W., Suite 700, Washington, D. C., 20005.
4. Iguchi, M. (1984), 'Earthquake response of embedded cylindrical foundation to SH and SV waves', Proc. 8th World Conf. Earthq. Engrg., San Francisco, California, 1081-1088.
5. Kausel, E., and Pais, A. (1987), 'Stochastic deconvolution of earthquake motions', J. Engrg. Mech., ASCE, 113(2), 266-277.

6. Luco, J. E., and Mita, A. (1987), 'Response of a circular foundation to spatially random ground motion," J. Engrg. Mech., ASCE, 113(1), 1-15.
7. Meek, J. W. and Veletsos, A. S. (1973), 'Simple models for foundations in lateral and rocking motion', Proc. 5th World Conf. Earthq. Engrg., Rome, Italy, 2, 2610-2613.
8. Nair, V. V. D. (1974), Dynamics of Certain Structure-Foundation Systems, Ph.D. thesis submitted to Rice Univ., Houston, Texas.
9. Scanlan, R. H. (1976), 'Seismic wave effects on soil-structure interaction', Earthq. Engrg. and Struct. Dyn. 4, 379-388.
10. Veletsos, A. S. (1977), 'Dynamics of structure-foundation systems', in W. J. Hall (ed.) Structural and Geotechnical Mechanics, Prentice-Hall, Inc., Englewood Cliffs, New Jersey, pp. 333-361.
11. Veletsos, A. S. (1978), 'Soil-structure interaction for buildings during earthquakes', Proc. 2nd Int. Conf. on Microzonation, San Francisco, California, 1, 111-133.
12. Veletsos, A. S. (1981), 'Seismic design provisions for soil-structure interaction', in O. Ergunay and M. Erdik (eds.) State of the Art in Earthquake Engineering, Turkish Natl. Com. Earthq. Engrg., pp. 236-248.
13. Veletsos, A. S. (1986), 'Some perspectives on dynamics of soil-structure interaction' in H. L. Graves and A. J. Philippacopoulos (eds.) Proc. Workshop on Soil-Structure Interaction, NUREG/CP-0054, BNL-NUREG-52011, Superintendent of Documents, U. S. Printing Office, Washington, D. C., pp. 213-232.
14. Veletsos, A. S., and Meek, J. W. (1974), 'Dynamic behaviour of building-foundation systems', Earthq. Engrg. and Struct. Dyn. 3, 121-138.
15. Veletsos, A. S. and Nair, V. V. D. (1974), 'Torsional vibration of viscoelastic foundations', J. Geotech. Engrg., ASCE, 100(3), 225-246.
16. Veletsos, A. S. and Nair, V. V. D. (1974), 'Response of torsionally excited foundations', J. Geotech. Engrg., ASCE, 100(4), 476-482.
17. Veletsos, A. S. and Nair, V. V. D. (1975), 'Seismic interaction of structures on hysteretic foundations', J. Struct. Div., ASCE, 101(1), 109-129.
18. Veletsos, A. S. and Prasad, A. M. (1988), 'Seismic interaction of structures and soils: stochastic approach', J. Struct. Eng., ASCE, 115(4), 935-956.
19. Veletsos, A. S., Prasad, A. M. and Tang, Y. (1988), 'Design approaches for soil-structure interaction', Proc. 9th World Conf. Earthq. Engrg., Tokyo, Japan, VIII, 341-352.
20. Veletsos, A. S. and Tang, Y. (1990), 'Deterministic assessment of effects of ground motion incoherence', J. Engrg. Mech., ASCE, 116(5), 1109-1124.
21. Veletsos, A. S. and Verbic, B. (1973), 'Vibration of viscoelastic foundations', Earthq. Engrg. and Struct. Dyn. 2(1), 87-102.
22. Veletsos, A. S. and Verbic, B. (1974), 'Basic response functions for elastic foundations', J. Engrg. Mech., ASCE, 100(2), 189-202.
23. Veletsos, A. S. and Wei, Y. T. (1971), 'Lateral and rocking vibration of footings', J. Soil Mech. and Found. Div., ASCE, 97, 1227-1248.
24. Verbic, B. (1972), Analysis of Certain Structure-Foundation Interaction Systems, Ph.D. thesis submitted to Rice Univ., Houston, Texas.
25. Wolf, J. P. (1988), Soil-Structure-Interaction Analysis in Time Domain, Prentice-Hall, Englewood Cliffs, N.J.
26. Wolf, J. P. and Somaini, D. R. (1986), ' Approximate dynamic model of embedded foundation in time domain ', Earthq. Engrg. and Struct. Dyn. 14(5), 683-703.

DYNAMIC CHARACTERISTICS OF QUATERNARY SAND-GRAVEL DEPOSITS

M. WATABE[*1)], K. MUROI[*2)], K. FUKUTAKE[*3)]
*1) Senior Managing Director,
*2) Nuclear Power Division, *3) Ohsaki Research Institute
Shimizu Corporation
Seavans South No.1-2-3 Shibaura Minatoku Tokyo, 105-07
Japan

ABSTRACT. There had been few researches and experiments concerned with the dynamic characteristics of sand-gravel, so far. But the number of report concerned with the dynamic characteristics of sand-gravel are increasing gradually now. This text provides a brief introduction to the basic dynamic characteristics of sand-gravel : initial shear modulus, cyclic deformation property, liquefaction resistance, settlement of ground, based on the results of large-scale field tests and on the recent researches of the in-situ frozen undisturbed sand-gravel samples, both of which have recently been conducted in Japan. And this text describes the dynamic properties of sand-gravel deposit in comparison with those of sand and clay.

1. Introduction

In case a large, heavy, and important structure such as a nuclear reactor building is to be constructed on relatively soft soil as compared with rock site, it is necessary to determine the dynamic characteristics, including the non-linear properties, of that soil, in order to anticipate the dynamic behavior of soil and structure during large earthquakes. Many data have been accumulated concerned with the dynamic characteristics of sand and clay, but there have been very few data concerned with those of sand-gravel. This is because in sand-gravel the particle diameter is far larger than that of sand, and therefore, large-sized laboratory test apparatus are required so that the dynamic characteristics of sand-gravel may be estimate with satisfactory accuracy. It is also technically difficult to obtain undisturbed samples of sand-gravel.

Recently, some undisturbed samples can be obtained in Japan, by the in-situ freezing sampling method. These samples were tested using a large-sized triaxial test apparatus. Furthermore, large-scale field test of sand-gravel deposits has also been carried out, and the quantitative evaluation of the dynamic characteristics of sand-gravel is now underway.

This text describes the large-scale field tests recently carried out in Japan, and gives a brief outline of the basic dynamic characteristics (shear modulus, damping factor, liquefaction resistance, and volumetric strain) determined through the recent research on sand-gravel.

P. Gülkan and R. W. Clough (eds.), Developments in Dynamic Soil-Structure Interaction, 327–348.

2. Large Scale Field Tests of Sand-Gravel Deposit

2.1. PENETRATION TEST

The type of soil investigation most commonly carried out in Japan is the standard penetration test (SPT). The N-values obtained from SPT are said to indicate the basic characteristics of the soil. When this SPT is applied to the investigation of sand-gravel, however, the test-result valued for physical characteristics are considered to be greater than those of the actual physical characteristics of sand-gravel. Because the large particles of gravel sometimes collide with the sampler, of which inner diameter is as small as 35 mm.

The inner diameter of the sampler was therefore increased to 50 mm, and the percussion energy was increased to approximately 1.5 times the previous value, then the large penetration test (LPT) was carried out using this larger sampler. Fig.1 [1] shows the results of the comparison of N-values obtained from the SPT and the Nd- values obtained from the LPT. As shown in Fig. 1, the Nd-value from the LPT has changed even in the layer in which the N-value from SPT is 50 or more, so it can be considered that the Nd-value becomes the index which shows the hardness of the sand-gravel deposit.

A few formulas have been proposed for estimating the engineering characteristics of sand and clay from the N-values obtained by SPT, but no conversion formula for estimating those of sand-gravel from the Nd-values obtained by LPT, has yet been established from the quantitative viewpoint, because of the insufficient amounts of data obtained so far. In future, when many data regarding the relationship between the Nd-values and the engineering characteristics of sand-gravel have been accumulated, it should be possible to estimate the engineering characteristics of sand-gravel from the Nd value.

Fig. 1 also shows a comparison of the test results for the surface layer, before and after excavation, clearly indicating the tendency for the N-values to become smaller after excavation. This result means that the effective confined pressure of the sand-gravel deposit has been decreased by the excavation of the surface layer, and that the bearing characteristics of the sand-gravel have been changed. These matters will be discussed later.

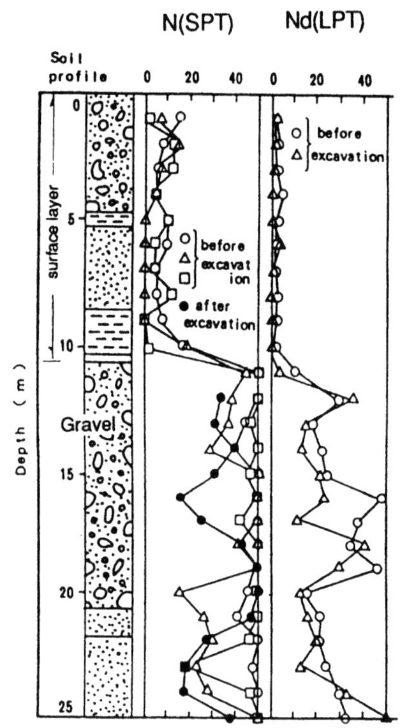

Fig.1 Comparison with SPT and LPT

2.2. IN-SITU FREEZING SAMPLING METHOD

To obtain undisturbed samples is desirable, in order to evaluate the realistic engineering characteristics of sand and sand-gravel ground. Also, the diameter of the individual particle forming gravel is larger than that of sand, and therefore, the more accurate evaluation can be realized if the size of the sample obtained is larger one. There are some researches [1), 2), 3), 4)] in Japan to evaluate the characteristics of soil by conducting a series of soil tests, using undisturbed samples which are thawed out in the test laboratory, after the obtaining of each undisturbed sample by freezing the soil in the test field (in-situ freezing sampling method), although each of these research cases has adopted uniquely-designed details for its method of freezing the soil, the main procedure in the freezing sampling method is that the ground is frozen using liquid nitrogen, and a column-shaped frozen sample is obtained, using a sampler which has a diamond cutter attached to its tip.

Fig. 2[4], 3[3], and 4[1] show comparisons of the test results for samples obtained by the ordinary sampling method, and those for samples obtained by the in-situ freezing sampling method, in the relationship between liquefaction resistance and repeating frequency, where σ_m' is the mean effective confining pressure and σ_d is the amplitude of cyclic loading axial stress. Fig. 2 shows the test results for sand, and Figs. 3 and 4 show the test results for sand-gravel. It is apparent that the results for disturbed samples are much different from those obtained from undisturbed samples. Accordingly, it is obvious that the obtaining of undisturbed samples is important, so that appropriate evaluation of the engineering characteristics of sand and sand-gravel can be carried out.

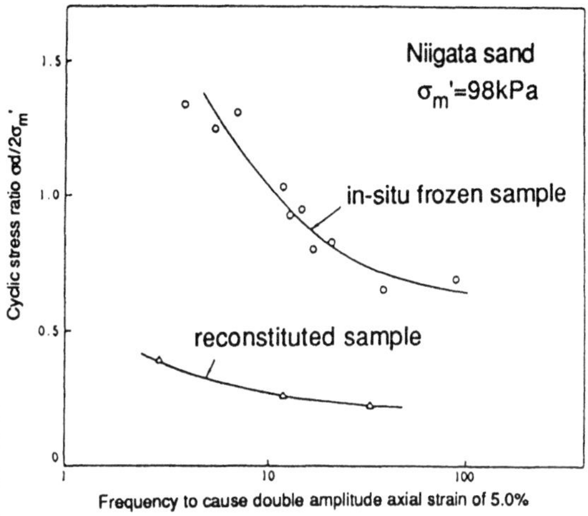

Fig.2 Undrained Cyclic Strength of Sand

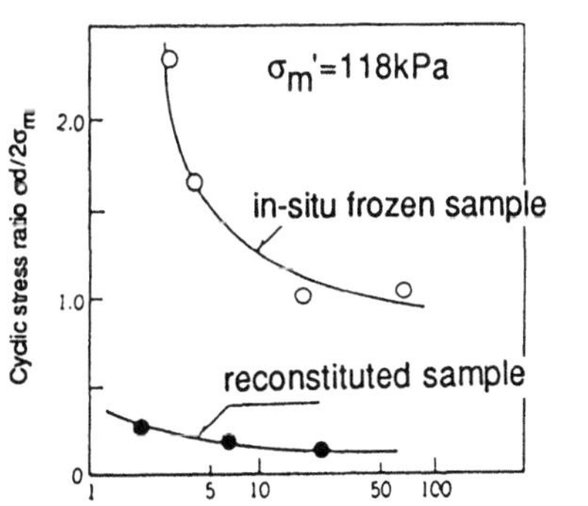

Fig.3 Undrained Cyclic Strength of Gravel against Liquefaction

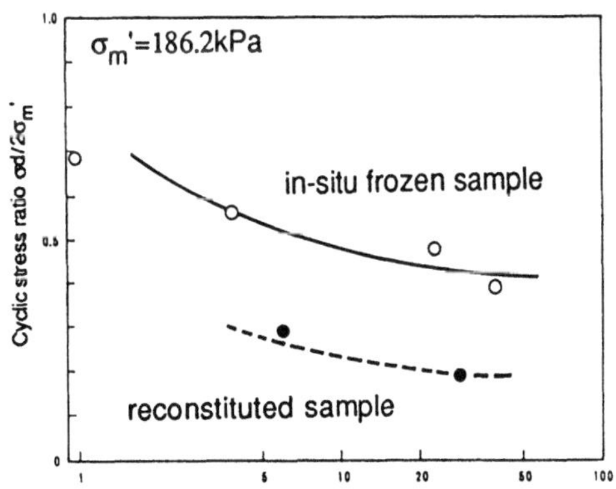

Fig4. Undrained Cyclic Strength of Gravel against Liquefaction

2.3. IN-SITU SOIL COLUMN TEST

In previous section, it has been pointed out that laboratory test using undisturbed samples obtained by the in-situ freezing sampling method is an effective technique for investigating the realistic engineering characteristics of sand-gravel. Due to the restriction imposed by the size of the apparatus, however, each in-situ frozen sample is a column with a diameter of 30 cm and a height of 60 cm.

Therefore, a full scale in-situ soil column tests, in which the sand-gravel ground shown in Fig. 5 were tested directly by applying pressure and shear force using a huge sand-gravel samples and apparatuses was implemented for the purpose of obtaining the realistic engineering characteristics of sand-gravel. Dynamic and static torsional shearing stress directly to the sand-gravel ground in its original condition can be obtained through these full-scale tests. Therefore, these series of tests can be regarded as to be a huge triaxial test. Two huge samples, each with a diameter of 10 m, and with heights of 9 m and 5 m respectively, were tested.

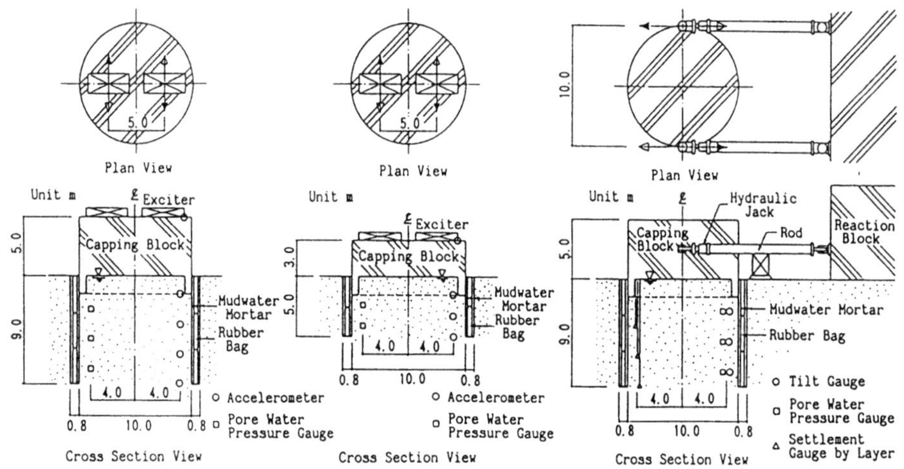

(a) 9m dynamic loading (b) 5m dynamic loading (c) 9m static loading

Fig.5 In-situ Soil Column Test

The following procedures were used to form up these samples.

(1) Dig and remove the surface layer (11m) to expose the sand-gravel deposit which is to be subjected to testing.
(2) Excavate 2 ring-shaped trenches, each with a diameter of 10 m and a width of 80 cm, to depths of 5m and 9m respectively, to prepare the columnar sand-gravel samples.
(3) Protect the sides of the sand-gravel samples with mortar which has been prepared in an appropriate strength and which may effect very little the rigidity of the sand-gravel samples. This mortar is used not only to prevent the collapse of the samples but also to prevent water draining from the sides of the samples.
(4) Insert a rubber bag around the entire circumference of each sand-gravel sample, in the ring-shaped trench.
(5) Inflate the rubber bag, using water, to apply pressure to the sand-gravel samples through the mortar from the side of each sample. The pressure simulates the lateral pressure which acted on

the sand-gravel samples when these were in their original condition.

(6) Put reinforced-concrete capping blocks on top of each sample. The weight of each capping block simulates the overburden pressure which so acted on each sand-gravel sample as in its original condition."

To apply to these soil column specimens the cyclic torsional-shearing stress through the entire specimens, the capping block was twisted dynamically or statically. According to the report, no considerable slipping occurred between the capping block and the soil sample, so the shearing stress may have been applied to the entire specimen. And it was reported that the torsional shearing stress generated had exceeded the maximum of 10^{-4} when dynamic stress was applied, and had exceeded the maximum of 10^{-3} when static stress was applied. It had also been reported that compressive deformation of samples had been observed because the excess pore water pressure in the sample had changed, linked with the pressure-application pattern, as shown in Fig. 6, at the level where the shearing strain was in the order of 10^{-3} when static stress was applied to the sample.

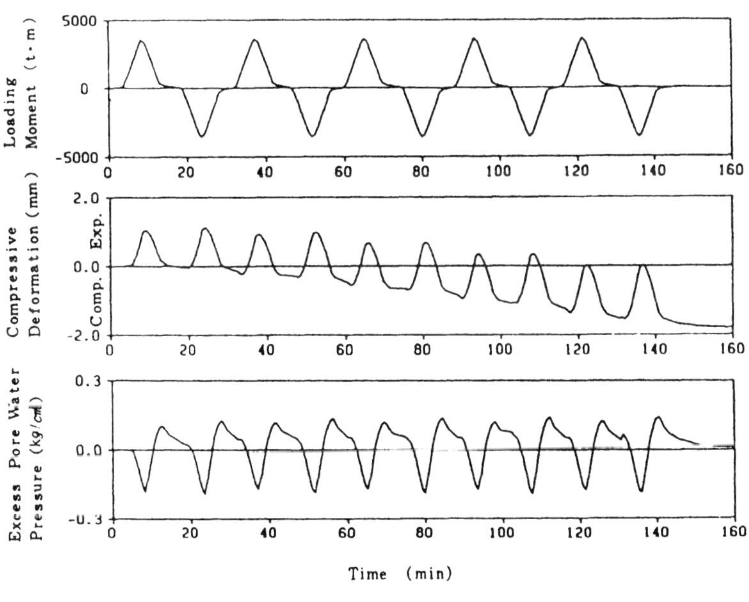

Fig.6 Time History of Loading Moment, Compressive Deformation and Excess Pore Water Pressure for Static Load Test of 9m Soil Column

2.4. CONCRETE BLOCK TEST

In order to evaluate of the dynamic soil-structure interaction, for Nuclear Power Plant, a concrete block is placed directly on the ground, and dynamic vibration is applied to the block, occasionally in Japan.

The report state that the concrete block with which weight to provide soil pressure of 470KPa, which was approximately the equivalent pressure to the ground as that for a standard nuclear reactor building in Japan, as shown in Fig. 7[1], was placed directly on the surface layer of the same sand-gravel deposit where the in-situ soil column testing was implemented. Dynamic and static cyclic loadings were applied to the large-scale concrete blocks.

Fig. 8[1] shows the Fourier spectrum for the microtremor record of block A. Accordingly, it was found that the predominant frequency of the ground is clearly shown to be 1.0 Hz, and the natural frequency of the block is shown to be 3.4 Hz.

Fig. 9[1] shows the resonance curve obtained from the dynamic vibration test. The resonance frequency is 3.25 Hz, and the stiffness of ground has become slightly softer than that estimated by the microtremor record. The report[1] also says that the nonlinear properties of the ground become more obvious as the vibration force applied is increased.

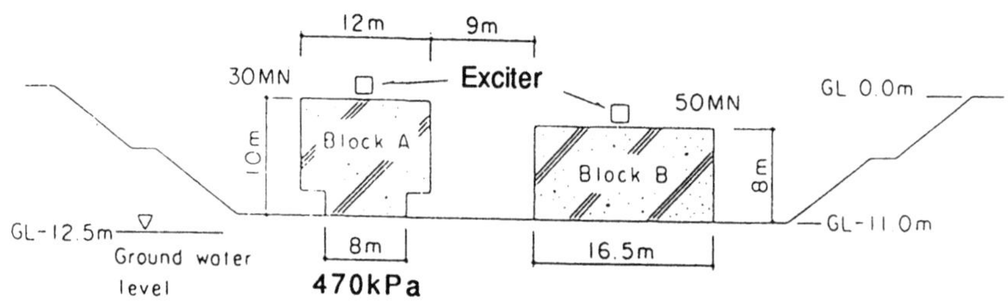

Fig.7 Section of Concrete Block Test

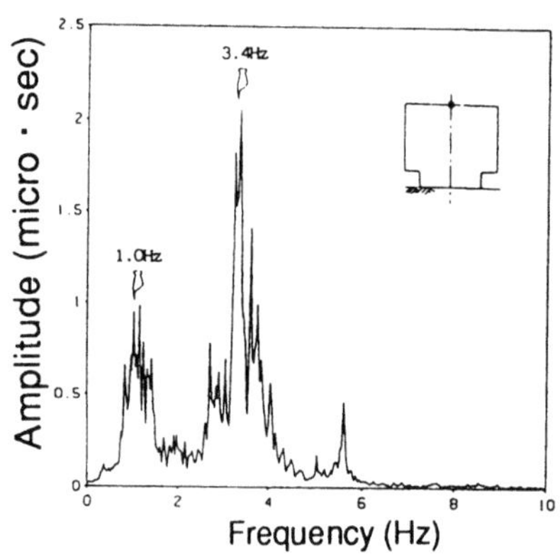

Fig.8 Fourier Spectrum of Microtremor

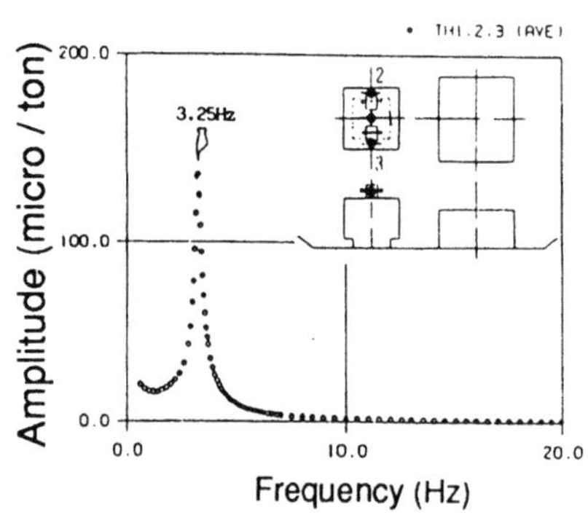

Fig.9 Resonance Curve of Block A

The report also stated that horizontal slip and setting were measured clearly by the static loading test, as shown in Fig. 10. Ground slip and settlement were measured when the strain level was as large as 10^{-3}, according to this report. When simulating the dynamic interaction between a sand-gravel deposit and a structure from the results of concrete block testing, the non-linear properties of the sand-gravel was one of the indispensable factors.

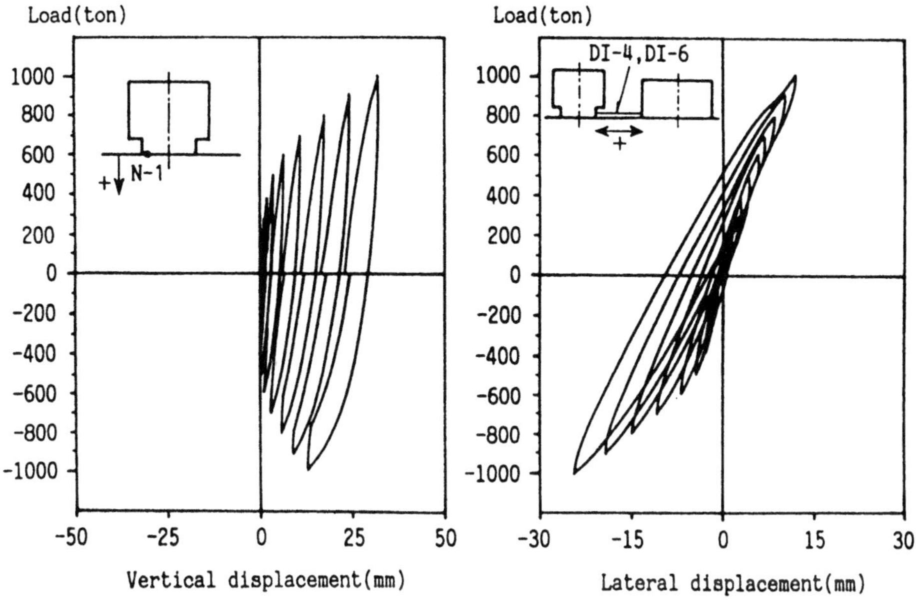

Fig.10 Load - Displacement Orbit of Concrete-block
(Fifth Cycle of Each Loading Level)

3. Dynamic Characteristics of Sand-Gravel

When carrying out dynamic analysis with consideration of the nonlinear properties of the ground, the parameters of soil properties specific to nonlinear analysis must be determined. These parameters are: the soil variety (sand, sand-gravel, clay, etc.), grain size distribution, density (relative density and void ratio), and stiffness (N-value or shear wave velocity). Also, when estimating the liquefaction of soil, test data for the liquefaction resistance curve etc. are needed, in addition to the above-mentioned data.

Through some recent researches for sand-gravel, data of dynamic characteristics of sand-gravel is being accumulated. And the comparison of dynamic characteristics of sand-gravel with those of sand and clay may also become possible as introduced following ;

3.1. INITIAL SHEAR MODULUS, Go

Factors which affect the shear modulus, G, of soil, are mainly: the sharing strain level, γ,the void ratio, e, the effective confined pressure, σ_m', soil variety (grain size distribution), and shear strength. Generally, G is evaluated by dividing it into a non-linear property portion dependent on γ, and an initial-condition portion, depending on σ_m'' and e. The initial shearing modulus, Go, where the strain level is 10^{-5} or less, will be mentioned here. At this lower strain level, the soil may be considered linear.

3.1.1. Influence of void ratio, and effective confined pressure . It gradually dawned that the relationship between the effective confined pressure, σ_m, void ratio, e, and the initial shear modulus, Go, of sand, can be defined by the following suggested experimental formula.

$$Go = A \cdot F(e) \cdot (\sigma_m')^n$$

Where, A and n are constants, and F(e) is the function of "e". Also, for normally-consolidated clay that has an over consolidated ratio of 1.0, the relation can be defined using the same formula. According to recent researches, for definition of the relationship between the above-mentioned factors for sand-gravel this formula in the same form can be applied. The parameters of formula for standard sand are shown as in Table 1, that for sand-gravel, as shown in Table 2, and that for normally-consolidated clay, as shown in Table 3. According to these tables, when the influence of the effective confined pressure on initial shear modulus is considered, most of the constants, n, for sand and clay are around 0.5. Meanfile, for sand-gravel, the value of the constant, n, seems to be slightly greater, around 0.6.

Fig. 11[1] and Fig. 12[3] show the relationship between effective confined pressure and Go, using reconstituted sample of sand-gravel, and undisturbed sample of sand-gravel obtained by the in-situ freezing-sampling method. According to the figures, the constant, n, shown for the non-disturbed sample of sand-gravel, are around 0.8 to 0.85, higher than 0.5, which was the constant, n, of the dependency of the effective confined pressure resulted from the disturbed sample. As shown in Fig. 1 of soil profile, the results for the N-value of the sand-gravel underneath the surface soil have changed, from before the excavation to after the excavation. From these results, it can be assumed that the influence of the effective confined pressure cannot be ignored for evaluation of the initial shear modulus of sand-gravel, Go.

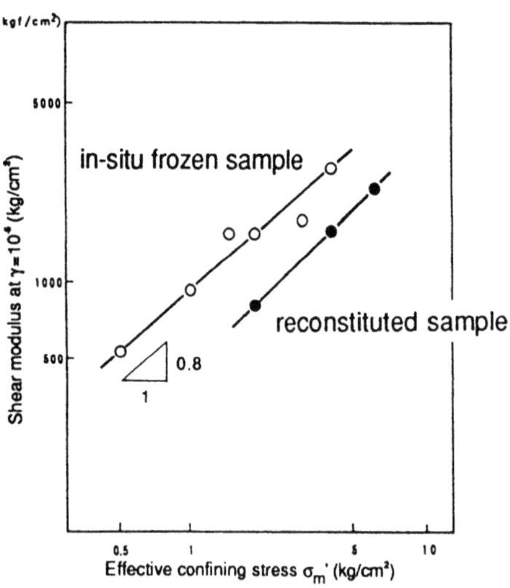

Fig.11 Relationship between Go and σ_m' of Gravel

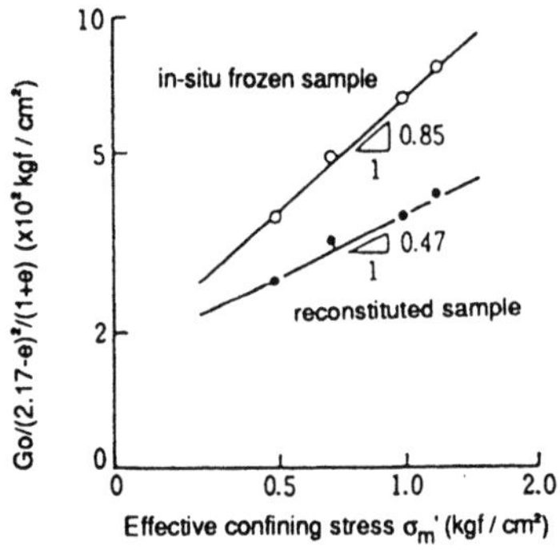

Fig.12 Relationship between Go and σ_m' of Gravel

Table-1 Initial shear modulus $Go = A\,F(e)\,(\sigma_m')^n$ for standard sand

Year	Specimen	A	F(e)	n	reference strain	Unit	researcher
1963	spherial particle Ottawa sand (e<0.8)	700	$(2.17-e)^2 / (1+e)$	0.5	$<10^{-4}$	kgf/cm^2	Hardin - Richart
	squarish particle silica sand	330	$(2.97-e)^2 / (1+e)$	0.5			
1965	Ottawa sand (σ_m' >2000psf)	1	$(32.17-14.8e)^2 / (1+e)$	0.5	$2.5\text{x}10^{-4}$	Go:psi σ_m' :psf	Hardin
	Ottawa sand (σ_m' <2000psf)	1	$(22.52-10.6e)^2 / (1+e)$	0.6			
1970	Ottawa sand	1	$(32.17-14.8e)^2 / (1+e)$	0.5	$<10^{-5}$	Go:psi σ_m' :psf	Drnevich - Richart
1972	sand	1230	$(2.97-e)^2 / (1+e)$	0.5	$<2.5\text{x}10^{-4}$	psi	Hardin - Drnevich
1975	3 kinds of sand	4200	$0.67-e / (1+e)$	0.5	10^{-5} to 10^{-6}	kgf/cm^2	Shibata - Soelarno
1977	11 kinds of small uniformity coefficient sand	900	$(2.17-e)^2 / (1+e)$	0.38	10^{-6}	kgf/cm^2	Iwasaki - Tatsuoka - Yoshida
		850	$(2.17-e)^2 / (1+e)$	0.44	10^{-5}		
		700	$(2.17-e)^2 / (1+e)$	0.5	10^{-4}		
1978	Monterey No.0 sand	1230	$(2.97-c)^2 / (1+e)$	0.5	$<10^{-5}$	psi	Drnevich
1978	sand	$Pa^{0.5}$	$625 / (0.3+0.7e^2)$	0.5	10^{-5}	(any)	Hardin
1978	Monterey No.0 sand	$Pa^{0.426}$	$428.2 / (0.3+0.7e^2)$	0.574	$<10^{-5}$	(any)	Sexena-Reddy
1980	Toyoura sand, Gifu sand	840	$(2.17-e)^2 / (1+e)$	0.5	10^{-6}	kgf/cm^2	Kokusho
1984	Monterey No.0 sand	$Pa^{0.52}$	$523 / (0.3+0.7e^2)$	0.48	10^{-5}	(any)	Chungt et.al.

Pa : atmosphere pressure

Table-2 Initial shear modulus Go = A F(e) (σ_m')^n for sand gravel

Year	Specimen	A	F(e)	n	reference strain	Unit	researcher
1978、1979	makadam of sand-stone (Dmax=9.5mm)	538	$(2.97\text{-}e)^2/(1+e)$	0.5		kgf/cm^2	Watanabe, et.al
		400	$(2.97\text{-}e)^2/(1+e)$	0.55			
1980	macadam	1300	$(2.17\text{-}e)^2/(1+e)$	0.55	10^6	kgf/cm^2	Kokusho, et.al
	spherical gravel	840	$(2.97\text{-}e)^2/(1+e)$	0.6			
1983	gravel (e=0.283 to 0.373)	500	$(2.17\text{-}e)^2/(1+e)$	0.66	10^{-6} to 10^{-5}	kgf/cm^2	Nishi, et.al
1984	macadam of andesite (Dmax=63.5,38.1mm)	649	$(2.17\text{-}e)^2/(1+e)$	0.65	$3\text{x}10^{-5}$	kgf/cm^2	Matsumoto et.al.
	gravel of riverbed (Dmax=38.1mm)	435	$(2.17\text{-}e)^2/(1+e)$	0.66			
1985	gravel fraction 25%	471	$(2.17\text{-}e)^2/(1+e)$	0.62	10^{-6} to 10^{-5}	kgf/cm^2	Tanaka, et.al.
	gravel fraction 50%	488	$(2.17\text{-}e)^2/(1+e)$	0.6			
1985	gravel fraction 25%	745	$(2.17\text{-}e)^2/(1+e)$	0.58	10^{-6} to 10^{-5}	kgf/cm^2	Kudoh, et.al.
	gravel fraction 50%	827	$(2.17\text{-}e)^2/(1+e)$	0.6			
1985	macadam (D_{50}=10.7mm)	710	$(2.17\text{-}e)^2/(1+e)$	0.44	10^{-5}	kgf/cm^2	Nishio,et.al

Table-3 Initial shear modulus Go = A F(e) (σ_m')^n for normaly consolidated clay

Year	Specimen	A	F(e)	n	Unit	researcher
1968	kaolinite in-situ clay	330	$(2.973\text{-}e)^2/(1+e)$	0.5	kgf/cm^2	Hardin-Black
1972	kaolinite	450	$(2.973\text{-}e)^2/(1+e)$	0.5	kgf/cm^2	Marcuson - Wahls
	bentonite	45	$(4.4\text{-}e)^2/(1+e)$	0.5		
1978	disturbed marine clay	200 - 400	$(2.17\text{-}e)^2/(1+e)$	0.5	kgf/cm^2	Kokusho, et.al
1982	Teganuma clay	9	$(7.32\text{-}e)^2/(1+e)$	0.5	kgf/cm^2	Kokusho, et.al.

3.1.2. Influence of grain size distribution. The experimental formulas in Table 1 are the results obtained from constituted sand samples which has fairly uniform grain size. The grain sizes of natural sand, however, are not uniform but contain a factor of irregularity. Fig. 13[5] shows an example of the relationship of the Go of sand with uniform grain size, to that of natural sand, indicating that the value for the Go of natural sand becomes smaller than that of sand constituted of grains of uniform size, under the same void ratio and confined pressure conditions. However, there have been no research results to date through which quantitative value of the influence of the grain size distribution can be determined not only for gravel but also for sand.

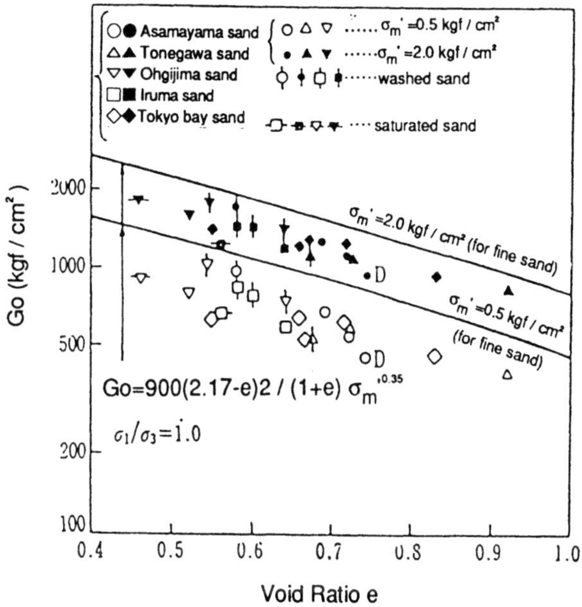

Fig.13 Relationship between Go and e

3.1.3. Quantitative determination of Go for Sand-Gravel. A high level of technique is needed to obtain the Go for sand-gravel by laboratory test. Because, Go is affected by the effective confined pressure, it is necessary to use undisturbed samples, and a high degree of accuracy is needed to measure Go at very small strain levels. Also, the freezing sampling method is also very expensive. Therefore, the method of estimating Go from the shear wave velocity, Vs, which can be obtained by elastic wave test, or the N-value from SPT, is frequently used quantitatively to calculate the Go for sand or clay. According to the past researches, the experimental formula for estimating the shear wave modulus from the N-value, can be defined following.

$$Go = a \cdot N^b$$

Where, a and b are constants. The parameters of the formula are shown in Table 4.
Go can be theoretically obtained by the following formula using Vs.

$$Go = \rho \cdot Vs^2$$

Where, ρ is the density of the soil.
Some formulae have been suggested by some researchers for estimating the Vs of sand-gravel from the N-value. When the SPT is applied to sand-gravel, however, the N-value itself tends to be excessively large, as explained in Fig. 1. Therefore, it is considered to be more reliable that when testing sand-gravel, Vs should be first measured directly by the elastic wave test, and then obtain Go than using Go estimated from the N-value.

Fig. 14 compares the G_{OL} value obtained by laboratory test with the G_{OF} obtained from the elastic wave test at the field in the case of sand. According to this report, it is found that when G_{OF} is larger than 500 kgf/cm², G_{OL}/G_{OF} becomes less than 1.0, and the major cause is disturbance of the sample. Because, applying G_{OL} obtained by in-situ freezing sample, G_{OL}/G_{OF} is nearly constant value about 1.0.

Table 4 $Go = a N^b$ kgf/cm²

scholar		a (kgf/cm²)	b
Imai	alluvial clay	159	0.626
	alluvial sand	94	0.715
	diluvium clay	242	0.552
	diluvium sand	170	0.656
	general	120	0.737
Hara		158	0.668
Ohsaki, Iwasaki		120	0.80
Ohta, Hara, Tanba, Sakano		139	0.72
Imai, Tonouchi		144	0.68
Ohsaki, Toriumi		122	0.62

3.2. NORMALIZED SHEAR MODULUS G/Go

The non-linear property of the shear modulus ,G, of soil, is generally evaluated from the relationship between the shear strain level, γ, and normalized shear modulus, G/Go, after establishing the value of Go, as introduced in previous section. The relationship between G/Go and γ will be considered here.

3.2.1. Disturbed samples and undisturbed samples. Fig. 15[3)] shows the results for the relationship of G-γ and G/Go-γ using a reconstituted gravel sample and using an undisturbed sand-gravel sample obtained by in-situ freezing sampling. According to these figures, each G-γ relationship shows the difference from disturbed and undisturbed, each G/Go-γ relationship, however, shows almost the same result regardless whether or not the sample has been disturbed. This fact has been

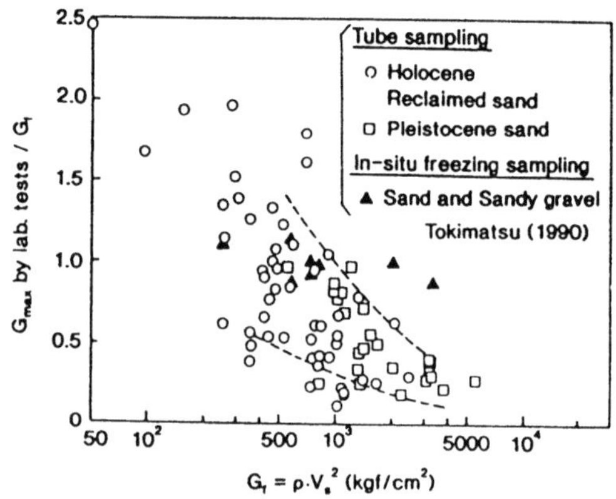

Fig.14 Comparison with G_{OL} from Laboratory Test and G_{OF} from Field Test

generally applied in the case of sand.

Therefore, it can be assumed that the effect of disturbance of the sample is not so great, on the G/Go relationship for sand-gravel.

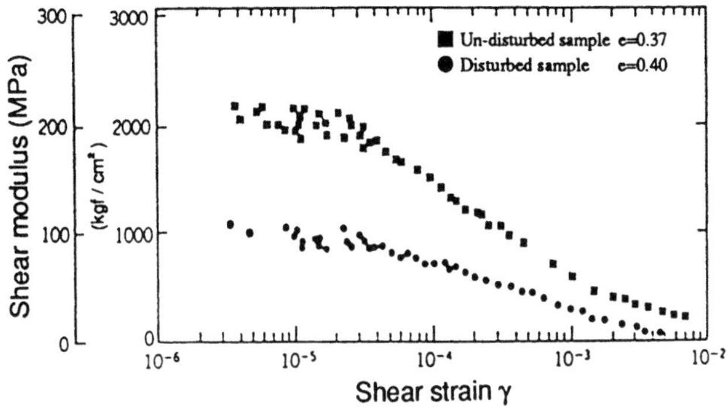

(a) Relationship between G and γ

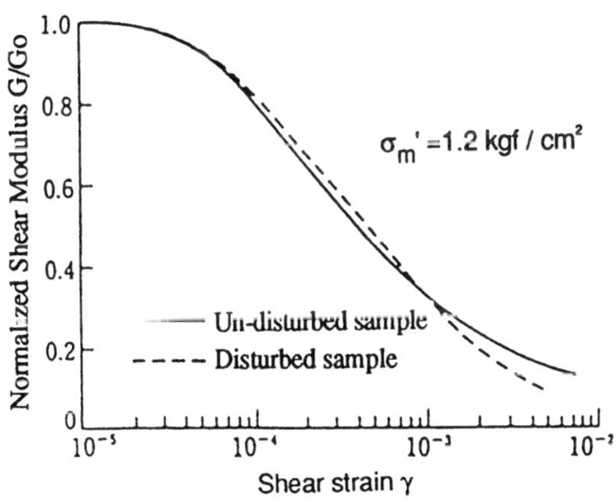

(b) Relationship between G/Go and γ

Fig.15 Shear Modulus Difference In-situ Frozen Sample from Reconstituted Sample

3.2.2. G/Go-γ, relationship for Sand-Gravel. The G/Go value decreases as the shear strain value increases. Fig. 16[8] shows a comparison of the various aspects of G/Go-γ relationships which have been statistically processed for each of sand, sand-gravel, and clay soils. These relationships were compared within one standard deviation range. From this comparison, it can be found that the decrease of normalized shear modulus of sand-gravel is most remarkable, followed by sand and clay. This means that there is a strong tendency for the normalized shear modulus to start decreasing while the shear strain ,γ, is still small, when the grain size of the soil becomes larger. Fig. 17[8] shows the G/Go - γ relationship which has become the basis of the statistical processing

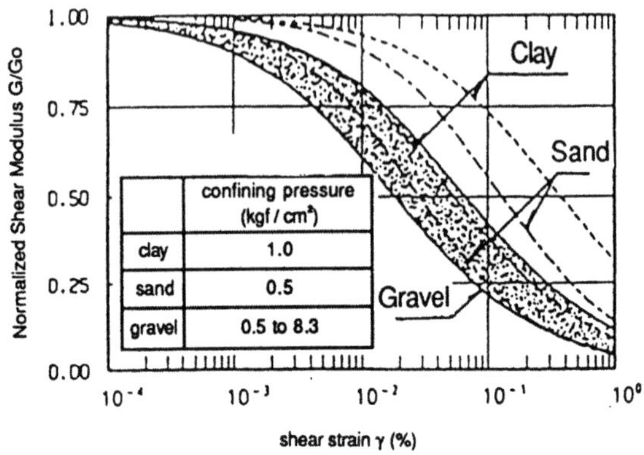

Fig.16 Variation of Soil on Statistical Relationship between G/Go and γ

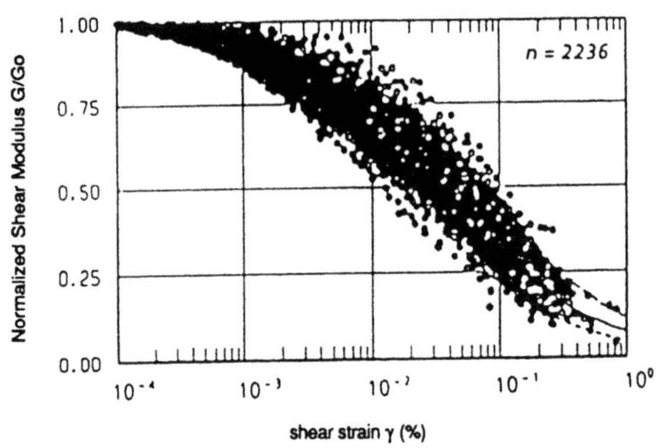

Fig.17 Relationship between G/Go and γ of Gravel (N=2236)

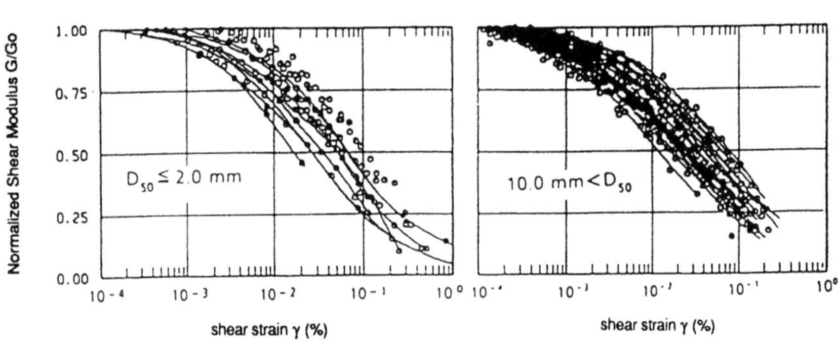

Fig.18 Variation of D_{50} on Relationship between G/Go and γ of Gravel

D_{50}: mean grain size

for the sand-gravel, showing that $\gamma_{0.5}$ is in the range of 2 x 10^{-4} to 9 x 10^{-4}. Also, Fig. 18[8] shows that the G/Go - γ relationship for sand-gravel is related to the mean grain size, D_{50}, indicating that when the grain size is larger, the G/Go value comes to be smaller. Fig. 19[9] shows an example of the results of test regarding the influence of the effective confined pressure σ_m' to the G/Go - γ relationship for sand, and Fig. 20[10] shows and example of the above mentioned influence for sand-gravel. It can be found that the G/Go - γ curve shifts to the right side as the effective confined pressure becomes greater. Upon this fact, it can be assumed that there is a tendency for the standard strain of sand and sand-gravel to become greater with an increase of the effective confined pressure,at the condition that shear strain at G/Go is equal to 0.5, $\gamma_{0.5}$.

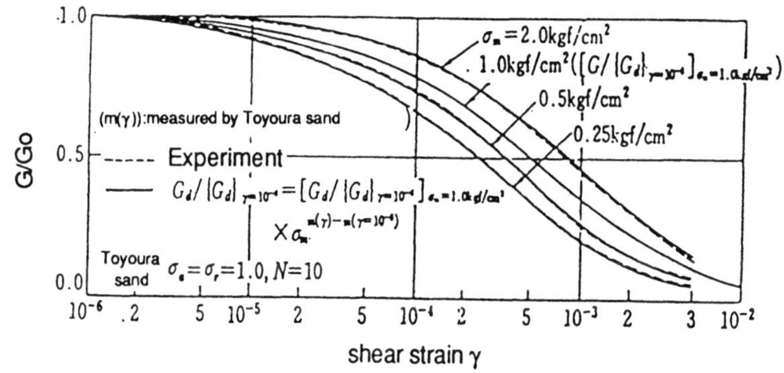

Fig.19 Variation of σ_m' on Relationship between G/Go and γ of Fine Sand

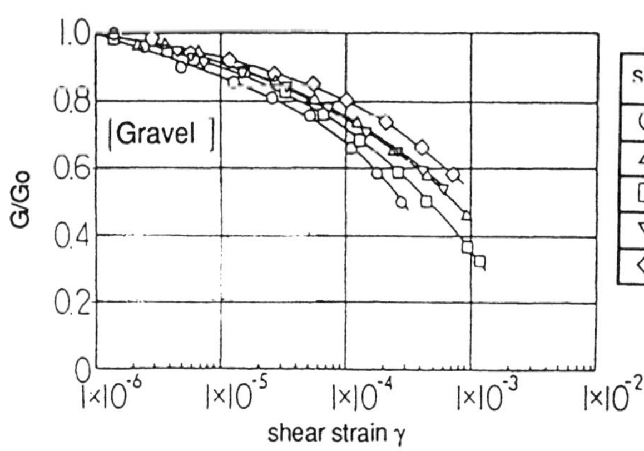

Sign	Lateral Pressure (kgf / cm²)	e after consolidation
○	0.5	0.358
△	1.0	0.383
□	2.0	0.363
▽	4.0	0.359
◇	8.0	0.364

Fig.20 Variation of σ_m' on Relationship between G/Go and γ of Gravel

3.3. DAMPING FACTOR h

Material damping factor of soil is shown as the damping factor, h. The damping factor, h, has a tendency to become greater with the increase of the shear strain , γ, mainly due to the friction among soil grains.

3.3.1. Influence of Effective Confined Pressure. Fig. 21 [11] shows the influence of the effective confined pressure ,σ_m', to the damping factor, h, for sand, spherical sand-gravel, and macadam, using γ as the parameter.. It may be seen that the damping factor decreases as σ_m' becomes larger, and the relationship between these figures forms an almost straight line on the logarithmic axis. This tendency is common regardless the differences of grain size, and the conditions of sampling ; disturbed or undisturbed. The research example in Fig. 22 [13] shows that this tendency may be applicable for sand-gravel. No research clearly show, however, the influence of the effective confined pressure on the damping factor for sand-gravel,yet.

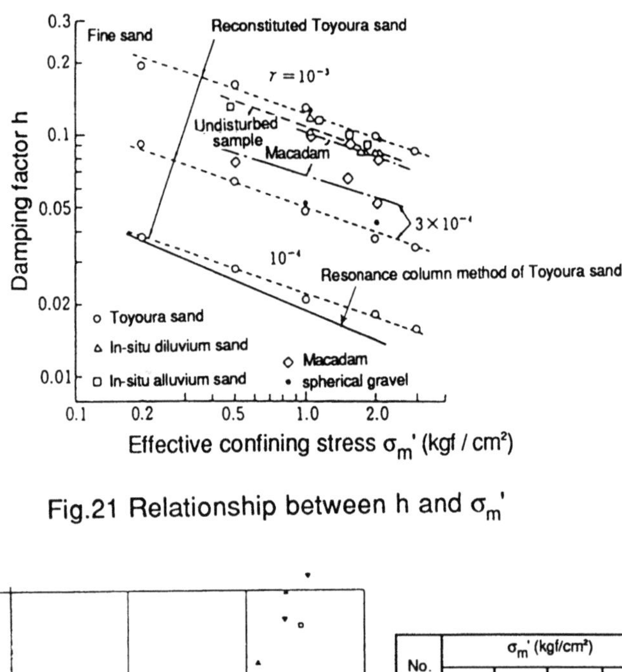

Fig.21 Relationship between h and σ_m'

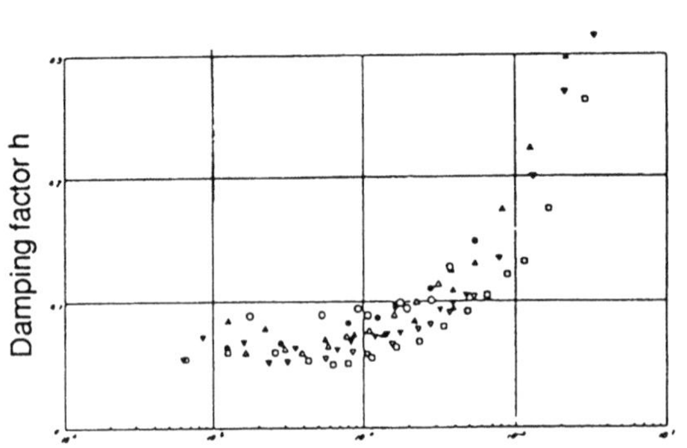

No.	σ_m' (kgf/cm²)			
	1.0	2.0	3.0	5.0
1		△		
2	○		▽	
3			▼	□
4		▲		
5	●			

Fig.22 Variation of effective confining pressure in Relationship between h and γ

3.3.2. h - γ relationship. Many h - γ relationships by the past research results have been statistically processed for each of sand, sand-gravel, and clay soil, and have been compared within the range of one standard deviation. The results of these comparisons are shown in Fig. 23[8]. According to this figure, it can be found that the maximum damping factor h_{max} is in the range of 0.12-0.18 for clay, in the range of 0.2 -0.35 for sand and in the range of 0.15- 0.25 for sand-gravel. A distinctive attribute of sand-gravel is that damping factor is rather large even for the smaller shear strain range.

Mean grain size, D_{50}, is divided into two areas, as shown in Fig. 24[8], for observation of the influence of grain size in sand-gravel. The influence of grain size on the value of the damping factor, h, is not clearly shown by this figure, but it can be assumed that there is some tendency for the fluctuation of the damping factor to be greater as the grain size is smaller.

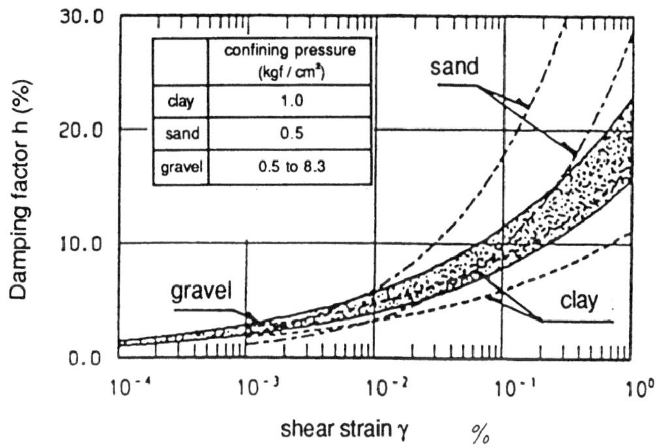

Fig.23 Variation of Soil on Statistical Relationship between h and γ

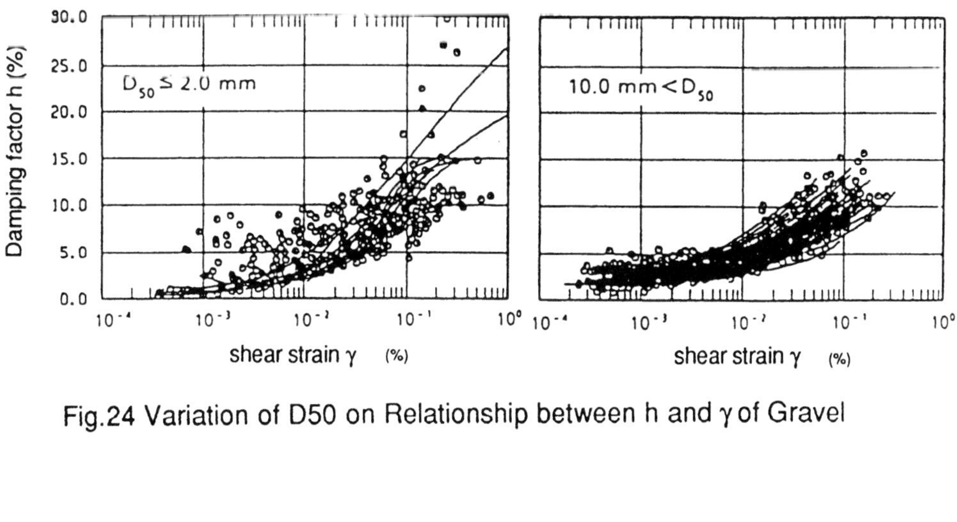

Fig.24 Variation of D50 on Relationship between h and γ of Gravel

3.4. LIQUEFACTION RESISTANCE

Liquefaction resistance can be defined by the stress ratio, which is obtained by dividing a fixed shearing stress amplitude, τd, by the initial effective confined pressure, σ_m', or by the effective overburden pressure, σvo', when the soil sample is loaded by repeated shear. The liquefaction resistance, however, is defined as $\sigma d/(2\sigma_m')$ when it is assumed that the difference between the amplitudes of the maximum principal stress and the minimum principal stress is σd, in triaxial testing in which axial stress is applied instead of shearing stress.

As shown in Figs. 3 and 4, there is a great difference between the liquefaction resistance determined based upon undisturbed samples and based upon disturbed samples. Therefore, the influence of disturbance is quite obvious consequently, to evaluate liquefaction resistance, to use undisturbed samples in the laboratory test is highly recommended.

A simple method of estimation from the N-value has been suggested here, for determining the liquefaction resistance of sand, but no simple method of estimation has been established yet for sand-gravel, because the accuracy of its N-values is poor and because the permeability of sand-gravel is generally higher than that of sand. There are also very few examples of research on the liquefaction resistance of sand-gravel. Therefore, the extensive data accumulation for liquefaction is required hereafter.

Under these circumstances, it has suggested, however, to estimate the liquefaction resistance of sand-gravel deposits from the Nd-value obtained from LPT, instead of using N-value obtained by SPT. The report[2] has also developed the following experimental formula for obtaining liquefaction resistance, using the effective stress, σm', and based on the Nd value.

$$\sigma d/(2\sigma_m') = f\{(\sigma_m'/P_1)^{-1}\cdot Nd\} \quad ; \; P_1{:}1.0\text{kgf/cm}^2\text{,const.}$$

Fig. 25[2] shows this within the limited range of the effective confined pressure. The Nd value obtained for sand-gravel by LPT is more reliable than the N-value obtained by SPT.

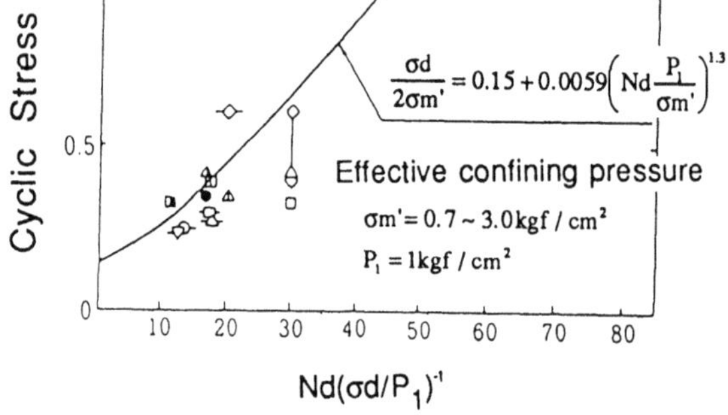

Fig.25 Relationship between Cyclic Stress Ratio and $Nd(\sigma_m'/Pa)^{-1}$

3.5. SETTLEMENT OF GROUND

It is important to estimate the residual settlement of ground caused by large earthquake. It can be calculated by dynamic analysis based on effective stress method, but the analysis requires enormous amount of complicated calculation. Moreover since its result depends on the soil constitutive equation, it is required to estimate the settlement value roughly beforehand.

Fig. 26[14] indicates the relationship between peak pore pressure ratio and volumetric strain after consolidation. It seems that the volumetric strain increases as the grain size increases, as peak pore pressure ratio increases, relative density, D_r, decreases.

In order to estimate the residual settlement value of ground caused by earthquake, some dynamic triaxial tests were carried out using in-situ freezing sand-gravel samples. In this test, the volumetric strain caused by scattered excess pore water pressure in the specimen was measured after undrained cyclic shear test. Therefore it was found out that the residual settlement by earthquake was correlated with shear strain during earthquake and soil density. And it is indicated that the relation as shown in Fig. 27[2] and following formula, are not affected by grain size distribution and effective confining pressure.

$$\varepsilon_v = 45\, DA_f / D_r\, (e_{max} - e_{min}) \quad (\%)$$

Where ε_v is volumetric strain, DA_f is the axial strain amplitude at the final state of undrained cyclic triaxial test, e_{max}, e_{min} is maximum and minimum void ratio, and D_r is relative density. And authors proposed the simple method through which estimate the settlement of ground caused by earthquake as shown in Fig. 28[2].

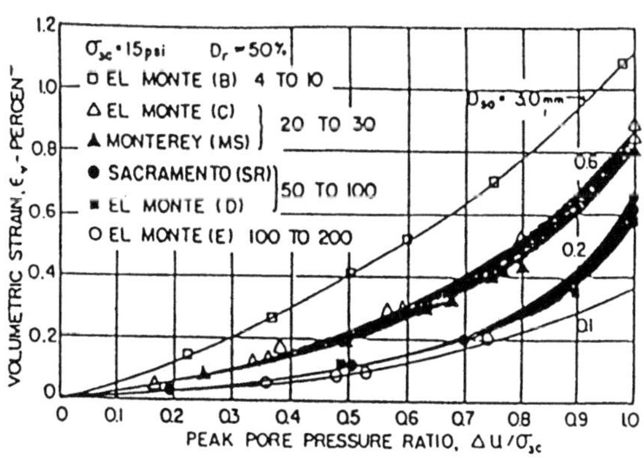

Fig.26 Relationship between Peak Pore Pressure Ratio and Volumetric strain

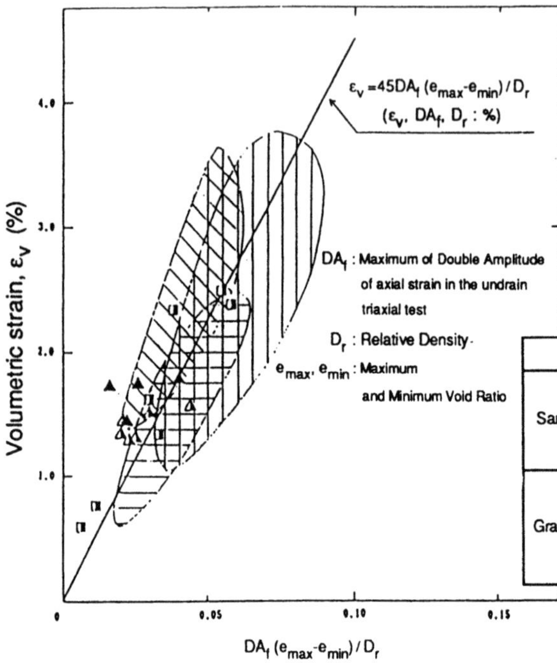

Sample		σm' (kgf/cm²)	Sign
Sand	Toyoura sand	1.0	
	Tonegawa sand	0.3, 0.5, 2.0	
	Narita sand	0.5, 1.0, 2.0	
Gravel	T site Upper deposit	1.8	△
		4.5 or 5.0	▲
	T site Lower deposit	1.2	□

Fig.27 Relationship between Volumetric strain and $DA_f\ (e_{max}-e_{min})\ /\ D_r$

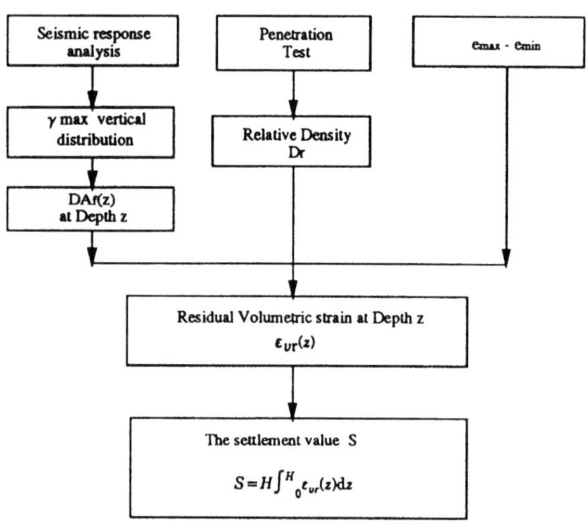

Fig.28 Simple Method which Estimate the Settlement of Ground caused by Earthquake

4. Conclusions

This paper has briefly introduced the large-scale field tests which has been carried out in Japan to understand the characteristics of sand-gravel, and has also mentioned recent studies of the dynamic characteristics of sand-gravel, which include initial shear modulus, cyclic deformation characteristics, liquefaction resistance, and settlement of ground.

The conclusions can be summarized as follows.

1 The initial shear modulus of sand-gravel is subjected to the much greater effect by the effective confined pressure than in the cases of sand of clay.
2 To obtain undisturbed samples is essential, in order to determine the initial shear modulus by laboratory test, and the liquefaction resistance.
3 There are no major differences between the G/Go - γ curve obtained for disturbed samples and for undisturbed ones, so the similar curve was obtained using undisturbed samples.
4 The G/Go - γ curve for sand-gravel shows a higher ratio of decreasing with the increase of shear strain than that for sand or clay.
5 The h - γ curve for sand-gravel shows a larger damping factor within a small range of shear strain than that for sand or for clay.
6 There are a few research reports which state that the liquefaction resistance of sand-gravel can be estimated from the Nd value obtained by large penetration test.
7 There are a research report which states that the settlement of ground caused by large earthquake is simply estimated by axial strain, void ratio, and relative density.

Data relating to the dynamic characteristics of sand-gravel are relatively fewer in number than those concerning the dynamic characteristics of sand or clay. It is therefore necessary to accumulate more data in future, in order to evaluate quantitative dynamic characteristics of sand-gravel.

5. References

1) M.Watabe, et.al.: "Large Scale Field Tests on Quaternary Sand and Gravel Deposits for Seismic Siting Technology", Second International Conference on Report Advances in Geotechnical Earthquake Engineering and Soil Dynamics, 1991
2) T.Kokusho, et.al.: "Study on Quaternary Ground Siting of Nuclear Power Plant-Part 1. Geological/Geotechnical Investigation Methods and Seismic Stability Evaluation Methods of Foundation Ground", Abiko Research Laboratory Rep. No. U19, 1991
3) Shamoto,Y., Nishio,S., Baba,K., Goto,S., Tamaoki,K., and Akagawa,S. Cyclic Stress Strain Behavior and Liquefaction Strength of Diluvial Gravels Utilizing Freezing Sampling, pp89~94 (1986)
4) Yoshimi,Y., et.al.: "Undrained cyclic shear strength of a dense Niigata sand", Soil and Foundation, Vol. 24, No. 4, pp131~145 (1984)
5) Iwasaki,T., F.Tatsuoka and Y.Takagi : Shear modulus of sands under cyclic torsional shear loading, Soils and Foundation, Vol. 18, No. 1, pp.39~56, (1978)
6) Iwasaki, T. and F. Tatsuoka : Effects of grain size and grading on dynamic shear modulus of sands, Soils and Foundations, Vol. 17, No. 3, pp.19~35, (1977)
7) Yasuda, Ymazaki:Dynamic Shear Modulus obtained by laboratory test and field test: The Japanese Society of Soil Mechanics and Foundation Engineering (1984)
8) Masanori Imazu, Kiyoshi Fukutake:Dynamic Shear Modulus and Damping of Gravel Materials The 21th Japan National Conference on Soil Mechanics and Foundation Engineering, Sapporo, Japan June, 10th-12th, 1986, ISSN 0285-7340

9) Iwasaki, T., et.al.: "Shear modulus of sands under cyclic torsional shear loading, Soils and Foundations, Vol. 18, No. 1, pp39~56.

10) Tatsuru Mizokami, et.al.: Correlation Between Dynamic Deformation Properties and Static Shear Strength of Various Materials

11) Zen, K. et.al.: "Laboratory tests and in-situ seismic surveyor vibratory shear modulus of clay soils with various elasticities, Proc. of 5th Japan Earthquake Engineering Symp., pp721~728 (1478)

12) JSCE : Dynamic analysis and earthquake-proof design, Vol. 1 1989, Gihodo.

13) Nobuhide Ogata, Masayoshi Yasuda
Dynamic Properties of Undisturbed Samples Containing gravels
The 17th Japan National Conference on Soil Mechanics and Foundation Engineering, Nara, Japan June, 8th-10th, 1982, ISSN 0285-7340

14) Lee,K.L. and Albasia, A. : Earthquake induced settlement in saturated sand, Jour. GE Div., Proc. ASCE, Vol. 100, No.GT$, pp387-406 (1974)

SEISMIC ANALYSIS OF RECTANGULAR TUNNELS IN SOFT GROUND

JOSEPH PENZIEN
International Civil Engineering Consultants, Inc.
Berkeley, California, U.S.A.

C. H. CHEN
National Taiwan University
Taipei, Taiwan, R.O.C.

W. Y. JEAN
D&E Services Company
Taipei, Taiwan, R.O.C.

Y. J. LEE
National Taiwan University
Taipei, Taiwan, R.O.C.

ABSTRACT: Practical analytical procedures are presented for assessing the seismic performance of rectangular concrete-lined tunnels located in soft soils as follows: (1) generating free-field ground motions in a horizontally-layered soil medium, (2) calculating cross-sectional deformations and overall global response of the lining to the free-field motions, including soil-structure interaction effects, and (3) evaluating separations in the lining's transverse joints. Numerical results are given to show the general nature of seismic performance for selected tunnel linings.

1. Introduction

To provide for the flow of automobile and train traffic in metropolitan areas, rectangular underground tunnels having reinforced concrete linings are commonly constructed by the cut-and-cover method. Often, these tunnels are located in soft ground which can experience high intensity motions due to the occurrence of large earthquakes. It has been common practice to design the linings on the basis that they will deform in a manner totally compatible with the free-field ground displacements (Kuesel 1969; DORTS 1988); even though, methods of analysis have been suggested which consider the effects of soil-structure interaction (Aoki, Y. 1973; Okamoto, S. and Tamura, C. 1973; Penzien, J. and Tseng, W. S. 1981). The purpose of this paper is to advance some of the more practical procedures for (1) predicting possible free-field ground displacements along the tunnel alignment which are critical to the lining response, (2) evaluating lining deformations produced by these displacements, including soil-structure interaction effects, and (3) estimating separations of transverse joints in the lining, and to present numerical results generated for the Sungshan Railway Extension project in Taipei, Taiwan, R.O.C..

P. Gülkan and R. W. Clough (eds.), Developments in Dynamic Soil-Structure Interaction, 349–359.

2. Free-Field Ground Motions

As is common practice, one set of horizontal free-field ground motions can be developed assuming vertically propagating shear waves in a horizontally-layered soil medium with the surface motions used for control purposes (Seed, H. B. and Idriss, I. M. 1970; Schnabel, P. B., et al. 1972). These control motions, which are usually specified to be compatible with a design response spectrum, can be obtained by systematically adjusting field-recorded or synthetically-generated accelerograms.

Assuming upward travelling incident shear-waves in the uniform half-space below the layered medium, which can be represented as a series of harmonics, one can calculate the corresponding upward and downward travelling harmonics in each layer, and the corresponding downward travelling harmonics in the half-space resulting from their interactions with the layered system. Decomposing the specified free-field surface acceleration time-history into its harmonics and satisfying the surface zero stress condition and the displacement and stress compatibility conditions at each interface of the layered system, one can calculate the amplitude and phase angle of each upward travelling harmonic and each downward travelling harmonic in each layer and in the half-space, thus providing time-histories of ground acceleration and displacement at all elevation levels in the layered media which are compatible with the soil modelling and the specified free-field surface motions. In carrying out these calculations, it is necessary to reduce the small-strain ($\gamma = 10^{-6}$) shear-modulus values and increase the corresponding damping ratios representing the soil in each layer in accordance with the standard equivalent linearization procedure and the appropriate shear-modulus and damping-ratio relations $G(\gamma)/G(\gamma = 10^{-6})$ and $\xi(\gamma)$ expressed as functions of shear strain (γ). For the seismic response calculations referred herein, the effective shear strain at each elevation was taken as 65% of the corresponding maximum strain. In the equivalent linearization procedure, iteration must be used in finalizing the strain-compatible shear-modulus and damping-ratio values.

Having obtained the horizontal free-field ground displacement time-histories at discrete elevations over the height of the lining cross-section by the above procedure, the shear (or racking) deformation of the cross-section can be evaluated by the finite-element procedure described subsequently, which includes the effects of soil-structure interaction.

To calculate overall global response of the lining, i.e., its combined axial and bending deformation, components of free-field ground displacement in a vertical plane at positions along the tunnel's longitudinal axis, under the most critical ground motion condition possible, are required. In calculating these displacement time-histories, use is made of the horizontal free-field ground displacement time history $V(t)$ at the mid-height position of the tunnel cross-section as generated by the above-described vertically-propagating shear-wave model.

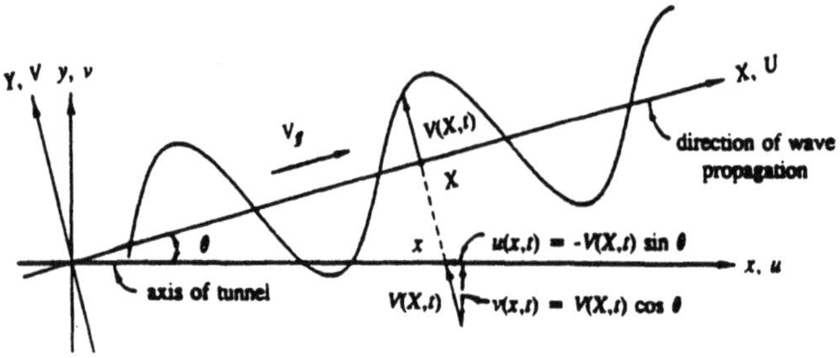

Figure 1. Free-field ground displacements for global analysis

However, in this application, it is assumed that displacement $V(t)$ is produced by a single train of shear waves travelling at velocity V_{ff} at an angle θ from the longitudinal axis of the tunnel as shown in Fig. 1; thus, the free-field soil displacements along the X-axis, but normal to it, are given by

$$V(X,t) = \sum_n V(i\bar{\omega}_n)\exp[i\bar{\omega}_n(t - \frac{X}{V_{ff}})] \tag{1}$$

in which $V(i\bar{\omega}_n)$ are the complex amplitudes of the discrete harmonics contained in $V(t)$ as generated by the FFT (Fast Fourier Transform) algorithm. The corresponding components of displacement along the x-axis in the x- and y-directions are

$$\begin{aligned} u(x,t) &= -\sin\theta \sum_n V(i\bar{\omega}_n)\exp[i\bar{\omega}_n(t - \frac{x\cos\theta}{V_{ff}})] \\ v(x,t) &= \cos\theta \sum_n V(i\bar{\omega}_n)\exp[i\bar{\omega}_n(t - \frac{x\cos\theta}{V_{ff}})] \end{aligned} \tag{2}$$

These free-field ground displacements are used subsequently to evaluate overall global response of the lining, including soil-structure interaction effects.

3. Cross-Section Racking Analysis

The most severe free-field ground motion environment causing the lining cross-section to undergo racking (shear-type) deformation is that condition produced by the vertically travelling shear waves in the layered medium. The most critical time when maximum racking occurs can be taken as that instant when the relative horizontal free-field soil displacement between the top and bottom positions of the lining cross-section is a maximum. At this instant, the horizontal free-field ground displacement profile over the entire depth of the lining is known.

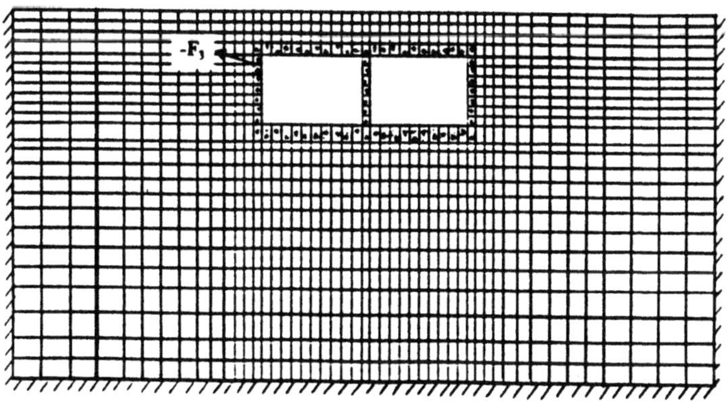

Figure 2. Finite-element modelling of soil and lining

Since the lining interacts with the surrounding soil in essentially a quasi-static fashion, the racking analysis can be carried out in three steps using finite-element modelling of the lining and soil as shown in Figure 2 and static loadings. Step 1--Using the finite element model of the lining cross-section only, static forces (represented by vector $\mathbf{F}_1$) are calculated, which, when applied to its outer-boundary nodes, will deform the cross-section so that these nodes displace exactly the same as do the corresponding points in the free-field soil (no lining present). The stresses in the cross-section under this deformed shape are calculated by the standard finite-element procedure. Step 2--The free-field shear stress distribution acting on the outer boundary of the rectangular soil element to be displaced by the cross-section is discretized, giving static forces (represented by vector $\mathbf{F}_2$) acting in the same nodal coordinates as those of the outer-boundary of the cross-section. If this soil element is now replaced by the lining cross-section and full bond is assumed at its outer-boundary with the soil, the external static forces (represented by vector $\mathbf{F}_3$) required in the coordinates of the cross-section outer-boundary nodes to cause these same nodes of the combined soil/lining model to displace in a manner totally compatible with the corresponding free-field soil displacements are the components in vector F_3 as defined by $F_3 = F_1 - F_2$. Step 3--Since the external static forces contained in F_3 do not actually exist, they must be canceled out by applying them in opposite directions to the combined soil/lining model as shown in Figure 2. The resulting stresses in the lining must be superimposed on the corresponding stresses generated in Step 1 to obtain the desired critical racking stresses in the lining.

4. Global Response Analysis

The horizontal and vertical free-field soil displacements along the longitudinal axis position of the tunnel as expressed by Eqs. (2) will be used to evaluated the overall global response of the lining. Because the mass of the lining is much less than the mass of the soil it displaces, the global response of the lining can be carried out in a quasi-static fashion using the corresponding equations which govern its axial and bending deformation, namely

$$\begin{aligned} \frac{\partial^2 u_t(x,t)}{\partial x^2} - \frac{k_x}{AE} u_t(x,t) &= -\frac{k_x}{AE} u(x,t) \\ \frac{\partial^4 v_t(x,t)}{\partial x^4} + \frac{k_y}{EI} v_t(x,t) &= +\frac{k_y}{EI} v(x,t) \end{aligned} \tag{3}$$

in which $u(x, t)$ and $v(x, t)$ are the free-field ground displacements given by Eqs. (2), u_t and v_t are the lining displacements along its centrordal axis, k_x and k_y are subgrade moduli in the x- and y-directions, and AE and EI are the axial and flexural stiffnesses of the lining.

Substituting Eqs. (2) into Eqs. (3) and obtaining the steady state solutions gives

$$\begin{aligned} u_t(x,t) &= -\sin\theta \sum_n \frac{V(i\bar{\omega}_n)}{(1 + \phi_{un})} \exp[i\bar{\omega}_n(t - \frac{x\cos\theta}{V_{ff}})] \\ v_t(x,t) &= \cos\theta \sum_n \frac{V(i\bar{\omega}_n)}{(1 + \phi_{vn})} \exp[i\bar{\omega}_n(t - \frac{x\cos\theta}{V_{ff}})] \end{aligned} \tag{4}$$

in which

$$\phi_{un} \equiv \frac{AE\bar{\omega}_n^2 \cos^2\theta}{k_x V_{ff}^2} \quad ; \quad \phi_{vn} \equiv \frac{EI\bar{\omega}_n^4 \cos^4\theta}{k_y V_{ff}^4} \tag{5}$$

Note that when ϕ_{un} and ϕ_{vn} equal zero which corresponds to k_x and k_y being equal to infinity, the tunnel displacements given by Eq. (4) are identical to the free-field soil displacements given by Eqs. (2), i.e., no soil-structure interaction effects are present.

The axial and bending normal strains in the lining at coordinate position y of the cross-section can be obtained by substituting Eqs. (4) into the relations

$$\epsilon_a(x,t) = \frac{\partial u_t(x,t)}{\partial x} \quad ; \quad \epsilon_b = -y\frac{\partial^2 v_t(x,t)}{\partial x^2} \tag{6}$$

giving

$$\begin{aligned} \epsilon_a(x,t) &= +\frac{\sin\theta\cos\theta}{V_{ff}} i \sum_n \frac{\bar{\omega}_n V(i\bar{\omega}_n)}{(1+\phi_{un})} \exp[i\bar{\omega}_n(t - \frac{x\cos\theta}{V_{ff}})] \\ \epsilon_b(x,t) &= +y\frac{\cos^3\theta}{V_{ff}^2} \sum_n \frac{\bar{\omega}_n^2 V(i\bar{\omega}_n)}{(1+\phi_{vn})} \exp[i\bar{\omega}_n(t - \frac{x\cos\theta}{V_{ff}})] \end{aligned} \tag{7}$$

The combined normal strain $\epsilon(x, t)$ at position y is now obtained by summing Eqs. (7), i.e., using

$$\epsilon(x,t) = \epsilon_a(x,t) + \epsilon_b(x,t) \tag{8}$$

Upon examining numerical results given by Eqs. (7) for the maximum value of $|y|$ on a cross-section and for the critical value of θ, which is close to 45°, one finds that $|\epsilon_a(x,t)|_{max}$ is much greater than $|\epsilon_b(x,t)|_{max}$. Further, since each harmonic in $\epsilon_a(x,t)$ is 90° out-of-phase with the corresponding harmonic in $\epsilon_b(x,t)$, $|\epsilon_a(x,t)|_{max}$ will occur at an instant of time quite different from that at which $|\epsilon_b(x,t)|_{max}$ occurs; thus, the square-root-of-the-sum-of-squares (SRSS) method could be used to estimate $|\epsilon(x,t)|_{max}$, i.e.,

$$|\epsilon(x,t)|_{max} = [|\epsilon_a(x,t)|^2_{max} + |\epsilon_b(x,t)|^2_{max}]^{\frac{1}{2}} \tag{9}$$

In view of the above observations, the bending contribution to $|\epsilon(x,t)|_{max}$ is negligible for the critical value of θ which is very close to 45°. Therefore, one can estimate the critical value of normal strain using the first of Eqs. (7) alone with θ set equal to 45°, i.e., using

$$|\epsilon(x,t)|_{max} = \frac{1}{2V_{ff}} \left| \sum_n i \frac{\bar{\omega}_n V(i\bar{\omega}_n)}{(1+\phi_{un})} \exp[i\bar{\omega}_n(t - \frac{x}{\sqrt{2}\, V_{ff}})] \right|_{max} \tag{10}$$

in which $\phi_{un} = AE\bar{\omega}_n^2 / 2k_x V_{ff}^2$. The subgrade modulus k_x cannot be evaluated rigorously for a given soil/liner system; however, for practical solutions, it can be approximated using $k_x \doteq 3G = 3\rho V_{si}^2$ in which ρ is the mass density of the soil adjacent to the lining, G is its effective shear modulus in the dominate region controlling soil-structure interaction, and V_{si} is the corresponding shear wave velocity. Note that V_{ff}, as defined earlier, is that shear wave velocity controlling the train of free-field travelling waves approaching the tunnel alignment; while V_{si} is the effective shear wave velocity controlling soil/structure interaction. These two shear-wave velocities may be significantly different from each other. Obviously, considerable judgement must be used in assigning their numerical values, after taking into consideration known factors such as geometry of soil layers relative to tunnel location, results of soil tests, and levels of soil shear strain produced by the free-field motions and by the soil-lining interaction. In considering the latter factor, one should note that the shear strains produced by the free-field ground motions and by soil-structure interaction, at the outer-boundary location of the lining cross-section, are given, respectively, by

$$\gamma_{ff}(X,t) = \frac{\partial V(X,t)}{\partial X} \;\; ; \;\; \gamma_{si}(x,t) = \frac{AE}{p\rho V_{si}^2} \frac{\partial \epsilon_a(x,t)}{\partial x} \tag{11}$$

in which p is the outside perimeter dimension of the cross-section. Making use of $X = x\cos\theta$, $\theta = 45°$, Eq. (1), and the first of Eqs. (7), these shear-strain relations become

$$\begin{aligned} \gamma_{ff}(x,t) &= -\frac{i}{V_{ff}} \sum_n \bar{\omega}_n V(i\bar{\omega}_n) \exp[i\bar{\omega}_n(t - \frac{x}{\sqrt{2}\,V_{ff}})] \\ \gamma_{si}(x,t) &= \frac{AE}{2\sqrt{2}\,p\rho V_{si}^2 V_{ff}^2} \sum_n \frac{\bar{\omega}_n^2 V(i\bar{\omega}_n)}{(1+\phi_{un})} \exp[i\bar{\omega}_n(t - \frac{x}{\sqrt{2}\,V_{ff}})] \end{aligned} \tag{12}$$

5. Separation of Transverse Joints

The axial strains described in the previous section were evaluated assuming no transverse joints to be present in the tunnel lining. Placement of such joints in the lining would reduce these strains to zero at the joints and would also reduce them at intermediate locations between adjacent pairs of joints. Using the first of Eqs. (3) and assuming the dominant free-field soil wave lengths producing $|\epsilon(x,t)|_{max}$ are long compared with the joint interval distance L, it can be shown that the ratio of the maximum absolute strain at the midway-point ($x = 0$) between an adjacent pair of joints to the maximum absolute strain given by Eq. (10) is

$$|\epsilon(0,t)|_{max} / |\epsilon(x,t)|_{max} = [1 - \frac{2}{\exp(\beta L/2) + \exp(-\beta L/2)}] \tag{13}$$

in which $\beta \equiv \sqrt{k_x/AE}$. Further, it can be shown that the strain reductions in a segment of lining tributary to a particular joint will allow it to separate a distance Δ_j as expressed by

$$\frac{\Delta_j}{L\,|\epsilon(x,t)|_{max}} = \frac{2}{\beta L} \left[\frac{\exp(\beta L/2) - \exp(-\beta L/2)}{\exp\beta L/2) + \exp(-\beta L/2)} \right] \tag{14}$$

6. Case Study

A case study is presented to demonstrate use of the above analytical procedures and to show the general nature of seismic response for selected tunnel linings.

6.1. LINING CROSS-SECTIONS

The basic lining cross-section used in this case study is shown in Figure 3.

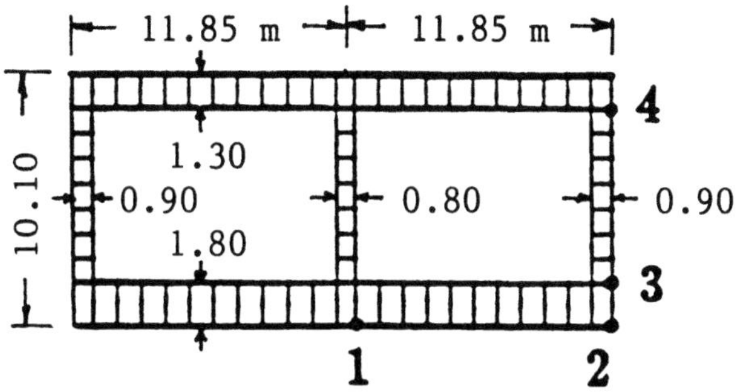

Figure 3. Basic lining cross-section

In this exact form, it is representative of one type of cross-section designed for the Sungshan Railway Extension. Variations on the wall and slab thicknesses, but not on the outside dimensions, of this cross-section were introduced to investigate the influence of tunnel stiffness on global response. The Young's modulus E and Poisson's ratio ν used for the concrete of the lining were 22,800 MN/m^2 and 0.20, respectively. The top of the cross-section is located approximately 3 meters under the ground surface.

6.2. SOIL PROPERTIES

The soil profile used for this case study consists of 19 layers of clay and sand materials resting on the half-space below. The total depth of the layered system is 48 m. The soil properties are quite variable over the entire depth consistent with the soft site-condition of the Taipei basin.

6.3. FREE-FIELD GROUND MOTIONS

The control free-field ground acceleration time-history representing surface motion was first generated to be compatible with the normalized pseudo-acceleration response spectrum for 5 percent damping as given by

$$\begin{aligned} S_a &= 1g && T \leq 0.054\,\text{sec} \\ S_a &= (0.238 + 14.11T)g && 0.054 \leq T \leq 0.20\,\text{sec} \\ S_a &= 3.06g && 0.20 \leq T \leq 0.168\,\text{sec} \\ S_a &= 8.661\,g/T^2 && T \geq 1.68\,\text{sec} \end{aligned} \tag{15}$$

which represents the Taipei basin site conditions (DORTS 1988). Then it was scaled to a PGA (peak ground acceleration) level equal to 0.18 g giving the resulting accelerogram shown in Figure 4.

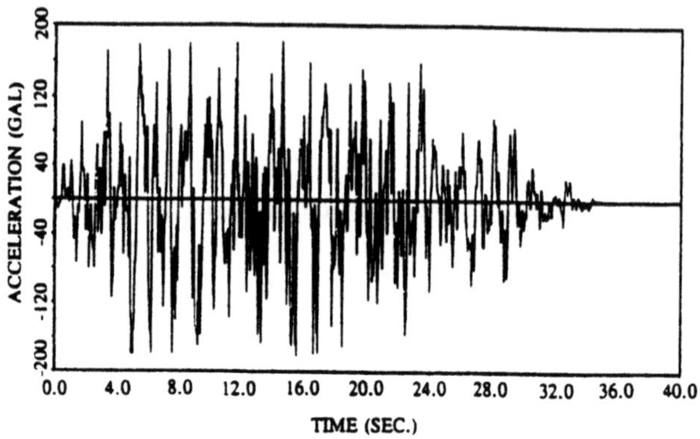

Figure 4. Free-field surface accelerogram

Using this control motion and the vertically-travelling shear-wave model of the 19-layer soil system, the time-histories of horizontal ground displacement at discrete values of elevation from top to bottom of the lining cross-section were obtained.

6.4. RACKING ANALYSIS RESULTS

A racking analysis as described in Sec. 3 was carried out for the lining cross-section shown in Figure 3 using the 19-layer soil system described above and the control acceleration time-history shown in Figure 4. The critical free-field soil displacement profile obtained in this analysis is shown in Figure 5.

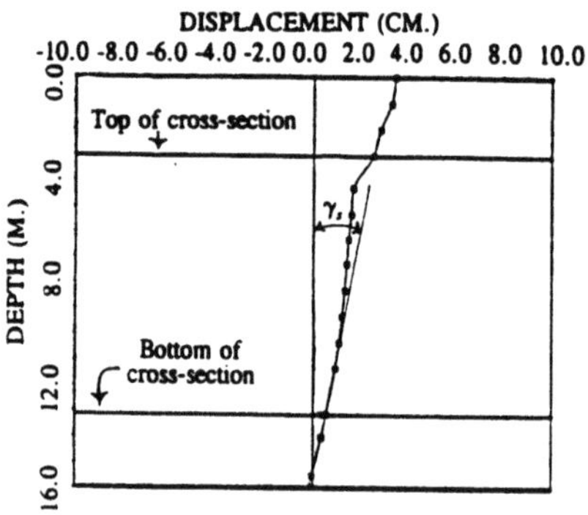

Figure 5. Critical free-field soil profile

Dividing the difference between the soil displacements at the top and bottom cross-section locations in this profile by the depth of the cross-section, one obtains an average free-field soil strain over this depth equal to 11.0×10^{-4} radians, which will be designated subsequently by γ_s.

Completing Steps 1, 2, and 3 as described in Sec. 3 and using superposition, the combined critical racking stresses in the cross-section and the corresponding nodal displacements were obtained. From the nodal displacements, a racking angle for the lining cross-section was obtained which is defined herein as the angle change between the straight line connecting Nodes 1 and 2 and the straight line connecting Nodes 3 and 4; see Figure 3. The numerical value of this angle change (γ_t) was found to be 4.6×10^{-4} radians. Dividing this angle by the critical free-field soil racking angle, one obtains $\gamma_t/\gamma_s = 4.6/11.0 = 0.42$ showing that for this basic cross-section, soil-structure interaction effects greatly reduce the racking deformations.

To study the importance of soil-structure interaction effects on the racking deformations of cross-sections when the relative stiffnesses of lining and surrounding soils are changed, parametric studies were carried out using (1) the same critical free-field soil displacement profile shown in Figure 5, (2) the same 19-layer geometry in the soil-structure interaction analysis but with all soil shear moduli (G) changed by a factor α from those values (G_0) used in the above case study ($\alpha \equiv G/G_0$), (3) the same outside geometry of cross-section as shown in Figure 3 but with all wall and slab thicknesses (t) increased by a factor β from those values (t_0) shown in the figure ($\beta \equiv t/t_0$) and (4) the same concrete Young's modulus (E) and Poisson's ratio given previously. The results of these parametric studies are shown in Figures 6 and 7.

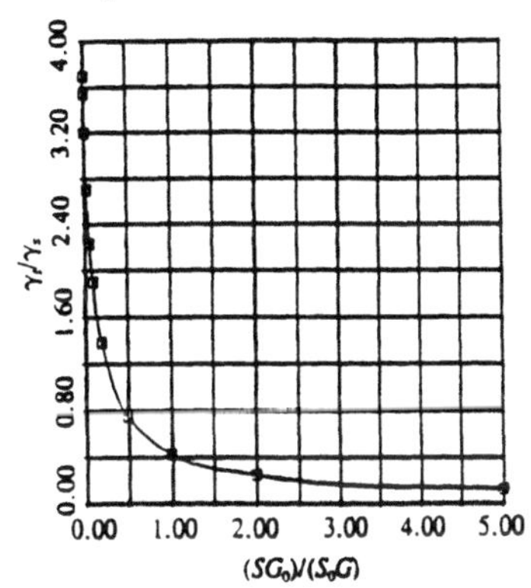

Figure 6. Racking angle ratio vs. lining/soil stiffness ratio

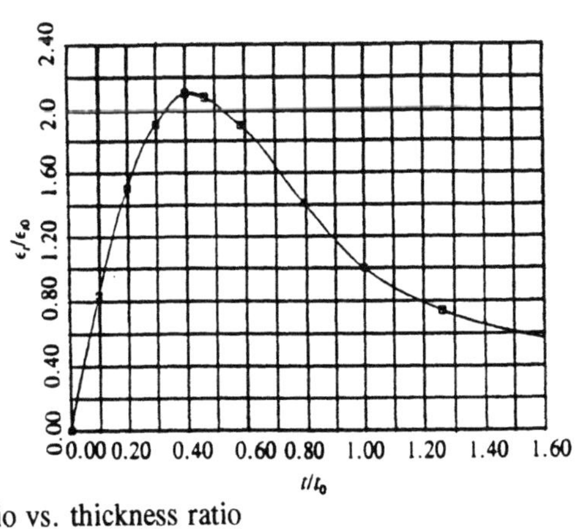

Figure 7. Strain ratio vs. thickness ratio

Figure 6 shows the ratio of cross-section racking angle to the free-field soil racking angle (γ_t/γ_s) plotted as the ordinate with the ratio $(SG_0)/(S_0G)$ plotted as the abscissa, in which S_0 represents the racking stiffness of the basic cross-section shown in Figure 3 and S is the corresponding variable stiffness. As seen in this figure, when $S = S_0$ and $G = G_0$, i.e., $(SG_0)/(S_0G) = 1$, $(\gamma_t/\gamma_s) = 0.42$ as stated above. When $(SG_0)/(S_0G)$ reduces towards zero, the ratio (γ_t/γ_s) increases to the value 3.64 corresponding to $\gamma_t = (3.64)(11.0)(10^{-4}) = 40 \times 10^{-4}$ radians. Note that this angle corresponds to the racking angle of a rectangular tunnel having no lining which is greater than the racking angle of the free-field soil having no tunnel by the factor 3.64. As $(SG_0)/(S_0G)$ increases from zero, the ratio (γ_t/γ_s) continues to decrease showing a corresponding increase in soil-structure interaction effects.

Letting $G = G_0$ so that the ratio $(SG_0)/(S_0G)$ reduces to the cross-section stiffness ratio (S/S_0), the graph of Figure 6 can be converted to that shown in Figure 7, where ϵ_0 refers to the critical flexure strain in the basic cross-section of Figure 3 due to racking and ϵ refers to the corresponding critical strain in the modified cross-section due to changing all wall, roof, and floor slab thicknesses from their basic values (t_0) to their varied values (t). This conversion is accomplished by recognizing that

$$(S/S_0) \doteq (t/t_0)^3 \tag{16}$$

and

$$(\gamma_t/\gamma_s) = (\gamma_{t0}/\gamma_s)(\gamma_t/\gamma_{t0}) \doteq (\gamma_{t0}/\gamma_s)(\epsilon/\epsilon_0)(t_0/t) \tag{17}$$

in which γ_{t0} is the racking angle of the basic cross-section in Figure 3 equal to 4.6×10^{-4} radians, γ_t is the corresponding racking angle of the modified cross-section, and γ_s is the critical free-field racking angle equal to 11×10^{-4} radians. Using these equations, the graph of Figure 7 is obtained directly from the graph of Figure 6. Note that as the thickness ratio reduces from the basic value $t/t_0 = 1$, the flexure strain ratio ϵ/ϵ_0 increases. Two partially offsetting effects are taking place. If the cross-section's racking-angle was held constant, the strain ratio would decrease with decreasing values of t/t_0 and, if the thickness ratio was held constant, the strain ratio would increase with increasing racking angles. The latter effect is dominant in this case. However, below the value $t/t_0 = 0.4$, decreasing values of t/t_0 result in decreasing values of ϵ/ϵ_0; thus, showing the former effect is dominating. Certainly, as the ratio t/t_0 goes to zero, the tunnel has no stiffness whatsoever so the racking angle is the same as the racking angle of the rectangular tunnel having no lining. While this particular racking angle is large, the strain goes to zero because the thicknesses t goes to zero. As the thickness ratio increases beyond the basic value $t/t_0 = 1$, the ratio ϵ/ϵ_0 continues to reduce due to soil-structure interaction effects.

6.5. GLOBAL RESPONSE RESULTS

Maximum axial strains were calculated by Eq. (10) for the cross-section shown in Figure 3 using the 19-layer soil medium described earlier, along with its associated soil properties, and the surface control motion presented in Figure 4. Unfortunately, limited space does not permit a detailed presentation of the results herein. Let it suffice to simply report the most critical result, namely the maximum normal stress on the cross-section which was evaluated to be approximately one-third the concrete compressive strength, i.e.,

$$|\sigma(x,t)|_{max} = E\,|\epsilon(x,t)|_{max} \doteq \frac{1}{3}f_c' \tag{18}$$

Considering that the free-field soil motion was assumed to be a single wave train moving at the most critical angle $\theta = 45°$, which is an extremely conservative assumption, the maximum stress calculated by Eq. (18) can be reduced substantially in estimating a more realistic value. Obviously, considerable judgement is required in making this final assessment.

6.6. EFFECTS OF TRANSVERSE JOINTS

Once the joint spacing L and the factor $\beta = \sqrt{k_x/AE}$ are known, Eqs. (13) and (14) can be used to obtain the mid-point axial strain and the corresponding joint separation. Since the maximum strain $|\epsilon(x,t)|_{max}$ is usually relatively low, placement of transverse joints for the purpose of reducing seismically induced strains is unnecessary. Should they be present for other reasons, their separations as caused by seismic effects can be tolerated using appropriate joint seals.

7. Concluding Statement

In assessing the seismic performance of tunnel linings, it is necessary to include soil-structure interaction effects. Once this is done, the deformations produced will usually be within tolerable limits.

8. Acknowledgement

The numerical results of the CASE STUDY were generated by the authors under a contract by and between Eastern International Engineers, Inc. and China Engineering Consultants, Inc.

9. References

Aoki, Y. (1973) 'Seismic design spectra for trench type tunnel', *Proc.* of Japan Society of Civil Eng., No. 211.

Department of Rapid Transit Systems (DORTS) (1988) 'Civil engineering design manual', Taipei Municipal Government, R.O.C.

Kuesel, T. R. (1969) 'Earthquake design criteria for subways', *Proc.* of ASCE, 95, ST6, 1213-1231.

Okamoto, S. and C. Tamura (1973) 'Behavior of subaqueous tunnels during earthquakes', *Earthquake Engineering and Structural Dynamics* 1, 253-166.

Penzien, J. and W. S. Tseng (1981) 'Seismic analysis of subaqueous tunnels', Workshop on Earthquake Engineering, The Chinese Institute of Civil and Hydraulic Engineering, Taipei.

Schnabel, P. B., et al. (1972) 'SHAKE - A computer program for earthquake response analysis of horizontally layered sites', Report No. EERC 72-12, University of California, Berkeley, California.

Seed, H. B. and I. M. Idriss (1970) 'Shear moduli and damping factors for dynamic response analysis', Report No. EERC 70-10, University of California, Berkeley, California.

PILE-SOIL-PILE INTERACTION UNDER SMALL AND LARGE DISPLACEMENTS

M. Novak
Faculty of Engineering Science
The University of Western Ontario
London, Ontario N6A 5B9
Canada

ABSTRACT. The paper describes a practical approach to the evaluation of stiffness and damping of pile groups based on dynamic interaction factors. Kinematic interaction under seismic excitation and the effect of large displacements are also discussed.

1. Introduction

Studies of pile behaviour under dynamic loads have two principal objectives: one is the evaluation of all possible hazards the piles may be exposed to; the other is the formulation of rational approaches to soil-pile-structure interaction. The hazards to pile safety include soil liquefaction, particularly when associated with large soil displacements, embankment movement, slope failure and resulting slides, and also inertia (seismic) forces generated by the vibration of the superstructure and deformations in the soil deposit during its vibration. The hazards mentioned have caused numerous pile failures, but the focus of this paper is on interaction of pile groups with the structure they support. The emphasis is on seismic excitation but the approaches treated are applicable to other types of loading too.

In order to establish the seismic forces expected to act on the piles and the structure, the evaluation of soil-pile-structure interaction is needed. This can be done experimentally or theoretically, but for design purposes the theoretical analysis is more readily applicable. The whole soil-pile-structure system can be analyzed either directly in one step or by means of substructuring. The direct one step analysis is desirable if strong nonlinearities are to be accounted for but for most practical cases, the substructuring technique, with the pile foundation being the substructure, is more readily applicable. In this procedure, the complete solution to the dynamic response is achieved in two steps, i.e. the kinematic interaction analysis, which accounts for the effect of seismic waves on pile motion, and inertial interaction analysis, which deals with the inertia (earthquake) forces generated by the motion of the building. This type of analysis is schematically depicted in Figure 1. Such an approach is essentially linear but nonlinearity can be accounted for approximately using equivalent linearization, adjusting soil properties to the expected level of strain, and other measures discussed later herein.

P. Gülkan and R. W. Clough (eds.), Developments in Dynamic Soil-Structure Interaction, 361–380.

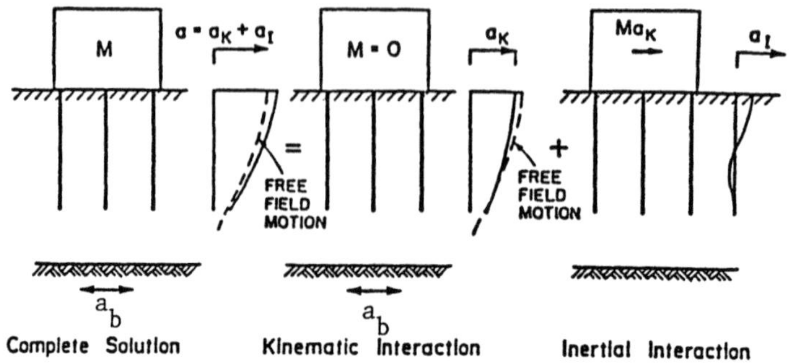

Figure 1 Schematic of complete seismic response analysis and separation of kinematic interaction from inertial interaction.

2. Kinematic Interaction

Kinematic interaction is a term used to describe the seismic response of piles, without the superstructure, relative to the free field ground motion. The difference between these two motions occurs because due to their stiffness and mass, the piles interact with the soil and interfere with the passage of seismic waves. Kinematic interaction is a complex phenomenon, particularly for pile groups whose behaviour is complicated by pile-soil-pile interaction.

Kinematic interaction can be quantified by expressing it as dimensionless ratios, known as transfer functions or kinematic interaction, relating the absolute values of pile head amplitudes to the free field horizontal amplitudes at ground surface, u_g. These ratios depend primarily on the type of the seismic wave, dimensionless frequency, a_o, the stiffness ratio E_p/E_s, the soil profile and for groups, also on pile number, their spacing ratio, s/d, and fixity to the cap. The dimensionless frequency

$$a_o = \omega d/V_s \tag{1}$$

where ω = circular frequency, d = pile diameter and V_s = soil shear wave velocity; E_p and E_s are Young's moduli of the pile and soil, respectively, and s = the distance between the piles. The basic types of seismic waves are the body waves, i.e. the SH and SV shear waves and the primary or P waves, and Rayleigh or surface waves. The body waves can act under different angles of incidence, θ, resulting in qualitatively and quantitatively different effects.

Most of the data available have been obtained by numerical models of the boundary element or finite element type assuming harmonic excitation. Examples of kinematic interaction factors obtained using the boundary element method are shown for single piles in Figure 2, displaying data for the most often assumed SH waves in which the soil particles oscillate horizontally, parallel to the propagating wave front. The data are for floating, vertical, single piles in homogeneous viscoelastic halfspace with slenderness ratio L/d=20, where L = pile length and ρ_p/ρ_s=3/2, where ρ_p and ρ_s are mass densities of the pile and soil, respectively. As Figure 2b indicates, the horizontal motion of single piles is accompanied by rocking of the pile head. At low dimensionless frequencies, a_o, the translation interaction

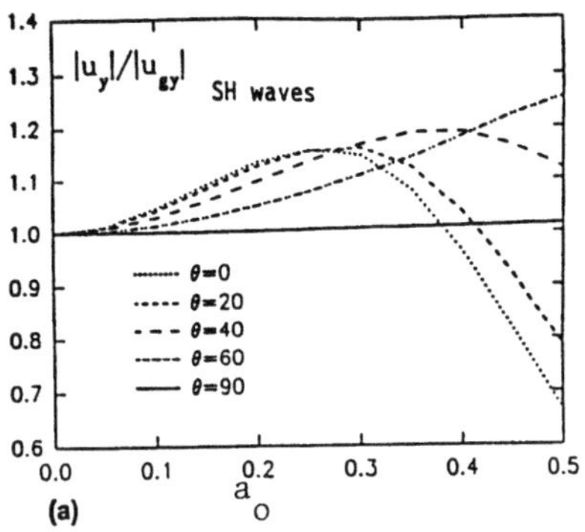

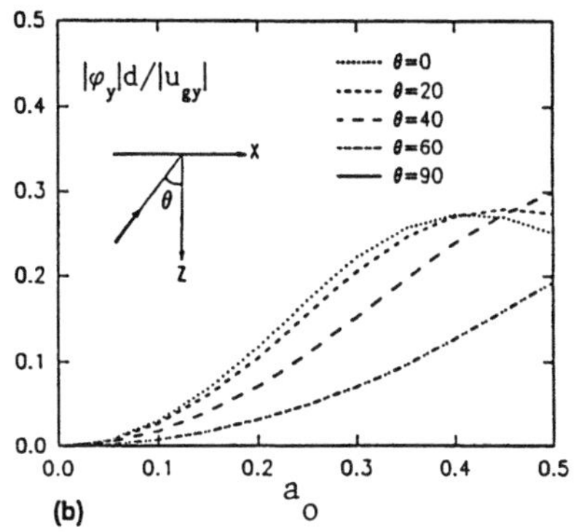

Figure 2 Kinematic interaction factors for single piles under SH shear waves acting with different angles of incidence θ: (a) - horizontal translation u_y and (b) - rocking ϕ_y (E_p/E_s = 1000; Kaynia and Novak, 1992).

factors are close to unity and the rocking interaction factors are close to zero, both indicating little kinematic interaction. At higher frequencies, kinematic interaction becomes significant and is strongly dependent on the angle of incidence, θ, with the greatest interaction effects produced by upward propagating shear waves (θ=0). The pile head translations are first amplified and then de-amplified; for smaller stiffness ratios than 1000 assumed in Figure 2, only a modest amplification occurs. For single endbearing piles under upward propagating shear waves, Gazetas (1984) generalized the results of a large parametric study for three soil profiles, i.e. linear, parabolic and constant and provided universal kinematic interaction factors.

For pile groups, kinematic interaction effects of SH waves are quite different. This is exemplified by Figure 3, displaying kinematic interaction factors for a 3x3 group with pile heads fixed to a massless, rigid cap, and can be summarized as follows: Kinematic interaction is weakest for upward propagating shear waves (θ=0°) and strongest for θ=90°; torsional pile group response is generated except for upward propagating shear waves; and rocking is negligible. Rocking is, however, generated by SV shear waves and Rayleigh waves. Further data on kinematic interaction in pile groups can be found for different types of waves in Kaynia and Novak (1992) and for shear waves propagating upwards in nonhomogeneous media in Fan et al. (1991).

For design purposes, the kinematic interaction factors can be applied in two ways: approximately, the free field response spectrum can be multiplied by the pertinent kinematic interaction curve to yield the design response spectrum for the motions at the pile heads. More rigorously, the free field motion can be Fourier transformed and each Fourier amplitude corrected by the kinematic interaction factor for the corresponding frequency. Inverse transform yields the design motion at the pile heads.

With the design motions at the pile heads established in this way, the inertial interaction analysis can proceed as is usual in seismic response analysis. To account for pile-structure interaction in this analysis, pile foundation stiffness and damping constants have to be evaluated. Basic data on this are given in the next section.

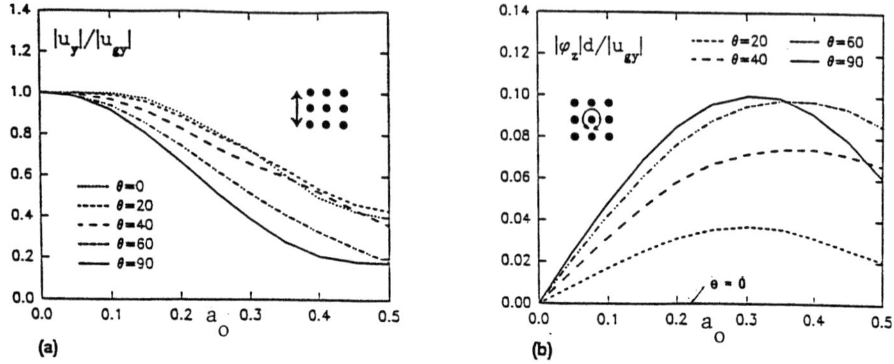

Figure 3 Kinematic interaction factors for 3x3 pile groups under SH shear waves: (a) - horizontal translation and (b) - torsion about vertical axis; (E_p/E_s=1000, pile spacing s/d=5, fixed head piles; Kaynia and Novak, 1992).

3. Stiffness and Damping Constants of Pile Foundations

When piles vibrate under dynamic loads, they interact with the surrounding soil. This interaction affects pile stiffness making it, in general, frequency dependent and dissipates energy which manifests itself as pile damping. The damping is caused by the radiation of elastic stress waves in the soil medium, hence radiation or geometric damping, and by material hysteresis, i.e. imperfect elasticity, in the soil and to a lesser degree in the pile. In groups of closely spaced piles, the interaction process is further complicated by interaction between individual piles known as pile-soil-pile interaction or the group effect.

Impedance functions are defined as amplitudes of harmonic forces (or moments) that have to be applied to the pile head in order to generate a harmonic motion with a unit amplitude in the specified direction. Because the generation of a specific impedance calls for a sole displacement amplitude, the horizontal impedance K_u, being an external force, must be accompanied by a coupling impedance $K_{\psi u}$ representing a moment acting in the direction ψ when there is a unit amplitude in the direction X. Because of the reciprocity principle, $K_{\psi u}=K_{u\psi}$.

The impedance functions can be expressed in any of the following ways, i.e.

$$K = K_1 + iK_2 \tag{2a}$$

$$= k + i\omega c \tag{2b}$$

$$= \bar{k}(k' + i\omega c') \tag{2c}$$

in which K_1 = ReK and K_2 = ImK are the real and imaginary parts of the complex stiffness, respectively and $i=\sqrt{-1}$; $k = K_1$ = true stiffness, $c = K_2/\omega$ = coefficient of equivalent viscous damping, and ω = circular frequency; $\bar{k}$ = static stiffness and $k' = k/\bar{k}$, $c' = c/k'$ = dimensionless stiffness and damping constants. All the parameters in Eqs. 2 depend on frequency ω or the dimensionless frequency a_o by Eq. 1 or

$$a_o = R\omega/V_s \tag{3}$$

where R=pile radius. Notice, this a_o is one half of that expressed in terms of d by Eq. 1.

3.1 SINGLE PILES

Extensive data on single pile impedances are available in the literature. They were obtained using the theory of elasticity in which the soil properties are described by the shear modulus $G_s = V_s^2\rho_s$, mass density ρ_s, Poisson's ratio ν and material damping. The latter can be characterized either by material damping ratio β or by $D=\tan\delta=2\beta$ where δ is the loss angle. Viscous, i.e. proportional to vibration velocity and thus frequency, material damping is sometimes assumed but frequency independent hysteretic damping is more realistic.

A schematic of the impedance of an endbearing pile, expressed in the alternative ways by Eqs. 2 is shown in Figure 4 in which the difference between viscous material damping and hysteretic, frequency independent damping is emphasized. For floating piles, the role of material damping is smaller as the layer resonances are absent and geometric damping due to progressive waves acts at all frequencies > 0. A schematic of floating pile impedances is shown in Figure 5 in which the effect of hysteretic, frequency independent damping is also depicted.

A few salient features of the pile impedances follow from the theoretical solutions: pile dynamic stiffness varies little with frequency, except for very heavy piles or very weak soils; for endbearing piles vibrating below the fundamental frequency of the soil layer, the geometric damping is absent because no progressive waves are generated in perfectly elastic medium, leaving soil and pile material damping as the only sources of energy dissipa- tion. Soil material damping reduces pile stiffness somewhat at a rate increasing with frequency and increases the total damping; with hysteretic damping the equivalent viscous damping constant grows ever more abruptly as the frequency approaches zero. The latter feature results from the definition of the equivalent viscous damping constant according to which $c = K_2/\omega$. Detailed data on single pile impedances can be found in Novak and El Sharnouby (1983), Gazetas (1984), Rajapakse and Shah (1989) and elsewhere.

3.2 PILE GROUPS

Piles are usually used in groups. When the distance between the piles is large, say twenty diameters or more, the piles do not affect each other. If, however, the piles are closely spaced, they interact with each other and this pile-soil-pile interaction or group effect exerts considerable influence on the properties and behaviour of the group. These two basic situations may be treated separately.

3.2.1 *Pile Interaction Neglected.* When the spacing between the piles is large, their interaction can be neglected and the stiffness and damping of the group are determined by the summation of stiffness and damping constants of the individual piles. In the vertical and horizontal directions this is straight-forward; for coupled sliding and rocking as well as torsion, the position of the reference point such as the cap centre of gravity, CG, and the arrangement of the piles in plan come into play. The group stiffness and damping in rotation derive from the horizontal, vertical and moment resistances of individual piles because the unit displacement $\psi=1$ occurs at the reference point (Figure 6). With the notations of

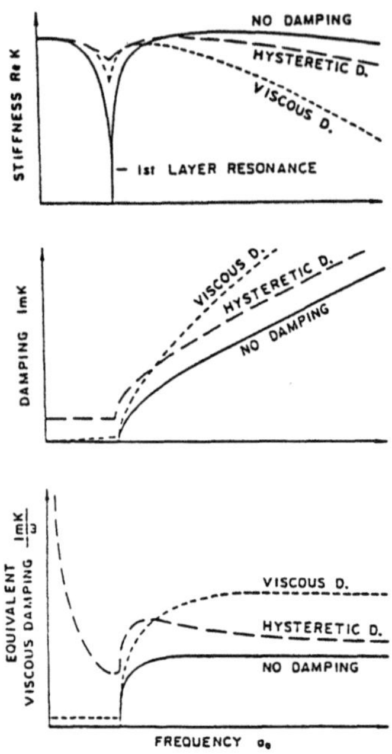

Figure 4 Schematic of endbearing pile impedance for three cases of soil material damping.

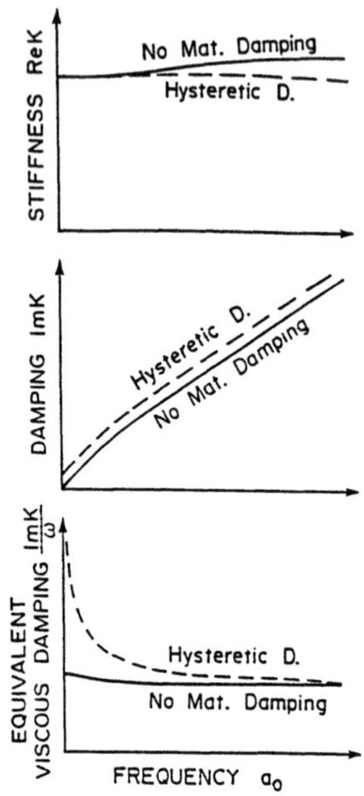

Figure 5 Schematic of floating pile impedance showing the effect of hysteretic material damping.

Figure 6, the stiffness constants of the pile group are for the individual directions as follows:

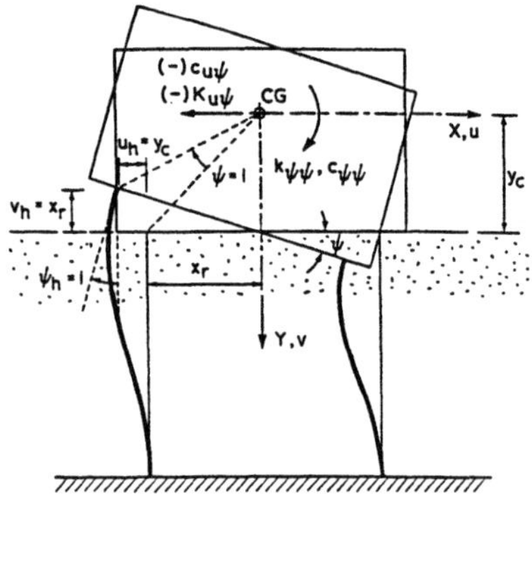

Figure 6 Generation of pile group impedances associated with cap rotation.

Vertical and horizontal translations

$$k_{vv} = \sum_r k_v \; ; \; k_{uu} = \sum_r k_u \tag{4}$$

Rotation of the cap in the vertical plane:

$$k_{\psi\psi} = \sum_r (k_\psi + k_v x_r^2 + k_u y_c^2 - 2k_c y_c) \tag{5}$$

Coupling between horizontal translation and rotation:

$$k_{u\psi} = k_{\psi u} = \sum_r (k_c - k_u y_c) \tag{6}$$

Torsion about vertical axis:

$$k_{\eta\eta} = \sum_r [k_\eta + k_u(x_r^2 + z_r^2)] \tag{7}$$

The summation extends over all the piles and the single subscripts indicate stiffness constants of single piles with k_c representing coupling between translation, u, and rocking, ψ. The distances x_r, z_r and y_c refer to the reference point as indicated in Figure 6. The torsional constant k_η related to pile twisting about its own axis can usually be neglected. The rotation and coupling impedances simplify if the cap is thin and y_c is taken as zero. For damping, formulae analogous to Eqs. 4 to 7 can be written with single pile damping constants replacing the stiffness constants.

3.2.2 *Pile Interaction Considered.* When piles are closely spaced, they interact with each other because the displacement of one pile contributes to the displacements of the others. Under *static* loads, pile interaction increases group settlement, redistributes the loads on individual piles and reduces bearing capacity, unless this reduction is counteracted by densification of the soil within the group due to pile driving. The investigation of static group effects, based on continuum mechanics, provided extensive data available in Poulos and Davis (1980), Butterfield and Douglas (1981), El Sharnouby and Novak (1985, 1990) and elsewhere. The static data are useful even in dynamics because at low frequencies, and particularly below the fundamental frequency of a stratum, the dynamic stiffness is usually quite close to the static stiffness.

Dynamic investigations of pile groups, reviewed in Novak (1991), indicate very complex behaviour: dynamic group effects are profound and differ considerably from static group effects; dynamic stiffness and damping of pile groups vary with frequency and these variations are much more pronounced than with single piles; group stiffness and damping can be either reduced or increased by pile-soil-pile interaction, may exhibit very sharp peaks and are affected even for very large pile spacings; and finally, group response cannot be deduced from a single pile response without due regard for pile-soil-pile interaction.

The analytical and numerical models available for direct group analysis are mostly of the boundary element or finite element type and their application requires considerable computing time and effort. A practical alternative is the superposition technique. The superposition technique widely used for approximate static analysis of pile groups, was extended to the dynamic case by Kaynia and Kausel (1982). This technique is implemented through a flexibility approach in conjunction with the dynamic interaction factors. The interaction factors define the normalized displacements of a load-free pile due to dynamic loads on an adjacent pile when no other piles are present. Interaction factors not only can simplify the dynamic analysis of pile groups but also contain features which elucidate the group behaviour.

The applicability of the superposition method depends on its accuracy, which can be assessed by comparing the results of direct analysis with those of the superposition method for typical pile group parameters. Such a comparison is displayed in Figures 7 and 8, showing the horizontal and vertical group stiffness and damping normalized by the sum of the single pile stiffnesses for the static case. Thus, the normalized group stiffness = $k/nk^s(0)$ and the normalized group damping = $c/nk^s(0)$ where $k^s(0)$ is the static single pile stiffness of the vibration mode considered and n is the number of piles. The results shown were calculated using the boundary integral approach formulated by Kaynia and Kausel (1982).

Figures 7 and 8 show the normalized stiffness and damping of a 3x3 pile group with different spacings for floating and endbearing piles in homogeneous soil. The direct analysis results are shown as continuous curves while the results from the superposition method are plotted as discrete points. The normalized single pile stiffness and damping are also given in order to show the group effect on the frequency variation. The parameters shared by all cases are as follows: $L/d=20$, $\rho_p/\rho_s=1.25$, $\nu_s=0.4$ and $\beta_s=0.05$.

Figure 7 demonstrates the typical group effects, the most prominent one being the strong fluctuations of group stiffness and damping with frequency as opposed to the corresponding single pile quantities. At low frequencies the pile groups, with the enclosed soil mass, behave similarly to a rigid footing, indicating a decrease in stiffness due to inertial effects. At some higher frequencies the forces carried by one pile tend to displace the closest piles in the opposite direction, hence the stiffness is increased. This dynamic group behaviour has been recognized for quite some time. The location and magnitude of the peaks in stiffness and damping depend on the pile spacing and the soil profile.

Figure 7 shows the comparison of the results from the two analyses for the case of a floating pile in homogeneous halfspace. In the horizontal case, the superposition method gives very good results for all frequencies. In the vertical case, significant deviations occur for the closely spaced pile group for frequencies higher than $a_o=0.25$. The moderately spaced pile group shows good agreement. The agreement is even better for stiffer piles than those assumed in Figures 7 and 8.

Figure 8 shows the comparisons for the case of endbearing piles embedded in a homogeneous soil underlain by a rigid base. Again, the horizontal case is in very good agreement. For the horizontal case the endbearing group stiffness and damping are very close to the corresponding homogeneous halfspace values except for the vanishing of radiation damping below the natural frequency of the layer and for the local peaks and troughs at subsequent resonance frequencies. The vertical group stiffness of the closely spaced pile group shows poor agreement between the two methods while for moderately spaced piles ($s/d=5$) the trend of the frequency variation is reproduced although differences occur within the regions of peaks in both stiffness and damping. The relative discrepancies are greater for stiffer pile groups (not shown) although the overall group effect is less pronounced compared to the homogeneous halfspace soil.

For piles in soils with parabolic and linear variations of the soil shear modulus with depth analogous comparisons are made in El Marsafawi et al. (1992). From all these comparisons it may be concluded that the superposition method works on the whole quite well except for the vertical response of endbearing piles and the higher frequency range in nonhomogeneous

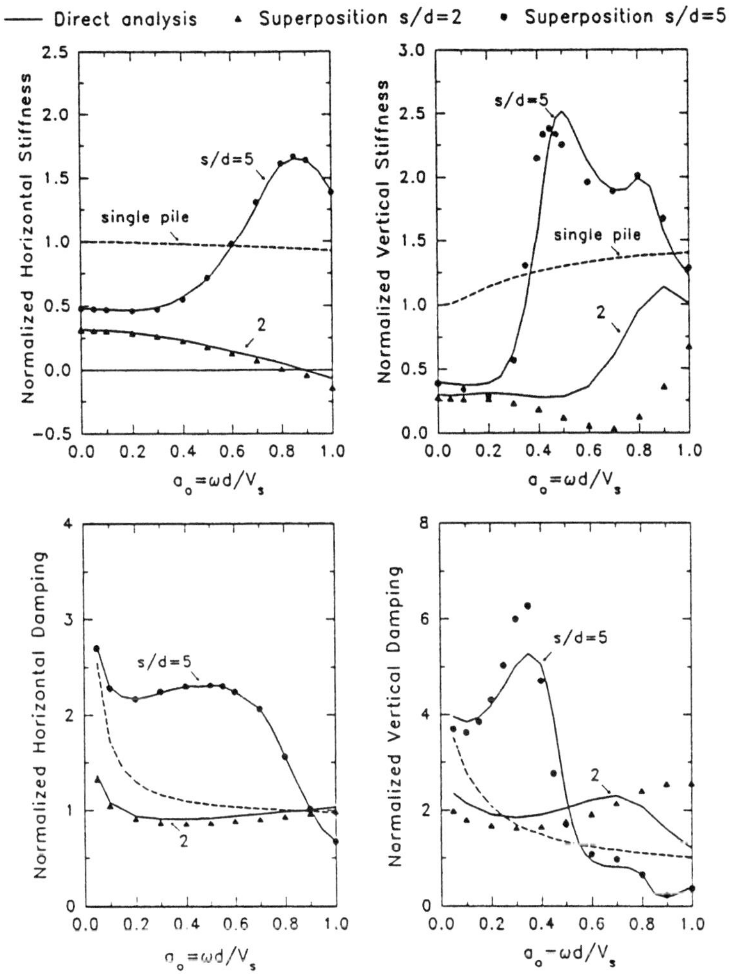

Figure 7 Horizontal and vertical normalized dynamic stiffness of a 3x3 floating pile group in homogeneous halfspace using direct analysis and the superposition method ($E_p/E_s=100$, L/d=20; El Marsafawi, Kaynia and Novak, 1992).

media. Thus, the superposition method based on dynamic interaction factors may be quite adequate for many practical applications.

3.2.3 *Dynamic Interaction Factors*. Dynamic interaction factors are dimensionless, frequency dependent complex numbers, defined as

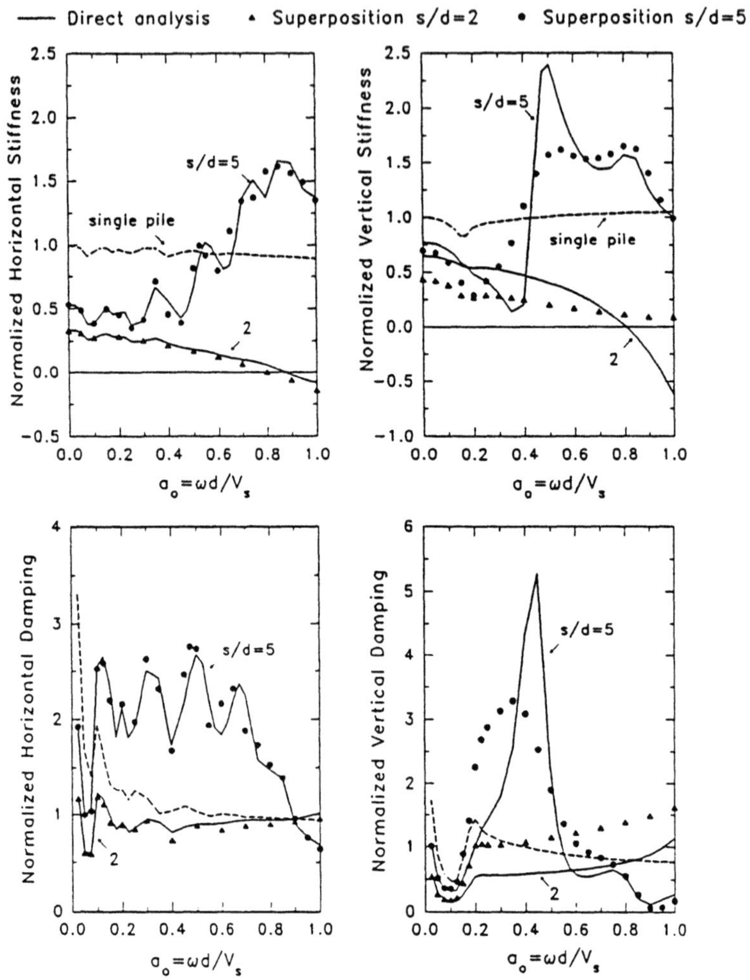

Figure 8 Horizontal and vertical normalized dynamic stiffness of a 3x3 endbearing pile group using direct analysis and the superposition method ($E_p/E_s=100$, L/d=20; El Marsafawi, Kaynia and Novak, 1992).

$$\alpha = \frac{\text{Dynamic displacement of pile 2}}{\text{Dynamic displacement of pile 1}} \tag{8}$$

in which the displacement of pile 2 is caused by a unit harmonic load on pile 1 and the dynamic displacement of pile 1 is established for an isolated pile. The displacement is either translation or rotation. (Static displacement can also be used in the denominator of Eq. 8.)

An example of the real and imaginary parts of the vertical interaction factors, calculated using the Kaynia and Kausel (1982) method for different spacing ratios s/d is plotted for homogeneous soil in Figure 9a. The interaction factors are oscillatory in character, i.e. negative as well as positive. Negative values of the imaginary part indicate a possible increase in group damping.

The interaction factors are commonly displayed in terms of their real and imaginary parts as in Figure 9a. This is a usual form but it makes interpolation for intermediate spacings difficult, especially at higher frequencies. This difficulty can be circumvented if the interaction factors are expressed in terms of amplitude, $|\alpha|$, and phase, ϕ. Thus, the dynamic interaction factor can be written as

$$\alpha = \alpha_1 + i\alpha_2 \tag{9a}$$

or

$$\alpha = |\alpha| e^{i\phi} \tag{9b}$$

in which the absolute value $|\alpha|$ and the phase angle ϕ are

$$|\alpha| = \sqrt{\alpha_1^2 + \alpha_2^2} \tag{10a}$$

$$\phi = \tan^{-1} (\alpha_2/\alpha_1) \tag{10b}$$

The vertical interaction factors from Figure 9a are presented in the latter form in Figures 9b and 9c. Also plotted in Figure 9c is the phase divided by a_o, ϕ/a_o. The latter form, shown in dashed lines, facilitates more accurate reading of the phase at low frequencies and is, therefore, used in the presentation of the vertical interaction factors together with the horizontal interaction factors for two basic load directions and fixed head piles in Figure 10. These data for floating piles are complemented by interaction factors for endbearing piles in Figure 11. For flexible piles with slenderness ratio greater than the L/d=20 used in Figure 11, the interaction factors are close to those of floating piles. With the amplitude and phase read from these figures, the real and imaginary parts of the interaction factors are

$$\alpha_1 = |\alpha| \cos\phi \ , \ \alpha_2 = |\alpha| \sin\phi \tag{10c}$$

The fixed-head horizontal interaction factors, plotted in Figures 10 and 11, are calculat- ed for piles whose heads do not rotate. They are adequate for the response which is dominated by horizontal translations, i.e. sliding, as is the case with a broad pile cap supporting a low structure. If significant cap rocking is anticipated, pile group analysis should consider coupling between all pile head motions, as indicated in Figure 6, and free-head interaction factors are needed. Such dynamic interaction factors are available in Kaynia and Kausel (1982), Fan et al. (1991) and El Marsafawi et al. (1992). These include interaction factors for pile head rotations and coupling and nonhomogeneous soil profiles.

The horizontal interaction factors for angles of load incidence, β, between 0° and 90° can be calculated from those for the two limit cases of $\beta=0$ and $\beta=\pi/2$ using the formula

$$\alpha(\beta) = \alpha(0^\circ)\cos^2\beta + \alpha(90^\circ)\sin^2\beta \tag{11}$$

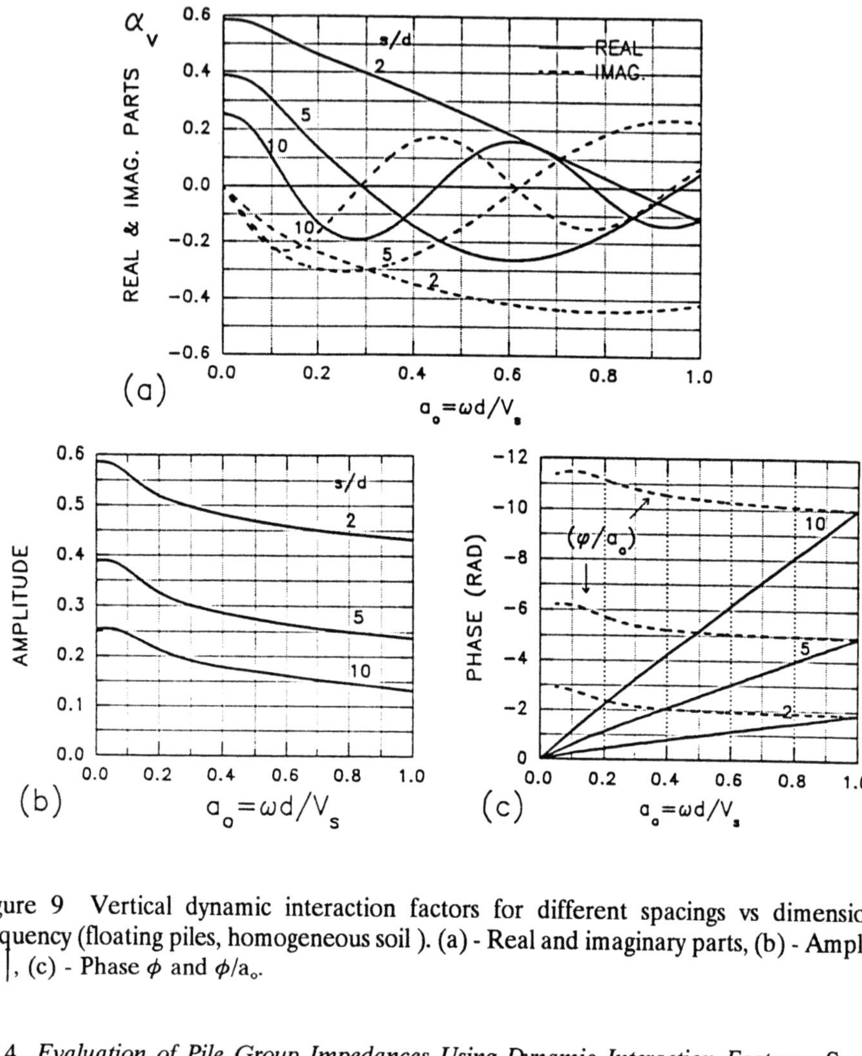

Figure 9 Vertical dynamic interaction factors for different spacings vs dimensionless frequency (floating piles, homogeneous soil). (a) - Real and imaginary parts, (b) - Amplitude $|\alpha|$, (c) - Phase ϕ and ϕ/a_o.

3.2.4 *Evaluation of Pile Group Impedances Using Dynamic Interaction Factors.* Superimposing the contributions from individual piles by means of the interaction factors and imposing the pertinent boundary conditions, the impedance for each vibration mode can be obtained. The basic vibration mode is horizontal translation of a group with a rigid cap. The evaluation of this case is simple if the loads on all piles of the group are the same, as in a symmetrical group of four piles, or are approximately assumed to be so. Consider fixed-head piles and denote the single pile complex impedance, K. Then, the complex stiffness of a group of n piles becomes

$$K^G = \frac{nK}{1 + \sum_{j=2}^{n} \alpha_j} \tag{12}$$

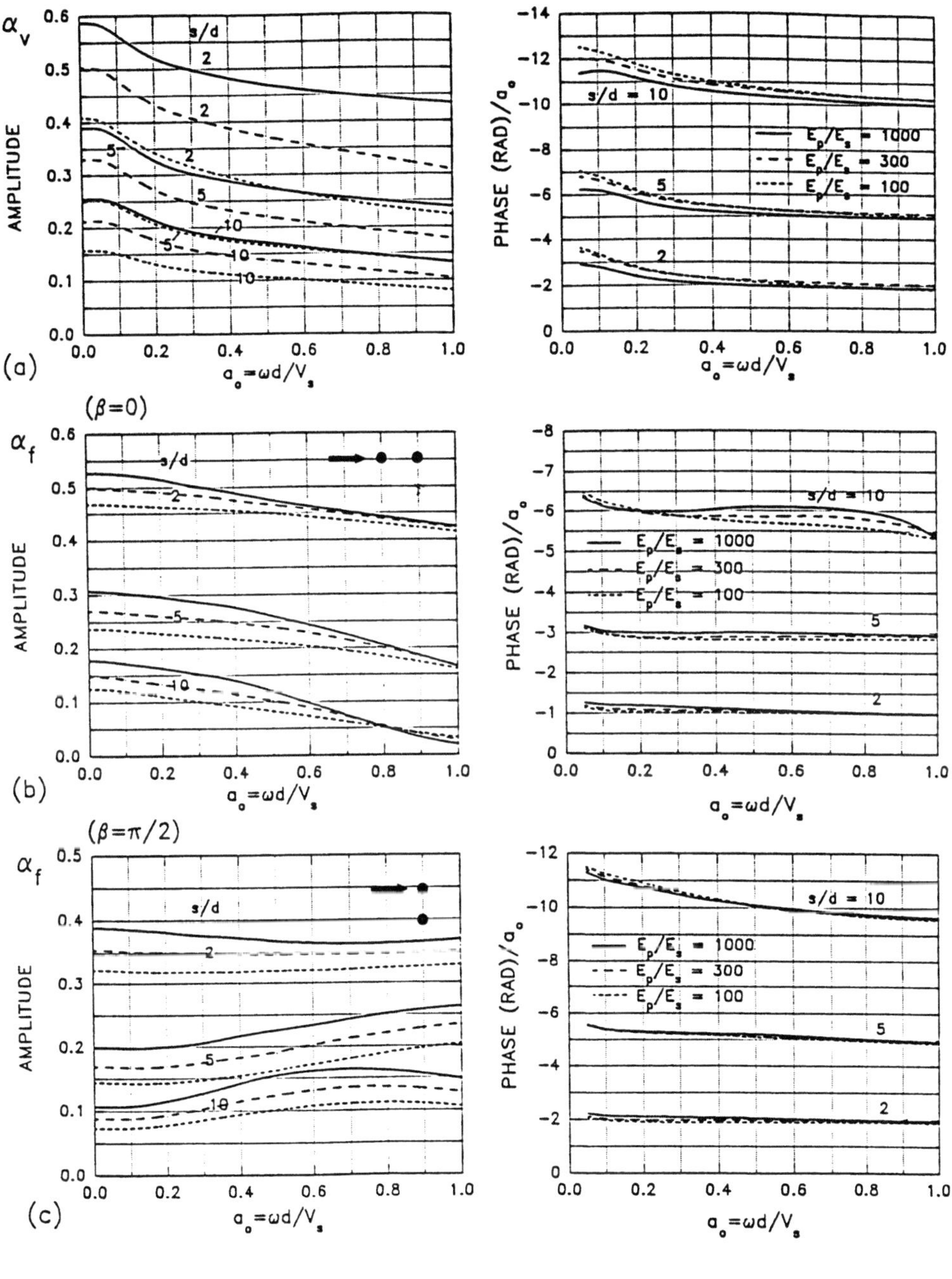

Figure 10 Interaction factors for floating, fixed-head piles in terms of amplitude and phase: (a) - vertical, (b) - horizontal with piles in line with load, $\beta=0$, (c) - horizontal with $\beta=90°$ (homogeneous soil, L/d=20, $\rho_p/\rho_s=1.25$, $\beta_s=0.05$, $\nu_s=0.4$; El Marsafawi, Kaynia and Novak, 1992).

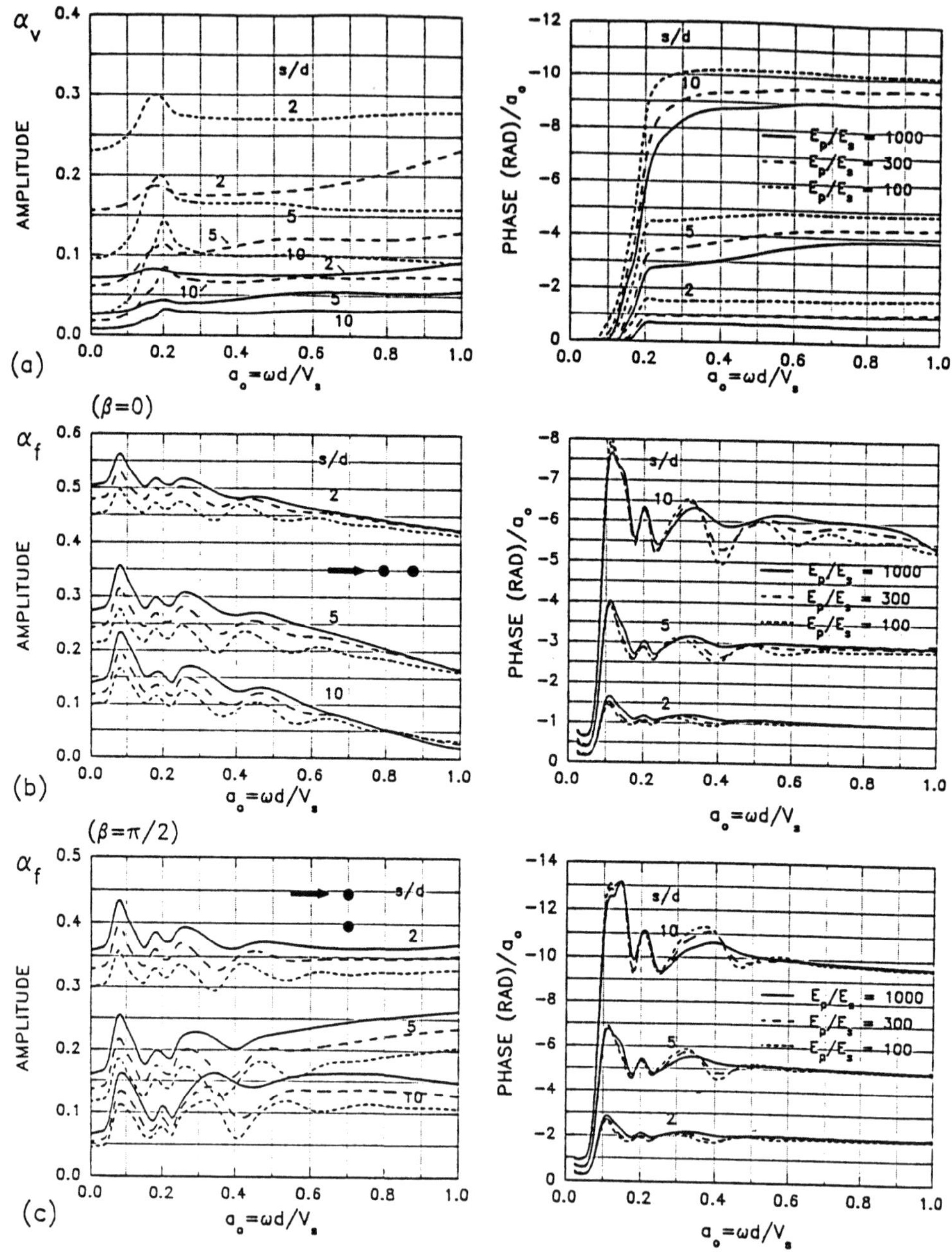

Figure 11 Interaction factors for fixed-head endbearing piles: (a) - vertical, (b) - horizontal with piles in line with load, $\beta=0$, (c) - horizontal with $\beta=90°$ (homogeneous soil, L/d=20, $\rho_p/\rho_s=1.25$, $\beta_s=0.05$, $\nu_s=0.4$; El Marsafawi, Kaynia and Novak, 1992).

To extract real stiffness and damping from this complex impedance denote

$$a = 1 + \sum_{j=2}^{n} \alpha_{j(1)} , \qquad b = \sum_{j=2}^{n} \alpha_{j(2)} \tag{13}$$

in which the subscripts (1) and (2) indicate respectively the real and imaginary parts of the interaction factors, by Eq. 10, between a reference pile, number one, and the other piles. Denoting further,

$$A = \frac{a}{a^2 + b^2} , \qquad B = \frac{-b}{a^2 + b^2} \tag{14}$$

the pile group real horizontal stiffness follows from Eq. 12 as

$$k^G = k_{uu} = n(k_u A - c_u B\omega) \tag{15a}$$

and the group coefficient of equivalent viscous damping as

$$c^G = c_{uu} = n(c_u A + k_u B/\omega) \tag{15b}$$

where k_u and c_u are single pile stiffness and damping coefficients. Vertical group impedance is obtained in analogous fashion. Similar formulae can be written for rocking and torsion of the group when the loads on individual piles are assumed to be proportional.

More rigorous formulae, not limited by the assumptions of equality or proportionality of the loads on individual piles, and including coupling between rocking and horizontal translation, as well as flexible caps are given in Novak and Mitwally (1990). For example, the horizontal or vertical dynamic group impedance is more generally defined as

$$K^G = K \sum_i \sum_r \varepsilon_{ir} \tag{16}$$

in which ε_{ir} are the elements of the inverted matrix $[\alpha]^{-1} = [\varepsilon]$ in which the matrix $[\alpha]$ lists all the complex dynamic interaction factors α_{ir} between any two piles in the group. For larger groups, these calculations are tedious unless suitable software is available. One such code is DYNA3. Dynamic pile-soil-pile interaction is a complication but its inclusion is important. In most cases, it reduces group stiffness and increases damping with corresponding effect on foundation response.

3.3 RESPONSE OF PILE SUPPORTED STRUCTURES

Once the impedances of the pile group are available, they can be included in the analysis of the superstructure just as in the case of shallow foundations. (See, e.g., the paper by Veletsos in these Proceedings.) For earthquake response analysis, kinematic interaction should be considered. For the horizontal response approached in terms of a response spectrum this can be done in an approximate way by multiplying the response spectrum coordinate, say for spectra acceleration S_a, by a kinematic interaction factor, $I_u = |u_y| / |u_{gy}|$, established by means of Figures 2a, 3a and similar data from the literature. Then, the horizontal seismic force on a rigid body with mass m vibrating in one degree of freedom is simply

$$Q_u = m\ S_a\ (\omega_1, \zeta_1)\ I_u\ (\omega_1) \tag{17}$$

in which ω_1 and ζ_1 are the natural frequency and total damping ratio of the system evaluated considering soil-structure interaction. For multi-degree-of-freedom systems, Eq. 17 is extended accordingly.

4. Nonlinear Response of Pile Foundations

Under large displacements, pile foundations behave in a nonlinear fashion. This manifests itself through a nonlinear relation between pile loading and deflection and associated with it hysteretic loops of complex shapes, or by the variation in resonance frequencies with shaking intensity and the lack of proportionality between vibration amplitudes and excitation forces. The reasons for such behaviour are soil nonlinearity at high strain, pile separation (gapping), soil degradation during cycling, as well as slippage and friction. The effects of these factors on pile impedance is to reduce both stiffness and radiation damping.

4.1 NONLINEAR RESPONSE OF SINGLE PILES

To incorporate the nonlinear factors into a continuum theory is extremely difficult and therefore, lumped mass models are most often used when nonlinear analysis is required. Such models comprise a number of lumped masses, interconnected by massless beam elements, with attached discrete nonlinear springs, nonlinear dampers, gaps and Coulomb friction blocks. The combination of these elements makes it possible to generate a variety of nonlinear force-displacement relationships and complex hysteretic loops associated with gapping observed in experiments. Models of this type were formulated by Penzien (1970), Matlock et al. (1978) and others. Such models are very versatile but it is difficult to relate their characteristics to standard geotechnical parameters of soil.

To help overcome this difficulty, various nonlinear soil resistance-deflection relationships known as p-y curves, for horizontal response, and t-z curves, for vertical response, have been recommended in the literature, e.g. by The American Petroleum Institute (1991).

Cyclic loading with low frequency provides basic insight into the material degradation due to soil plasticity and mechanic degradation due to gapping associated with large displacements. Many studies were devoted to this subject. Summarizing their observations, Swane and Poulos (1982) postulate that during cyclic lateral loading the two forms of degradation lead to the increase in pile deflection and bending stresses; but if this degradation stabilizes, the pile is said to "shakedown" to a state of permanent strains and residual stresses and will react elastically to any further cyclic loading of the same intensity. When the pile does not stabilize into an elastic or inelastic response, the pile deflections continue to increase and incremental collapse may result.

The shakedown phenomenon is favourable from the point of view of the applicability of the linear theories outlined above. It explains why, with adequate adjustments, particularly for pile separation, such theories may give reasonable results, even in cases where overall strong nonlinearity of the response is manifested.

4.2 NONLINEAR RESPONSE OF PILE GROUPS

Nonlinear dynamic analysis of pile groups is very difficult. Akiyoshi and Fuchida (1982) formulated an approximate solution for vertically vibrating endbearing piles considering imperfect adhesion between the pile and the soil and Nogami and Konagai (1987) developed a similar analysis. Nogami et al. (1988) and Otani et al. (1991) extended the concept of the dynamic Winkler medium further to include horizontal response, slippage, gapping and inelastic soil behaviour being able to generate a variety of degrading hysteresis loops. To account for the nonlinear effects in a very approximate way, Novak and Sheta (1980) introduced a weak cylindrical zone around the pile thus allowing for radial nonhomogeneity. For single piles this zone reduces pile stiffness somewhat but reduces radiation damping substantially. In pile groups, the weak zone dulls the sharp peaks in the impedances and reduces radiation damping at all frequencies quite markedly (Sheta and Novak, 1982).

Some insight into pile group behaviour under large displacements can be obtained from static experiments. Figure 12 shows the results of field tests conducted on free-headed test piles being steel pipes 0.1016 m in outer diameter and 3.05 m in length. The interaction factor diminishes with increasing deflection, dropping to about one half of the original value at the deflection of about 3.5 percent of the diameter, and then levels off. This reduction varies with spacing and the angle of incidence. If the pile is unloaded and reloaded, the interaction factors for small displacements become much smaller than the original ones, apparently due to gaps generated by the preceding large displacements, and then asymptotically approach the values from the first loading. This behaviour obviously makes

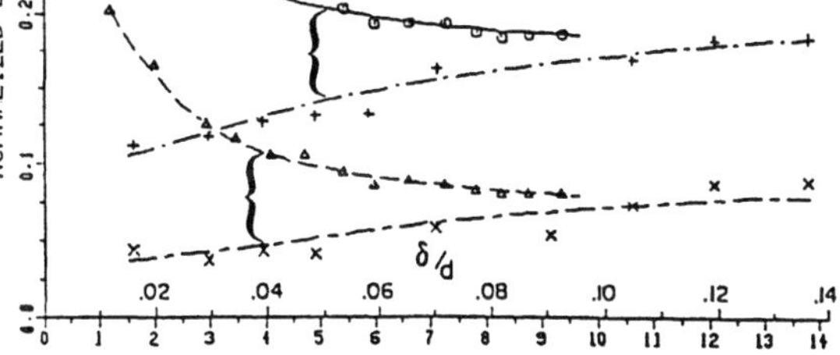

Figure 12 Horizontal static interaction factors for first loading and reloading (Janes & Novak, 1989).

the analysis of irregular transient response very difficult. More research in this area is needed.

The main conclusions of the theoretical and experimental investigations are that under large displacements pile-soil-pile interaction is diminished, but not eliminated, and both stiffness and radiation damping are substantially reduced.

The primary cause of the reduction in stiffness and damping is gapping which produces an effect similar to that of a free length. This makes it possible to account approximately for pile gapping, the major source of pile response overall nonlinearity, by representing the gap by a free length of pile sticking out of the ground. The reduction of interaction factors to a degree counteracts the reduction in the single pile impedance. The net result of these two trends depends on pile spacing, stiffness, the free length, e, and under dynamic loads, also on frequency. Figure 13 shows an example of the effect of the pile free length on static stiffness of a 3x3 group. While the group stiffness decreases with increasing e/d, faster for more flexible piles, group efficiency increases indicating a diminishing effect of pile-soil-pile interaction. Dynamic interaction factors for piles with free length are given by Mitwally and Novak (1987).

With the pile properties being so sensitive to the gap, assessment of its depth is very important. Considerations concerning the development of the gap have to include lateral capacity of the piles and the actual level of oscillatory loading. At larger force levels, the elastic range of soil stresses will be exceeded in the layer close to ground surface and the depth of the region so affected will increase with load amplitude. In sands, and particularly

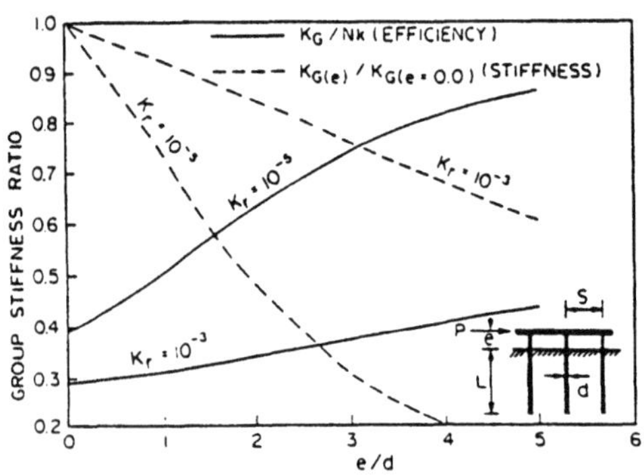

Figure 13 Effect of pile free length on horizontal group stiffness and efficiency of 3x3 group ($L/d \geq 25$, $S/d=3$, $K_r=E_pI_p/E_sL^4$; El Sharnouby & Novak, 1985).

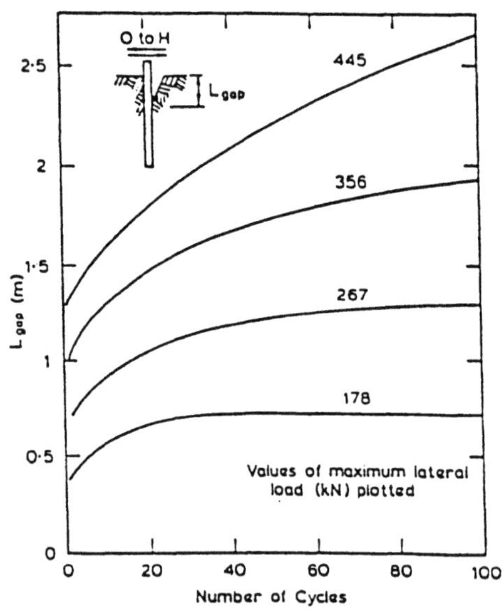

Figure 14 Development of gap separating pile from soil with number of cycles for different load amplitudes (Swane & Poulos, 1984).

saturated sand, this will cause nonlinear softening and a decrease in soil resistance to pile motion. In clays, the exceedance of the pile lateral capacity may result in the opening up of a gap between the pile and adjacent soil. Some data on this phenomenon were provided by Swane and Poulos (1984) who demonstrated how the gap grows with the number of cycles and the amplitude of the load. For a 610 mm diameter steel tube pile in stiff clay they produced a theoretical relationship between the number of load cycles, load amplitude and gap depth displayed in Figure 14. At a lower level of loading, the gap stabilizes after a number of cycles depending on the force amplitude; after that, the pile response is essentially elastic, in accordance with the shakedown phenomenon. At a high level of loading, the gap depth may keep growing or even lead to incremental collapse, obviously a condition to avoid in design. Further references on nonlinear pile behaviour can be found in Novak (1991).

5. Conclusions

For practical purposes, dynamic pile group analysis can be conducted by means of superposition and dynamic interaction factors. On the whole, this technique works quite well except for vertical response of endbearing piles and higher frequencies in nonhomogeneous media.

Nonlinearity under large strains reduces group interaction effects but does not eliminate them. Under steady-state conditions, some of these effects can be allowed for in an approximate way by considering a free length of the piles to account for pile separation due to gapping.

6. References

Akiyoshi, T. and Fuchida, K. (1982). "An approximate solution of vertical vibration of end-bearing piles with frictional interface," Proc. JSCE, No. 324, August, 31-40 (in Japanese).

American Petroleum Institute (1991). "Recommended practice for planning, designing and constructing fixed offshore platforms," (19th edition), 1220 L St. N.W., Washington, D.C. 20005.

Butterfield, R. and Douglas, R.A. (1981). "Flexibility coefficients for the design of piles and pile groups," Construction Industry Res. and Inf. Assoc., 6 Storey's Gate, London, England, Technical Notes, 108.

DYNA3, A Computer Program for Calculation of Foundation Response to Dynamic Loads; M. Novak et al., Geotechnical Research Centre, University of Western Ontario, London, Ontario N6A 5B9, Canada.

El Marsafawi, H., Kaynia, A.M. and Novak, M. (1992). "Superposition method for pile group dynamic analysis," Proc. ASCE Session on Piles Under Dynamic Loads, New York, N.Y.; also, Res. Rep. GEOT-1-1992, Fac. of Eng. Sci., U.W.O., London, Ont.

El Sharnouby, B. and Novak, M. (1985). "Static and low frequency response of pile groups," Canadian Geotech. J., Vol. 22, No. 1, 79-94.

El Sharnouby, B. and Novak, M. (1990). "Stiffness constants and interaction factors for vertical response of pile groups," Canadian Geotech. J., Vol. 27, No. 6, 813-822.

Fan, K., Gazetas, G., Kaynia, A., Kausel, E. and Ahmad, S. (1991). "Kinematic seismic response of single piles and pile groups," J. Geotech. Eng., ASCE, Vol. 117, No. 12, December, 1860-1879.

Gazetas, G. (1984). "Seismic response of end-bearing single piles," Soil Dyn. and Earthq. Eng., Vol. 3, No. 2, 82-93.

Janes, M. and Novak, M. (1989). "Pile group response to large lateral loads," Proc. 42nd Canadian Geotech. Conf., Winnipeg, Manitoba, Oct. 23-25, 216-222.

Kaynia, A.M. and Kausel, E. (1982). "Dynamic behavior of pile groups," 2nd Int. Conf. on Num. Methods in Offshore Piling, Austin, TX, 509-532.

Kaynia, A.M. and Novak, M. (1992). "Response of pile foundations to Rayleigh waves and obliquely incident body waves," J. Earthq. Eng. and Struct. Dyn., Vol. 21, No. 4, April.

Matlock, H., Foo, H.C. and Bryant, L.M. (1978). "Simulation of lateral pile behaviour under earthquake motion," Proc. Am. Soc. Civ. Engrs. Specialty Conf. on Earthq. Eng. & Soil Dyn., Pasadena, CA, II, 600-619.

Mitwally, H. and Novak, M. (1987). "Response of offshore towers with pile interaction," J. of Eng. Mech., July, Vol. 113, No. 7, 1065-1084.

Nogami, T. and Konagai, K. (1987). "Dynamic response of vertically loaded nonlinear pile foundations," J. Geotech. Eng., Vol. 113, No. 2, 147-160.

Nogami, T., Konagai, K. and Otani, J. (1988). "Nonlinear pile foundation model for time-domain dynamic response analysis," 9th WCEE, Tokyo, Vol. III, 593-598.

Novak, M. (1991). "Piles under dynamic loads," 2nd Int. Conf. Recent Advances in Geotech. Earthq. Eng. & Soil Dyn., St. Louis, Missouri (USA), March, Proc. Vol. 3, 250-273.

Novak, M. and El Sharnouby, B. (1983). "Stiffness constants of single piles," J. Geotech. Eng. Div. ASCE, Vol. 109, No. 7, 961-974.

Novak, M. and Mitwally, H. (1990). "Random response of offshore towers with pile-soil-pile interaction," J. Offshore Mech. and Arctic Eng., Feb., Vol. 112, 35-41.

Novak, M. and Sheta, M. (1980). "Approximate approach to contact problems of piles," Proc. Geotech. Eng. Div. ASCE National Convention "Dynamic Response of Pile Foundations: Analytic Aspects," Florida, Oct. 30, 53-79.

Otani, J., Nogami, T. and Konagai, K. (1991). "Non-linear time domain numerical model for pile group under transient dynamic forces," Proc. 2nd Int. Conf. on Recent Advances in Geotech. Earthq. Eng. & Soil Dyn., St. Louis, Missouri.

Penzien, J. (1970). "Soil-pile foundation interaction," in Earthquake Engineering, Ed. R.L. Wiegel, Prentice-Hall, Englewood Cliffs, New Jersey, 349-381.

Poulos, H.G. and Davis, E.H. (1980). Pile Foundations Analysis and Design, John Wiley & Sons, 397 p.

Rajapakse, R.K.N.D. and Shah, A.H. (1989). "Impedance curves for an elastic pile," Soil Dyn. and Earthq. Eng., Vol. 8, No. 3, 145-152.

Sheta, M. and Novak, M. (1982). "Vertical vibration of pile groups," J. Geotech. Eng. Div. ASCE, 112(11), 1016-1032.

Swane, I.C. and Poulos, H.G. (1982). "A theoretical study of the cyclic shakedown of laterally loaded piles," Res. Rep. No. R415, Univ. of Sydney, School of Civil and Mining Eng., July.

Swane, I.C. and Poulos, H.G. (1984). "Shakedown analysis of laterally loaded pile tested in stiff clay," Proc. 4th Australia-New Zealand Conf. on Geomech., Perth, Vol. I, 165-169.

THE SOIL-STRUCTURE INTERACTION PROBLEM AS IT IS DEALT WITH BY VARIOUS CODES FOR EARTHQUAKE RESISTANT DESIGN OF STRUCTURES

P. G. CARYDIS
National Technical University
Laboratory for Earthquake Engineering
Zografos 15700, Athens
Greece

ABSTRACT. In the present communication the various earthquake resistant regulations which deal with the subject of soil-structure interaction have been reviewed and presented. The respective parts, which deal with the subject under consideration, of Eurocode 8 and ATC-3-06 are also reviewed. Due to the soil-structure interaction, reduction factors of the seismic design forces are provided by the various codes, while an increase of the natural period of the structure and of the overall seismic design displacements are provided as well. Mainly, the most codes dealing with the problem, provide reduction factors for the overturning moments of the structure along the height. This reduction is higher in the base. Depending on the existing seismotectonic regime in combination to the prevailing soil conditions that happen to occur during destructive past earthquakes, some of the codes, increase or decrease the reduction of the design seismic forces due to the Soil-Structure Interaction. It has been observed, also, that between two successive editions of the same code, the soil-structure interaction problem is dealt with, differently.

1. Introduction

The present paper deals with the problem of structures resting on the ground, and not for structures which are partially or totally embeded into the ground.

There are two basic approaches that an Engineer may follow during the design, in order to take under consideration the soil-structure interaction phenomenon: The first is to consider the structure with its ground down to the bed rock, or down to the adequate depth and to handle it as a unique model (Fig.1a), and the second is to consider the structure fixed on the ground and to use correction factors in order to compensate the influence of the ground (Fig.1b). The correction factors are not always justified, but for simple structures, or for structures of common importance the latter approach is the most popular.

The first approach needs the determination of all parameters influencing the response of the compound model, the various assumptions used (structure and foundation: mass distribution, stiffness, damping, geometry; ground profile: layers, density, shear modulus, geometry, boundaries, damping; input motion etc.) and a more rigorous analytical procedure

P. Gülkan and R. W. Clough (eds.), Developments in Dynamic Soil-Structure Interaction, 381–399.

in the linear or non linear domain, taking into consideration the uncertainties involved. The quantity and quality of these uncertainties, and the departures of the assumptions used from the reality make some time the Designer Engineer to prefer a simpler model, like the one shown in Fig.(1b) and to carry out the appropriate corrections, rather than a more sophisticated, like the one shown in Fig.(1a). The fixed on the base model, with the appropriate correction factors, is the one followed by the various codes, who deal with the Soil-Structure Interaction.

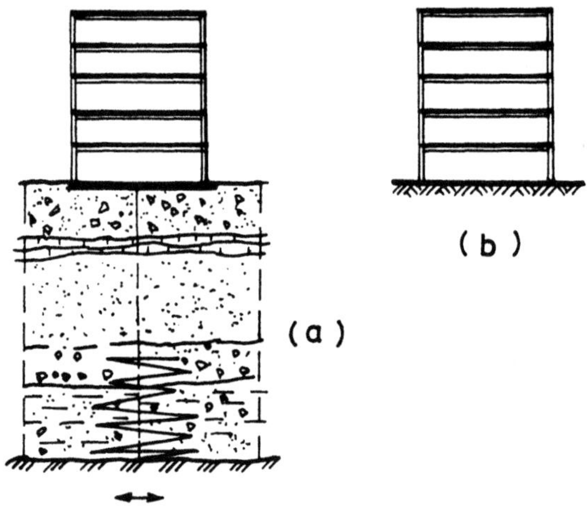

Figure 1. Two basic approaches are followed: (a) the structure with the soil profile down to the bed rock and (b) the structure fixed on the ground, with the appropriate correction factors.

It is worth while to mention here that the most popular assumption about the input excitation, is to have an horizontal ground motion equally exciting the lower parts of a structure, even in the case in which a more sophisticated procedure is followed. But this, it is

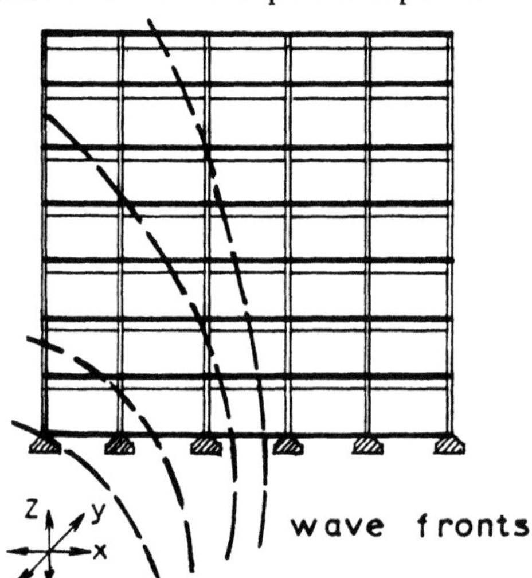

Figure 2. The earthquake excitation of the structure is a time and space varying function.

not always the case (see Fig.2), when we have a rather long structure, for example 40m, resting on a rather soft ground with a shear wave velocity of the order of 300 m/sec. The phase difference of the incident wave motion between the first and the last footing is of the order of $\pi/4$.

The Designer Engineer for the every day practice, prefers to minimize the various uncertainties and parameters involved, which influence the results of his computations in an uncontrollable way.

In general, the structure-foundation-ground integrated model, (see Fig.3a) creates another earthquake motion at the foundation-soil interface different from that of the free field (with a spectrum of the horizontal component moved towards the longer periods), the overall damping may be higher, the various natural periods are increased, while the respective modal shapes and henceforth the participation factors are accordingly modified (Eurocode 8, Part 5). For example, in Fig.($3b_1$ and $3b_2$) the first and second modal shapes are shown for the structure fixed on the ground. When the flexibility of the ground is increased (case of Fig.$3c_1$ and $3c_2$), the deformation of the above ground structure is minimized in expense of the deformation of the ground. In the case of a softer ground and relatively stiffer structure, the structure may be deformed as a rigid body (Fig.$3d_1$ and $3d_2$).

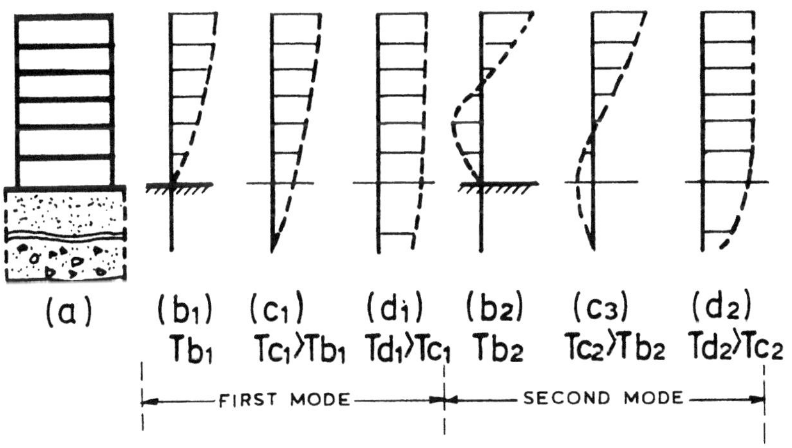

Figure 3. First and second normal modes of a structure with the ground. The cases b_1 and b_2 are the first and second modes for rigid foundation-structure fixed on the ground. Cases c_1 and c_2 are the first and second modes respectively for softer ground, while d_1 and d_2 are the respective modes for even a softer ground (Matsushima and Carydis (1971)).

Most of the codes who deal with the Soil-Structure Interaction (S.S.I.) phenomenon diminish and misinterpreting it, mainly, to a reduction of the overturning moments of the structure. The reduction is higher from the top to the bottom of the structure. Also, some codes give some increase functions of the fundamental period of the structure. The codes have to combine two contradicting parameters: the assumption under which the structures are considered as fixed on the ground, and the fact that the ground under the foundation has zero tensile strength.

Accurate calculation of the axial forces, which are developed on the vertical load carrying members of a structure is of primary importance. A reduction, or an increase of the design axial force of a column, during its seismic response, may drasticly decrease its strength and ductility.

The number of codes explicitly dealing with the S.S.I. problem is relatively small. But, almost all codes deal with the modification of the input motion according to the prevailing soil conditions. The latter case has not been considered as a result of the S.S.I., and therefore it is not analysed in the present communication.

2. Earthquake Resistant Design Codes reviewed

The basic assumption prevailing at the various codes is that of the building model fixed on the foundation. If some other assumption is followed, it will be respectively stated in the following evaluation. It should be stated here that each code takes under consideration the prevailing seismotectonic conditions, the existing construction practice, the socio-economic framework, the quality of materials, the various building codes and specifications other than the seismic codes, the level of the technological development, the available computing tools and the recent earthquake experience from destructive events. The return period of the strong earthquakes compared to the human life span plays an important role on the implementation of the regulations for the construction of earthquake resistant structures. In the case that one wants to compare the various codes, the abovementioned parameters should be taken into account.

2.1 ATC-3-06 REPORT

The Tentative Provisions for the Development of Seismic Regulations for Buildings (1984) are the most elaborated among the other codes and for this reason are preceded in the present communication. The relevant content of this report is presented in a concise way.

2.1.1. *S.S.I. for use with the Equivalent Lateral Force Procedure.* The lateral seismic force, F_x, applied at any level, x, is equal to:

$$F_x = C_{vx} V \tag{1}$$

where

$$C_{vx} = \frac{w_x h_x^k}{\sum_{i=1}^{n} w_i h_i^k} \tag{2}$$

k is related to the period of the building
$V = C_s W$, the seismic base shear
C_s = the seismic design coefficient given by the formula:

$$C_s = \frac{1.2 A_v S}{R T^{2/3}} \tag{3}$$

A_v = coefficient representing the Effective Peak Velocity-Related Acceleration,
S = coefficient for the soil
R = response modification factor
T = fundamental period of the building, fixed at the base
W = the total gravity load of the whole building.

The **Overturning Moment** in each story shall be distributed to the vertical load carrying members in the same proportion as for the distribution of the horizontal seismic forces, and is given in accordance to the formula:

$$M_x = \kappa \sum_{i=x}^{n} F_i (h_i - h_x) \quad (4)$$

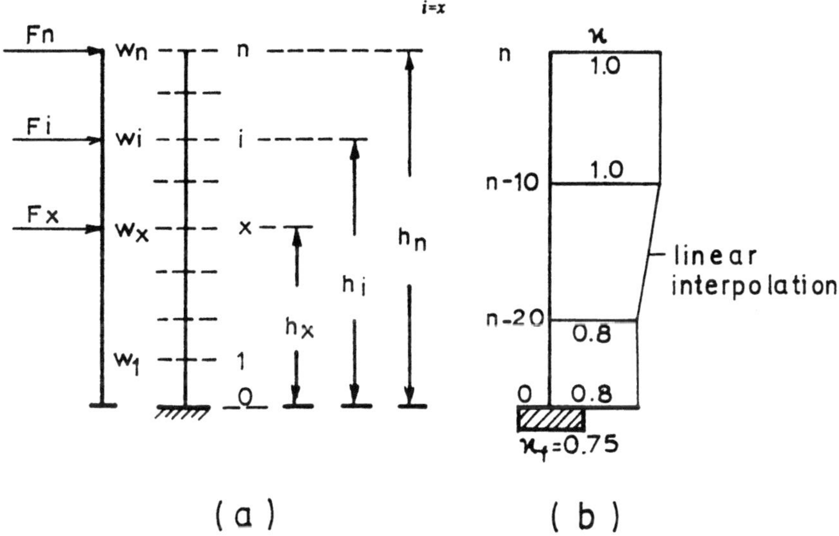

Figure 4. (a) The lateral seismic forces F_x. (b) The associated overturning moment reduction factor κ.

The variation of the reduction factor κ is given in Fig.4. Between the 10th and 20th story from the top, the reduction factor varies linearly. At the level of the soil-foundation interface, the reduction factor $\kappa_f = 0.75$. The overturning moment reduction factor is justified according to the following three reasons: (i) the horizontal seismic forces do not attain their maximum values simultaneously; (ii) the horizontal forces are towards the conservative side, henceforth their summation for story shears and the statically consistent to them resulting overturning moments are overestimated; (iii) in reality, the foundation of the building it is not fixed on the ground, therefore it can not undertake any tensile force and the foundation may up-lift for short time instances.

Only the third of the above mentioned three reasons for the reduction of the overturning moment is attributed to the S.S.I. effect.

The S.S.I. effect is further expressed by the increase of both the fundamental natural period of the compound structure $\tilde{T}$, and its effective damping $\tilde{\beta}$. Due to the change of the natural period and damping, the modified values of the base shear, of the story seismic forces, the resulting overturning moments, and tortional effects are calculated according to the following:

The **Base Shear** V given above may be reduced to an amount ΔV in order to account for the S.S.I. effects:

$$\tilde{V}=V-\Delta V \tag{5}$$

$$\Delta V=[C_s-\tilde{C}_s(\frac{0.05}{\tilde{\beta}})^{0.4}]\bar{W} \tag{6}$$

where

$\tilde{C}_s$ = the seismic design coefficient due to the flexibility of the ground. This derives from the use of formula (3), but with a period $\tilde{T}$ given below.

$\tilde{\beta}$ = the damping ratio for the soil-structure integrated system

$\bar{W}$ = the effective gravity load of the whole building, $\bar{W}=0.7$ W. Except for one concentrated mass: $\bar{W}=W$

The maximum value of ΔV can not be more than 0.3V. Therefore, the base shear $\tilde{V}$ can not be less than 0.7V.

The **Effective Building Period** due to the effect of the flexibility of the soil shall be determined from the use of the formula:

$$\tilde{T}=T\sqrt{1+\frac{\bar{K}}{K_y}(1+\frac{K_y\bar{h}^2}{K_\theta})} \tag{7}$$

where

$\bar{K}$ = the "stiffness" of the building fixed at base, calculated from the formula:

$$\bar{K}=4\pi^2\frac{\bar{W}}{gT^2} \tag{8}$$

$\bar{h}$ = the effective height of the building. Generally $\bar{h}=0.7\,h_n$, except for the case where the mass is effectively concentrated in one only level n, it shall be taken as $\bar{h}=h_n$

K_y = the lateral stiffness of the foundation. (Static force per produced deflection parallel to the force applied at the level of foundation, and that of the analysis of the structure).

K_θ = the rocking stiffness of the foundation. (Static moment per produced average rotation at the level of foundation in the plane of the applied moment and that of the analysis of the structure).

g = acceleration of gravity

In order to calculate the foundation stiffnesses K_y and K_θ, for various strain levels of the foundation soil,its average shear modulus may be determined using the values given in Table 1.

TABLE 1. Values of the ratios of shear modulus G/G_o and shear wave velocity υ_s/υ_{so} for various ground Acceleration Coefficients

Ground Acceleration Coefficient, A_v	$\leq$0.10	0.15	0.20	$\geq$0.30
Ratio of Shear Modulus G/G_o	0.81	0.64	0.49	0.42
Ratio of Shear Wave Velocity υ_s/υ_{so}	0.9	0.8	0.7	0.65

where

υ_{so}=the average soil shear wave velocity beneath the foundation at small strains ($\leq 10^{-3}\%$)

G=the average soil shear modulus beneath the foundation

$G_o=\gamma\upsilon_{so}^2/g$, the respective value of G at small strains and

γ=the average soil unit weight

The **Effective Damping**, $\tilde{\beta}$, for the soil-structure integrated system. The damping ratio $\tilde{\beta}$ shall be computed from the formula:

$$\tilde{\beta}=\beta_o+\frac{0.05}{(\tilde{T}/T)^3} \tag{9}$$

where

β_o=the foundation damping ratio given in Fig.5

The quantity r in Fig.5 is a characteristic foundation length, which shall be determined according to the formulas:

$$\textit{For}\quad \bar{h}\leq 0.5L_0:\quad r=r_a=\sqrt{A_o/\pi} \tag{10a}$$

$$\textit{For}\quad \bar{h}\geq L_o:\quad r=r_m=\sqrt[4]{4I_o/\pi} \tag{10b}$$

where

L_o=the overall length of the side of the foundation, parallel to the direction of the analysis of the structure

A_o=the area of the load bearing foundation

I_o=the static moment of inertia of the load bearing foundation

Exception: For buildings supported on point bearing piles and when the foundation soil consists of a soft stratum of reasonably uniform properties underlain by a much stiffer, rock-like deposit with an abrupt increase in stiffness, the factor β_o in formula (9) shall be replaced by:

$$\beta_o' = (\frac{4D_s}{\upsilon_s T})^2 \beta_o \quad if \quad \frac{4D_s}{\upsilon_s \tilde{\bar{T}}} < 1 \tag{9a}$$

where

D_s = the thickness of the soft stratum underlain the foundation of the building

υ_s = the shear wave velocity

In this case of the foundation conditions β_o' is less than β_o, but in any case should be taken β_o' less than 5%.

The **Vertical Distribution of Seismic Forces,** will be the same as for the building without interaction, but the Base Shear in the relation (1) will be equal to $\tilde{V}$.

The **Modified Story Shears, Overturning Moments and Tortional Effects** about a vertical axis, shall be determined using the same procedures as for structures without interaction, but

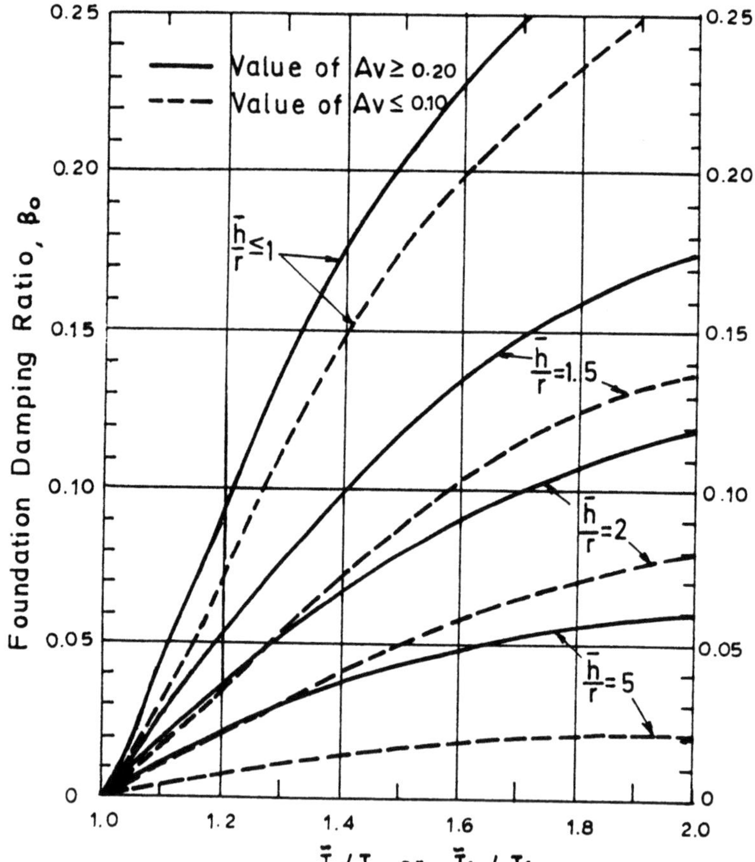

Figure 5. Foundation Damping Ratio.For values of A_v=0.15, the ratios β_o shall be taken by averaging the values from the solid and dashed lines.

with the reduced lateral forces.

The **Modified Deflections,** $\tilde{\delta}_x$, shall be determined according to the formula:

$$\tilde{\delta}_x = \frac{\tilde{V}}{V}\left[\frac{M_o h_x}{K_\theta} + \delta_x\right] \tag{11}$$

where

M_o = the overturning moment at the base (level 0, Fig.4a), as expressed in formula (4), using the unmodified lateral seismic forces (see formulas (1) and (2)).

h_x = the height above the base (level 0, Fig.4a) to the level under consideration

δ_x = the deflections of the structure fixed at the base, using the unmodified seismic forces, after an elastic analysis, and the appropriate deflection amplification factor (C_d).

The **Modified Story Drifts,** $\tilde{\Delta}$, shall be calculated as the difference of the modified deflections $\tilde{\delta}_x$, at the top and bottom of the story under consideration.

The **Modified Story P-Δ** effects, shall be evaluated by multiplying in each story the modified story diflection $\tilde{\delta}$, the story drift $\tilde{\Delta}$ and the modified story shear $\tilde{V}$ by the factor $(1+a_d)$ and recompute the story shears, overturning moments, and other seismic force effects corresponding to these augmented story shears. The coefficient $a_d=\theta/(1-\theta)$, where θ is the stability coefficient, given in formula (4-10) of the report ATC-3-06, using the modified story shears and modified deflections determined previously. These P-delta effects need not be considered when the stability coefficient θ is equal or less than 0.10.

2.1.2. *S.S.I. for use with the Modal Analysis Procedure.* First, the basic quantities without considering the S.S.I. effects are exposed in the following.

The portion of the base shear contributed by the mth mode, V_m, is equal to:

$$V_m = C_{sm}\bar{W}_m \tag{12}$$

where

C_{sm} = the modal seismic design coefficient, given below

$\bar{W}_m$ = the effective modal gravity load determined in accordance to the formula:

$$\bar{W}_m = \frac{\left[\sum_{i=1}^{n} w_i\phi_{im}\right]^2}{\sum_{i=1}^{n} w_i\phi_{im}^2} \tag{13}$$

where

φ_{im} = the modal displacement at the ith level in its mth mode, for the building fixed at its base.

The modal seismic design coefficient C_{sm} shall be determined in accordance to the formula:

$$C_{sm} = \frac{1.2\ A_v S}{R\ T_m^{2/3}} \tag{14}$$

The **Modal Forces, F_{xm}**, applied at each level, x, shall be determined according to the formula:

$$F_{xm}=C_{vxm}\,V_m \tag{15}$$

where

$$C_{vxm}=\frac{w_x\phi_{xm}}{\sum_{i=1}^{n} w_i\phi_{im}} \tag{16}$$

Formulas (15) and (16) are analogous to formulas (1) and (2).

The **Modal Deflections, δ_{xm}**, at each level x, shall be determined according to the formula:

$$\delta_{xm}=C_d\delta_{xem} \tag{17}$$

where

C_d = the deflection amplification factor

and

$$\delta_{xem}=\frac{g}{4\pi^2}\frac{T_m^2F_{xm}}{w_x} \tag{18}$$

The **Modal Story Drifts, Δ_m**, shall be computed as the difference of the deflections, δ_{xm}, at the top and bottom of the story under consideration.

The **Modal Story Shears, Story overturning Moments and the Corresponding Member Forces and Moments** due to the modal seismic forces determined from the formulas (15),(16) and (17) shall be computed for each mode by linear static methods.

The **Overturning Moment** at the foundation-soil interface may be reduced by 10%. Overturning moments along the height of the structure are calculated mode by mode and recognizing the fact that the direction of forces F_{xm} is controlled by the algebraic sign of φ_{xm}, one cocludes that there is no reason for reducing these moments. This is in contrast with reductions permitted in overturning moments calculated from the Equivalent Lateral Force procedure. The above mentioned reduction of the overturning moment by 10% is attributed to the third reason justifying the overturning moment reduction factor κ (see formula 4)-which is mainly due to the S.S.I. effect.

In order to account for the S.S.I. effects, in the ATC-3-06 report it is further commended that studies of the dynamic response of elastically supported multi-degree-of-freedom systems reveal that, within the ranges of parameters that are of interest in the design of building structures subjected to earthquakes, Soil-Structure Interaction affects substantially only the response component contributed by the fundamental mode of vibration of the superstructure.

Interaction effects are considered only in evaluating the contribution of the fundamental natural mode. The contributions of the higher modes are computed as if the structure were fixed at the base, and the maximum value of a response quantity is determined, as for rigidly supported structures, by taking the square root of the sum of the squares of the maximum modal contributions.

The **Modal Base Shear**, corresponding to the fundamental mode of vibration, V_1 may be reduced to

$$\tilde{V}_1 = V_1 - \Delta V_1 \tag{19}$$

where

ΔV_1 = the reduction computed according to the formula (6) in which

$\bar{W}$ = the gravity load $\bar{W}_1$ defined by formula (13)

C_s shall be calculated from the formula (14), using the fundamental period T_1 of the fixed at the base building

C_s shall be calculated from the same formula (14), but using the effective building period $\tilde{T}_1$ determined from formula (7). In this formula it must be put $T = T_1$, $\bar{K}$ will be evaluated from formula (8) with $\bar{W} = W_1$, $T = T_1$, while $\bar{h}$ will be computed from the formula:

$$\bar{h} = \frac{\sum_{i=1}^{n} w_i \phi_{i1} h_i}{\sum_{i=1}^{n} w_i \phi_{i1}} \tag{20}$$

The above designated values of T_1 and $\tilde{T}_1$ shall be used in order to evaluate the damping ratio $\tilde{\beta}$ and β_o from Fig.5.

No reduction is permitted in shear components contributed by the higher modes of vibration due to S.S.I. effect.

The reduced base shear, $\tilde{V}_1$, shall not be taken less than 0.7 V_1.

The **Modified Modal Seismic Forces, Story Shears and Overturning Moments**, shall be determined as for buildings without S.S.I. effect, using the modified base shear force, $\tilde{V}_1$, instead of V_1.

The **Modified Modal Deflections**, $\tilde{\delta}_{xm}$, shall be determined according to the formulas:

$$\tilde{\delta}_{x1} = \frac{\tilde{V}_1}{V_1}\left[\frac{M_{01} h_x}{K_\theta} + \delta_{x1}\right] \tag{21}$$

and

$$\delta_{xm} = \delta_{xm} \; for \; m = 2,3,...... \tag{21b}$$

where

M_{01} = the overturning base moment for the fundamental mode of the fixed-base building, as determined in the description of the Modal Analysis Procedure, using the unmodified modal base shear V_1.

δ_{xm} = the modal deflections at level x of the fixed-base building, as determined by formulas (17) and (18), using the unmodified modal shears, V_m.

The **Modified Story Modal Drifts,** Δ_m, shall be computed as the difference of the deflections, δ_{xm}, at the top and bottom of the story under consideration.

The **Design Values of the Modified Shears, Moments, Deflections, and Story Drifts** shall be determined as for structures without S.S.I. effects by taking the square root of the sum of the squares of the respective modal contributions. In the design of the foundation, the overturning moment at the foundation-soil interface determined in this manner may be reduced by 10 percent, as for structures without S.S.I. effects.

A reference paper documenting most of the above presented material is given by Veletsos (1978 and 1982).

2.2 EUROCODE 8-PART 5

This Part,5, of the Eurocode 8, Faccioli, Cuellar, Gazetas et al.(1992), is dealing with Foundations, Retaining Structures and Geotechnical Aspects. For the Soil-Structure Interaction it is mentioned the following:

2.2.1 General Effects and Significance. A flexibly-supported structure (on soil profiles B and C) will respond differently, in several ways, compared to the response of the same structure, founded on rigid ground (fixed base) and excited by the same free field motion due to:

1. The foundation motion of the flexibly supported structure will differ from the free field and may include an important rocking component
2. The fundamental period will be longer than that corresponding to the fixed-base
3. The natural periods, mode shapes and mode participation factors of the flexibly supported structure will be different than that corresponding to the fixed base structure.
4. The overall damping associated with a flexibly supported structure will include contributions from the "radiation" and "hysteretic" damping generated at the soil foundation interface, in addition to the structural damping generated in the superstructure.

For the majority of usual building structures, the effect of S.S.I. tend to be beneficial, except for structures sensitive to rocking and to large relative displacements, and for structures founded on soft soil deposits for which the S.S.I. effect may be detrimental.

2.2.2 Methods of analysis. Dependence of the soil properties on strain-amplitude and frequency, and on local changes in lithology, must be taken into account by parametric variation. In general, a variation of shear modulus G will be sufficient. Values as those recommended in Table 1 are adopted, also, in this code.

For regular building structures, for which the code allows a fundamental mode simplified dynamic analysis, the effects of S.S.I. can be accounted for, through simplified procedures.

For structures other than regular buildings, and requiring an evaluation of S.S.I. effects, these must be assessed using more rigorous procedures including:

a) "Substructuring" methods, in which the soil is replaced by springs and dashpots.
b) "One-step" methods, in which the structure, its foundation, and the supporting soil are modeled as one system, subjected to an excitation consistent with the free field response spectrum.

Such rigorous procedures are also necessary if:
- the soil is arbitrarily layered
- damping has to be considered in detail (by separation of material damping and radiation damping)
- stiffness and damping depend on frequency
- the flexibility of the foundation and of the embedded building parts has to be considered in detail
- the type of the foundation is not uniform
- the analysis of nonlinearities, such as gaps between base and soil, is necessary.

2.3. ARGENTINA

The overturning moment at any level of the structure shall be calculated from the above that level statically acting seismic forces, without reduction.

2.4. CANADA

The overturning moment M_x at any level, x, of the building, shall be multiplied by the coefficient, J_x, calculated from the formula:

$$J_x = J + (1 - J)\left(\frac{h_x}{h_n}\right)^3 \tag{22}$$

where
h_x and h_n is the height from the level zero of the x and n levels respectively (see Fig.4a).

The overturning moment, M, at the base of the structure shall be multiplied by a reduction coefficient, J, according to the formulas:

$$J = 1, \quad \textit{for } T < 0.5 \text{ (sec)} \tag{23a}$$

$$J = (1.1 - 0.2T), \quad \textit{for } 0.5 \leq T \leq 1.5 \text{ (sec)} \tag{23b}$$

$$J = 0.8, \quad \textit{for } T > 1.5 \text{ (sec)} \tag{23c}$$

2.5. CHILE

The overturning moment M_x at any level, x, of the building, shall be multiplied by the reduction factor, J_x, calculated from the formula:

$$J_x=0.8+0.2\frac{h_x}{h_n} \tag{24}$$

This reduction is applied only for buildings having 4 or more stories and for the "static method of analysis".

2.6 FRANCE

The seismic coefficient is given by the formula:

$$C=\alpha\ \beta\ \gamma\ \delta \tag{25}$$

where
α = intensity coefficient
β = response spectrum coefficient (a function of natural period, damping and soil quality)
γ = distribution coefficient along the height of the structure
δ = foundation coefficient, which takes into account the effect of the foundation on the behaviour of the structure. This coefficient is given in Table 2.

TABLE 2. Values of the foundation coefficient δ

Type of Foundation	Soil			
	A Rocky Ground	B Ground of Average Consistency	C Loose Ground with a fairly large water content	D Water-logged muds and silts
Surface Pads	1.00	1.15	1.25	
Isolated Surface Pads	0.90	1.15	1.25	
Basement foundations	0.90	1.00	1.25	
Raft		1.00	1.10	1.20
Bearing Piles passing through soils B,C or D		1.10	1.15	1.30
Friction Piles		1.10	1.30	

2.7. GERMANY

The natural period T_1 of the fundamental mode, takes into consideration the flexibility of the ground, under the following way. First, a value of the flexibility of the above ground structure is calculated with the formula:

$$f_1=\frac{h_n}{3EI} \tag{26a}$$

the flexibility of the ground, with the formula:

$$f_2=\frac{1}{C_k I_f} \tag{26b}$$

where

E = the modulus of elasticity of the structure (MN/m^2)
I = the moment of inertia of the substitute bar of the structure (m^4)
C_k = the dynamic overturning modulus of foundation support (MN/m^3), given from the formula:

$$C_k=\frac{dyn\ E_s}{0.25\sqrt{A}} \tag{26c}$$

A = the area of the foundation (m^2)
$dynE_s$ = the dynamic modulus of the subsoil (MN/m^2)
I_t = the moment of inertia of the plan of the foundation (m^4)

The natural period T_1 of a substitute bar will be calculated according to the formula:

$$T_1=1.5\sqrt{(f_1+f_2)\sum_{i=1}^{n} w_i h_i^2} \quad \text{(sec)} \tag{27}$$

This formula is a nice application of the Southwell-Dunkerley method (see Jacobsen and Ayre (1958), and Richart, Hall and Woods (1970)), combining the two periods (one of the building and the other of the foundation).

2.8. IRAN

2.8.1. *Iranian Code, previous to 1984 Edition.* The overturning moment, M_x at any level, x, shall be multiplied by the reduction factor, J_x, calculated from the formula:

$$J_x=\frac{h_n-h_x}{h_n}=1-\frac{h_x}{h_n} \tag{28}$$

The overturning moment at the base of the building, or structure, shall be determined in accordance to the formula:

$$M=J\sum_{i=1}^{n}F_i h_i \tag{29a}$$

where

$$J=\frac{0.6}{\sqrt[3]{T}}, \qquad 0.45\lesssim J\lesssim 1.0 \tag{29b}$$

2.8.2. *Iranian Code, 1988 Edition.* The building must be stable against overturning. The overturning moment at any level is given by formula (29a) with J = 1.0. Factor of safety against overturning, at the interface between the foundation and the supporting ground, shall be at least 1.75.

One may observe that the 1988 edition is more conservative than the previous one as far as the S.S.I. effect is concerned.

2.9. MEXICO

2.9.1. *Mexico Code, 1977 Edition.* The overturning moment may be reduced by a factor given in formula (24), but it shall not be taken smaller than the design seismic force at the elevation considered times its distance to the centre of gravity of all the corresponding masses above.

2.9.2. *Mexico Code, 1987 Edition.* For the static analysis yields the same requirement as for the 1977 edition. A new, quite extended chapter, has been added dealing with the S.S.I. problem. Additional information are given with microzonation maps where various characteristic parameters are given, as for example, about the discrimination of the zones, about the corresponding predominant period of the ground, and about the effective depth of the soft soil material down to the firm soil layer.

The fundamental period of the structure due to the flexibility of the soil, shall be calculated according to the formula:

$$T_1=\sqrt{T_o^2+T_x^2+T_r^2} \quad \text{(sec)} \tag{30}$$

where

T_o = the fundamental period of the fixed at the rigid base structure, in (sec)

T_x = the fundamental period of the structure as if it were a rigid body supported by the soil, along the direction under consideration (sec)

T_r = the fundamental period of the structure as if it were a rigid body supported by the soil, around an horizontal axis perpendicular to the direction under consideration and passing through the centroid of the above described rigid body, in (sec)

For the calculation of the periods T_x and T_r in the text of the code are given the adequate formulas and other miscellaneous information. Also, formula (30) is an application of the

Southwell-Dunkerley method mentioned in paragraph 2.7.

2.9. NEW ZEALAND

The seismic overturning moments shall be derived from the horizontal seismic forces without reduction.

2.10. NICARAGUA

The overturning moment at any level, x, shall be determined according to the formula:

$$M_x = \sum_{i=x+1}^{n} F_i(h_i - h_x) \tag{31}$$

2.11. PERU

The overturning moment acting on the foundation and any story, x, shall be calculated according to the formula (31).

For buildings having more than 10 stories, it is possible to make a reduction at the lower stories. The reduction will be 2% for each story increasing up to a maximum value 20%.

In the foundation design, it can be applied a further reduction of 10% to the already reduced overturning moment in the interface between the soil and the foundation body.

For inverted pendulum structures the reduction, mentioned above, it is not permitted.

The overturning safety factor shall be equal to, or greater than 1.5.

2.12. PHILIPPINES

2.12.1. *Philippines Code, 1972 Edition.* A reduction factor for the overturning moment at each level, x, is permitted according to the formula:

$$J_x = J + (1 - J)\left(\frac{h_x}{h_n}\right)^3 \tag{32}$$

where

$$J = \frac{0.6}{\sqrt[3]{T}} \tag{33}$$

The overturning moment at the base of the building or structure, shall be reduced by the factor J of formula (33).

2.12.2. *Philippines Code, Previous to 1988 Edition.* Not any reduction of the overturning is permitted.

2.13. U.S.A.

Every structure shall be designed to resist the overturning effects caused by earthquake forces, specified in another section of the Code.

At any level, x, the overturning moments to be resisted shall be determined using those seismic forces derived from the force F_t, applied at the top of the structure, and the forces F_i which act on levels above the level x.

At any level, the incremental changes of the design overturning moment shall be distributed to the various resisting elements in the manner prescribed in another section of the code. Overturning effects on every element shall be carried down to the foundation.

2.14. VENEZUELA

The overturning moment at any level, x, shall be calculated according to the formula:

$$M_x = J \sum_{i=x}^{n} F_i(h_i - h_x) \tag{34}$$

where

$$J = 1 - 0.04\frac{T}{T^*} \tag{35}$$

T = the fundamental period of the structure
T^* = 0.4(sec) for soil type S_1
0.6(sec) for soil type S_2
1.0(sec) for soil type S_3

The reduction of the overturning moment is equal to 1.0 for the top 5 stories and equal to J for all levels under the 8 topmost levels. For the levels between 5 and 8 from the top a linear interpolation shall be made.

3. Conclusions

Not any design guidelines may be proposed to account for the S.S.I. effects except for the general notes about the modification of the input motion, the increase of the natural period of the building, the modal shape and the damping. Expect for the damping, which is, generally, increased by the S.S.I. effects, and which is acting always to the safety side, it is nevertheless difficult to say, for the other parameters mentioned above, whether or not the S.S.I. effects are in favor or against the overall safety of the structure. This depends very much on the prevailing seismotectonic regime in combination with the soil conditions and the structural characteristics. If, for example, we have a seismic activity from strong distant foci, the effects of the S.S.I. will be quite different from those anticipated from the effects of near field earthquakes.

4. Acknowledgents

The help of Messrs G.Ioakimidis and E.A.Vougioukas is acknowledged.

5. References

Applied Technology Council, (1984) Tentative Provisions for the Development of Seismic Regulations for Buildings, Second Printing, ATC-3-06 report, Palo Alto, California

Bossenmayer, H., Carvalho, E.C., Giuffre, A., et al (1988) Eurocodo No 8, Structures in Seismic Regions, Part 1, General and Building, Commission of the European Communities, Brussels

Faccioli, E., Cuellar, V., Gazetas, G., et al, (1992) EC8-Part 5, Foundations, Retaining Structures and Geotechnical Aspects, 1st redrafting

I.A.E.E., (1988) Earthquake Resistant Regulations, A World List-1988, Gakujutsu Bunken Fukyu-kai, Tokyo

Jacobsen, L.S., Ayre, R.S.,(1958) "Engineering Vibrations", Mc Graw-Hill Book Co., Inc., pp.566

Matsushima, Y. and Carydis, P.,(1971) "Study on the seismic interaction problem between ground and shear structures", in Bulletin of International Institute of Seismology and Earthquake Engienering, Tokyo, Japan, Vol.8, pp.237-252

Richart, F.E.Jr., Hall,J.R.Jr., and Woods, R.D.,(1970) "Vibrations of Soils and Foundations", Prentice-Hall, Inc., Englewood Cliffs, New Jersey, U.S.A., pp.416

Veletsos, A.S.,(1982) "Foundation dynamics and soil-structure interaction", EERI (eds.), A Regional Seminar on Earthquake Fundamentals for Alaska, Alaska Pacific University, p.p.80

Veletsos, A.S.,(1978) "Soil Structure Interaction for Buildings During Earthquakes", Proceedings of the Second International Conference on Microzonation for Safer Construction-Research and Application, San Francisco, California, November 26-December 1, p.p.111-134.

ESTIMATION AND GENERATION OF DESIGN INPUT MOTION

M. WATABE,
Shimizu Corporation
Seavans South
1-2-3 *Shibaura, Minatoku*
Tokyo, 105-07
Japan

T. WATANABE, T. SATO, and K. DAN
Ohsaki Research Institute, Shimizu Corporation
Fukoku Seimei Bldg.
2-2-2 *Uchisaiwaicho, Chiyodaku*
Tokyo, 100
Japan

ABSTRACT. This paper presents estimation and generation of design input motion, based on the physical understanding of earthquake ground motions. First, principal parameters to characterize earthquake ground motions from the view points of seismic source, path, and local site conditions are described. Then, the procedure and methods to generate design input motions are introduced. The methods are classified into three groups: the method using strong-motion records, the method based on fault model, and the method using standard spectra. Their applicabilities to actual problems are also discussed. The methods using strong-motion records or standard spectra are conventional and practical but a great attention has to be paid to a range of the applicability, which depends on the quality and quantity of the records. In a theoretical approach based on the fault model, the heterogeneity of the fault rupture has to be taken into account for the evaluation of design input motion in the period range of the engineering interest. A semi-empirical approach using small-event records, based on the fault model, is useful at a site in the vicinity of a large fault.

1. Introduction

Large earthquakes have caused severe damage in many countries unfortunately. In these countries, structures should be designed and constructed to be safe during the earthquakes. It is very important to evaluate design input motions, i.e. earthquake ground motions which are to be used in the dynamic response analysis of structures for earthquake-resistant design, based on the physical understanding of earthquake ground motions.

The purpose of this paper is to present estimation and generation of the design input motions by prospective earthquakes for earthquake-resistant design of structures. It has been established that earthquake ground motions mainly result from three factors, that is, seismic source, propagation path of seismic waves, and local site conditions (e.g., Iwata and Irikura (1988); Kato, *et al.* (1992)). Therefore, first, principal parameters to characterize earthquake ground motions from the view points of the above three factors are described (e.g., Bolt (1983); Joyner and Boore (1988); Aki and Irikura (1991)). Then, the procedures and methods to generate the design input motions are introduced. These methods are classified into three groups: the method using strong-motion records, the method based on fault model, and the method using standard spectra. The applicability of these methods to actual problems is also discussed.

P. Gülkan and R. W. Clough (eds.), Developments in Dynamic Soil-Structure Interaction, 401–434.

2. Principal Parameters to Characterize Earthquake Ground Motions

In this section, principal parameters to characterize earthquake ground motions are presented from the view points of seismic source, propagation path of seismic waves, and local site conditions as shown in Figure 1.

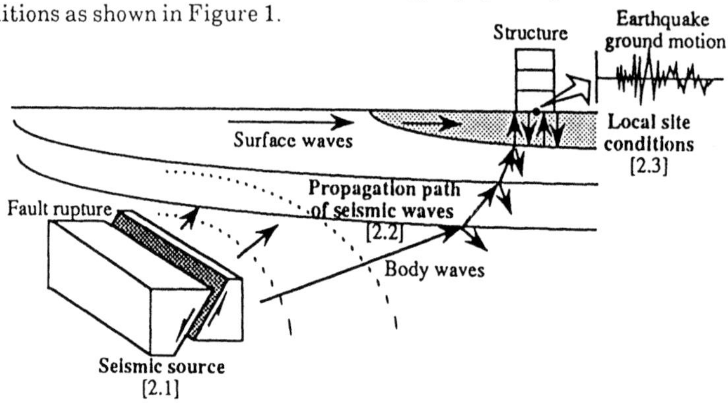

Figure 1. Generation of earthquake ground motion.The sign of [ ] denotes the section in which the item is described.

2.1. SEISMIC SOURCE

2.1.1. *Cause of Earthquake*. The Earth's lithosphere consists of several plates. The plates move and interact with each other (see Figure 2). Plate interactions occur at the following places: ① Mid-ocean ridges where two plates split apart, ② Transform faults where two plates slide horizontally, and ③ Subduction zones where high density oceanic plates subduct beneath low density continental plates. Many earthquakes have occurred in these places.

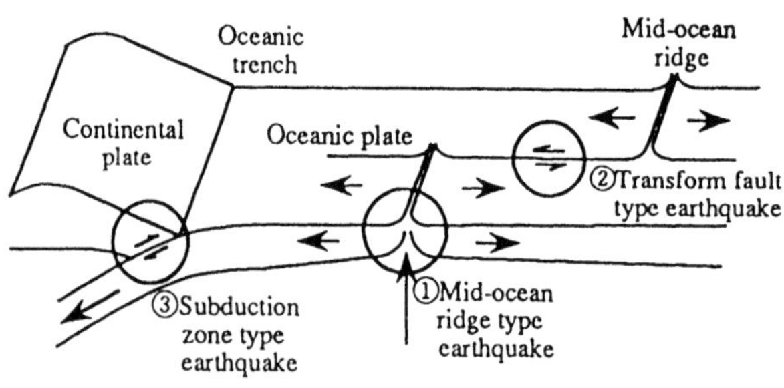

Figure 2. Movements and interactions of plates of the Earth's lithosphere. Many earthquakes occur at places where the plates interact.

The physical concept of earthquake phenomena is based on the elastic rebound theory derived by Reid (1911). Figure 3 shows a mechanism of an earthquake in the subduction zone. Due to interaction between two plates, strains in the lithosphere build up until a failure point (see Figure 3 (b)). Then, rupture takes place along a fault in the lithosphere and the strained portion rebounds on each side of the fault so as to relieve accumulated strains (see Figure 3 (c)). Namely, the fault rupture is a seismic source. Fault parameters used to characterize the fault rupture for the synthesis of earthquake ground motions (see section 5).

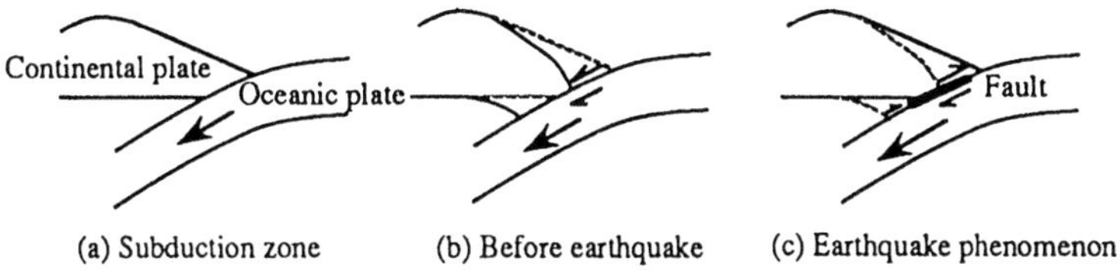

Figure 3. Mechanism of an earthquake in subduction zone based on the elastic rebound theory.

2.1.2. *Earthquake Magnitude.* Earthquake ground motions depend on the size of the earthquake. An earthquake magnitude M is widely used to measure the size of earthquake.

The earthquake magnitude M was originally defined by Richter (1935) as the common logarithm of the maximum amplitude A (in microns) recorded by a standard torsion Wood-Anderson Seismograph with a natural period of 0.8 second, critical damping ratio of 0.8, and magnification of 2,800, located at a site 100 km away from the epicenter. A calibration curve was constructed to reduce the amplitude observed at an arbitrary epicentral distance Δ (see Figure 4) to that expected at 100 km, given in the following form:

$$M = \log_{10}A + \alpha \log_{10}\Delta + \gamma, \tag{1}$$

where α and γ are constants.

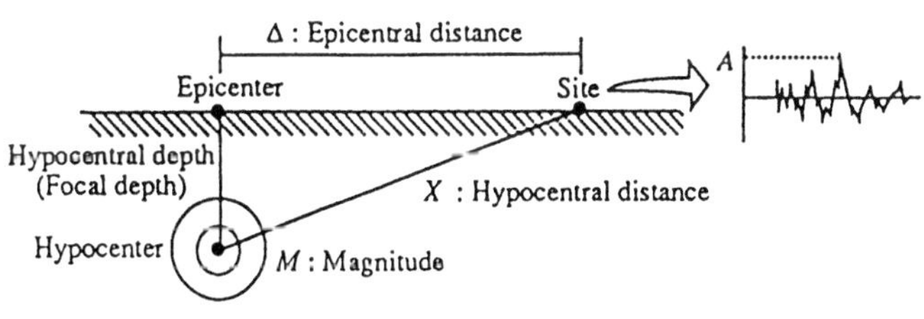

Figure 4. Earthquake magnitude M and epicentral distance Δ.

Based on the above definition, Richter (1935) determined M of local shallow earthquakes ($\Delta \leqq 600$ km) in South California, the U.S.A. This magnitude is called the Richter local magnitude M_L. The surface wave magnitude M_s, which is determined from the ground motion associated with surface waves with a period of 20 second recorded at stations in the world, is also commonly used in earthquake engineering (Gutenberg (1945); Vaněk *et al.* (1962)). Besides, there are many different kinds of magnitude such as the moment magnitude M_w (see section 5), the body wave magnitude m_b, and the Japan Meteorological Agency magnitude M_{JMA}. Therefore, a great attention has to be paid to what kind of magnitude is used. The theoretical and empirical relationships between M and some of fault parameters have been studied (e.g., Kanamori and Anderson (1975); Sato (1979)). These relationships can be used to estimate the fault parameters when only the magnitude of an earthquake is given.

2.2. PROPAGATION PATH OF SEISMIC WAVES

2.2.1. *Types of Seismic Waves.* There are mainly two types of seismic waves in the Earth: body waves that propagate through the Earth's interior and surface waves that propagate along the Earth's free surface or along other discontinuity in the Earth's interior.

From an earthquake source, body waves are radiated into the Earth's interior. There are two types of body waves, that is, compressional waves and shear waves. The compressional waves travel faster than the shear waves and are often called *P*-waves or primary waves. Shear waves are called S-waves or secondary waves. *P*-wave velocity V_p and *S*- wave velocity V_s are expressed by $V_p=\{(\lambda+2\mu)/\rho\}^{1/2}$ and $V_s=(\mu/\rho)^{1/2}$, respectively, where λ and μ are the Làme's constants and ρ is density of the medium. The relationship between them is given by $V_p/V_s=\{2(1-\nu)/(1-2\nu)\}^{1/2}$, where ν is a Poisson's ratio.

At a site far from the epicenter, surface waves are generally predominant. The energy of the surface waves are always concentrated near the surface. There are two types of surface waves, that is, Rayleigh waves and Love waves. The Earth's structure roughly consists of multi-layered media whose *P*- and *S*- wave velocities are larger with depth. The velocities of the surface waves for these inhomogeneous media vary with their frequencies, called velocity dispersion. These velocities are also characterized by *P*- and *S*- wave velocities.

Therefore, *P*- and *S*- wave velocity structures between the source and the site are required in order to know the behavior of the body waves and the surface waves. Figure 5 illustrates the motions of *P*-wave, *S*-wave, Rayleigh wave, and Love wave.

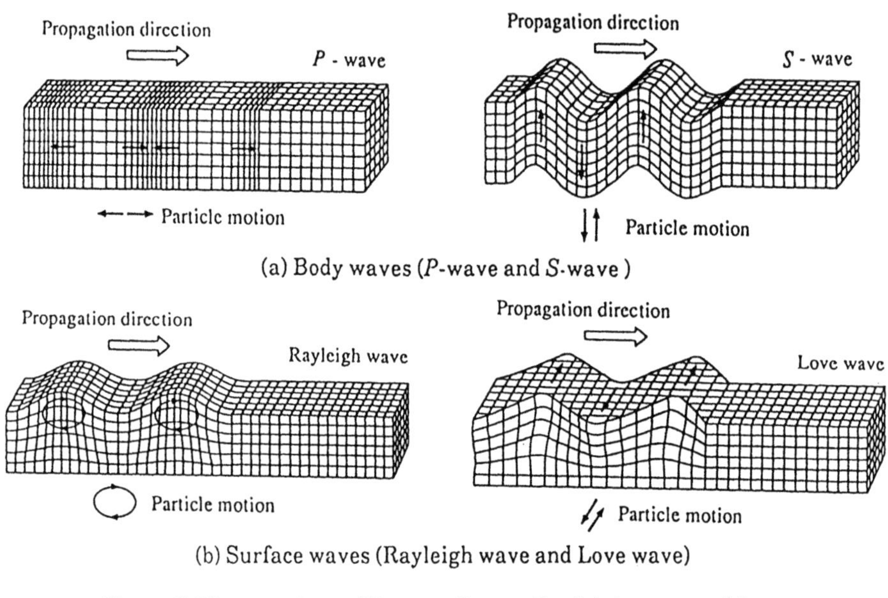

Figure 5. Wave motions of *P*-wave, *S*-wave, Rayleigh wave, and Love wave.

Figure 6 shows an example of long-period ground motion records at distance of a few hundreds of kilometers. These seismograms show long-period output of the STS-1 seismometer with broad-band and wide dynamic range at Inuyama station, Aichi Prefecture in Japan, for the 1987 Chibaken-Toho-Oki earthquake of M_{JMA} 6.7 and the focal depth 58 km. (Yamada, *et al.* (1989)). The epicentral distance is 316 km. These seismograms have the period contents longer than a few seconds. It is clearly seen that the several types of seismic waves arrive at different times because of their different velocities and that the surface waves are more predominant than the body waves.

Figure 7 shows an example of acceleration strong ground motion records within a few tens of kilometers from the source. These accelerograms were obtained at Echujima station, Tokyo ($\Delta = 55$ km) for the same earthquake. The strong parts of horizontal accelerograms have the predominant period contents less than a few seconds and they mainly consist of S-waves. Many studies have been done on the identification of seismic wave types from observation (e.g., Ishii, *et al.* (1989)). From these results, it is pointed out that the S-waves are significant for the engineering purpose.

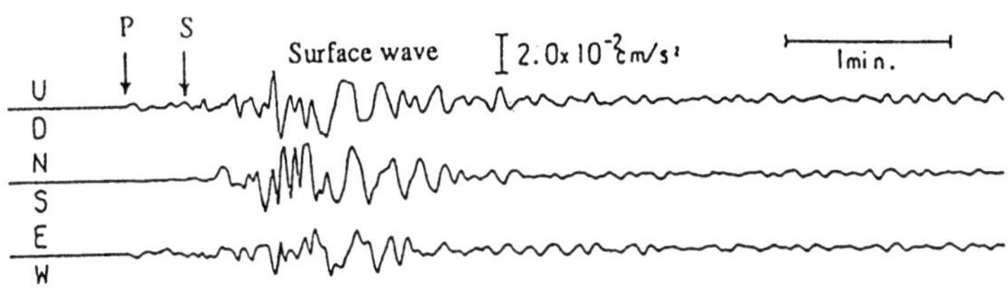

Figure 6. An example of long-period ground motion records at a distance of a few hundreds of kilometers from the source. The three components are long-period (more than a few seconds) output of the STS-1 seismometer with broad-band and wide dynamic range at Inuyama station, Aichi Prefecture in Japan, for the Chibaken-Toho-Oki earthquake on December 17, 1987 (M_{JMA} 6.7, focal depth 58 km, and $\Delta = 316$ km), (after Yamada. *et al.* (1989)).

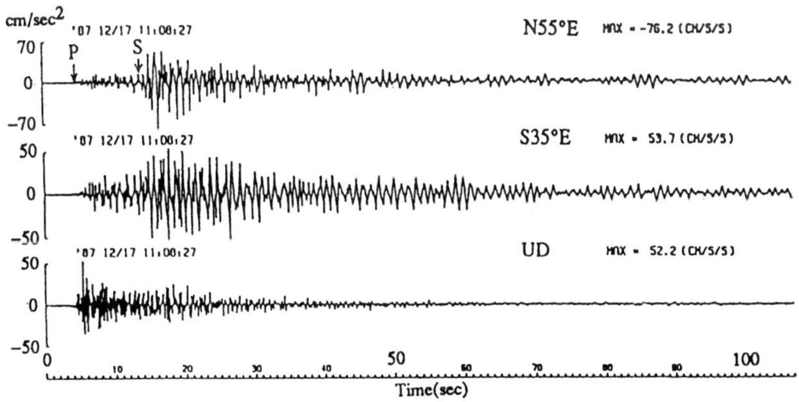

Figure 7. An example of strong ground motion records within a few tens of kilometers from the source, obtained at Echujima station, Tokyo in Japan, for the same earthquake described in Figure 6 ($\Delta = 55$ km).

2.2.2. *Attenuation of Seismic Waves*. The wave amplitudes decrease by geometrical spreading, by intrinsic attenuation, and by losses due to inhomogeneity in the Earth.

The geometric spreading is caused by the increase of the wavefront area with the travel distance increasing. For body waves, the reduction in amplitude due to geometrical spreading can be expressed by $1/x$, where x is the travel distance. For surface waves, the corresponding reduction factor can be expressed by $(1/x)^{1/2}$.

The intrinsic attenuation is caused by the internal friction of materials. The gross effect of the intrinsic attenuation is summarized by the dimensionless value Q called a quality factor, $1/Q(\omega) = -\delta E/2\pi E$, where E is the peak strain energy stored in a volume of material and $-\delta E$ is the energy loss in each cycle at a circular frequency ω. The relationship between the quality factor Q and the damping factor h is given by $h = 1/(2Q)$. The reduction in amplitude due to this attenuation is expressed by $\exp(-\omega x/(2V_c Q))$,where V_c is the phase velocity of a seismic wave.

The loss due to inhomogeneity in the Earth is involved in reflection, refraction, mode conversion, diffraction, and scattering by the existence of the boundaries or velocity fluctuation in the Earth's medium (e.g., Aki and Wu (1988); Sato (1990); Sato and Kawase (1992)).

It is difficult to evaluate these attenuations from the information on the Earth's structure at present. Therefore, the following empirical attenuation relations obtained from observation have been studied.

2.2.3. *Empirical Attenuation Relations for Earthquake Ground Motions.* The empirical attenuation relations for peak acceleration, peak velocity, and response spectra of earthquake ground motions have been obtained (e.g., Kanai (1958); Kobayashi and Nagahashi (1977); Joyner and Boore (1981); Joyner and Boore (1982); Watabe and Tohdo (1982); Cambell (1985); Kamiyama and Yanagisawa (1986); Fukushima and Tanaka (1990)).

The typical equation form of empirical relationships between the amplitude A of an earthquake ground motion and the source distance x is generally expressed by

$$\log_{10} A = aM + b\log_{10} x + c. \tag{2}$$

Here, a, b, and c are regression coefficients determined from many records.

In the equation, there is ambiguity in defining the source distance x because the fault rupture of earthquakes may extend over tens or hundreds of kilometers. Variation of measuring distance from the source are illustrated in Figure 8 (Shakal and Bernreuter (1981)). In the past, the earthquake source was considered as a point and earthquake ground motions were estimated by using epicentral or hypocentral distance between the point source and the site because of their readily available distances. However, most recent works have used some variation on the closest distance to the fault rupture area.

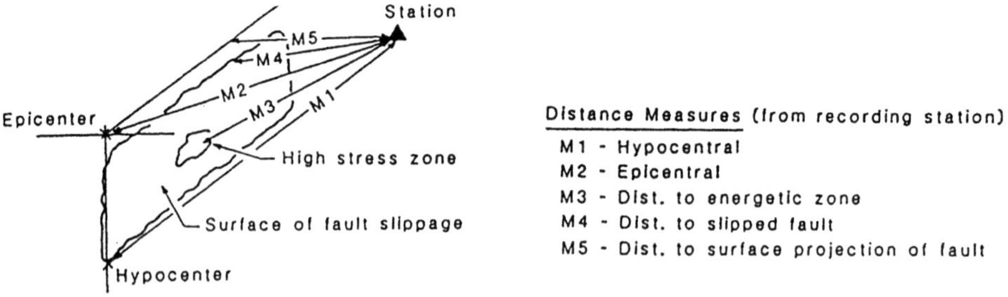

Figure 8. Different source distance measures used in empirical attenuation relationships (after Shakal and Bernreuter (1981)).

Figure 9 shows an example of the empirical attenuation relationships for peak horizontal accelerations obtained from the records of 28 earthquakes in Japan and 15 earthquakes in the U.S.A. and other countries (Fukushima and Tanaka (1990)). The resultant attenuation curve of peak horizontal accelerations is given by

$$\log_{10}\mathrm{PGA} = 0.41M_s - \log_{10}(X + 0.032 * 10^{0.41M_s}) - 0.0034X + 1.30. \qquad (3)$$

Here, PGA (cm/sec^2) is the mean peak acceleration of two horizontal components at each site, X (km) is the shortest distance between the site and the fault rupture area, and M_s is the surface-wave magnitude. In equation (3), the terms of $\log_{10}X$ and $0.0034X$ represent the attenuation due to geometrical spreading and other attenuations, respectively.

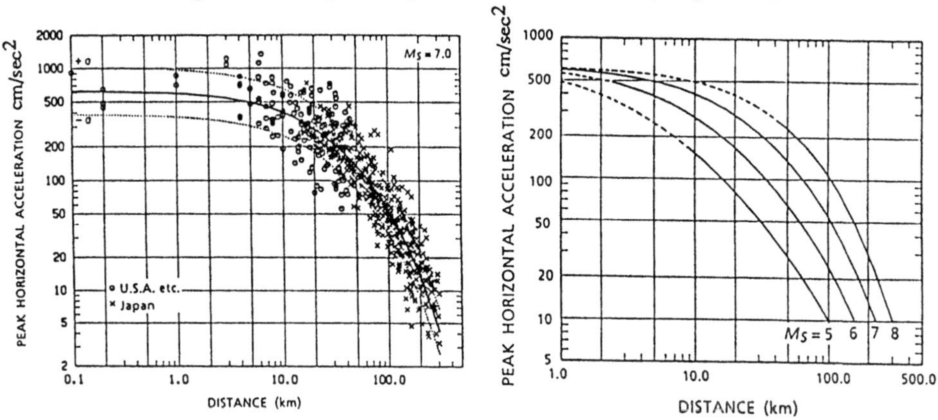

(a) Attenuation curve with data points for M_s 7.0 (b) Attenuation curves for M_s of 5 to 8

Figure 9. Empirical attenuation relation for horizontal peak accelerations. The broken lines in figure (b) indicate extrapolation range due to the lacks of data (after Fukushima and Tanaka (1990)).

These empirical attenuation relationships are widely used for estimating strong motions. However, a great attention has to be paid to the limited applicability of them to the estimation of strong-motions, because of the lacks of data for large earthquakes or in the vicinity of faults (see the broken lines in Figure 9 (b)).

2.3. LOCAL SITE CONDITIONS

2.3.1. *Multiple Reflection of Body Waves in Flat Soft Surface Layers.* In general, there exist soft sediment layers near the surface. When body waves are impinging upon the surface layer, successive reflections of the body waves occur within these surface layers. This phenomenon, which is called multiple reflection, causes the resonant amplification and it is highly sensitive to frequency . Therefore, the effect of soft surface layers on ground motions is often very significant (Kanai (1983)).

In order to estimate this effect, the soft sediment layers are modeled flat and analyzed by one-dimensional wave propagation theory. So-called flat layer assumption means that a geological structure underneath the site consists of laterally homogeneous layers. One-dimensional wave propagation theory gives responses at an arbitrary position in the flat layers due to vertically incident body waves.

Figure 10 shows the theoretical amplification factor of surface ground motion due to an S-wave impinging vertically from the below on a single flat soft surface layer with and without damping over a base stratum, where the incident wave is harmonic with unit amplitude and circular frequency ω. H, V_{s1}, ρ_1, h_1 are the thickness, S-wave velocity, density, and damping factor of the surface layer, respectively. V_{s2} and ρ_2 are the S-wave velocity and density of the base stratum.

The amplification has peaks at incident wave length of 4H, 4/3H, 4/5H, etc. Namely, the predominant periods T_H (or the predominant frequency $f_H=\omega_H/2\pi=1/T_H$) are given by $T_H=4H/V_{s1}$, $4H/3V_{s1}$, $4H/5V_{s1}$,etc. In the case of the surface layer without damping ($h_1=0.00$), the amplification factor at the peaks is equal to twice of the impedance ratio between the base stratum and the surface layer, i.e. $2\rho_2V_{s2}/\rho_1V_{s1}$, which is 4 in this case. The dotted line in Figure 10 (b) indicates the response in the case of the damping factor $h_1=0.05$, whose value is independent of frequency. It can be seen that the amplification factor is decreasing with frequency increasing.

Thus, S-wave velocity, density, thickness, and damping factor of local geology near the site are parameters to characterize local site conditions. S-wave velocity and density can be obtained by explorations and soil tests. But it is difficult to obtain the information on damping factor by these explorations. Recently, frequency-dependent characteristics of damping factor for sediments have been studied from borehole array observation (e.g., Kinoshita (1992)).

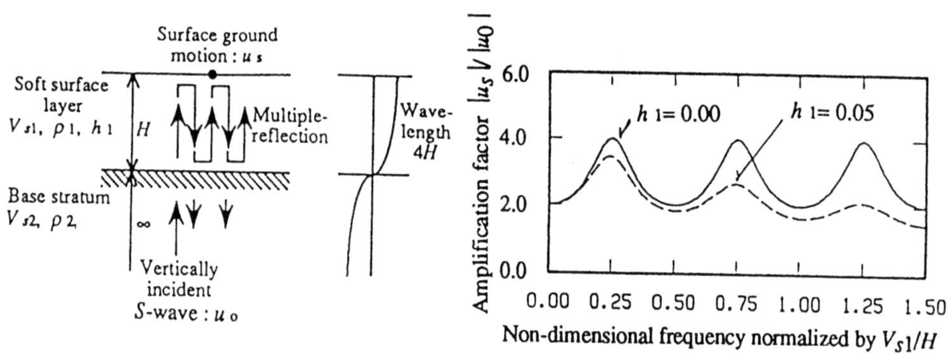

(a) One-dimensional wave propagation theory (b) Amplification factor $|u_s|/|u_0|$ v.s. frequency (Impedance contrast $\rho_2V_{s2}/\rho_1V_{s1}=2$)

Figure 10. One-dimensional wave propagation theory and amplification factor of a surface layer for vertically incident S-waves.

2.3.2. *Effect of Nonlinearity of Soft Soil.* In the soft surface layer subjected to very strong incident waves, the layer shows the strain-dependent nonlinear characteristics. The nonlinearity of the soil is characterized by a shear modulus reduction curve and a damping curve as a function of strain (e.g., Seed and Idriss, 1970). The predominant period is longer and the amplification factor is smaller for large strain during strong motion than those for small strain during weak motion (e.g., Tazoh, *et al.* (1988); Satoh, *et al.* (1992)). SHAKE (Schnabel, *et al.* (1972)) is the best known computer program for calculating this effect based on the one-dimensional wave propagation theory.

2.3.3. *Effect of Topographic or Geologic Irregularity.* Actual soft sediments are not infinitely flat but are laterally confined often in the form of a sediment-filled basin. There have been many theoretical or numerical studies on responses of a sediment-filled basin (e.g., Bard and Bouchon (1980); Sasatani (1988); Horike, *et al.* (1990); Ohori,*et al.* (1992)). Figure 11 shows the surface response of a sediment-filled basin due to vertically incident *SH* type Ricker wavelet by Kawase and Aki (1989). It is clear that the duration of the wave trains becomes longer in the basin because the surface waves generate at the edges of the basin and propagate horizontally. Recently, there have been some studies on comparing observed seismograms with synthetic ones (e.g., Ohtsuki, *et al.* (1984); Kawase and Sato (1992)). Several researchers have pointed out that this effect may be one of the reasons of the severe damage in Mexico city during the 1985 Michoacan, Mexico, earthquake (e.g., Sánches-Sesma, *et al.* (1988); Kawase and Aki (1989)).

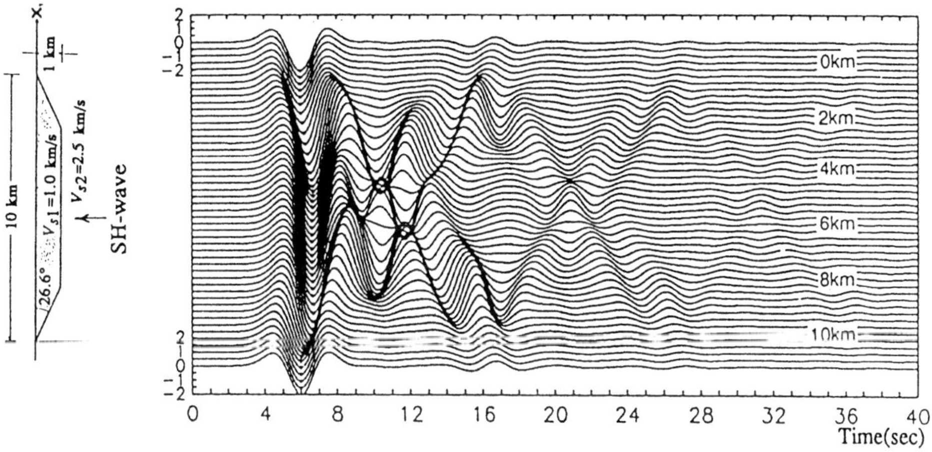

Figure 11. Response of sediment-filled basin due to vertically incident *SH* type Ricker wavelet with the characteristic frequency of 0.25 Hz. Surface waves generate at the edges of the basin that propagate horizontally inside the basin (after Kawase and Aki (1989)).

3. Basic Concept of Estimation and Generation of Design Input Motion

After a construction site is chosen, design earthquakes are selected for the assessment of earthquake-resistant design. Then, the design input motions at the site due to the design earthquakes are estimated by several methods as indicated in Figure 12.

3.1. DESIGN EARTHQUAKES

In Japan and the U.S.A., the following two kinds of earthquakes are considered as the design earthquakes for important structures such as high rise buildings and nuclear power plants.

One is called 'Design basis earthquake' which is thought to occur at least once during the life span of the structure. When the structure is subjected to this earthquake, no significant damage to the structure should occur.

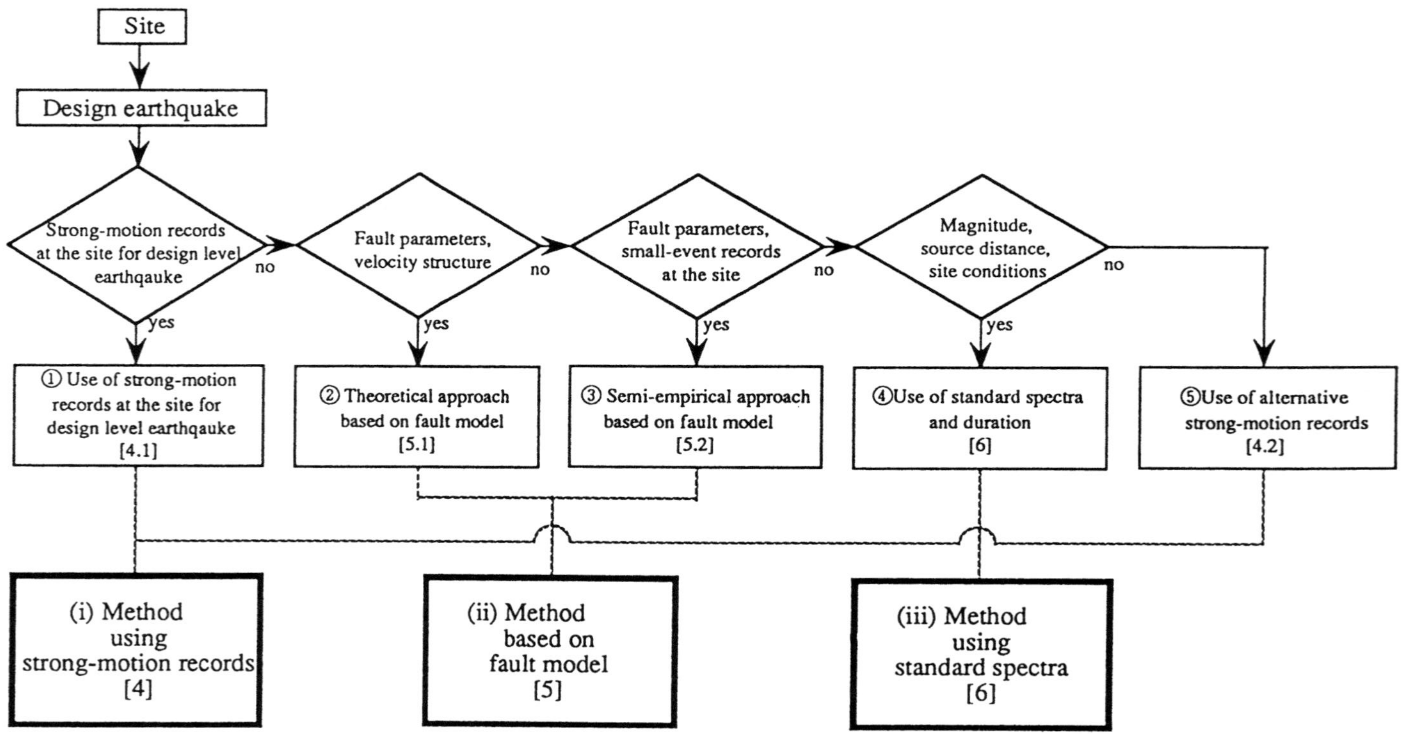

Figure 12. Flow-chart of estimation and generation of design input motion. The sign of [ ] denotes the section in which the method is described.

The other earthquake is called 'Maximum earthquake' which is thought to occur once in hundreds of years, more than life span years of the structures and generates the strongest ground motion at the site. When the structure is subjected to this earthquake, some damage can be tolerated, but the structure should not have severe damage that would result in collapse of the structure or loss of life.

This concept is often called 'Dual mode design.' The design earthquake is determined primarily based on the records of historic earthquakes, highly active faults, and seismo-tectonic structure lying under the site region (e.g., Itoh, *et al.* (1987)). Then, the parameters characterizing the design earthquakes, such as the magnitude, the source distance, or the fault parameters, will be examined and fixed.

3.2. DEFINITION OF DESIGN INPUT MOTION

The design input motions are basically defined as the ground motions at the free surface of the base stratum which supports the foundation of the structure as shown in Figure 13. Here, the free surface of the base stratum is defined to be a nearly flat surface of the base stratum extending over a considerable area, above which neither surface layers nor structures are assumed to be present. The base stratum should be relatively hard soil or rock. In Japan, the base stratum is considered to exhibit S-wave velocity greater than 400 ~ 500 m/sec for high rise buildings and 700 m/sec for nuclear power plants.

Suppose there exist soft surface layers over the base stratum, the soil-structure system including the surface layers should be employed as a response analysis model with the design input motion.

When ground motion records obtained at a different point from the free surface of the base stratum are utilized, it is desirable to calculate ground motions at a hypothetical free surface of the base stratum by deconvolution or convolution technique such as a one-dimensional wave propagation theory.

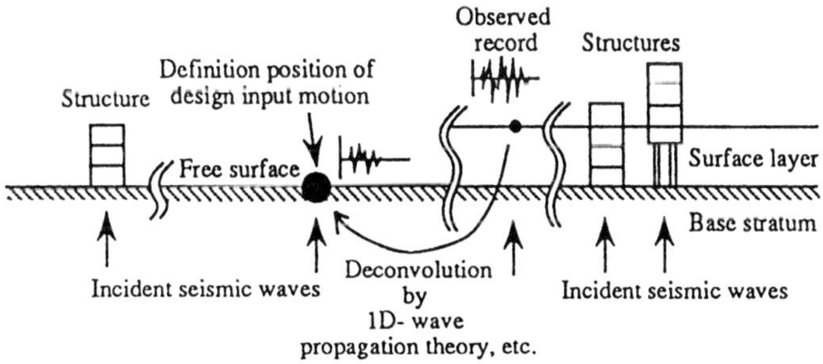

Figure 13. Definition of design input motion.

3.3. METHODS FOR ESTIMATING AND GENERATING DESIGN INPUT MOTION

At present, the following five approaches can be used to generate the design input motions:

① Use of strong-motion records at the site for the design level earthquake
② Theoretical approach based on fault model
③ Semi-empirical approach based on fault model
④ Use of standard spectra and duration
⑤ Use of alternative strong-motion records.

These approaches will be selected based on the following idea.

When there exist strong-motion records obtained at the site for the design earthquake, the approach ① will be given the priority because of all the information defining the design input motions being included in those records. When there exist no records and both information on the fault parameters and the velocity structure of the path between the earthquake source to the site can be specified, the approach ② will be used. When there exist appropriate small-event records at the site and the information on the fault parameters of the design earthquake, the approach ③ will be employed. When the magnitude of the design earthquake and the source distance are known, the approach ④ will be used. When there is no specific information of both the site and the design earthquake, the approach ⑤ will have to be employed.

From the view point of analogy of methodology, the five approaches can be classified into the following three methods:

(i) Method using strong-motion records: Approaches ① and ⑤
(ii) Method based on fault model: Approaches ② and ③
(iii) Method using standard spectra: Approach ④.

The details of these methods are described in the following sections 4, 5, and 6.

4. Method Using Strong-Motion Records

This method consists of two cases. In the first case, strong-motion records have been obtained at or near the site for the design level earthquakes. In the second case, alternative strong-motion records are used, after normalized by peak values, because no available records have been obtained at or near the site.

4.1. USE OF STRONG-MOTION RECORDS AT OR NEAR THE SITE FOR THE DESIGN EARTHQUAKES

After the site is chosen, the design earthquakes are selected for assessment of earthquake-resistant design. It is the best way to use the strong-motion records observed at the site, if there are any from the selected earthquakes.

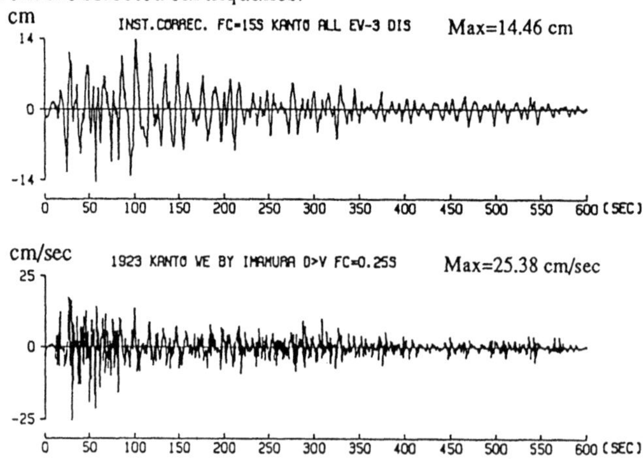

Figure 14. Mended displacement motion and the differentiated velocity motion (east-west component) at Hongo, Tokyo, from the 1923 Kanto, Japan, earthquake.

For instance, the 1923 Kanto, Japan, earthquake of M_{JMA} 7.9 is considered one of the design earthquakes in Tokyo. During this earthquake, the ground-displacement seismograms were recorded at Hongo in Tokyo by Imamura strong-motion seismograph with a magnification factor of 2. The records were saturated. Yokota *et al.* (1989) mended the saturated parts by the method of Noda *et al.* (1988) and corrected the characteristics of the seismograph. Figure 14 shows the mended seismogram and the differentiated velocity motion. The peak displacement is 14.5 cm and the velocity 25.4 cm/sec. The reliable period is from 2 to 15 seconds and the motion is the east-west (N77°E) component. The north-south (N347°E) and vertical components were too saturated to mend. After examining their coda parts and the aftershock records, Yokota *et al.* estimated that the north-south component should have been 1.5 times larger than the east-west component.

Consequently, it is appropriate to enlarge the mended seismogram by 1.5, equivalent to the north-south component, and use it as one of the design input motions for the structures with a natural period of 2 to 15 seconds in Tokyo. The peak displacement of the motion becomes 21.7 cm and the velocity 38.1 cm/sec.

4.2. USE OF ALTERNATIVE STRONG-MOTION RECORDS

In an actual design, we can hardly expect strong-motion records obtained at or near the site for the design earthquakes. Hence, in the U.S.A., Canada, and Japan, for instance, representative strong-motion records such as the record obtained at El Centro, California, during the 1940 Imperial Valley earthquake of M_L 7.1 are used as one of the design input motions after normalized by peak values.

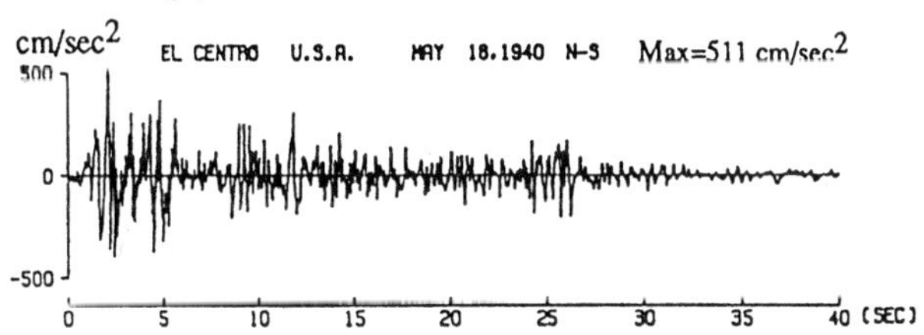

Figure 15. The El Centro, California, record normalized by the peak velocity of 50 cm/sec.

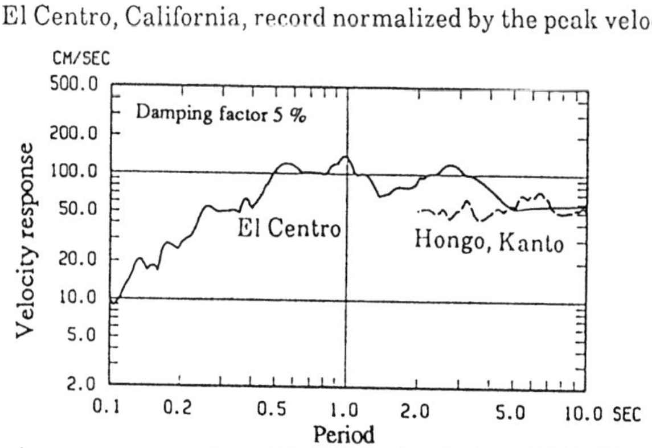

Figure 16. Velocity response spectra with a damping factor of 5 %. The solid line is for the El Centro record normalized by the peak velocity of 50 cm/sec and the broken line is for the north-south component estimated from the mended seismogram at Hongo, Tokyo, from the 1923 Kanto, Japan, earthquake.

The peak values for normalization are determined according to the seismicity around the site. For high-rise buildings in Tokyo, for instance, the peak velocity for normalizing the records is 25 cm/sec for the 'Design basis earthquake', and 50 cm/sec for the 'Maximum earthquake'.

Figure 15 shows the acceleration motion of El Centro whose peak velocity is normalized to be 50 cm/sec and Figure 16 shows its velocity response spectrum by a solid line, with a damping factor of 5 %. Figure 16 also shows, by a broken line, the velocity response spectrum of the north-south component estimated at Hongo in Tokyo during the 1923 Kanto earthquake. Since the response spectrum of the El Centro record is flat in the period range of 5 to 10 seconds, the reliable period of the record is 0.1 to 5 seconds. When the strong-motion records are used, a great attention has to be paid to the reliable period range of them.

This approach is so easy to apply that we can see many examples in the actual design. However, the alternative records does not reflect well the source, the path, and the local site conditions as specific ones to the site.

5. Method Based on Fault Model

Representation of the fault rupture and the seismic wave radiated into the Earth's interior accompanying the fault rupture is needed in order to calculate the earthquake motions theoretically. The former is called a 'fault model' and the latter a 'Green's function'. This method consists of two approaches. The fist one is a strictly theoretical approach, in which every physical phenomenon is described mathematically. The second one is a semi-empirical approach, in which some parts of the theory are replaced by records.

5.1. THEORETICAL APPROACH

The 'fault model' can be classified into two types: a 'kinematic model' and a 'dynamic model' (Aki and Richards (1980)). The 'kinematic model' is based on elastic theory of dislocation and also called a 'dislocation model.' In this method, the temporal and spatial distribution of the dislocation at both sides of a crack is given a priori as the boundary condition and the response of the elastic medium is calculated. The 'dynamic model' is based on fracture mechanics. In this method, the fracture criteria of the fault and the condition of the stress are given and then a process of the appearance, growth, and stop of a crack is solved. The temporal and spatial distribution of the fault dislocation is calculated and the response of the elastic medium is obtained. Up to now, the 'kinematic model' has been used in most cases, because the 'dynamic model' is difficult to deal with and exact solutions are limited to simple problems.

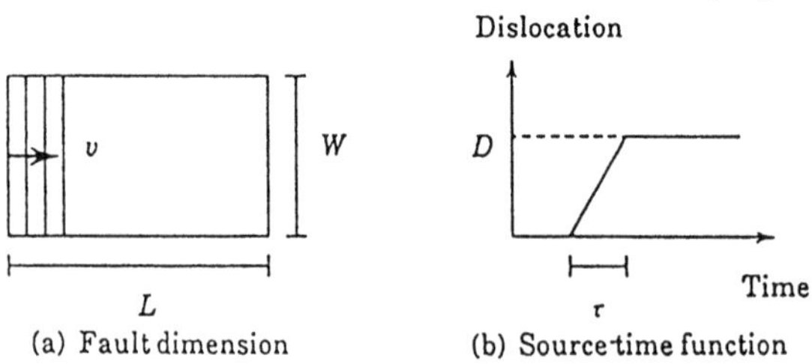

Figure 17. The five source parameters of Haskell model. L, W, D, τ, and v indicate the fault length, the fault width, the averaged dislocation, the rise time of the dislocation, and the rupture velocity, respectively.

Haskell model (1964) is the most popular model among the 'kinematic model.' He described the fault model by five source parameters; the fault length L, the fault width W, the averaged dislocation D, the rise time of the dislocation τ, and the rupture velocity v (See Figure 17). By using the 'kinematic model', Aki (1969) successfully reproduced displacement seismograms recorded at the site whose distance from the fault rupture for the Parkfield earthquake of 1966 was about 60 m.

Aki (1966) introduced a seismic moment M_0, which is defined by $M_0 = \mu SD$. Here, μ is the rigidity of the medium and S is the fault area ($L \times W$ for a rectangular fault). The seismic moment is an important physical parameter, indicating the size of the earthquakes. Kanamori and Anderson (1975) gave a theoretical basis of some relations among fault parameters, the seismic moment, and the surface-wave magnitudes. For huge earthquakes with magnitudes of over 8, M_L or M_s described in section 2.1.2 is not suitable to measure the size of earthquakes and instead Kanamori (1977) defined a moment magnitude M_w by

$$\log_{10} M_0 \,[\text{dyne}\cdot\text{cm}] = 1.5\, M_w + 16.1. \tag{4}$$

As the Earth's interior has been investigated by the analyses of earthquake waves, the propagation of the long-period body waves and surface waves and the oscillation of the Earth have been described mathematically. At present, the direction and the amplitude of the seismic moment (moment tensor) is determined in a routine work for all earthquakes with magnitudes of over 6 in the world (Kanamori and Given (1981); Dziewonski, *et al.* (1983)).

The Green's function represents a response at an arbitrary point in the medium when a unit force is applied to a point source. It is obtained by solving the differential equations and the boundary conditions. Hence, many studies have been performed to calculate the Green's function representing the inhomogeneous medium and the complex boundary conditions in order to model the real phenomenon. The study began to deal with a half space medium because of easy mathematical operation (e.g., Lamb (1904); de Hoop (1960); Kawasaki *et al.* (1973)). The theoretical solutions for the multilayered half-space medium were then obtained (e.g., Harkrider (1964); Helmberger (1968); Swanger and Boore (1978); Bouchon (1979); Luco and Apsel (1983)). Recently, the numerical methods such as the finite difference method (Vidale and Helmberger (1987); Dreger and Helmberger (1990); Yamanaka, *et al.* (1992); Yomogida and Etdgen (1992)), the finite element method (Satoh, *et al.* (1988); Sato (1990)), the boundary element method (Hisada, *et al.* (1988)), and the paraxial method (Graves and Clayton (1992)) as well as the analytical methods (e.g., Kudo (1980); Yomogida (1985); Kohketsu (1987); Uebayashi *et al.* (1992)) have been applied to obtain the Green's function for difficult problems like a laterally heterogeneous medium.

An example of the simulation carried out by Sato (1990) is shown here. Figure 18 shows the seismograms of the west-east direction at different stations of Kobe and Osaka during the 1961 Kita-Mino, Japan, earthquake of M_{JMA} 7.0. The two observation stations are located at almost same distance and direction from the epicenter. Kobe is located at the edge of the sediment-filled Osaka Basin and Osaka is located in the center of it. This figure shows a clear effect of the basin. The synthetic motions are calculated by combining the Haskell type dislocation theory with the thin layer finite element method (Kausel *et al.* (1975); Tajimi (1980)) by Sato *et al.* (1988, 1990), in which the basin is modeled as shown in Figure 19. Figure 20 shows that the synthetic displacement motions agree well with the observed ones.

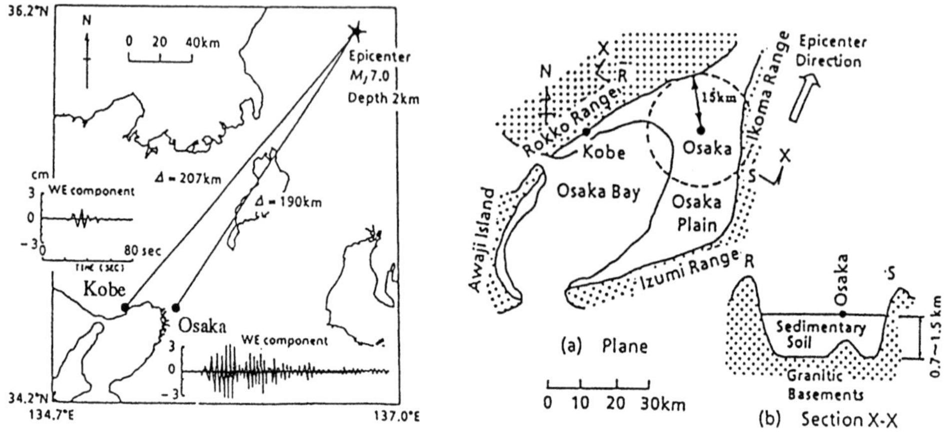

(a) Location of the epicenter and the seismograms observed at Kobe and Osaka stations

(b) Diagrammatic profile of the Osaka basin and stations of Kobe and Osaka.

Figure 18. Location of the epicenter of the 1961 Kita-Mino, Japan, earthquake of M_{JMA} 7.0, the seismograms observed at Kobe and Osaka stations, and diagrammatic profile of the sediment-filled Osaka basin (after Sato (1990)).

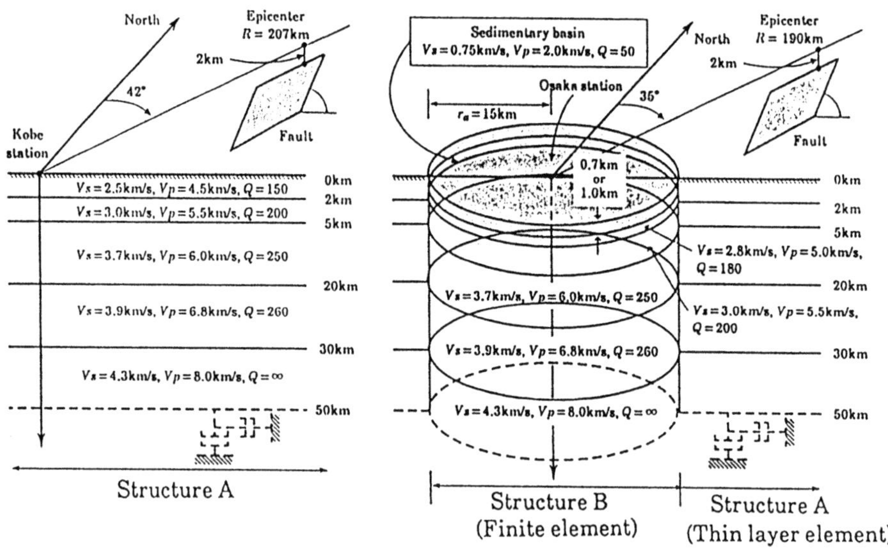

(a) Layered model used to calculate synthetic seismograms at Kobe

(b) Sediment-filled basin model used to calculate synthetic seismograms at Osaka

Figure 19. Velocity structure model used to calculate synthetic seismograms by the thin layer finite element method (after Sato (1990)).

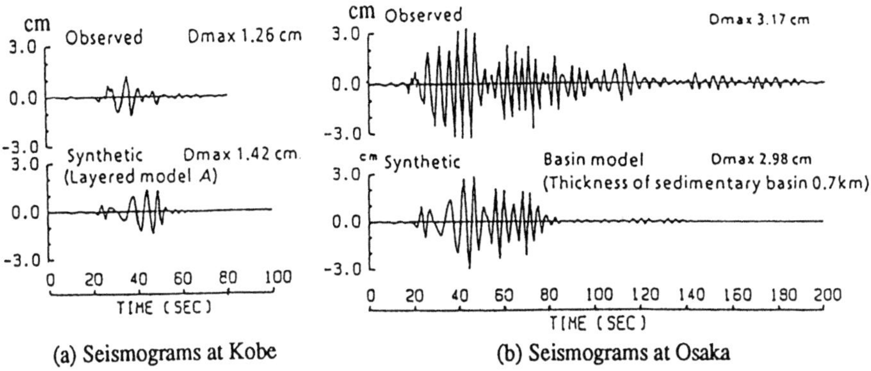

Figure 20. Comparison between the observed and synthetic seismograms (west-east component).

Now it is possible to simulate earthquake ground motions with a period longer than a few seconds for an earthquake with a magnitude smaller than about 7 by the 'kinematic model' with smooth and uniform rupture. However, the natural period of the structures is, in general, less than several seconds in actual design. Information on the inhomogeneous micro-fracture mechanics of the source and on the detailed underground structures is necessary to simulate the short-period motions. Moreover, for an earthquake with a magnitude over 7, the seismic waves even with a period longer than 10 seconds is affected by the inhomogeneous rupture of the source. Hence, studies on the inhomogeneous rupture of the source have been performed by many researches (e.g., Aki (1967); Brune (1970); Wyss and Brune (1967); Hanks and Wyss (1972); Madariaga (1976); Das and Aki (1977); Mikumo and Miyatake (1978); Kikuchi and Kanamori (1982); Olson and Apsel (1982); Papageorgiou and Aki (1983); Takeo (1987)). For the application of the results to the actual design, further results of the studies must be accumulated because of the large uncertainty of the parameters at present.

5.2. SEMI-EMPIRICAL APPROACH

The semi-empirical approach was developed by Hartzell (1978). In this approach, small-event records are used as Green's functions, as shown in Figure 21, instead of calculating the Green's functions in the theoretical approach .

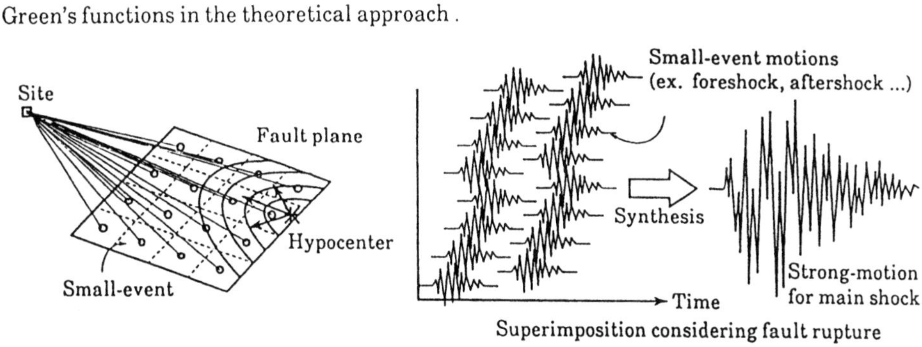

Figure 21. Schematic concept of the semi-empirical approach.

This approach is very useful for estimating the short-period motions because the small-event motions include not only the local site effect near the site and the three-dimensionally inhomogeneous structures of the path but also the complex rupture mechanism of the source. Kanamori (1979) synthesized the Love waves with a period of 2 to 10 seconds considering the difference of the seismic moment of the large and small events. Irikura (1983) proposed a method based on the similarity relation of the source parameters, insisting that the fault length, the fault width, and the averaged dislocation are proportional to each other and the stress drop and the slip velocity are constant independently of the source size. He applied the method to the velocity motions with a predominant period of about 10 seconds. In his method, the number of the superposition of the small-event records is also the ratio of the seismic moment between the two events. On the other hand, in order to simulate the short-period motions, Tanaka *et al.* (1982) synthesized the acceleration records and showed that the number of the superposition of the small-event records should be the ratio of the seismic moment to the power of two thirds, $(M_{0l}/M_{0s})^{2/3}$, here, l and s denote the indexes for the large and small events, respectively. Dan *et al.* (1990), based on the ω-square source spectrum model (Brune (1970); Hanks and Wyss (1972); Sato and Hirasawa (1973); Izutani (1981); Hanks and McGuire (1981), Boore (1983)), explained consistently the difference between the synthesis methods for the long-period and short-period motions. There have been several variations of the semi-empirical approach (e.g., Takemura and Ikeura (1988); Wald, *et al.* (1988); Akao, *et al.* (1988); Izutani and Katagiri (1992)).

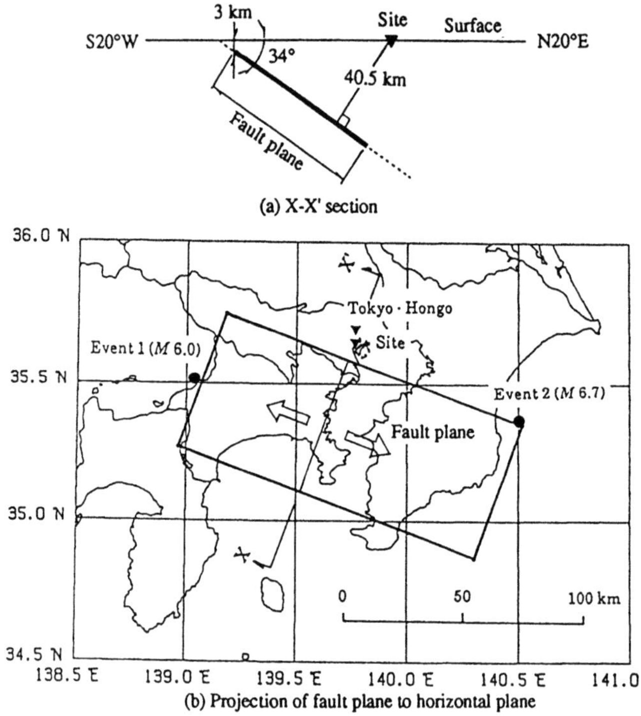

Figure 22. Location of the site, the fault model of the 1923 Kanto, Japan, earthquake, and the epicenters of the small-events for the Green's functions.

Here, an example of the simulation in Tokyo by the method of Dan *et al.* is shown for the 1923 Kanto earthquake. Figure 22 shows the location of the site, the fault model of the Kanto earthquake, and the epicenters of the small events for the Green's functions. Figure 23 shows small-event accelerograms used as the Green's functions in the synthesis. Figure 24 shows the synthesized acceleration motion and the integrated velocity motion on the base stratum supporting the structures (the *S*-wave velocity is 400 to 500 m/sec). Figure 25 shows the velocity response spectra of the synthesized motion with a damping factor of 5 % by the dotted line. The reliable period is in the whole period range of 0.1 to 10 seconds for the synthesized motion, while it is longer than 2 seconds for the mended seismogram from the Kanto earthquake (the broken line) and shorter than 5 seconds for the El Centro record (the solid line).

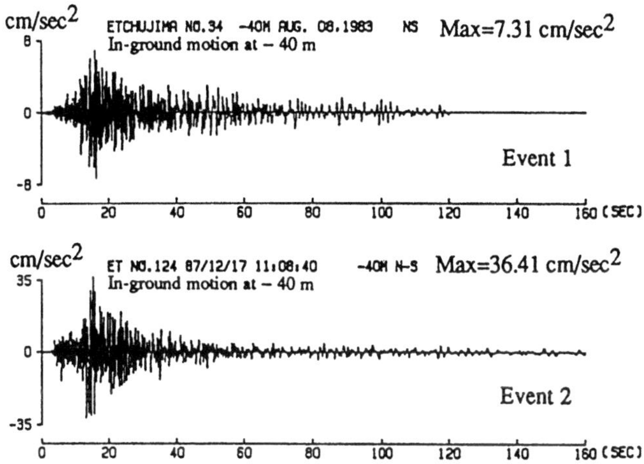

Figure 23. Small-event records used as the Green's functions in synthetics. The upper seismogram is from the Event 1 and the lower one is from the Event 2.

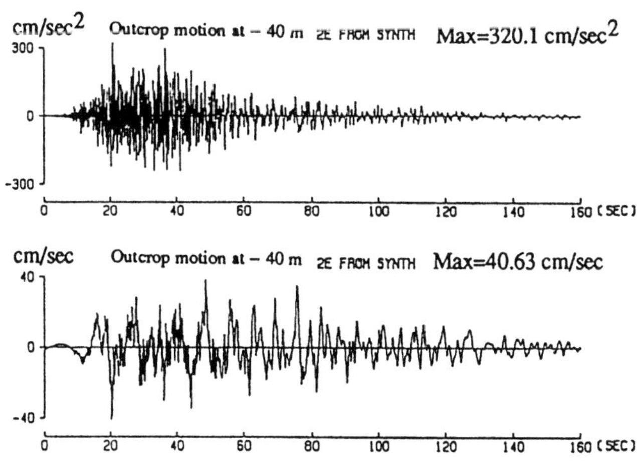

Figure 24. Synthesized acceleration motion and the integrated velocity motion on the base stratum in Tokyo from the 1923 Kanto, Japan, earthquake.

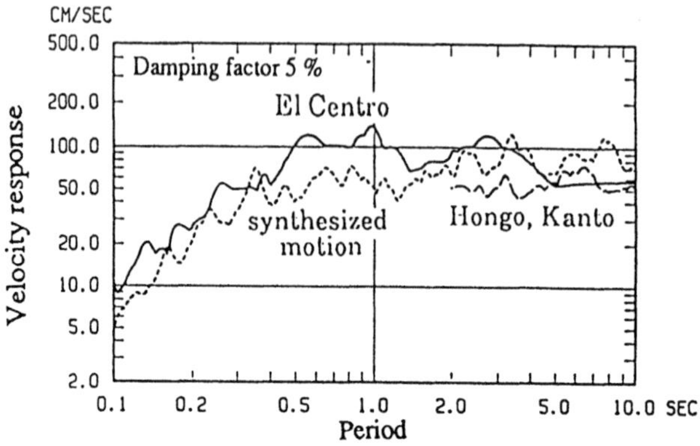

Figure 25. Velocity response spectra with a damping factor of 5 %. The dotted line is for the synthesized acceleration motion in Figure 24, the broken line is for the estimated north-south component of the seismogram at Hongo, Tokyo, from the 1923 Kanto, Japan, earthquake, and the solid line is for the El Centro record normalized by the peak velocity of 50 cm/sec.

The semi-empirical approach is the most practical method for evaluating the input motions based on the fault model, but it needs at least the small-event records, the value of the seismic moment of the small event, and the fault model of the large event. The key of this approach is whether the small-event records include enough characteristics representing to the natural periods of the structures to be analyzed.

6. Method Using Standard Spectra

This method generates a design input motion by the superposition of sinusoidal waves in such a manner that its spectrum matches the target spectrum. In this method, the standard response spectrum or power spectrum, duration, and time-dependent variation of amplitude envelop curve (intensity function) are determined as a function of the earthquake magnitude, distance, and local site conditions. The information on the phase of ground motions can be used instead of duration and time-dependent variation of amplitude envelop curve. These values are usually given by regression analyses based on many strong-motion records. Therefore, an attention has to be paid to a range of applicability, which depends on the quality and quantity of the records.

6.1. FREQUENCY CONTENTS OF EARTHQUAKE GROUND MOTIONS

6.1.1. *Normalized Spectra.* The normalized spectra by the peak amplitude of ground motions such as peak acceleration express just only their spectral shapes, that is, the frequency-dependent characteristics. For design earthquake motions, it is quite practical to apply this normalized spectra with the aid of the maximum amplitudes of ground motions which are evaluated by the seismic hazard analysis.

Tajimi (1960) expressed a power spectral density function by

$$S(\omega)=S_0\cdot\frac{1+4h_g{}^2(\omega/\omega_g)^2}{\{1-(\omega/\omega_g)^2\}^2+4h_g{}^2(\omega/\omega_g)^2}. \tag{5}$$

Here, S_0 is the power spectrum at $\omega = 0$, ω_g is the predominant frequency, and h_g is the damping factor of subsoil layers. This spectrum, called 'Kanai-Tajimi spectrum', is convenient for expressing the predominant frequencies caused by the resonance of subsoil layers.

The normalized response spectra have been used for seismic design of nuclear power plants in the U.S.A. and other countries (U.S. Atomic energy commission (1973)).

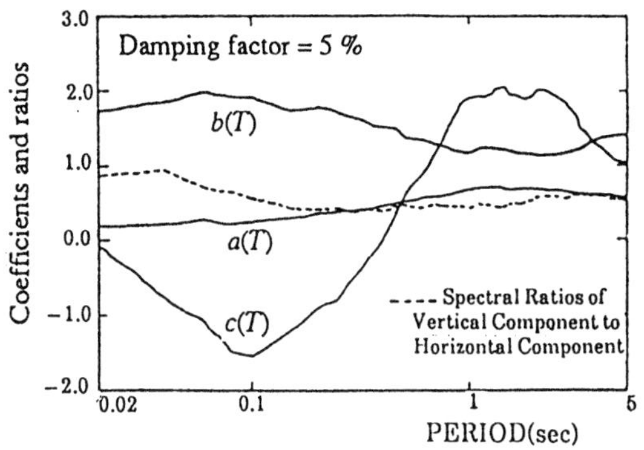

Figure 26. The coefficients $a(T)$, $b(T)$, and $c(T)$ of the response spectrum (solid lines) and the ratios of the response spectrum of vertical component to that of horizontal component (broken line) (after Watabe and Tohdo (1982)).

6.1.2. *Response Spectra.* The spectral shape of the motion recorded on the rock surface may be influenced less than the motion recorded on soil deposit by the local site conditions. The velocity response spectrum with a damping factor of 5 % on the exposed rock surface is assumed to have the same form as the equation (2). The coefficients $a(T)$, $b(T)$, and $c(T)$ are obtained by regression analysis for each period as shown in Figure 26 (Watabe and Tohdo (1982)). Some examples of the response spectra are shown in Figure 27. The solid lines indicate the results using the coefficients $a(T)$, $b(T)$, and $c(T)$ defined in Figure 26 and the broken lines are modified smooth spectra.

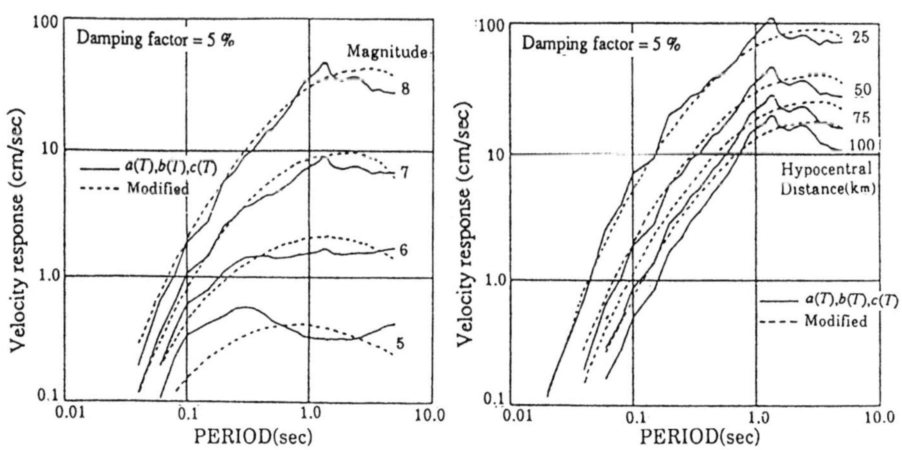

(a) Hypocentral distance of 50 km

(b) Magnitude(M_{JMA}) of 8

Figure 27. Velocity response spectra on rock site (after Watabe and Tohdo (1982)).

Hisada *et al.* (1978) proposed the standard response spectra for the seismic design of stiff structures, such as nuclear power plant facilities, after analyzing peak accelerations of the records obtained in the near-field and about 50 accelerograms recorded on outcropped rock surface in Japan. They determined spectral shapes of the standard spectra by the control periods for three different epicentral distances, referred to as near, intermediate, and far, for each of three different magnitudes 6, 7, and 8. Figure 28 shows the response spectra for nine categories.

6.1.3. *Response Spectrum of Vertical Components.* As for vertical component, the ratios of the vertical spectra to the horizontal spectra have been investigated with limited data of vertical components. The ratio for 8 accelerograms are shown by the broken line in Figure 26.

The characteristics of peak accelerations and acceleration response spectra of vertical ground motions have been investigated for the near-field records obtained in U.S.A. (Watabe *et al.* (1990)).

The mean value of the ratios of the peak vertical accelerations during the strong part of horizontal ground motions to the peak horizontal ones is equal to 0.53. The predominant period of vertical component are shorter than that of horizontal component. The

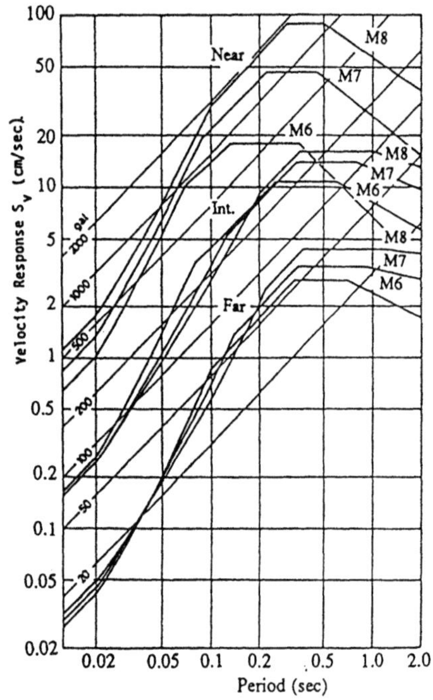

Figure 28. Standard response spectra for stiff structures on rock (after Hisada *et al.* (1978)).

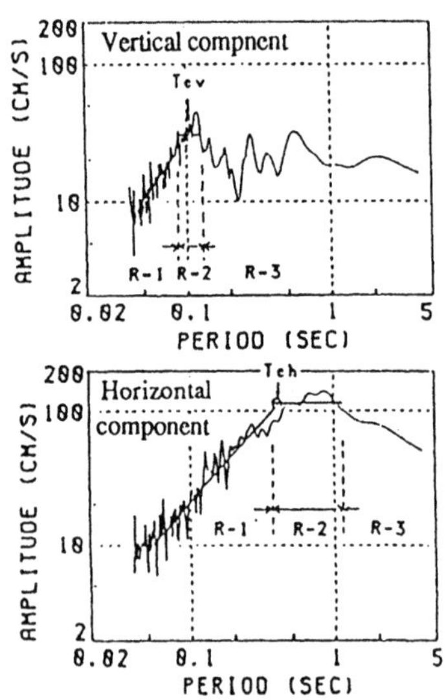

Figure 29. Fourier amplitudes and lower-side corner periods (Tcv, Tch) of vertical and horizontal components for El Centro 1940 (after Watabe *et al.* (1990)).

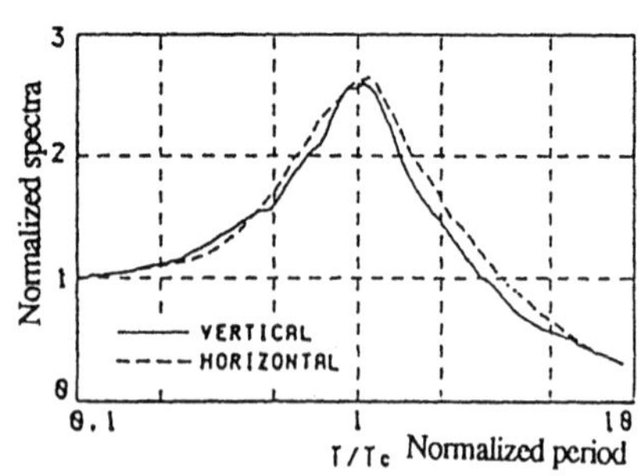

Figure 30. The overall average spectra normalized by both the peak acceleration and the lower-side corner period (after Watabe *et al.* (1990)).

spectral shapes of vertical and horizontal acceleration response spectra can be regarded as almost identical after normalized by both the peak acceleration and the lower-side corner period of the Fourier amplitude spectra of the accelerogram.

Figure 29 shows examples of the Fourier amplitudes and lower-side corner periods. The Fourier amplitudes rapidly decrease in shorter periods from the lower-side corner periods. The mean values of the lower-side corner periods for the horizontal and vertical components are 0.21 sec and 0.13 sec, respectively. The overall average spectra normalized by both the peak acceleration and the lower-side corner periods are compared for the horizontal and vertical components in Figure 30.

6.1.4. *The effect of damping factor on response spectra.* The spectral characteristics of earthquake ground motions are often represented by the velocity response spectra with a damping factor of 5 %. The response spectra with a damping factor other than 5 % can be obtained by multiplying the amplification coefficient η described in equation (6) (Hisada *et al.* (1978)). The amplification coefficient was derived from stationary random vibration theory.

$$\eta = \begin{cases} 1/\sqrt{1+17(\xi-0.05)\exp(-2.5T/T_0)}, & \text{for } T \geqq 0.10\text{ sec} \\ 1, & \text{for } T=0.02\text{ sec} \end{cases} \tag{6}$$

Here, ξ is a damping factor, T is a period, and T_0 is an effective duration of motion ($T_0 = 10^{0.31M-1.2}$). The equation (6) was verified by utilizing 135 horizontal accelerograms.

6.2. DURATION AND ENVELOPE FUNCTION

As an earthquake becomes larger, the length of a fault becomes longer, whereas the rupture velocity does not change so greatly. Therefore, the duration of earthquake ground motions is closely related with the size of earthquake or the earthquake magnitude.

According to the research work based on the fault model done by Sato (1979), the length of a fault is obtained by $\log_{10}L[\text{km}] = 0.5M - 1.88$. By using this relation, the total rupture duration T_r of the earthquake motion can be calculated by $\log_{10}T_r[\text{sec}] = 0.5M - 2.36$, with the assumption of a rupture velocity 3 km/sec of the unilateral fault. The various definitions of duration are proposed for observed accelerograms.

One of the relations between duration T_d and the magnitude M is given by $\log_{10}T_d[\text{sec}] = 0.3M - 0.77$ (Hisada and Ando (1973)). The duration in this equation was defined as the time after which peak value doesn't exceed 10 percent of the peak acceleration.

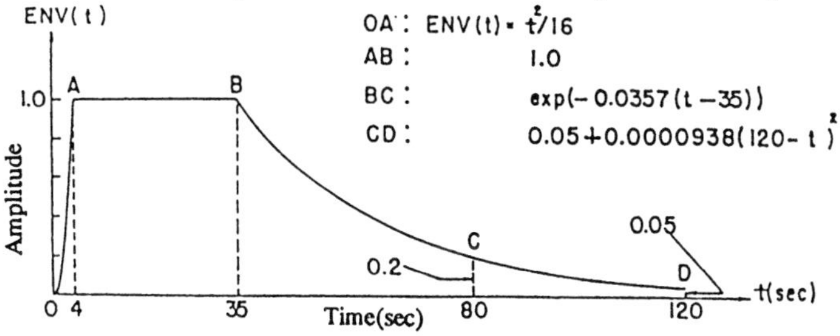

Figure 31. Time-dependent variation of amplitude envelop function for the strong motion due to an earthquake of M 8 (after Jennings *et al.* (1969)).

The amplitude envelope function is represented by three parts: build-up, nearly constant with high intensity level, and decaying (Jennings *et al.* (1969)). Figure 31 shows an example of the envelope functions for an earthquake of M 8. The envelope function is also called the deterministic intensity function.

6.3. CHARACTERISTICS OF PHASES OF EARTHQUAKE GROUND MOTIONS

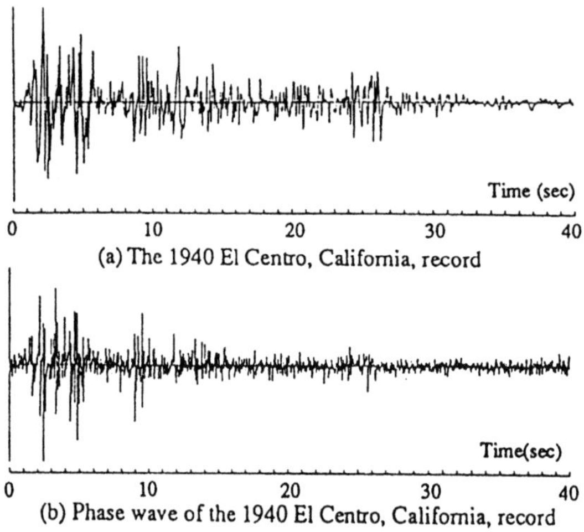

Figure 32. Record of El Centro 1940 (a) and its phase wave (b) (after Ohsaki (1979)).

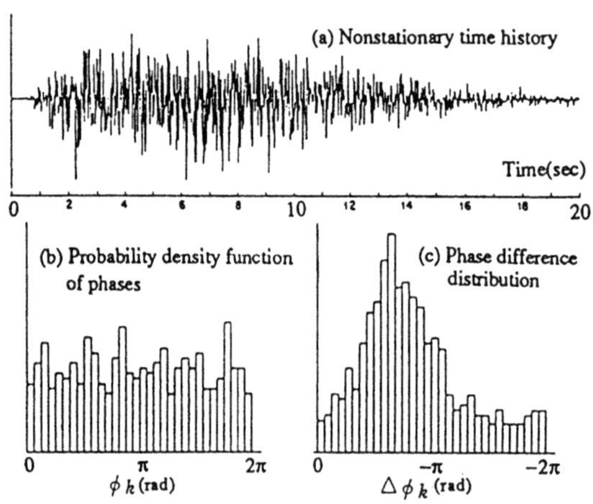

Figure 33. Nonstationary time history (a), its probability density functions of phases (b), and the phase difference distribution (c) (after Ohsaki (1979)).

Most researches suggest the randomness and the uniform probability distribution of phases of earthquake ground motion. It is also the fact, however, that the non-stationarity is composed not only of Fourier amplitudes but also of phase angles.

Figure 32 shows the record of El Centro 1940, the north-west component, and synthetic one which has uniform amplitudes in each frequency component with the same set of phases of the record. Let this kind of synthetic wave be called as a 'phase wave'. It is clearly shown in Figure 32 that the envelope function of the phase wave is quite similar to original one.

It was pointed out by Ohsaki(1979) and Katukura *et al.*(1978,1984) that the characteristics of non-stationarity of records can be expressed by the phase properties. The phase properties are expressed by the distribution of phase difference $\Delta\phi$, that is given by $\Delta\phi_k = \phi_{k+1} - \phi_k$, where ϕ_k is Fourier phase and $\Delta\phi$ is defined in the range of $-2\pi < \Delta\phi < 0$ (Ohsaki (1979)).

Figure 33 shows the nonstationary time history, the probability density function of phases and the phase difference distribution, respectively. The figure shows that the probability density function of phase is nearly uniform, while the probability density function of phase differences has a quite similar feature to the envelope function of time history normalized by the duration.

6.4. SYNTHESIS OF EARTHQUAKE MOTION

6.4.1 *Synthetic earthquake motions fitting to the design response spectra.* A simple model of synthetic earthquake motion is expressed as follows:

$$y(t) = E(t)\, g(t), \qquad (7)$$

where $g(t)$ is the stationary random process and $E(t)$ is the envelope function.

The stationary process is synthesized by the superposition of sinusoidal waves as follows:

$$g(t) = \Sigma\, A_j \cdot \cos\,(\omega_j t + \phi_j), \qquad (8)$$

where ω_j is the frequency equally spaced by $\Delta\,\omega$ and ϕ_j is random variables for phases.

The synthetic motion is generated in an iterative procedure to fit the target spectra. Figure 34 shows the flow chart of the synthetic earthquake motion fitting to the target spectra. When the set of phases or the phase differences of a recorded motion is used for the Fourier phase of a synthetic earthquake motion, it is not necessary to multiply the envelop function. Because the synthetic earthquake motion, $g(t)$, would have the similar envelop as the recorded one.

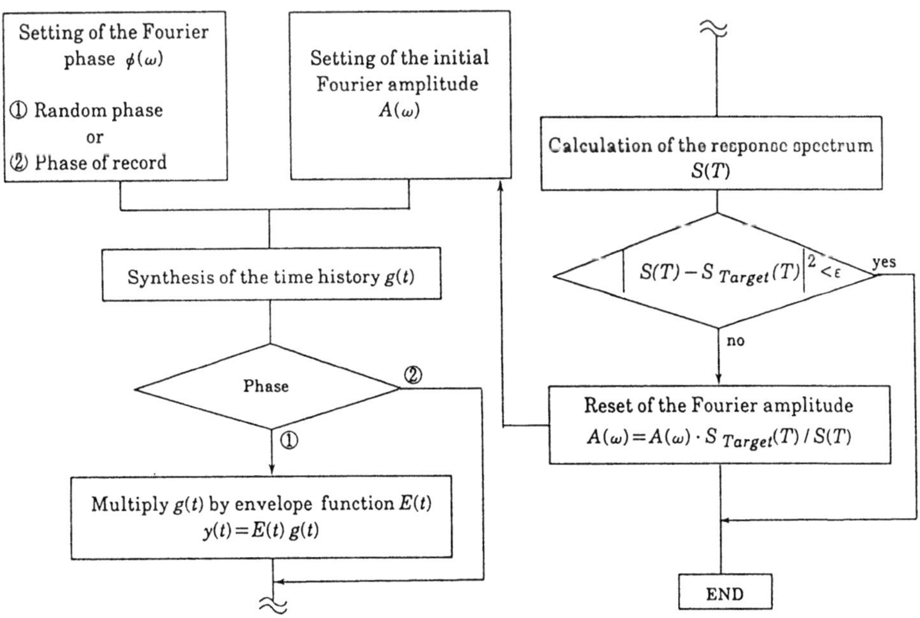

Figure 34. Flow chart of the synthetic earthquake ground motions fitting to the response spectra $S_{TARGET}(T)$.

6.4.2. *Synthetic earthquake motions fitting to multi-damping response spectra.* In generating synthetic earthquake motion, the iterative calculation is usually carried out to fit the design response spectrum with a damping factor of 5 %. In this procedure, the response spectra with other values of damping factor, for example 1 %, are not satisfactorily controlled. When the structures have low damping components, it is desired that the synthetic motion should fit to the response spectra of different damping factors simultaneously. From this point of view, several researches were carried out.

Hirasawa and Watabe (1992) presented an effective method for generation of a time history compatible with two or three damping target spectra (see Figure 35). It is an extension of the method developed by Lilhanand and Tseng (1987).

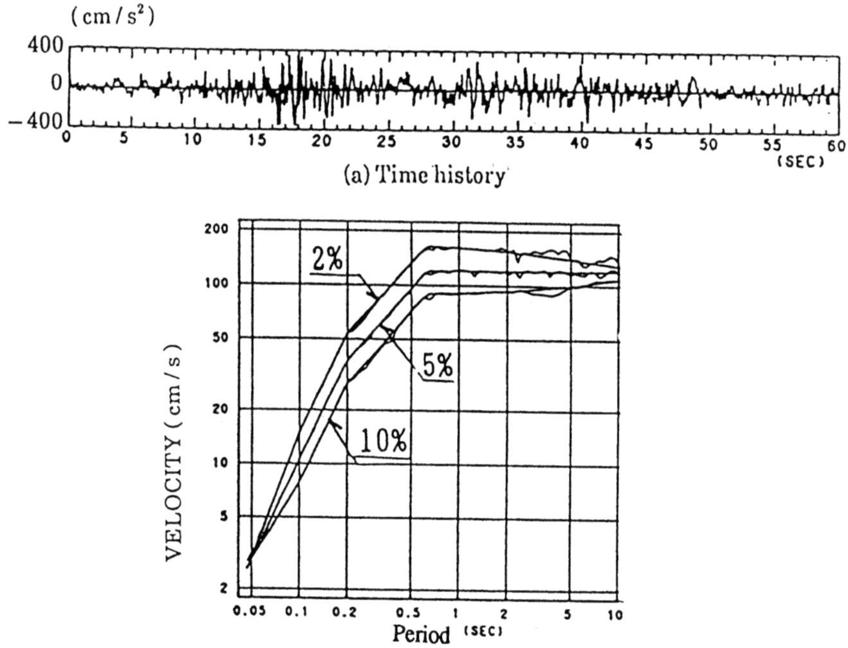

(b) Velocity response spectra

Figure 35. The results of the synthetic earthquake motion compatible with multi-damping response spectra. (a) time history, (b) comparison of the target spectra with the simulated ones (after Hirasawa and Watabe (1992)).

6.4.3. *Synthetic nonstationary earthquake motions using phase gradients.* In general, earthquake motions have a time change of frequency contents, i.e. nonstationary characteristics as shown in Figures 6 and 7. When a nonstationary earthquake motion is band-pass filtered, the filtered time histories have their different amplitude envelopes depending on the frequency ranges. Mean and standard deviation of the distribution of Fourier phase gradient $t_{gr}(\omega_k)$ $(=\Delta\phi_k/\Delta\omega$ where $\Delta\phi_k$ is a phase difference and $\Delta\omega$ is a circular frequency interval, (Katukura, *et al.* (1984)) of an earthquake motion are related to the mean and standard deviation of the amplitude envelope of the motion, respectively. Therefore, it is possible to control the nonstationarity of the earthquake motion by giving the distribution of the Fourier phase gradients to each range of frequencies.

Naraoka and Watanabe (1987) described the nonstationary earthquake ground motion $y(t)$ by

$$y(t)=\sum_{k=1}^{N/2} A_k \sin(\omega_k t+\phi_{k-1}+t_{gr}(\omega_k)\Delta\omega), \tag{9}$$

where $\omega_k(=k\Delta\omega)$, A_k, and $t_{gr}(\omega_k)$ $(=\Delta\phi_k/\Delta\omega)$ are a circular frequency of the k-th component, a Fourier amplitude at ω_k, and a Fourier phase gradient at ω_k, respectively. The Fourier phase gradients were assumed to be a log-normal distribution with the same mean and standard deviation as those of the band-pass filtered record. The energy response time histories of a single-degree-of-freedom system with a natural circular frequency ω_k and with 5 % damping factor for the motion was adopted as a band-filter.

Figure 36 shows an example of a synthetic nonstationary motion using phase gradients. Figure 36 (a) indicates the mean and standard deviation of the Fourier phase gradients of an earthquake motion record at Niigata for the 1964 Niigata, Japan, earthquake as shown in Figure 36 (b). Figure 36 (c) shows a simulated motion and it is recognized that the record is well reproduced (Naraoka and Watanabe (1987)).

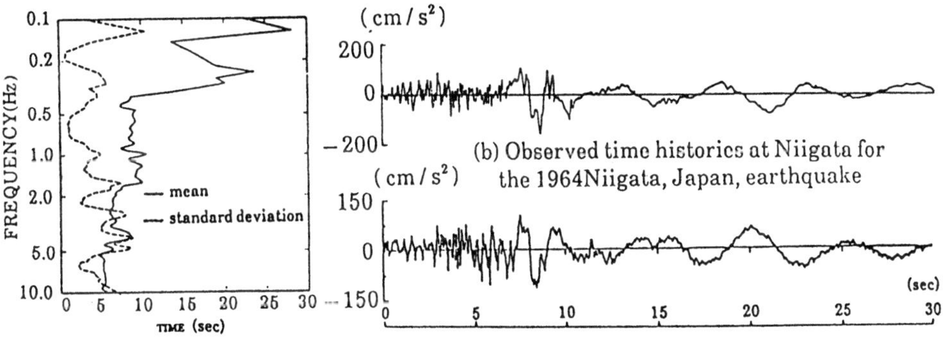

(a) Fourier phase gradient and standard deviation of recorded motion (b)

(c) Simulated time histories by using Fourier phase gradient and standard deviation of recorded motion (b)

Figure 36. Simulation of nonstationary earthquake motion using Fourier phase gradients (after Naraoka and Watanabe (1987)).

7. Conclusions

The conclusions can be summarized as follows:

(1) The important parameters to characterize earthquake ground motions are ① the earthquake magnitude or the fault parameters for seismic source, ② the source distance or the velocity structures of the path between the source and the site, and ③ the S-wave velocity, density, damping factor, and thickness of layers near the site for local conditions.

(2) In the procedure of the estimation and generation of design input motion, the concept of two different kinds of design earthquakes, namely, the 'Design basis earthquake' and the

'Maximum earthquake' are introduced. The design input motions are defined at the surface of the base stratum which supports the foundation of the structure.

(3) The five approaches can be used to generate the design input motions at present. They are classified into the following three groups:

(i) The method using strong-motion records
Use of strong-motion records at the site for the design level earthquake (Approach ①)
Use of alternative strong-motion records (Approach ⑤)

(ii) The method based on fault model
Theoretical approach (Approach ②)
Semi-empirical approach (Approach ③)

(iii) The method using standard spectra
Use of standard spectra and duration (Approach ④)

(4) When there exist strong-motion records obtained at the site for the design level earthquake, the approach ① will be given the priority, because of all the information being included in those records. It is noted that a great attention has to be paid to the reliable period range of the records. But there have been very few such records at present.

(5) When there exist no records and both of the detailed information on the fault parameters and the velocity structure of the path between the earthquake source to the site can be specified, the approach ② will be used. But it is difficult to get both information at present.

(6) When there are appropriate small-event records at the site and the information on fault parameters of the design earthquake, the approach ③ will be employed. This approach is very useful and practical for estimating the short-period motions, because the small-event motions include not only the local site effect near the site and the inhomogeneous structures of the path but also the complex rupture mechanism of the source.

(7) When the magnitude of the design earthquake and the source distance are known, the approach ④ will be employed. In this approach, the design input motions are generated by the standard response spectrum or power spectrum, duration, and time-dependent variation of amplitude envelop curve or phase of ground motions. These values are usually given by regression analyses using many strong-motion records. Therefore, a great attention has to be paid to a range of applicability, which depends on the quality and/or quantity of the records.

(8) When there is no specific information of both the site and the design earthquake, the approach ⑤ will have to be employed. When this approach is employed, a great attention has to be paid to the reliable period range of the records. But this does not reflect well the source, path, and local site conditions.

In the estimation and generation of design input motions, there are substantially three kinds of uncertainty as follows (Abrahamson *et al.*): The first one is the parametric uncertainty which represents uncertainty in the values of source parameters for future earthquakes. The second one is the random uncertainty which represents detailed aspects of the earthquake source and wave propagation that cannot be modeled deterministically (e.g., Somerville, *et al.* (1988); Loh and Yeh (1988), Katayama, *et al.* (1990)). The third one is the modeling uncertainty which represents differences between the actual process that generates the strong ground motion and the simulation of those processes in the procedure.

To generate more accurate and reliable design input motions, future works should be directed to the quantitative evaluation on these kinds of uncertainty.

8. Acknowledgement

The authors are grateful to Dr. T. Itoh, Ohsaki Research Institute, Shimizu Corporation for his review of the manuscript.

9. References

Abrahamson, N. A., Somerville, P. G., and Cornell, C. A., 'Uncertainty in numerical strong motion predictions for engineering applications', (submitted to Bull. Seism. Soc. Am.).

Akao, Y., Noda, S., and Inoue, R. (1988) "Theoretical and semi-empirical approaches by using normal mode solution for reproducing seismic motions at intermediate periods', Proc. 9th World Conference on Earthquake Engineering, Tokyo-Kyoto, II, 769-774.

Aki, K. (1966) 'Generation and propagation of G waves from the Niigata earthquake of June 16, 1964', Bull. Earthq. Res. Inst, Univ. of Tokyo, 44, 73-88.

Aki, K. (1967) 'Scaling law of seismic spectrum', J. Geophys. Res., 72, 1217-1231.

Aki, K. (1968) 'Seismic displacement near a fault', J. Geophys. Res., 73, 5359-5376.

Aki, K., and Irikura, K. (1991) 'Characterization and mapping of earthquake shaking for seismic zonation', Proc. Fourth Int. Conf. Seismic Zonation, Stanford, California, 1, 61-110.

Aki, K. and Richards, P. G. (1980) Quantitative Seismology, W. H. Freeman and Company, San Francisco, Chapters 14 and 15.

Aki, K. and Wu, R. S. (eds.) (1988) Scattering and attenuation of seismic waves, Part I, Pure and Applied Geophysics, 128, No. 1/2.

Bard, P. Y. and Bouchon, M. (1980) 'Seismic response of sediment-filled valleys, Part 1: The case of incident SH waves', Bull. Seism. Soc. Am., 70, 1263-1286.

Bolt, B. A. (1983) 'Necessary and sufficient conditions for strong motion parameters', Seismic Design of Embankments and Caverns, Proc. of a Symp. sponsored by the ASCE Geotech.Eng. Div. in Conjunction with the ASCE National Convention, Philadelphia, Pennsylvania, May 16-20, Howard, T. R., editor, Am. Soc. Civil Eng., 10-22.

Boore, D. M. (1983) 'Stochastic simulation of high frequency ground motion based on seismological models of radiated spectra', Bull. Seism. Soc. Am., 73, 1865-1894.

Bouchon, M. (1979) 'Discrete wave number representation of elastic wave fields in three-dimensions', J. Geophys. Res., 84, 3609-3614.

Brune, J. N. (1970) "Tectonic stress and the spectra of seismic shear waves from earthquakes', J. Geophys. Res., 75, 4997-5009.

Cambell, K. W. (1985) 'Strong motion attenuation relations: ten years perspective', Earthq. Spectra, 1, 759-804.

Dan, K., Watanabe, T., Tanaka, T., and Sato, R. (1990) 'Stability of earthquake ground motion synthesized by using different small-event records as empirical Green's functions', Bull. Seism. Soc. Am., 80, 1433-1455.

Das, S. and Aki, K. (1977) 'Fault plane with barriers', J. Geophys. Res., 82, 5658-5670.

de Hoop, A. T. (1960) 'A modification of Cagniard's method for solving seismic pulse problems', Applied Scientific Research, Section B, 8, 349-356.

Dreger, D. S. and Helmberger, D. V. (1990) 'Broad band modeling of local earthquakes', Bull. Seism. Soc. Am., 80, 1162-1179.

Dziewonski, A. M., Friedman, A., Giardini, D., and Woodhouse, J. H. (1983) 'Global seismicity of 1982', Phys. Earth Planet. Interiors, 33, 76-90.
Fukushima, Y and Tanaka, T. (1990) 'A new attenuation relation for peak horizontal acceleration of strong earthquake ground motion in Japan', Bull. Seism. Soc. Am., 84, 757-783.
Graves, R. W. and Clayton, R W. (1992) 'Modeling path effects in three-dimensional basin structure', Bull. Seism. Soc. Am., 82, 81-103.
Gutenberg, B. (1945) 'Amplitude of surface waves and magnitudes of shallow earthquakes', Bull. Seism. Soc. Am., 35, 3-12.
Hanks, T. C. and McGuire, R. K. (1981) 'The character of high frequency strong ground motion', Bull. Seism. Soc. Am., 71, 2071-2095.
Hanks, T. C. and Wyss, M. (1972) 'The use of body-wave spectra in the determination of seismic-source parameters', Bull. Seism. Soc. Am., 62, 561-589.
Harkrider, D. G. (1964) 'Surface waves in multilayered elastic media', Bull. Seism. Soc. Am., 54, 627-679.
Hartzell, S. H. (1978) 'Earthquake aftershocks as Green's functions', Geophys. Res. Letters, 5, 1-4.
Haskell, N. A. (1964) 'Total energy and energy spectral density of elastic wave radiation from propagating faults', Bull. Seism. Soc. Am., 54, 1811-1841.
Helmberger, D. V. (1968) 'The crust-mantle transition in the Bering Sea', Bull. Seism. Soc. Am. 58, 179-214.
Hirasawa, M. and Watabe, M. (1992) 'Generation of simulated earthquake motions compatible with multi-danping response spectra', Proc. Tenth World Conf. Earthq. Eng., Madrid (to be published).
Hisada, T. and Ando, H. (1973) 'On relationship between duration and magnitude of earthquake motion at bedrock', Kajima Inst. Const. Tech., 1187-1198.
Hisada, T., Ohsaki, Y., Watabe, M., and Ohta, T. (1978) 'Design spectra for stiff structures on rock', Proc. Second Int. Conf. on Microzonation, San Francisco, III., 1187-1198.
Hisada, Y., Yamamoto, S., and Tani, S. (1988) 'Analysis of strong ground motion of plain and basin being composed of soft soil by fault model and boundary element method', Proc. Eighth World Conf. Earthq. Eng., San-Francisco, II, 701-706.
Horike, M., Uebayashi, H., and Takeuchi, Y. (1990) 'Seismic response in three-dimensional sedimentary basin due to plane S wave incidence', J. Phys. Earth, 33, 261-284.
Irikura, K. (1983) 'Semi-empirical estimation of strong ground motions during large earthquakes', Bull. Disast. Prev. Res. Instit., Kyoto Univ., 33, Part 2, No. 298, 63-104.
Ishii, T., Sato, T., and Satoh, T. (1989) 'A study on seismic wave propagation in Lotung, Taiwan', 7-12. Trans. of Tenth Int. Conf. on Struct. Mech. Reactor Tech., K1, 7-12.
Itoh, T., Ishii, K., Okumura, T., and Ishikawa, Y. (1987) 'Development of seismic hazard analysis in Japan', Trans. of Ninth Int. Conf. on Struct. Mech. Reactor Tech., K1, 69-74.
Iwata, T. and Irikura, K. (1988) 'Source parameters of the 1983 Japan Sea earthquake sequence', J. Phys. Earth, 36, 155-184.
Izutani, Y. (1981) 'A statistical model for prediction of quasi-realistic strong ground motion', J. Phys. Earth, 29, 537-557.
Izutani, Y. and Katagiri, F. (1992) 'Empirical Green's function corrected for source effect', Earthq. Eng. Struct. Dyn., 21, 341-349.
Jennings, P. C., Housner, G. W., and Tsai, N. C. (1969) 'Simulated earthquake motions for design purposes', Proc. Fourth World Conf. Earthq. Eng., 145-160.

Joyner, W. B. and Boore, D. M. (1981) 'Peak horizontal acceleration and velocity from strong-motion records including records from the 1979 Imperial Valley, California, earthquake', Bull. Seism. Soc. Am., 71, 2011-2038.
Joyner, W. B. and Boore, D. M. (1982) 'Prediction of earthquake response spectra', U. S. Geological Survey, Open-file report, 82-977, 1-16.
Joyner, W. B. and Boore, D. M. (1988) 'Measurement, characterization, and prediction of strong ground motion', Earthq. Eng. Soil Dyn. II, Proc. of the Specialty Conf. Sponcered by the Geotech. Eng. Div. of the Am. Soc. Civil Eng., Geotechnical Special Publication, No.20, 43-102.
Kamiyama, M. and Yanagisawa, E. (1986) 'A statistical model for estimating response spectra of strong earthquake ground motions with emphasis on local soil conditions', Soils and Foundations, 26, 16-32.
Kanai, K. (1958) 'A study of strong ground motions', Bull. Earthq. Res. Inst., Univ. of Tokyo, 36, 295-310.
Kanai, K. (1983) Engineering Seismology, The Univ.of Tokyo Press, Tokyo.
Kanamori, H. (1977) 'The energy release in great earthquakes', J. Geophys. Res., 82, 2981-2987.
Kanamori, H. (1979) 'A semi-empirical approach to prediction of long-period ground motions from great earthquakes', Bull. Seism. Soc. Am., 69, 1645-1670.
Kanamori, H. and Anderson, D. L. (1975) 'Theoretical basis of some empirical relations in seismology', Bull. Seism. Soc. Am., 65, 1073-1096.
Kanamori, H. and Given, J. W. (1981) 'Use of long-period surface waves for rapid determination of earthquake-source parameters', Phys. Earth Planet. Interiors, 27, 8-31.
Katayama, T., Yamazaki, F., Nagata, S., Lu, L., and Turker, T. (1990) 'A strong motion database for the Chiba seismometer array and its engineering analysis', Earthq. Eng. Struct. Dyn., 19, 1089-1106.
Kato, K., Takemura, M., Ikeura, T., Urao, K., and Uetake, T. (1992) 'Preliminary analysis for evaluation of local site effects from strong motion spectra by an inversion method', J. Phys. Earth, 40, 175-191.
Katukura, H., Watanabe, T., and Izumi, M. (1978) 'A study on the phase properties of seismic waves', Proc. Fifth Japan Earthq. Eng. Symp., 209-216.
Katukura, H., Watanabe, T. and Izumi, M. (1984) 'A study on the Fourier analysis of nonstationary seismic waves', Proc. Eighth World Conf. Earthq. Eng., San-Francisco, II, 525-532.
Kausel, E., Roesset, J. M., and Waas, G. (1975) 'Dynamic analysis of footings on layered media', J. Eng. Mech., Am. Soc. Civil Eng., EM5, 101, 679-693.
Kawasaki, I., Suzuki, Y., and Sato, R. (1973) 'Seismic waves due to a shear fault in a semi-infinite medium', J. Phys. Earth, 21, 251-284.
Kawase, H. and Aki, K. (1989) 'A study on the response of a soft basin for incident S, P, and Rayleigh waves with special reference to the long duration observed in Mexico city', Bull. Seism. Soc. Am., 79, 1361-1382.
Kawase, H. and Sato, T. (1992) 'Simulation analysis of strong motions in the Ashigara valley considering one- and two-dimensional geological structures', J. Phys. Earth, 40, 27-56.
Kikuchi, M. and Kanamori, H. (1982) 'Inversion of complex body waves', Bull. Seism. Soc. Am., 72, 491-506.
Kinoshita, S. (1992) 'Frequency dependent attenuation of shear waves in a sedimentary layers-basement system in the Kanto area, Japan', Proc. of Int. Symp. Earthq. Disast. Prev., CENAPRED, Mexico, May (to be published).

Kobayashi, H. and Nagahashi, S. (1977) 'Response spectra on seismic bedrock during earthquake' Proc. Sixth World Conf. Earthq. Eng., New-Delhi, 516-623.
Kohketsu, K. (1987) 'Synthetic seismograms in realistic media', Bull. Earthq. Res. Inst., Univ.of Tokyo, 62, 201-245.
Kudo, K. (1980) 'A study on the contribution of surface waves to strong ground motions', Proc. Seventh World Conf. Earthq. Eng., Istanbul, 2, 499-506.
Lamb, H. (1904) 'On the propagation of tremors over the surface of an elastic solid', Philosophical Trans.s of the Royal Society of London, Series A, 203, 1-42.
Loh, C. H. and Yeh, Y. T. (1988) 'Spatial variation and stochastic modelling of seismic differential ground movement', Earthq. Eng. Struct. Dyn., 16, 583-596.
Lilhanand, K. and Tseng, W. S. (1987) 'Generation of synthetic time histories compatible with multi-damping design response spectra', Trans. of Ninth Int. Conf. on Struct. Mech. Reactor Tech., K1,105-110.
Luco, J. E. and Apsel, R. J. (1983) 'On the Green's functions for Layered half-space', Bull. Seism. Soc. Am., 73, 909-929.
Madariaga, R. (1976) 'Dynamics of an expanding circular fault', Bull. Seism. Soc. Am., 66, 639-666.
Mikumo, T. and Miyatake, T. (1978) 'Dynamical rupture process on a three dimensional fault with non-uniform frictions and near-field seismic waves', Geophy. J. Roy. astr. Soc., 54, 417-438.
Naraoka, K. and Watanabe, T. (1987) 'Generation of nonstationary earthquake ground motions using phase characteristics', Trans. of Ninth Int. Conf. on Struct. Mech. Reactor Tech., K1, 37-42.
Noda, S., Shimada, S., and Yoshida, T. (1988) 'Relatively long-period ground motions obtained by mending of saturated records of a low-magnification seismograph', Proc. Ninth World Conf. Earthq. Eng., Tokyo-Kyoto, II , 263-268.
Ohori, M., Koketsu, K., and Minami, T. (1992) 'Seismic response of three-dimensional sediment-filled valleys due to incident plane waves', J. Phys. Earth, 40, 209-222.
Ohsaki, Y. (1979) 'On the significance of phase content in earthquake ground motions', Earthq. Eng. Struct. Dyn., 7, 427-439.
Ohtsuki, A., Yamahara, H., and Tazoh, H. (1984) 'Effect on lateral inhomogeneity on seismic waves. Part II: Observations and analysis', Earthq. Eng. Struct. Dyn., 12, 795-816.
Olson, A. H. and Apsel, R. J. (1982) 'Finite faults and inverse theory with applications to the 1979 Imperial Valley earthquake', Bull. Seism. Soc. Am., 72, 1969-2001.
Papageorgiou, A. S. and Aki, K. (1983) 'A specific barrier model for the quantitative description of inhomogeneous faulting and prediction of strong ground motion, Part I description of the model', Bull. Seism. Soc. Am., 73, 693-722.
Reid. H. F. (1911) 'The elastic-rebound theory of earthquakes', Bull. Dept. of Geol., 6, 413-444.
Richter, C. F. (1935) 'An instrumental earthquake magnitude scale', Bull. Seism. Soc. Am., 25, 1-32.
Somerville, P. G., Mclaren, J. P., Saikia, C., K., and Helmberger, D. V. (1988) 'Site specific estimation on spatial incoherence of strong ground motion', Earthq. Eng. Soil Dyn. II, Proc. of the Specialty Conf. Sponcered by the Geotech. Eng. Div. of the Am. Soc. Civil Eng., Geotechnical Special Publication, No.20, 188-202.
Sánches-Sesma, F. J., Chaves-Perez, S., Suarez, M., Bravo, M. A., and Perez-Rocha (1988) 'The Mexico earthquake of September 19, 1985-On the seismic response of the valley of Mexico', Earthq. Spectra, 4, No.3, 569-589.

Sasatani, T. (1988) 'SH-waves in laterally varying medium: (1) The seismic response of sediment-filled valleys to plane-incident SH waves', J. Fac. Sci., Hokkaido Univ., Ser. VII (Geophysics), 8, 191-208.
Sato, H. (1990) 'Unified approach to amplitude attenuation and coda excitation in the randomly inhomogeneous lithosphere', Pure and Applied Geophysics, 132, No1/2, 93-121.
Sato, R. (1979) "Theoretical basis on relationships between focal parameters and earthquake magnitude', J. Phys. Earth., 27, 353-372.
Sato, T. (1990) 'Simulation of observed seismograms on sedimentary basin using theoretical seismograms in the period range from 2 to 20 sec', Proc. 8th Japan Earthq. Eng. Symp., 193-198 (in Japanese with English abstract).
Sato, T. and Hasegaea, M. (1990) 'Formulation of hybrid approach for theoretical seismograms using thin layer element and axsymmetric finite element', J. Struct. Const. Eng. (Trans. of Archtectural Instit. Japan), 414, 55-69 (in Japanese with English abstract).
Sato, T., and Hirasawa, T. (1973) 'Body wave spectra from propagating shear cracks', J. Phys. Earth, 21, 415-431. '
Sato, T. and Kawase, H. (1992) 'Finite element simulation of seismic wave propagation in near-surface random media', Proc. of the Int. Symp. on the Effects of Surface Geology on Seismic Motion, Odawara, Japan, March, I, 257-262.
Satoh, T., Hasegawa, M., Nakai, S., Watanabe, T., and Tanaka, T. (1988) 'Analysis of theoretical seismograms by the thin layer finite element method and its application to the 1980 Izu-Hanto-Toho-Oki earthquake in Japan', Proc. Ninth World Conf. Earthq. Eng., Tokyo-Kyoto, II, 763-768.
Satoh, T., Sato, T., and Kawase, H. (1992) 'Local site effects on weak and strong ground motions', Proc. of Int. Symp. of Earthq. Disast. Prev., CENAPRED, Mexico, May (to be published).
Schnabel, P. B., Lysmer, J., and Seed, H. B. (1972) 'SHAKE: A computer program for earthquake response analysis of horizontally layered sites', Report EERC-72-12, Univ.of California, Berkeley.
Seed, H. B. and Idriss, I. M. (1970) 'Soil moduli and damping factors for dynamic response analysis', Earthquake Engineering Research Center, U. C. Berkeley, California, Report No. EERC 70-10.
Shakal, A. L. and Bernreuter, D. L. (1981) 'Empirical analysis of near-source ground motion', U. S. Nuclear Regulatory Commission Report NUREG/CR-2095.
Swanger, H. J. and Boore, D. M. (1978) 'Simulation of strong-motion displacements using surface-wave modal superposition', Bull. Seism. Soc. Am., 68, 907-922.
Tajimi, H. (1960) 'A statistical method of determining the maximum response of a building structure during an earthquake' Proc. 2nd World Conf. Earthq. Eng., Japan, 2, 781-797.
Tajimi, H. (1980) 'A contribution to theoretical prediction of dynamic stiffness of surface foundations', Proc. Seventh World Conf. Earthq. Eng., Istanbul, 105-112.
Takemura, M. and Ikeura, T. (1988) 'A semi-empirical method using a hybrid of stochastic and deterministic fault models: Simulation of strong ground motions during large earthquakes', J. Phys. Earth, 36, 89-106.
Takeo, M. (1988) 'Rupture process of the 1980 Izu-Hanto-Toho-Oki earthquake deduced from strong motion seismograms', Bull. Seism. Soc. Am., 78, 1074-1091.
Tanaka, T., Yoshizawa, S., Sakaue, M., and Osawa, Y. (1982) 'Estimation of acceleration characteristics of strong ground motion by synthesis of accelerogram obtained during a small earthquake', Bull. Earthq. Res. Inst., Univ.of Tokyo, 57, 561-579.

Tazoh, T., Sato, K., Shimizu, K., and Hatakeyama, A. (1988) 'Nonlinear seismic response analysis of soil deposit using strong seismic records', Proc. Ninth World Conf. Earthq. Eng., Tokyo-Kyoto, 507-512.

Uebayashi, H., Horike, M., and Takeuchi, Y. (1992) 'Seismic motion in a three-dimensional arbitrarily-shaped sedimentary basin, due to a rectangular dislocation source', J. Phys. Earth, 40, 223-240.

U.S. Atomic energy commission (1973) 'Regulatory Guide 1.60, Design response spectra for seismic design of nuclear power plants'.

Vaněk,J., Zatopek, A., Karnik, V., Kondorskaya, N. V., Riznichenco, Y. V., Savarensky, E. F., Solov'ev, S. L., and Shebalin, N. V. (1962) 'Standardization of magnitude scales', Izv. Acad. Sci.,USSR, Geophys. Ser., 108-111 (English translation by AGU).

Vidale, J. E. and Helmberger, D. V. (1988) 'Elastic finite difference modeling of the 1971 San Fernando, California earthquake', Bull. Seism. Soc. Am., 78, 121-141.

Wald, D. J., Burdick, L. J., and Somerville, P. G. (1988) 'Simulation of acceleration time histories close to large earthquakes', Earthq. Eng. and Soil Dyn. II, Geotechnical Special Publication, No. 20, 430-444.

Watabe, M. and Tohdo, M. (1982) 'Research on the design earthquake ground motions, Part-2', Trans. Archi. Inst. Japan, No. 312, 63-72 (in Japanese with English abstract).

Watabe, M., Tohdo, M., Chiba, O., and Fukuzawa, R. (1990) 'Peak accelerations and response spectra of vertical strong-ground motions from near-field records in USA', Proc. Eighth Japan Earthq. Eng. Symp., 1, 301-306, (in Japanese with English abstract).

Wyss, M. and Brune, J. N. (1967) 'The Alaska earthquake of 28 March 1964', Bull. Seism. Soc. Am., 57, 1017-1023.

Yamada, I., Fukao, Y., Ishihara, Y., and Aoki, H. (1989) 'Broad-band and wide dynamic range seismic observation by an STS-seismograph', Zisin, J. Seism. Soc. Japan, 42, 21-31 (in Japanese with English abstract).

Yamanaka, H., Seo, K., and Samano, T. (1992) 'Analysis and numerical modeling of surface-wave propagation in a sedimentary basin', J. Phys. Earth, 40, 57-71.

Yokota, H., Kataoka, S., Tanaka, T., and Yoshizawa, S. (1989) 'Estimation of long-period ground motion of the 1923 great Kanto earthquake', J. Struct. Const. Eng. (Trans. of Archtectural Instit. Japan), 401, 35-45 (in Japanese with English abstract).

Yomogida, K. (1985) 'Gaussian beams for surface waves in laterally slowly-varying media', Geophy. J. Roy. astr. Soc., 82, 511-533.

Yomogida, K. and Etdgen J. (1992) '3-D wave propagation in the Los Angeles basin for the Whittier-narrows earthquake' (submitted to Bull. Seism. Soc. Am.).

INDEX

P. Gülkan and R. W. Clough (eds.), Developments in Dynamic Soil-Structure Interaction, 435–439.

GPSR Compliance

The European Union's (EU) General Product Safety Regulation (GPSR) is a set of rules that requires consumer products to be safe and our obligations to ensure this.

If you have any concerns about our products, you can contact us on ProductSafety@springernature.com

In case Publisher is established outside the EU, the EU authorized representative is:

Springer Nature Customer Service Center GmbH
Europaplatz 3
69115 Heidelberg, Germany

Batch number: 10607320

Printed by Printforce, the Netherlands